Mechanics and Strength of Materials

Vitor Dias da Silva

Mechanics and Strength of Materials

 Springer

Vitor Dias da Silva

Department of Civil Engineering
Faculty of Science & Technology
University of Coimbra
Polo II da Universidade - Pinhal de Marrocos
3030-290 Coimbra
Portugal
E-mail: vdsilva@dec.uc.pt

ISBN-13 978-3-642-06423-4   e-ISBN-13 978-3-540-30813-3

Springer is a part of Springer Science+Business Media
springeronline.com
© Springer-Verlag Berlin Heidelberg 2010
Printed in The Netherlands

Cover design: *design & production* GmbH, Heidelberg

Printed on acid-free paper

Preface To The English Edition

The first English edition of this book corresponds to the third Portuguese edition. Since the translation has been done by the author, a complete review of the text has been carried out simultaneously. As a result, small improvements have been made, especially by explaining the introductory parts of some Chapters and sections in more detail.

The Portuguese academic environment has distinguished this book, since its first edition, with an excellent level of acceptance. In fact, only a small fraction of the copies published has been absorbed by the school for which it was originally designed – the Department of Civil Engineering of the University of Coimbra. This fact justifies the continuous effort made by the author to improve and complement its contents, and, indeed, requires it of him. Thus, the 423 pages of the first Portuguese edition have now grown to 478 in the present version. This increment is due to the inclusion of more solved and proposed exercises and also of additional subjects, such as an introduction to the fatigue failure of materials, an analysis of torsion of circular cross-sections in the elasto-plastic regime, an introduction to the study of the effect of the plastification of deformable elements of a structure on its post-critical behaviour, and a demonstration of the theorem of virtual forces.

The author would like to thank all the colleagues and students of Engineering who have used the first two Portuguese editions for their comments about the text and for their help in the detection of misprints. This has greatly contributed to improving the quality and the precision of the explanations.

The author also thanks Springer-Verlag for agreeing to publish this book and also for their kind cooperation in the whole publishing process.

Coimbra   *V. Dias da Silva*
March 2005

Preface to the First Portuguese Edition

The motivation for writing this book came from an awareness of the lack of a treatise, written in European Portuguese, which contains the theoretical material taught in the disciplines of the Mechanics of Solid Materials and the Strength of Materials, and explained with a degree of depth appropriate to Engineering courses in Portuguese universities, with special reference to the University of Coimbra. In fact, this book is the result of the theoretical texts and exercises prepared and improved on by the author between 1989-94, for the disciplines of Applied Mechanics II (Introduction to the Mechanics of Materials) and Strength of Materials, taught by the author in the Civil Engineering course and also in the Geological Engineering, Materials Engineering and Architecture courses at the University of Coimbra.

A physical approach has been favoured when explaining topics, sometimes rejecting the more elaborate mathematical formulations, since the physical understanding of the phenomena is of crucial importance for the student of Engineering. In fact, in this way, we are able to develop in future Engineers the intuition which will allow them, in their professional activity, to recognize the difference between a bad and a good structural solution more readily and rapidly.

The book is divided into two parts. In the first one the Mechanics of Materials is introduced on the basis of Continuum Mechanics, while the second one deals with basic concepts about the behaviour of materials and structures, as well as the Theory of Slender Members, in the form which is usually called Strength of Materials.

The introduction to the Mechanics of Materials is described in the first four chapters. The first chapter has an introductory character and explains fundamental physical notions, such as continuity and rheological behaviour. It also explains why the topics that compose Solid Continuum Mechanics are divided into three chapters: the stress theory, the strain theory and the constitutive law. The second chapter contains the stress theory. This theory is expounded almost exclusively by exploring the balance conditions inside the body, gradually introducing the mathematical notion of tensor. As this notion is also used in the theory of strain, which is dealt with in the third chapter, the explanation of this theory may be restricted to the essential physical aspects of the deformation, since the merely tensorial conclusions may be drawn by analogy with the stress tensor. In this chapter, the physical approach adopted allows the introduction of notions whose mathematical description would be too complex and lengthy to be included in an elementary book. The finite strains and the integral conditions of compatibility in multiply-connected bodies are examples of such notions. In the fourth chapter the basic phenomena which determine the relations between stresses and strains are explained with the help of physical models, and the constitutive laws in the simplest three-dimensional cases are deduced. The most usual theories for predicting the yielding and rupture of isotropic materials complete the chapter on the constitutive law of materials.

In the remaining chapters, the topics traditionally included in the Strength of Materials discipline are expounded. Chapter five describes the basic notions and general principles which are needed for the analysis and safety evaluation of structures. Chapters six to eleven contain the theory of slender members. The way this is explained is innovative in some aspects. As an example, an alternative Lagrangian formulation for the computation of displacements caused by bending, and the analysis of the error introduced by the assumption of infinitesimal rotations when the usual methods are applied to problems where the rotations are not small, may be mentioned. The comparison of the usual methods for computing the deflections caused by the shear force, clarifying some confusion in the traditional literature about the way as this deformation should be computed, is another example. Chapter twelve contains theorems about the energy associated with the deformation of solid bodies with applications to framed structures. This chapter includes a physical demonstration of the theorems of virtual displacements and virtual forces, based on considerations of energy conservation, instead of these theorems being presented without demonstration, as is usual in books on the Strength of Materials and Structural Analysis, or else with a lengthy mathematical demonstration.

Although this book is the result of the author working practically alone, including the typesetting and the pictures (which were drawn using a self-developed computer program), the author must nevertheless acknowledge the important contribution of his former students of Strength of Materials for their help in identifying parts in the texts that preceded this treatise that were not as clear as they might be, allowing their gradual improvement. The author must also thank Rui Cardoso for his meticulous work on the search for misprints and for the resolution of proposed exercises, and other colleagues, especially Rogério Martins of the University of Porto, for their comments on the preceding texts and for their encouragement for the laborious task of writing a technical book.

This book is also a belated tribute to the great Engineer and designer of large dams, Professor Joaquim Laginha Serafim, who the Civil Engineering Department of the University of Coimbra had the honour to have as Professor of Strength of Materials. It is to him that the author owes the first and most determined encouragement for the preparation of a book on this subject.

Coimbra   *V. Dias da Silva*
July 1995

Contents

Part I

Introduction to the Mechanics of Materials

I

Introduction

I.1 General Considerations

Materials are of a discrete nature, since they are made of atoms and molecules, in the case of liquids and gases, or, in the case of solid materials, also of fibres, crystals, granules, associations of different materials, etc. The physical interactions between these constituent elements determine the behaviour of the materials. Of the different facets of a material's behaviour, *rheological behaviour* is needed for the Mechanics of Materials. It may be defined as the way the material deforms under the action of forces.

The influence of those interactions on macroscopic material behaviour is studied by sciences like the Physics of Solid State, and has mostly been clarified, at least from a qualitative point of view. However, due to the extreme complexity of the phenomena that influence material behaviour, the quantitative description based on these elementary interactions is still a relatively young field of scientific activity. For this reason, the deductive quantification of the rheological behaviour of materials has only been successfully applied to some *composite materials* – associations of two or more materials – whose rheological behaviour may be deduced from the behaviour of the individual materials, in the cases where the precise layout of each material is known, such as plastics reinforced with glass or carbon fibres, or reinforced concrete.

In all other materials rheological behaviour is idealized by means of physical or mathematical models which reproduce the most important features observed in experimental tests. This is the so-called *phenomenological approach.*

From these considerations we conclude that, in Mechanics of Materials, a phenomenological approach must almost always be used to quantify the rheological behaviour of a solid, a liquid or a gas. Furthermore, as the consideration of the discontinuities that are always present in the internal structure of materials (for example the interface between two crystals or two granules, micro-cracks, etc.), substantially increases the degree of complexity of the problem, we assume, whenever possible, that the material is *continuous.*

From a mathematical point of view, the hypothesis of continuity may be expressed by stating that the functions which describe the forces inside the material, the displacements, the deformations, etc., are continuous functions of space and time.

From a physical point of view, this hypothesis corresponds to assuming that the macroscopically observed material behaviour does not change with the dimensions of the piece of material considered, especially when they tend to zero. This is equivalent to accepting that the material is a mass of points with zero dimensions and all with the same properties.

The validity of this hypothesis is fundamentally related to the size of the smallest geometrical dimension that must be analysed, as compared with the maximum dimension of the discontinuities actually present in the material.

Thus, in a liquid, the maximum dimension of the discontinuities is the size of a molecule, which is almost always much smaller than the smallest geometrical dimension that must be analysed. This is why, in liquids, the hypothesis of continuity may almost always be used without restrictions.

On the other side, in solid materials, the validity of this hypothesis must be analysed more carefully. In fact, although in a metal the size of the crystals is usually much smaller than the smallest geometrical dimension that must be analysed, in other materials like concrete, for example, the minimum dimension that must be analysed is often of the same order of magnitude as the maximum size of the discontinuities, which may be represented by the maximum dimension of the aggregates or by the distance between cracks.

In gases, the maximum dimension of the discontinuities may be represented by the distance between molecules. Thus, in very rarefied gases the hypothesis of continuity may not be acceptable.

In the theory expounded in the first part of this book the validity of the hypothesis of continuity is always accepted. This allows the material behaviour to be defined independently of the geometrical dimensions of the solid body of the liquid mass under consideration. For this reason, the matters studied here are integrated into Continuum Mechanics.

I.2 Fundamental Definitions

In the Theory of Structures, actions on the structural elements are defined as everything which may cause forces inside the material, deformations, accelerations, etc., or change its mechanical properties or its internal structure. In accordance with this definition, examples of actions are the forces acting on a body, the imposed displacements, the temperature variations, the chemical aggressions, the time (in the sense that is causes aging and that it is involved in viscous deformations), etc. In the theory expounded here we consider mainly the effects of applied forces, imposed displacements and temperature.

Some basic concepts are used frequently throughout this book, so it is worthwhile defining them at the beginning. Thus, we define:

- *Internal force* – Force exerted by a part of a body or of a liquid mass on another part. These forces may act on imaginary surfaces defined in the interior of the material, or on its mass. Examples of the first kind are axial and shear forces and bending and torsional moments which act on the cross-sections of slender members (bars). Examples of the second kind would be gravitational attraction or electromagnetic forces between two parts of the body. However, the second kind does not play a significant role in the current applications of the Mechanics of Materials to Engineering problems, and so the designation *internal force* usually corresponds to the first kind (internal surface forces).
- *External forces* – Forces exerted by external entities on a solid body or liquid mass. The forces may also be sub-divided into *surface external forces* and *mass external forces*. The corresponding definitions are:
  - *External surface forces* – External forces acting on the boundary surface of a body. Examples of these include the weight of non-structural parts of a building, equipment, etc., acting on its structure, wind loads on a building, a bridge, or other Civil Engineering structure, aerodynamic pressures in the fuselage and wings of a plane, hydrostatic pressure on the upstream face of a dam or on a ship hull, the reaction forces on the supports of a structure, etc.
  - *External mass forces* – External forces acting on the mass of a solid body or liquid. Examples of external mass forces are: the weight of the material a structure is made of (earth gravity force), the inertial forces caused by an earthquake or by other kinds of accelerations, such as impact, vibrations, traction, braking and curve acceleration in vehicles and planes, and external electromagnetic forces.
- *Rigid body motion* – displacement of the points of a body which do not change the distances between the points inside the body.
- *Deformation* – Variation of the distance between any two points inside the solid body or the liquid mass.

These definitions are general and valid independently of assuming that the material is continuous or not. In the case of continuous materials two other very useful concepts may be defined:

- *Stress* – Physical entity which allows the definition of internal forces in a way that is independent of the dimensions and geometry of a solid body or a liquid mass. There are several definitions for stress. The simplest one is used in this book, which states that stress is the internal force per surface unit.
- *Strain* – Physical entity which allows the definition of deformations in a way that is independent of the dimensions and geometry of a solid body or a liquid mass. As with stress, there also are several definitions for strain. The simplest one states that strain is the variation of the distance between two points divided by the original distance (longitudinal strain), or half the

variation of a right angle caused by the deformation (shearing strain). This strain definition is used throughout this book.

I.3 Subdivisions of the Mechanics of Materials

The Mechanics of Materials aims to find relations between the four main physical entities defined above (external and internal forces, displacements and deformations). Schematically, we may state that, in a solid body which is deformed as a consequence of the action of external forces, or in a flowing liquid under the action gravity, inertial, or other external forces, the following relations may be established

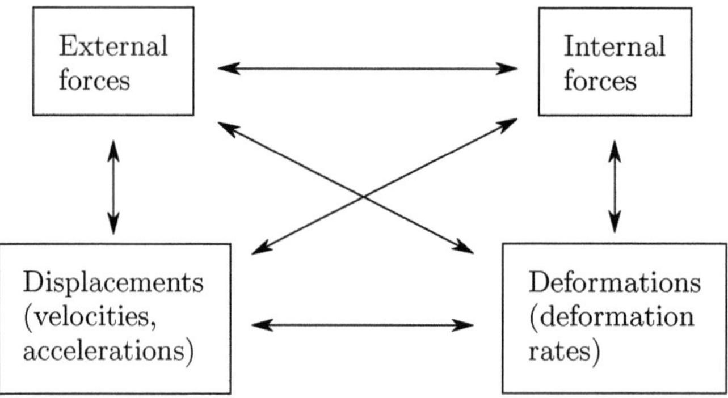

When the validity of the hypothesis of continuity is accepted, these relations may be grouped into three distinct sets

$$\text{force} \xleftrightarrow{\;\;①\;\;} \text{stress} \xleftrightarrow{\;\;③\;\;} \text{strain} \xleftrightarrow{\;\;②\;\;} \text{displacement}$$

1 – *Force-stress relations* – Group of relations based on force equilibrium conditions. Defines the mathematical entity which describes the stress – the stress tensor – and relates its components with the external forces. This set of relations defines the *theory of stresses*. This theory is completely independent of the properties of the material the body is made of, except that the continuity hypothesis must be acceptable (otherwise stress could not be defined).

2 – *Displacement-strain relations* – Group of relations based on kinematic compatibility conditions. Defines the strain tensor and relates its components to the functions describing the displacement of the points of the body. This set of relations defines the *theory of strain*. It is also independent of the rheological behaviour of material. In the form explained in more detail in Chap. III, the theory of strain is only valid if the deformations and the rotations are small enough to be treated as infinitesimal quantities.

3 – *Constitutive law* – Defines the rheological behaviour of the material, that is, it establishes the relations between the stress and strain tensors. As mentioned above, the material rheology is determined by the complex physical phenomena that occur in the internal structure of the material, at the level of atom, molecule, crystal, etc. Since, as a consequence of this complexity, the material behaviour still cannot be quantified by deductive means, a phenomenological approach, based on experimental observation, must used in the definition of the constitutive law. To this end, given forces are applied to a specimen of the material and the corresponding deformations are measured, or vice versa. These experimentally obtained force-displacement relations are then used to characterize the rheological behaviour of the material.

The constitutive law is the potentially most complex element in the chain that links forces to displacements, since it may be conditioned by several factors, like plasticity, viscosity, anisotropy, non-linear behaviour, etc. For this reason, the definition of adequate constitutive laws to describe the rheological behaviour of materials is one of the most extensive research fields inside Solid Mechanics.

II

The Stress Tensor

II.1 Introduction

Some physical quantities, like the mass of a body, its volume, its surface, etc., are mathematically represented by a scalar, which means that only one parameter is necessary to define them. Others, like forces, displacements, velocities, etc., are vectorial entities, which need three quantities to be defined in a three-dimensional space, or two in the case of a two-dimensional space. Other physical entities, like the *states of stress and strain* around a material point inside a body under internal forces, are tensorial quantities, which may be described by nine components in a three-dimensional space, or by four in a two-dimensional space.

In a more general and systematic way, a scalar may be defined as a tensor of order zero with $3^0 = 1$ components, and a vector as a first order tensor with $3^1 = 3$ components. A second order tensor, or simply, tensor, has $3^2 = 9$ components. Higher order tensors may also be defined. An nth order tensor will have 3^n components in a three-dimensional space (or 2^n in a two-dimensional space). As will be seen later, the tensor components are not necessarily all independent.

Below, the stress tensor is defined and some of its properties are analysed.

II.2 General Considerations

Consider a solid body under a system of self-equilibrating forces, as shown in Fig. 1-a. Imagine that the body is divided in two parts by the section represented in the same Figure. Internal forces act in the left surface of the section, representing the action of the right part of the body on the left part. Similarly, as a consequence of the equilibrium condition, in the right surface forces act with the same magnitude and in opposite directions, as shown in Fig. 1-b. The force F and the moment M represent the resultant of the internal forces distributed in the section, which generally vary from point to point. However, by considering an infinitesimal area, $d\Omega$, in the surface (Fig. 2-a), we may consider a homogeneous distribution of the internal force in this area. Dividing the infinitesimal force dF, which acts in the infinitesimal area $d\Omega$, we get the internal force per unit of area or *stress*.

$$T = \frac{dF}{d\Omega} \, .$$
(1)

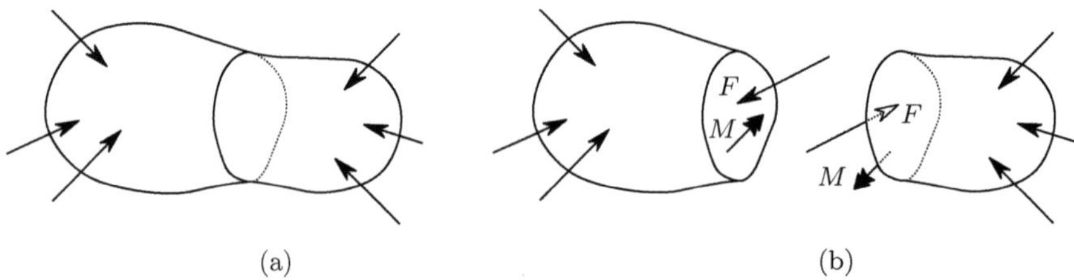

(a)   (b)

Fig. 1. Internal forces in a solid body under a self-equilibrating system of forces

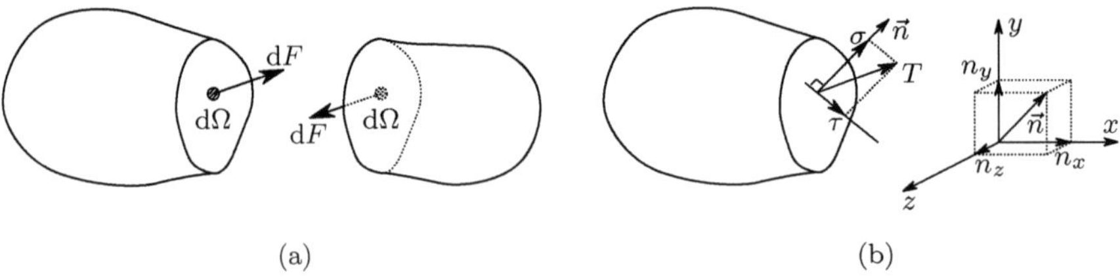

(a)   (b)

Fig. 2. Stress in an infinitesimal surface (facet)

The orientation of the infinitesimal surface of area $d\Omega$ (facet) in a rectangular Cartesian reference frame xyz may be defined by a unit vector $\vec{n}$, which is perpendicular to the facet and points to the outside direction in relation to the part of the body considered (Fig. 2-b). This vector $\vec{n}$, is the *semi-normal of the facet* and, as a unit vector, its components are the cosines of the angles between the vector and the coordinate axes – the *direction cosines* of the facet

$$\begin{cases} n_x = \cos(n, x) = l \\ n_y = \cos(n, y) = m \\ n_z = \cos(n, z) = n \, . \end{cases}$$

As the vector has a unit length, we have

$$l^2 + m^2 + n^2 = 1 \, .$$
(2)

The stress acting on the facet may be decomposed into two components: a *normal* one, with the direction of the semi-normal of the facet $\sigma = T \cos \alpha$, and a *tangential* or *shearing* component $\tau = T\sin\alpha$, where α is the angle between the semi-normal $\vec{n}$ and the total stress vector T (Fig. 2-b).

In the right surface of the section we may define a facet, which is coincident with the left one, but has an opposite semi-normal with direction cosines $-l, -m, -n$ and stresses σ and τ with the same magnitude as in the left facet, but opposite directions. In the case of a facet which is perpendicular to a coordinate axis, it will be a positive facet if its semi-normal has the same direction as the axis to which it is parallel, and it will be negative in the opposite case. As the normal stress σ in these facets is parallel to one of the coordinate axes, the shearing stress τ may be decomposed in the directions of the other two coordinate axes.

In the presentation that follows the Von-Karman convention will be used for the stresses. According to this convention, the stresses are positive if they have the same direction as the coordinate axis to which they are parallel, in the case of a positive facet. In the case of a negative facet, the stresses will be positive, if they have the direction opposite to the corresponding coordinate axis. We will denote the normal stresses parallel to the axes x, y and z by σ_x, σ_y and σ_z, respectively. The shearing stresses are represented by the notation τ_{ij}, where the first index represents the direction of the semi-normal of the facet and the second one the direction of the shearing stress vector. For example τ_{yz} denotes the shearing stress component which is parallel to the z coordinate axis and acts in a facet whose semi-normal is parallel to the y axis.

External force components are positive if they have the same direction as the coordinate axes to which they are parallel.

Figure 3 shows the stresses acting in a rectangular parallelepiped defined by three pairs of facets, which are perpendicular to the three coordinate axis and are located in an infinitesimal neighborhood of point P.

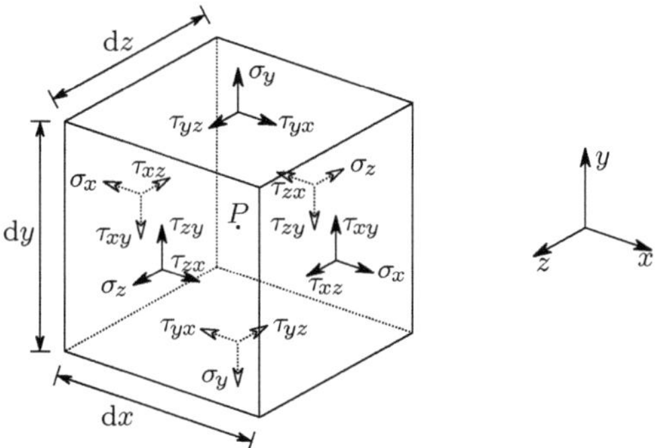

Fig. 3. Positive normal and shearing stresses

II.3 Equilibrium Conditions

Stresses and external forces must obey static and dynamic equilibrium conditions. Using these conditions, some relations may be established in the interior of the body, as well as in its boundary. These fundamental relations are deduced in the following two sub-sections.

II.3.a Equilibrium in the Interior of the Body

The static equilibrium of a body, or a part of it, under the action of a system of forces demands that both its resulting force and its resulting moment vanish. If the resulting moment is zero, we have rotation equilibrium; if the resulting force is zero, equilibrium of translation is attained.

The forces acting in the rectangular parallelepiped defined by the three pairs of facets in Fig. 3 are in equilibrium of translation, since the stress vectors in each pair of facets are equal (more precisely, the difference between them is infinitesimal) and have opposite directions. The external body forces are therefore equilibrated by the infinitesimal difference between the stresses in the negative and positive facets of the pair. The corresponding expressions are presented later. We will first analyse the rotation equilibrium conditions.

Equilibrium of Rotation

Assuming that the translation equilibrium is guaranteed, the resulting moment will be zero or a couple. The latter will vanish if the moments of the forces in relation to three axes, which have a common point, are non-parallel and do not lie along to the same plane, are zero. For simplicity, we consider axes, which are parallel to the reference system and contain the geometrical center of the infinitesimal parallelepiped (Fig. 3). Considering, for example, the axis x' parallel to x, the only forces which have a non-zero moment in relation to this axis are the resultants of τ_{yz} and τ_{zy}, as it can be confirmed by looking at Fig. 3 and as represented in Fig. 4.

The condition of zero moment of the forces which result from the stresses represented in Fig. 4, around the axis x', may be expressed by the equation

$$2\left(\tau_{yz}\,\mathrm{d}x\,\mathrm{d}z\,\frac{\mathrm{d}y}{2}\right) - 2\left(\tau_{zy}\,\mathrm{d}x\,\mathrm{d}y\,\frac{\mathrm{d}z}{2}\right) = 0 \Rightarrow \tau_{zy} = \tau_{yz}\,. \tag{3}$$

The conditions which express the equilibrium of rotation around the axes y' and z', parallel to the global axes y and z, respectively lead to the conclusion that $\tau_{xy} = \tau_{yx}$ and $\tau_{xz} = \tau_{zx}$. These expressions, together with expression 3, represent the so-called *reciprocity of shearing stresses* in perpendicular facets. Since the reference axes may have any spatial orientation, the reciprocity may be expressed in the following way, which is independent of reference axes: *considering two perpendicular facets, the components of the shearing stresses*

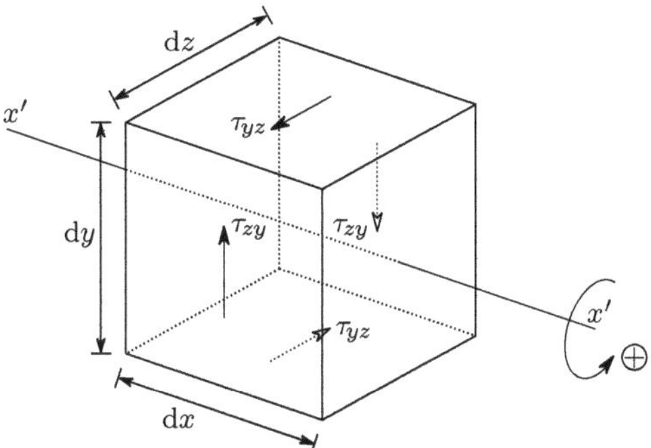

Fig. 4. Equilibrium of rotation around axis x'

which are perpendicular to the common edge of the two facets have the same magnitude and either both point to that edge or both diverge from it.[1]

Equilibrium of Translation

As stated above, the translation equilibrium, in terms of the forces, which act on the faces of the infinitesimal parallelepiped (Fig. 3) is verified. These forces are infinitesimal quantities of the second order: for example, the force corresponding to the stress σ_y is $\sigma_y \, dx \, dz$. The body forces acting in the parallelepiped are infinitesimal quantities of the third order: for example, the force corresponding to the body force per unit of volume in the direction x, X, is $X \, dx \, dy \, dz$. For these reasons, the body forces can be related to the forces corresponding to the variation of the stress, which are also infinitesimal quantities of third order. Since $\sigma_x, \ldots, \tau_{zy}$ are the mean values of the stresses in the facet, it is only necessary to compute the variation of the stress in the direction of the coordinate corresponding to the semi-normal of the facet, on which the stress acts. Figure 5 displays the forces acting on the infinitesimal parallelepiped, including the body forces and the variations of the stress functions.

The condition of equilibrium of the forces acting in direction x leads to the expression

$$\mathrm{d}\sigma_x \, \mathrm{d}y \, \mathrm{d}z + \mathrm{d}\tau_{yx} \, \mathrm{d}x \, \mathrm{d}z + \mathrm{d}\tau_{zx} \, \mathrm{d}x \, \mathrm{d}y + X \, \mathrm{d}x \, \mathrm{d}y \, \mathrm{d}z = 0 \, . \tag{4}$$

---

[1] If the external loading were to include moments M_X, M_Y, M_Z, distributed in the volume of the body, instead of equation (3) we would obtain the expression $\tau_{yz} - \tau_{zy} + M_X = 0$ and there would be no reciprocity of the shearing stresses. However, this kind of loading does not usually have physical significance, except in problems which are beyond the scope of this text, such as the case of the influence of a strong magnetic field on the stress distribution in a magnetized body. For this reason, in the discussion below, the reciprocity of the shearing stresses will will always be considered valid.

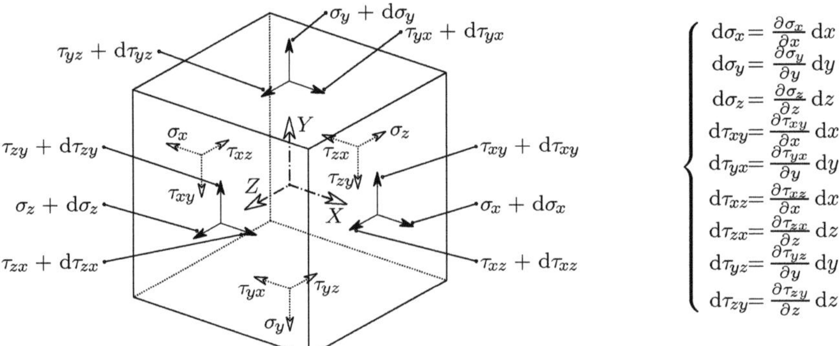

$$\begin{cases} \mathrm{d}\sigma_x = \dfrac{\partial \sigma_x}{\partial x}\,\mathrm{d}x \\[4pt] \mathrm{d}\sigma_y = \dfrac{\partial \sigma_y}{\partial y}\,\mathrm{d}y \\[4pt] \mathrm{d}\sigma_z = \dfrac{\partial \sigma_z}{\partial z}\,\mathrm{d}z \\[4pt] \mathrm{d}\tau_{xy} = \dfrac{\partial \tau_{xy}}{\partial x}\,\mathrm{d}x \\[4pt] \mathrm{d}\tau_{yx} = \dfrac{\partial \tau_{yx}}{\partial y}\,\mathrm{d}y \\[4pt] \mathrm{d}\tau_{xz} = \dfrac{\partial \tau_{xz}}{\partial x}\,\mathrm{d}x \\[4pt] \mathrm{d}\tau_{zx} = \dfrac{\partial \tau_{zx}}{\partial z}\,\mathrm{d}z \\[4pt] \mathrm{d}\tau_{yz} = \dfrac{\partial \tau_{yz}}{\partial y}\,\mathrm{d}y \\[4pt] \mathrm{d}\tau_{zy} = \dfrac{\partial \tau_{zy}}{\partial z}\,\mathrm{d}z \end{cases}$$

Fig. 5. Forces acting on the infinitesimal parallelepiped

By substituting the stress variations with their values as defined in Fig. 5 and eliminating the product $\mathrm{d}x\,\mathrm{d}y\,\mathrm{d}z$, which appears in every element of the resulting expression, we get the first of the *differential equations of equilibrium*, which are

$$\begin{cases} \dfrac{\partial \sigma_x}{\partial x} + \dfrac{\partial \tau_{yx}}{\partial y} + \dfrac{\partial \tau_{zx}}{\partial z} + X = 0 \\[8pt] \dfrac{\partial \tau_{xy}}{\partial x} + \dfrac{\partial \sigma_y}{\partial y} + \dfrac{\partial \tau_{zy}}{\partial z} + Y = 0 \\[8pt] \dfrac{\partial \tau_{xz}}{\partial x} + \dfrac{\partial \tau_{yz}}{\partial y} + \dfrac{\partial \sigma_z}{\partial z} + Z = 0\,. \end{cases} \tag{5}$$

The last two expressions are obviously obtained from the conditions of equilibrium of translation in directions y and z, respectively.

Expressions 5 have been obtained by using the equilibrium conditions in a solid body in static equilibrium or in uniform motion. But it is very easy to generalize them to solids or liquids in non-uniform motion, by including the *inertial forces* in the body forces.

To this end, let us consider the situation represented in Fig. 5, for the case of no static balance. In this case, the resulting force is not zero, but induces an acceleration, which, in the most general case, has components in the three coordinate axes. Taking the direction x, for example, instead of expression 4, the fundamental equation of dynamics yields the relation

$$\underbrace{\mathrm{d}\sigma_x\,\mathrm{d}y\,\mathrm{d}z + \mathrm{d}\tau_{yx}\,\mathrm{d}x\,\mathrm{d}z + \mathrm{d}\tau_{zx}\,\mathrm{d}x\,\mathrm{d}y + X\,\mathrm{d}x\,\mathrm{d}y\,\mathrm{d}z}_{\text{force}}$$

$$= \underbrace{\rho\,\mathrm{d}x\,\mathrm{d}y\,\mathrm{d}z}_{\text{mass}}\ \overbrace{a_x}^{\text{acceleration}}\ \Rightarrow\ \dfrac{\partial \sigma_x}{\partial x} + \dfrac{\partial \tau_{yx}}{\partial y} + \dfrac{\partial \tau_{zx}}{\partial z} + \underbrace{X - \rho a_x}_{X_i} = 0\,, \tag{6}$$

where a_x represents the acceleration component in direction x and ρ is the density of the material. If we define the inertial forces

$$\begin{cases} X_i = -\rho a_x \\ Y_i = -\rho a_y \\ Z_i = -\rho a_z \;, \end{cases}$$

these may be treated as body forces in a body in static equilibrium, as stated by expression 6.

II.3.b Equilibrium at the Boundary

The balance conditions of the forces acting in the infinitesimal neighborhood of a point belonging to the boundary of the body may be established by considering an infinitesimal tetrahedron defined by three facets, whose semi-normals are parallel to the coordinate axes and by a facet on the boundary. Figure 6 shows this tetrahedron and the stresses and boundary forces per area unit $(\overline{X}, \overline{Y}, \overline{Z})$ acting on its faces. Since stresses and boundary forces may be considered as uniformly distributed, their resultants act on the centroids of the facets.

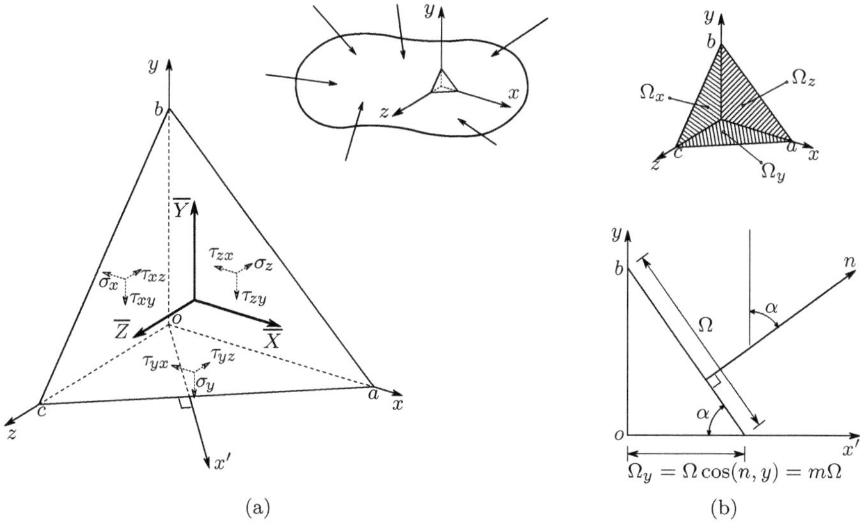

(a)   (b)

Fig. 6. Infinitesimal tetrahedron defined at the boundary of a body

The conditions expressing the rotation equilibrium around the axis of $\overline{X}, \overline{Y}$ and $\overline{Z}$ confirm the reciprocity of the shearing stresses, since the moments of the body forces acting on the tetrahedron do not need to be considered, because they are infinitesimal quantities of the fourth order, while the moments of the stress resultants are infinitesimal quantities of the third order (note that boundary forces and normal stresses are on the same lines).

The balance equation for the translation in direction x yields the expression (Fig. 6-a)

$$\overline{X}\Omega - \sigma_x\Omega_x - \tau_{yx}\Omega_y - \tau_{zx}\Omega_z = 0 \, , \tag{7}$$

where Ω_x, Ω_y, Ω_z, Ω represent the areas of the triangles obc, oac, oab, abc, respectively. Denoting the direction cosines of the semi-normal of the facet abc by l, m, n, the following relations are easily stated (cf. Fig. 6-b)

$$\Omega_x = l\Omega \qquad \Omega_y = m\Omega \qquad \Omega_z = n\Omega \, .$$

By substituting these relations in equation (7), we get the first of the *boundary balance equations*, which are

$$\begin{cases} l\sigma_x + m\tau_{yx} + n\tau_{zx} = \overline{X} \\ l\tau_{xy} + m\sigma_y + n\tau_{zy} = \overline{Y} \\ l\tau_{xz} + m\tau_{yz} + n\sigma_z = \overline{Z} \, . \end{cases} \tag{8}$$

The last two equations are obviously obtained from the conditions of equilibrium in the directions y and z, respectively. Expressions 8 are also valid in presence of inertial forces, since these, as body forces, lead to infinitesimal quantities of the higher order in the balance equations, so that they do not need to be considered.

II.4 Stresses in an Inclined Facet

The stresses acting on an inclined facet (a facet whose semi-normal is not parallel to any of the coordinate axes) may be obtained from the balance equations of the forces acting in an infinitesimal tetrahedron similar to the one in Fig. 6, with the difference that the triangle abc represents the inclined facet inside the body (Fig. 7-a).

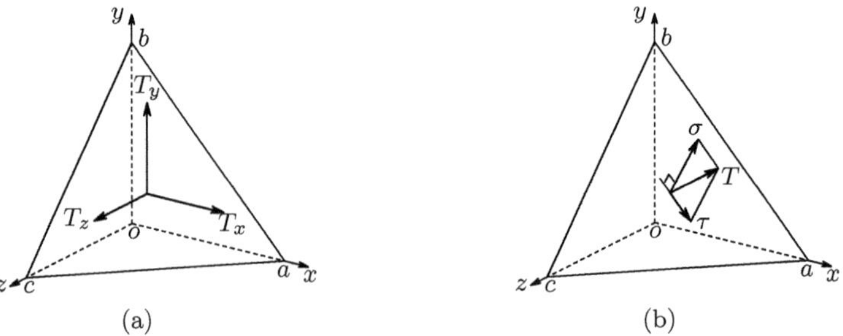

Fig. 7. Stresses in an inclined facet

Denoting by T_x, T_y, T_z the components in the reference directions of the stress vector acting on the facet abc and by l, m and n the direction cosines of its semi-normal, expression 8 directly gives the *Cauchy equations*

$$\begin{cases} T_x = l\sigma_x + m\tau_{yx} + n\tau_{zx} \\ T_y = l\tau_{xy} + m\sigma_y + n\tau_{zy} \\ T_z = l\tau_{xz} + m\tau_{yz} + n\sigma_z \ . \end{cases} \tag{9}$$

Using matrix notation, we may write

$$\underbrace{\begin{Bmatrix} T_x \\ T_y \\ T_z \end{Bmatrix}}_{\{T\}} = \underbrace{\begin{bmatrix} \sigma_x & \tau_{yx} & \tau_{zx} \\ \tau_{xy} & \sigma_y & \tau_{zy} \\ \tau_{xz} & \tau_{yz} & \sigma_z \end{bmatrix}}_{[\sigma]} \underbrace{\begin{Bmatrix} l \\ m \\ n \end{Bmatrix}}_{\{l\}} \ . \tag{10}$$

We may conclude that the elements of matrix $[\sigma]$ are sufficient to compute the stress in any inclined facet around point $\underline{o}$, which means that they completely define the *state of stress around point $\underline{o}$*. This matrix therefore defines the *stress tensor*. As a consequence of the reciprocity of the shearing stresses, only six of its nine components are independent, which means that six quantities are generally necessary (and sufficient) to define the stress state around a point.

The normal stress component is the projection of vector T in the direction of the semi-normal to the facet. Taking into consideration the reciprocity of the shearing stresses ($\tau_{xy} = \tau_{yx}$, $\tau_{xz} = \tau_{zx}$ and $\tau_{yz} = \tau_{zy}$), we get

$$\sigma = lT_x + mT_y + nT_z$$

$$= l^2\sigma_x + m^2\sigma_y + n^2\sigma_z + 2lm\tau_{xy} + 2ln\tau_{xz} + 2mn\tau_{yz} \ . \tag{11}$$

The magnitude of the shearing stress may be found by means of Pythagoras' theorem, $\tau^2 = T^2 - \sigma^2$ (Fig. 7-b). The components τ_x, τ_y and τ_z of the shearing stress in the reference directions may be obtained by subtracting the components of the normal stress σ to the components of the total stress T yielding

$$\begin{cases} \tau_x = T_x - l\sigma \\ \tau_y = T_y - m\sigma \\ \tau_z = T_z - n\sigma \ . \end{cases} \tag{12}$$

II.5 Transposition of the Reference Axes

Rotating the reference axes obviously causes a change in the components of the stress tensor. These are the stresses that act in facets, which are perpendicular to the new reference axes as shown in Fig. 8. Next we develop an expression to compute the new components of the tensor when the Cartesian rectangular reference system rotates.

Let us first consider the stress $T_{x'}$, which acts on the facets with a semi-normal x' and has the components $T_{x'x}$, $T_{x'y}$ and $T_{x'z}$ in the original reference

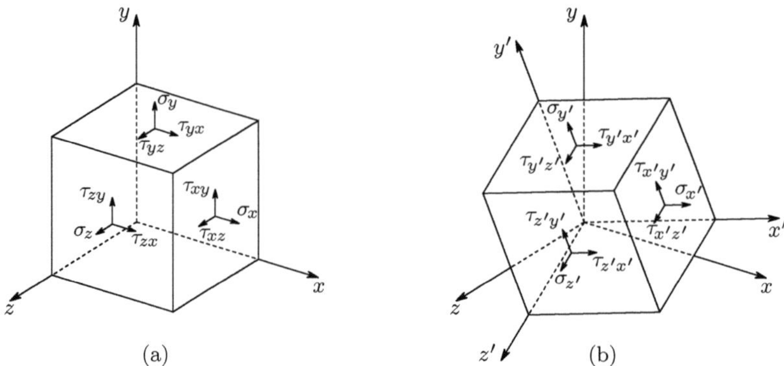

Fig. 8. Transposition of the reference axes

system (xyz). Changing the notation used for the direction cosines, expressions 10 give

$$
\left\{
\begin{array}{l}
l = (x',x) \\
m = (x',y) \\
n = (x',z)
\end{array}
\right\}
\Rightarrow
\left\{
\begin{array}{c}
T_{x'x} \\
T_{x'y} \\
T_{x'z}
\end{array}
\right\}
=
\begin{bmatrix}
\sigma_x & \tau_{yx} & \tau_{zx} \\
\tau_{xy} & \sigma_y & \tau_{zy} \\
\tau_{xz} & \tau_{yz} & \sigma_z
\end{bmatrix}
\left\{
\begin{array}{c}
(x',x) \\
(x',y) \\
(x',z)
\end{array}
\right\} .
$$

Proceeding in the same way in relation to the stresses acting in the facets with semi-normals y' and z', we get, in matrix notation

$$
\underbrace{
\begin{bmatrix}
T_{x'x} & T_{y'x} & T_{z'x} \\
T_{x'y} & T_{y'y} & T_{z'y} \\
T_{x'z} & T_{y'z} & T_{z'z}
\end{bmatrix}
}_{[T]}
=
\underbrace{
\begin{bmatrix}
\sigma_x & \tau_{yx} & \tau_{zx} \\
\tau_{xy} & \sigma_y & \tau_{zy} \\
\tau_{xz} & \tau_{yz} & \sigma_z
\end{bmatrix}
}_{[\sigma]}
\times
\underbrace{
\begin{bmatrix}
(x',x) & (y',x) & (z',x) \\
(x',y) & (y',y) & (z',y) \\
(x',z) & (y',z) & (z',z)
\end{bmatrix}
}_{[l]} .
$$

$$(13)$$

The elements of matrix $[T]$ are the stresses acting in the new facets (semi-normals x', y' and z'), but still represented by their components in the original xyz reference system. The components of the tensor in the new reference system $x'y'z'$ are the projections of the stresses $[T]$ in the directions $x'y'z'$. These components may be obtained by the matrix operation

$$
\underbrace{
\begin{bmatrix}
\sigma_{x'} & \tau_{y'x'} & \tau_{z'x'} \\
\tau_{x'y'} & \sigma_{y'} & \tau_{z'y'} \\
\tau_{x'z'} & \tau_{y'z'} & \sigma_{z'}
\end{bmatrix}
}_{[\sigma']}
=
\underbrace{
\begin{bmatrix}
(x',x) & (x',y) & (x',z) \\
(y',x) & (y',y) & (y',z) \\
(z',x) & (z',y) & (z',z)
\end{bmatrix}
}_{[l]^t}
\times
\underbrace{
\begin{bmatrix}
T_{x'x} & T_{y'x} & T_{z'x} \\
T_{x'y} & T_{y'y} & T_{z'y} \\
T_{x'z} & T_{y'z} & T_{z'z}
\end{bmatrix}
}_{[T]=[\sigma][l]} .
$$

$$(14)$$

Combining expressions 13 and 14, we get

$$
[\sigma'] = [l]^t [\sigma][l] . \tag{15}
$$

As the vectors in matrix $[l]$ are orthogonal and have unit length and since the scalar product of orthogonal vectors is zero, we get

$$[l]^{t}[l] = [I] = [l][l]^{t} \Rightarrow [\sigma] = [l][\sigma'][l]^{t} , \tag{16}$$

where $[I]$ represents the identity matrix.

II.6 Principal Stresses and Principal Directions

The stress tensor $[\sigma]$ may be seen as a linear operator, which transforms the unit vector represented by the semi-normal of the facet, with components l, m and n, in the vector of components T_x, T_y, T_z (the stress on the facet), as described by expression 10.

Since it is a symmetrical linear operator, it is known from the linear Algebra that it can always be diagonalized, that the three roots of its characteristic equation are all real and, if they are all different, its eigenvectors are orthogonal. Transposing these conclusions to the stress state around a point, this means that there are always three facets, perpendicular to each other, where the stress vector has the same direction as the normal to the facet. As a consequence, the shearing stress vanishes. The stresses in those *principal facets* are the *principal stresses* and their normals are the *principal directions* of the stress state.

In the following exposition, these notions are analysed and expressions for their computation from the components of the stress tensor in a rectangular Cartesian system are deduced. As far as possible, a physical analysis of the stress state will be preferred to a mathematical analysis of the linear operator $[\sigma]$, since, for the student of engineering, the physical understanding of the underlying phenomena is of crucial importance.

Let us consider a principal facet. The stress acting on it has only the normal component σ, so that the components of the stress vector are $T_x = l\sigma$, $T_y = m\sigma$ and $T_z = n\sigma$. Substituting these values in expression 10, we get the homogeneous system of linear equations

$$\underbrace{\begin{bmatrix} \sigma_x - \sigma & \tau_{xy} & \tau_{xz} \\ \tau_{xy} & \sigma_y - \sigma & \tau_{yz} \\ \tau_{xz} & \tau_{yz} & \sigma_z - \sigma \end{bmatrix}}_{[C]} \begin{Bmatrix} l \\ m \\ n \end{Bmatrix} = \begin{Bmatrix} 0 \\ 0 \\ 0 \end{Bmatrix} . \tag{17}$$

Such a system of equations has the trivial solution $l = m = n = 0$, and has other non-zero solutions only if there is a linear dependency between the equations, that is, if the determinant of the system matrix, $[C]$, vanishes. The direction cosines l, m and n cannot be zero simultaneously, since they are the components of a unit vector. Thus, the second possibility (zero determinant) must yield, as expressed by the condition

$$\begin{vmatrix} \sigma_x - \sigma & \tau_{xy} & \tau_{xz} \\ \tau_{xy} & \sigma_y - \sigma & \tau_{yz} \\ \tau_{xz} & \tau_{yz} & \sigma_z - \sigma \end{vmatrix} = -\sigma^3 + I_1\sigma^2 - I_2\sigma + I_3 = 0 . \tag{18}$$

In this expression the quantities I_1, I_2 and I_3 take the values

$$I_1 = \sigma_x + \sigma_y + \sigma_z$$

$$I_2 = \begin{vmatrix} \sigma_x & \tau_{xy} \\ \tau_{xy} & \sigma_y \end{vmatrix} + \begin{vmatrix} \sigma_x & \tau_{xz} \\ \tau_{xz} & \sigma_z \end{vmatrix} + \begin{vmatrix} \sigma_y & \tau_{yz} \\ \tau_{yz} & \sigma_z \end{vmatrix}$$

$$= \sigma_x\sigma_y + \sigma_x\sigma_z + \sigma_y\sigma_z - \tau_{xy}^2 - \tau_{xz}^2 - \tau_{yz}^2$$

$$I_3 = \begin{vmatrix} \sigma_x & \tau_{xy} & \tau_{xz} \\ \tau_{xy} & \sigma_y & \tau_{yz} \\ \tau_{xz} & \tau_{yz} & \sigma_z \end{vmatrix} = \sigma_x\sigma_y\sigma_z + 2\tau_{xy}\tau_{xz}\tau_{yz} - \sigma_x\tau_{yz}^2 - \sigma_y\tau_{xz}^2 - \sigma_z\tau_{xy}^2 .$$

The roots of equation (18) are the stresses, which satisfy equation (17), with non-simultaneous zero direction cosines l, m and n.[2] They represent the normal stresses in facets, where the shearing stress is zero, which means that they are principal stresses. The direction cosines of the normals to these facets – the principal directions – may be computed by substituting in Expression 17 σ for one of the roots of equation (18) and considering the supplementary condition $l^2 + m^2 + n^2 = 1$, since, with that substitution, equations (17) become linearly dependent ($|C| = 0$). Usually the principal stresses are denoted by σ_1, σ_2 and σ_3 with $\sigma_1 > \sigma_2 > \sigma_3$ (cf. example II.1).

The roots of equation (18) must not vary when the reference system is rotated, since they represent the principal stresses, which are intrinsic values of the stress state and therefore must not depend on the particular reference system used to describe the stress tensor. For this reason, equation (18) is designated as the *characteristic equation of the stress tensor*. The roots of this equation will be independent of the reference system if the coefficients I_1, I_2, I_3 are insensitive to coordinate changes. These coefficients are therefore *invariants of the stress tensor*.

Sometimes (for example in elasto-plastic computations) it is more convenient to define the invariants in the following way

$$J_1 = \sum_{i=1}^{3} \sigma_{ii} = I_1$$

$$J_2 = \frac{1}{2}\sum_{i=1}^{3}\sum_{j=1}^{3} \sigma_{ij}\sigma_{ij} = \frac{1}{2}I_1^2 - I_2$$

$$J_3 = \frac{1}{3}\sum_{i=1}^{3}\sum_{j=1}^{3}\sum_{k=1}^{3} \sigma_{ij}\sigma_{jk}\sigma_{ki} = \frac{1}{3}I_1^3 - I_1 I_2 + I_3 , \tag{19}$$

where $\sigma_{11} = \sigma_x$, $\sigma_{22} = \sigma_y$, $\sigma_{33} = \sigma_z$, $\sigma_{12} = \sigma_{21} = \tau_{xy}$, $\sigma_{13} = \sigma_{31} = \tau_{xz}$ and $\sigma_{23} = \sigma_{32} = \tau_{yz}$. These relations may be verified by direct substitution. The last verification is, however, rather time-consuming. Obviously, if I_1, I_2 and I_3 are invariant, J_1, J_2 and J_3 will also be.

---

[2] As components of a unit vector these direction cosines must obey the condition $l^2 + m^2 + n^2 = 1$.

II.6.a The Roots of the Characteristic Equation

The characteristic equation always has three real roots. In order to prove this statement, let us first remember that a third order polynomial equation always has at least one real root, since an odd-degree polynomial may take arbitrary high values, positive or negative, by assigning sufficiently high positive or negative values to the variable. Now, let us assume that one of the reference axes (for example axis z) is parallel to the principal direction, which corresponds to that real root. For simplicity, we will consider $\sigma_z \equiv \sigma_3$ (in this section we abandon the convention $\sigma_1 > \sigma_2 > \sigma_3$). In this case, the shearing stresses τ_{xz} and τ_{yz} will vanish and expression 17 takes the form

$$\begin{bmatrix} \sigma_x - \sigma & \tau_{xy} & 0 \\ \tau_{xy} & \sigma_y - \sigma & 0 \\ 0 & 0 & \sigma_z - \sigma \end{bmatrix} \begin{Bmatrix} l \\ m \\ n \end{Bmatrix} = \begin{Bmatrix} 0 \\ 0 \\ 0 \end{Bmatrix} . \tag{20}$$

The characteristic equation is therefore

$$(\sigma_z - \sigma) \begin{vmatrix} \sigma_x - \sigma & \tau_{xy} \\ \tau_{xy} & \sigma_y - \sigma \end{vmatrix} = 0$$

$$\Rightarrow (\sigma_z - \sigma) \left[x\sigma_y - (\sigma_x + \sigma_y)\sigma + \sigma^2 - \tau_{xy}^2 \right] = 0 . \tag{21}$$

One of the roots is obviously $\sigma = \sigma_z = \sigma_3$, as expected, since z is a principal direction. The other two roots may be obtained by solving the second degree equation

$$\sigma^2 - (\sigma_x + \sigma_y)\sigma + \left(\sigma_x\sigma_y - \tau_{xy}^2\right) = 0 .$$

The solution of this equation may be written as follows

$$\sigma = \frac{\sigma_x + \sigma_y}{2} \pm \frac{1}{2}\sqrt{\sigma_x^2 + 2\sigma_x\sigma_y + \sigma_y^2 - 4\sigma_x\sigma_y + 4\tau_{xy}^2}$$

$$= \frac{\sigma_x + \sigma_y}{2} \pm \frac{1}{2}\sqrt{\underbrace{(\sigma_x - \sigma_y)^2 + 4\tau_{xy}^2}_{\geq 0}} . \tag{22}$$

The roots of this equation are always real, since the binomial under the square root cannot take negative values. Therefore, there are always three real roots of the characteristic equation. The roots can, however, be double or even triple. For example, if the binomial is zero, we have for any pair of reference axes x, y of a plane perpendicular to axis z

$$(\sigma_x - \sigma_y)^2 + 4\tau_{xy}^2 = 0 \Rightarrow \begin{cases} \sigma_x = \sigma_y \\ \tau_{xy} = 0 \end{cases}$$

$$\Rightarrow \sigma_1 = \sigma_2 = \frac{\sigma_x + \sigma_y}{2} = \sigma_x = \sigma_y . \tag{23}$$

From this expression the conclusion may be drawn that, if two roots are equal (double root) and the third is different, then all the normal stresses of the plane, which is perpendicular to the principal direction corresponding to the third root (in this case the direction z and the plane x, y, respectively), are principal stresses and take the same value, since $\sigma_1 = \sigma_2 = \sigma_x = \sigma_y$ and $\tau_{xy} = 0$. We have, in this case, a stress state, which is axis-symmetric, i.e. symmetric in relation to an axis (the z axis, in this case).

If the three roots are equal (triple root), the shearing stress vanishes in every facet, as a similar analysis in any plane containing the z axis easily shows. Furthermore, the normal stress has the same value in every facet. Since the stresses do not vary with the orientation of the facet, we have an *isotropic stress state*. The components of this stress tensor are $\sigma_x = \sigma_y = \sigma_z = \sigma$ and $\tau_{xy} = \tau_{xz} = \tau_{yz} = 0$, regardless of the orientation of the reference system.

II.6.b Orthogonality of the Principal Directions

In the case of three different principal stresses, the corresponding principal directions are perpendicular to each other. This has already been implicitly demonstrated in the previous considerations, since the plane xy is perpendicular to direction z, which coincides with one of the principal directions. The orthogonality may, however be proved more clearly from expression 20.

The last equation in this expression is linearly independent of the other two, unless $\sigma = \sigma_z = \sigma_3$. In this last case, we must have

$$\begin{vmatrix} \sigma_x - \sigma & \tau_{xy} \\ \tau_{xy} & \sigma_y - \sigma \end{vmatrix} \neq 0 \, ,$$

since the value of σ, for which this determinant vanishes, is different from σ_3 (cf. (21)). Thus, the direction cosines must take the values $l = m = 0$ and $n = 1$, to obey equations (20). These are the direction cosines of direction z, as expected.

In the case of $\sigma \neq \sigma_3$, equations (20) are satisfied only if

$$n = 0 \qquad \text{and} \qquad \begin{vmatrix} \sigma_x - \sigma & \tau_{xy} \\ \tau_{xy} & \sigma_y - \sigma \end{vmatrix} = 0 \, ,$$

since, in this case, $l^2 + m^2 = 1$. As $n = 0$, the principal directions corresponding to the principal stresses σ_1 and σ_2 are in the plane xy, i.e. they are perpendicular to z. As axis z is parallel any of the three principal directions, they must be all be perpendicular to each other.

II.6.c Lamé's Ellipsoid

In the previous section we have demonstrated that there are always three orthogonal principal directions in a stress state. It is therefore always possible to choose a rectangular Cartesian reference system which coincides with the three principal directions. In this case, the shearing components of the stress tensor vanish and it takes the form

$$\begin{cases} \sigma_x = \sigma_1 \\ \sigma_y = \sigma_2 \\ \sigma_z = \sigma_3 \end{cases} \text{and} \quad \begin{cases} \tau_{xy} = 0 \\ \tau_{xz} = 0 \\ \tau_{yz} = 0 \end{cases} \Rightarrow [\sigma] = \begin{bmatrix} \sigma_1 & 0 & 0 \\ 0 & \sigma_2 & 0 \\ 0 & 0 & \sigma_3 \end{bmatrix}. \tag{24}$$

In an inclined facet, with a semi-normal defined by the direction cosines l, m, n, the relation between the components of the stress vector and the principal stresses may be deduced from expression 9, yielding

$$\begin{cases} T_1 = l\sigma_1 \\ T_2 = m\sigma_2 \\ T_3 = n\sigma_3 \end{cases} \Rightarrow \begin{cases} l = \dfrac{T_1}{\sigma_1} \\ m = \dfrac{T_2}{\sigma_2} \\ n = \dfrac{T_3}{\sigma_3}. \end{cases} \tag{25}$$

Since the direction cosines must obey the condition $l^2 + m^2 + n^2 = 1$, expression 25 gives

$$\frac{T_1^2}{\sigma_1^2} + \frac{T_2^2}{\sigma_2^2} + \frac{T_3^2}{\sigma_3^2} = 1 \,. \tag{26}$$

If we consider a Cartesian reference system T_1, T_2, T_3, this expression represents the equation of an ellipsoid, whose principal axes are the reference system and where the points on the ellipsoid are the tips P of the stress vectors $\overrightarrow{OP}$ (T_1, T_2, T_3) acting in facets containing the point with the stress state defined by expression 24 (point O, Fig. 9)

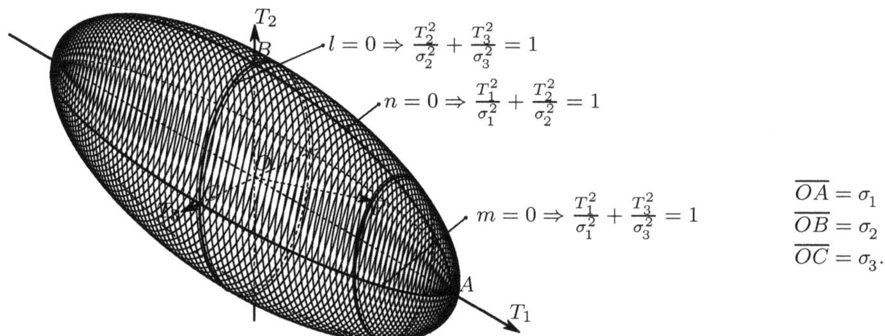

Fig. 9. Lamé's Ellipsoid or stress ellipsoid

This ellipsoid is a complete representation of the magnitudes of the stress vectors in facets around point O. It allows an important conclusion about the stress state: the magnitude of the stress in any facet takes a value between the maximum principal stress σ_1 and the minimum principal stress σ_3. It must be mentioned here that this conclusion is only valid for the absolute value of the stress, since in expression 26 only the squares of the stresses are considered.

From Fig. 9 we conclude immediately that if the absolute values of two principal stresses are equal the ellipsoid takes a shape of revolution around the third principal direction and if the three principal stresses have the same absolute value the ellipsoid becomes a sphere.

In the first case, the stress $\vec{T}$ acting in facets, which are parallel to the third principal direction have the same absolute value. Besides, if these two principal stresses have the same sign, we have an axisymmetric stress state, as concluded in Sect. II.6.a.

In the second case ($|\sigma_1| = |\sigma_2| = |\sigma_3|$), the stress $\vec{T}$ has the same magnitude in all facets. Furthermore, if $\sigma_1 = \sigma_2 = \sigma_3$, we have an isotropic stress state (cf. Sect. II.6.a).

II.7 Isotropic and Deviatoric Components of the Stress Tensor

The stress tensor may be considered as a system of forces in equilibrium, acting on an infinitesimal parallelepiped. Such a system may be decomposed in subsystems of forces in equilibrium.

When applying the stress theory to isotropic materials it is often necessary to separate the component of the stress tensor, which induces only volume changes in the material, from the component, which causes distortions. For example, as will be seen later in the study of the strain tensor and the constitutive law, the volume change in an isotropic material depends only on the isotropic component of the stress tensor

$$
\begin{bmatrix}
\sigma_m & 0 & 0 \\
0 & \sigma_m & 0 \\
0 & 0 & \sigma_m
\end{bmatrix}
\quad \text{with} \quad
\sigma_m = \frac{\sigma_x + \sigma_y + \sigma_z}{3} = \frac{I_1}{3} . \tag{27}
$$

The decomposition of the stress tensor may be described by the expression

$$
\begin{bmatrix}
\sigma_x & \tau_{yx} & \tau_{zx} \\
\tau_{xy} & \sigma_y & \tau_{zy} \\
\tau_{xz} & \tau_{yz} & \sigma_z
\end{bmatrix}
=
\underbrace{\begin{bmatrix}
\sigma_m & 0 & 0 \\
0 & \sigma_m & 0 \\
0 & 0 & \sigma_m
\end{bmatrix}}_{\substack{\text{isotropic tensor} \\ \text{component}}}
+
\underbrace{\begin{bmatrix}
\sigma_x - \sigma_m & \tau_{yx} & \tau_{zx} \\
\tau_{xy} & \sigma_y - \sigma_m & \tau_{zy} \\
\tau_{xz} & \tau_{yz} & \sigma_z - \sigma_m
\end{bmatrix}}_{\substack{\text{deviatoric tensor} \\ \text{component}}} .
$$

$$\tag{28}$$

In isotropic materials the deviatoric component of the stress tensor does not cause volume change, as will be seen later. In this tensor component the first invariant vanishes ($I_1' = \sigma_x + \sigma_x + \sigma_z - 3\sigma_m = 0$), which means that $J_2' = -I_2'$ and $J_3' = I_3'$ (cf. (19)).

II.8 Octahedral Stresses

Octahedral stresses are stresses acting in facets which are equally inclined in relation to the principal directions. Considering a reference system, where the axes lie in the principal directions of the stress state, the semi-normals of these facets have direction cosines with equal absolute values. Since there are eight facets obeying this condition (one in each of the eight trihedrons), they define one octahedron, which is symmetrical in relation to the principal planes (Fig. 10).

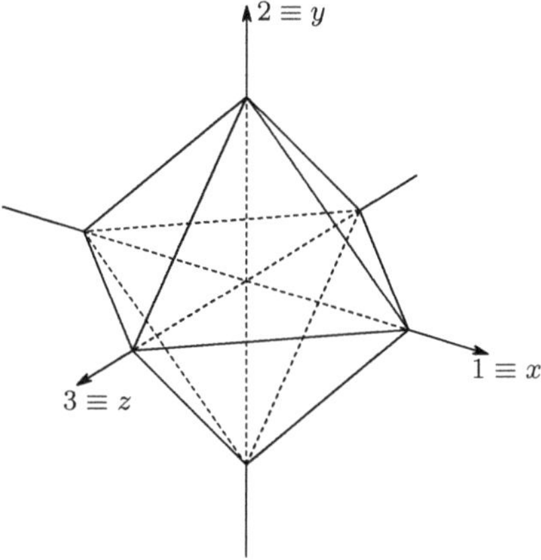

Fig. 10. Octahedron defined by equally inclined facets in relation to the principal directions 1, 2, 3

The direction cosines of the octahedral semi-normals take the values

$$
\begin{cases} |l| = |m| = |n| \\ l^2 + m^2 + n^2 = 1 \end{cases}
\Rightarrow
\begin{cases} l = \pm \dfrac{1}{\sqrt{3}} \\ m = \pm \dfrac{1}{\sqrt{3}} \\ n = \pm \dfrac{1}{\sqrt{3}} \, . \end{cases}
$$

As the reference system is a principal one, the shearing components of the stress tensor vanish. Therefore, the Cauchy equations (9) furnish the stress components

$$
\begin{cases} T_x = l\sigma_1 = \pm \dfrac{\sigma_1}{\sqrt{3}} \\ T_y = m\sigma_2 = \pm \dfrac{\sigma_2}{\sqrt{3}} \\ T_z = n\sigma_3 = \pm \dfrac{\sigma_3}{\sqrt{3}} \, . \end{cases}
$$

The normal component of the octahedral stress is then

$$\sigma_{oct} = lT_x + mT_y + nT_z = l^2\sigma_1 + m^2\sigma_2 + n^2\sigma_3$$

$$= \frac{\sigma_1 + \sigma_2 + \sigma_3}{3} = \frac{I_1}{3} = \frac{\sigma_x + \sigma_y + \sigma_z}{3} \, . \tag{29}$$

This stress coincides with the isotropic stress (cf. (27)).

The magnitude of the shearing component of the octahedral stress may be computed by using Pythagoras' theorem (cf. Sect. II.4), yielding

$$\tau_{oct}^2 = T_{oct}^2 - \sigma_{oct}^2 = T_x^2 + T_y^2 + T_z^2 - \sigma_{oct}^2 = l^2\sigma_1^2 + m^2\sigma_2^2 + n^2\sigma_3^2 - \sigma_{oct}^2$$

$$= \frac{1}{3}\underbrace{\left(\sigma_1^2 + \sigma_2^2 + \sigma_3^2\right)}_{I_1^2 - 2I_2} - \frac{I_1^2}{9} = \frac{2}{9}\left(I_1^2 - 3I_2\right) \, . \tag{30}$$

As the quantities I_1 and I_2 are insensitive to changes in the reference coordinates, the octahedral shearing stress may be expressed directly as a function of the components of the stress tensor in any rectangular Cartesian reference system xyz

$$\tau_{oct} = \frac{\sqrt{2}}{3}\sqrt{I_1^2 - 3I_2}$$

$$= \frac{\sqrt{2}}{3}\sqrt{\left(\sigma_x + \sigma_y + \sigma_z\right)^2 - 3\left(\sigma_x\sigma_y + \sigma_x\sigma_z + \sigma_y\sigma_z - \tau_{xy}^2 - \tau_{xz}^2 - \tau_{yz}^2\right)}$$

$$= \frac{1}{3}\sqrt{\left(\sigma_x - \sigma_y\right)^2 + \left(\sigma_y - \sigma_z\right)^2 + \left(\sigma_x - \sigma_z\right)^2 + 6\left(\tau_{xy}^2 + \tau_{xz}^2 + \tau_{yz}^2\right)} \, . \tag{31}$$

By substituting in the last expression σ_x, σ_y and σ_z for $\sigma_x - \sigma_m$, $\sigma_y - \sigma_m$ and $\sigma_z - \sigma_m$, respectively, we conclude immediately that the octahedral shearing stresses of the complete stress tensor and of its deviatoric component (28) are equal. As we shall see later (Sect. IV.7.b.v), the octahedral shearing stress plays an important role in one of the plastic yielding theories.

An even more simple expression of the octahedral shearing stress in terms of the invariants (cf. (30)) may be obtained by considering only the deviatoric tensor. For this purpose, we establish a relation between the second invariant of the deviatoric tensor, I_2', and the two first invariants of the complete stress tensor I_1 and I_2 (cf. (28))

$$I_2' = \begin{vmatrix} \sigma_x - \sigma_m & \tau_{xy} \\ \tau_{xy} & \sigma_y - \sigma_m \end{vmatrix} + \begin{vmatrix} \sigma_x - \sigma_m & \tau_{xz} \\ \tau_{xz} & \sigma_z - \sigma_m \end{vmatrix} + \begin{vmatrix} \sigma_y - \sigma_m & \tau_{yz} \\ \tau_{yz} & \sigma_z - \sigma_m \end{vmatrix}$$

$$= \begin{vmatrix} \sigma_x & \tau_{xy} \\ \tau_{xy} & \sigma_y \end{vmatrix} + \begin{vmatrix} \sigma_x & \tau_{xz} \\ \tau_{xz} & \sigma_z \end{vmatrix} + \begin{vmatrix} \sigma_y & \tau_{yz} \\ \tau_{yz} & \sigma_z \end{vmatrix}$$

$$\underbrace{+\,\sigma_m^2 - \left(\sigma_x + \sigma_y\right)\sigma_m + \sigma_m^2 - \left(\sigma_x + \sigma_z\right)\sigma_m + \sigma_m^2 - \left(\sigma_y + \sigma_z\right)\sigma_m}_{-3\sigma_m^2}$$

$$\Rightarrow \qquad I_2' = I_2 - \frac{I_1^2}{3} \qquad \Leftrightarrow \qquad I_1^2 - 3I_2 = -3I_2' \ .$$

Substituting in Expression 30, we get

$$\tau_{oct}^2 = -\frac{2}{3}I_2' \ . \tag{32}$$

From this expression we conclude that the second invariant of the deviatoric stress tensor always takes a negative value.

The third invariant of the deviatoric stress tensor, I_3', may also be expressed in terms of the invariants of the complete tensor, as follows

$$I_3' = \begin{vmatrix} \sigma_x - \sigma_m & \tau_{xy} & \tau_{xz} \\ \tau_{xy} & \sigma_y - \sigma_m & \tau_{yz} \\ \tau_{xz} & \tau_{yz} & \sigma_z - \sigma_m \end{vmatrix} = \begin{vmatrix} \sigma_x & \tau_{xy} & \tau_{xz} \\ \tau_{xy} & \sigma_y & \tau_{yz} \\ \tau_{xz} & \tau_{yz} & \sigma_z \end{vmatrix}$$

$$- \underbrace{\left(\sigma_y \sigma_z + \sigma_x \sigma_z + \sigma_x \sigma_y - \tau_{yz}^2 - \tau_{xz}^2 - \tau_{xy}^2 \right)}_{I_2} \sigma_m + \underbrace{\left(\sigma_x + \sigma_y + \sigma_z \right)}_{3\sigma_m} \sigma_m^2 - \sigma_m^3$$

$$= I_3 - I_2 \sigma_m + 2\sigma_m^3 = I_3 - \tfrac{1}{3}I_1 I_2 + \tfrac{2}{27}I_1^3 \ .$$

II.9 Two-Dimensional Analysis of the Stress Tensor

II.9.a Introduction

In many applications of the stress theory, one of the principal directions is known. As examples, we may refer the stress state at the surface of a body (in the very common case of no tangential surface loads), the stress state in a thin plate under in-plane forces, the stress states induced by the normal and shear forces and by the bending and torsion moments in bars, etc. In many cases, the principal stress, which corresponds to the known principal direction, is zero, as in the referred case of the thin plate, or in the surface of a body, where there are no external forces applied. In this case we have a *plane stress state*.

In any of these cases, a two-dimensional analysis of the stress tensor is enough to compute the remaining two principal stresses and directions. Since the three principal directions are perpendicular to each other, the remaining two principal directions act in facets, which are parallel to the known principal direction. Therefore, only this family of facets needs to be considered. As this two-dimensional analysis is considerably simpler than a three-dimensional one, a deeper insight into the stress state is possible.

The two-dimensional analysis could be performed by particularizing the expressions developed for the three-dimensional case and by developing them further in the simplified two-dimensional form. However, in the following account, the two-dimensional expressions will be deduced from scratch, i.e. without using the three-dimensional framework described in the previous sections. This option is useful because it allows the two-dimensional case to be understood, without first having to learn the more demanding three-dimensional one. As a side effect, some of the conclusions obtained in the general case will be repeated in the two-dimensional analysis, although they are obtained in a different way.

For simplicity, we will consider that the known principal direction is direction 3, and that that direction coincides with axis z. Thus the two-dimensional analysis is performed in plane xy, by considering facets which are perpendicular to this plane, and in which there are no shearing stresses with a z-component, since z is a principal direction.

II.9.b Stresses on an Inclined Facet

Let us consider a triangular prism, where two of the lateral faces are perpendicular to the coordinate axes x and y and the third lateral face has an orientation defined by the angle θ between its semi-normal and axis x. Figure 11 illustrates this prism and the stresses acting in its facets.

The equilibrium condition of the forces acting in direction θ yields

$$\sigma_\theta \, \mathrm{d}z \, \mathrm{d}s = \sigma_x \, \mathrm{d}z \, \mathrm{d}y \cos\theta + \sigma_y \, \mathrm{d}z \, \mathrm{d}x \sin\theta + \tau_{xy} \, \mathrm{d}z \, \mathrm{d}y \sin\theta + \tau_{yx} \, \mathrm{d}z \, \mathrm{d}x \cos\theta \ ,$$

or, as $\mathrm{d}x = \mathrm{d}s \sin\theta$ and $\mathrm{d}y = \mathrm{d}s \cos\theta$,

$$\sigma_\theta = \sigma_x \cos^2\theta + \sigma_y \sin^2\theta + 2\tau_{xy} \sin\theta \cos\theta \ . \tag{33}$$

Similarly, the equilibrium condition in the perpendicular direction $(\theta \pm \frac{\pi}{2})$ yields the relation

$$\tau_\theta \, \mathrm{d}z \, \mathrm{d}s + \sigma_x \, \mathrm{d}z \, \mathrm{d}y \sin\theta + \tau_{yx} \, \mathrm{d}z \, \mathrm{d}x \sin\theta = \tau_{xy} \, \mathrm{d}z \, \mathrm{d}y \cos\theta + \sigma_y \, \mathrm{d}z \, \mathrm{d}x \cos\theta \ .$$

Simplifying, we get

$$\tau_\theta = (\sigma_y - \sigma_x) \sin\theta \cos\theta + \tau_{xy} \left(\cos^2\theta - \sin^2\theta \right) \ . \tag{34}$$

Expressions 33 to 34 show that the stresses σ_x, σ_y and τ_{xy} allow the computation of the stresses in an arbitrary facet, whose orientation is defined by angle θ. They thus fully define the two-dimensional stress state around point P (Fig. 11). These stresses are the components of the stress tensor in the reference system xy.

The expressions 33 and 34 may be given another form, if we take into account the trigonometric relations

$$\sin\theta \cos\theta = \frac{\sin 2\theta}{2} \qquad \sin^2\theta = \frac{1 - \cos 2\theta}{2} \qquad \cos^2\theta = \frac{1 + \cos 2\theta}{2} \ .$$

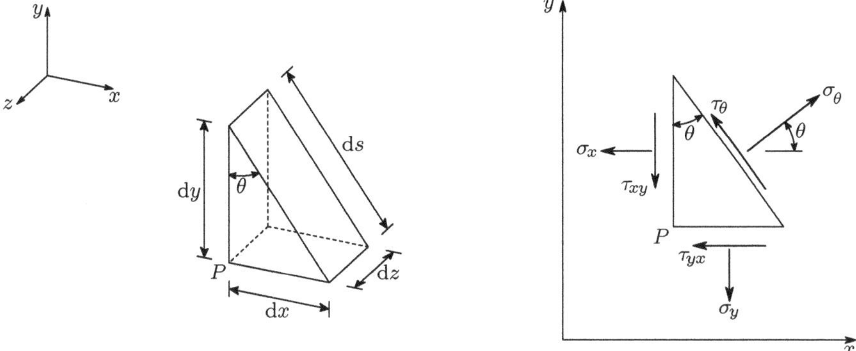

Fig. 11. Infinitesimal prism used in the two-dimensional analysis of the stress state

Substituting these relations in expression 33, we get

$$\sigma_\theta = \frac{\sigma_x + \sigma_y}{2} + \frac{\sigma_x - \sigma_y}{2}\cos 2\theta + \tau_{xy}\sin 2\theta \ . \tag{35}$$

In the same way, expression 34 becomes

$$\tau_\theta = -\frac{\sigma_x - \sigma_y}{2}\sin 2\theta + \tau_{xy}\cos 2\theta \ . \tag{36}$$

II.9.c Principal Stresses and Directions

Expressions 35 and 36 furnish the normal and shearing components of the stress acting in facet θ, as functions of the stress tensor components σ_x, σ_y and τ_{xy}. With these expressions, the evolution of σ_θ and τ_θ with the facet orientation θ may be analysed. Differentiating expression 35 in relation to θ and equating to zero gives

$$\frac{\partial \sigma_\theta}{\partial \theta} = -(\sigma_x - \sigma_y)\sin 2\theta + 2\tau_{xy}\cos 2\theta = 0$$
$$\Rightarrow \tan 2\theta = \frac{2\tau_{xy}}{\sigma_x - \sigma_y} \Rightarrow \theta = \frac{1}{2}\arctan\frac{2\tau_{xy}}{\sigma_x - \sigma_y} \ . \tag{37}$$

Expression 37 yields two values of θ (θ_1 and $\theta_2 = \theta_1 + \frac{\pi}{2}$), which correspond to a maximum and a minimum of σ_θ. By substituting expression 37 in expression 36, we get $\tau_\theta = 0$. This means that, in a two-dimensional stress state, there are always two orthogonal directions which define facets where the shearing stress takes a zero value and where the normal stress takes its minimum and maximum values. These directions are the *principal directions* and the corresponding values of the stress are the *principal stresses*. Usually, the maximum principal stress is denoted by σ_1 and the minimum by σ_2.[3]

---

[3] As we have seen in the three-dimensional analysis of the stress state, there is a third principal stress in a parallel facet to the plane xy. A descending ordering of the

These stresses may be computed by substituting in Expression 35 θ with the values of θ_1 and θ_2 given by (37). To this end, the following trigonometric relations are used

$$\cos 2\theta = \frac{1}{\pm\sqrt{1 + \tan^2 2\theta}} \qquad \sin 2\theta = \frac{\tan 2\theta}{\pm\sqrt{1 + \tan^2 2\theta}} \; .$$

By substituting the last but one of Equations 37 in these expressions and the result in Expression 35, after some manipulation we get

$$\left\{ \begin{array}{c} \sigma_1 \\ \sigma_2 \end{array} \right\} = \frac{\sigma_x + \sigma_y}{2} \left\{ \begin{array}{c} + \\ - \end{array} \right\} \sqrt{\left(\frac{\sigma_x - \sigma_y}{2}\right)^2 + \tau_{xy}^2} \; . \tag{38}$$

It should be noted here that σ_1 does not necessarily corresponds to the direction of θ_1, as defined above, since in Equation 38 the convention $\sigma_1 \geq \sigma_2$ is used. These values are not known when the directions θ_1 and θ_2 are obtained from Expression 37.

The value of θ, which corresponds to each of the principal stresses given by expression 38 may, however, be computed easily by using a relation which is deduced directly from the equilibrium condition, in direction x or y, of the forces acting in the prism shown in Fig. 11. By considering $\theta = \theta_1$ and, as a consequence, $\sigma_\theta = \sigma_1$ and $\tau_\theta = 0$, the equilibrium condition in direction x yields

$$\sigma_1 \, ds \cos\theta_1 = \sigma_x \, \overbrace{ds \cos\theta_1}^{dy} + \tau_{xy} \, \underbrace{ds \sin\theta_1}_{dx} \Rightarrow \tan\theta_1 = \frac{\sigma_1 - \sigma_x}{\tau_{xy}} \; .$$

The equilibrium condition in direction y gives the relation $\tan\theta_1 = \frac{\tau_{xy}}{\sigma_1 - \sigma_y}$.

If the reference system coincides with the principal directions, the shearing stress is zero and the normal and shearing stresses acting on an arbitrary facet are

$$\left\{ \begin{array}{l} \sigma_x = \sigma_1 \\ \sigma_y = \sigma_2 \\ \tau_{xy} = 0 \end{array} \right. \Rightarrow \left\{ \begin{array}{l} \sigma = \dfrac{\sigma_1 + \sigma_2}{2} + \dfrac{\sigma_1 - \sigma_2}{2} \cos 2\alpha \\[2ex] \tau = -\dfrac{\sigma_1 - \sigma_2}{2} \sin 2\alpha \; , \end{array} \right. \tag{39}$$

where α is the angle between the principal direction 1 and the semi-normal to the facet, as shown in Fig. 11 (with the principal direction 1 in the place of axis x).

---

principal stresses could demand a different ordering (for example, $\sigma_z = \sigma_1$ or $\sigma_z = \sigma_2$ instead of $\sigma_z = \sigma_3$). However, for the sake of simplicity, in the two-dimensional case we adopt the descending order only for the principal stresses lying in the plane xy.

II.9.d Mohr's Circle

In the last two sections we have implicitly adopted a sign convention for the shearing stresses, where a positive stress corresponds to the y-direction in a referential system, which is obtained by a rotation θ in the direct direction (counterclockwise), of the xy referential system represented in Fig. 11. This means that a positive shearing stress in the inclined facet corresponds to a rotation, in the direct direction, around point P.

If we adopt the opposite convention – the shearing stress is positive, when it defines a clockwise rotation – the negative sign in the second of Expressions 39 disappears. In this case, these expressions are the parametric equations of a circle in a rectangular Cartesian reference system σ-τ. This circle can be used to represent the whole stress state graphically, since each point in the circle represents the stress vector in a facet, whose orientation is defined by angle α (cf. (39)). This representation of the stress tensor was developed by the end of the 19th century by Otto Mohr and it still remains very popular, despite the decline of the graphic methods with the emergence of computational tools, because of its simplicity and capacity for visualizing the whole stress state.

Representing the normal stresses in the axis of abscissas (horizontal direction), and the shearing stress in the axis of ordinates (with the second sign convention defined above) Expressions 39 define a circle with radius $\frac{\sigma_1-\sigma_2}{2}$, whose center is the point of abscissa $\frac{\sigma_1+\sigma_2}{2}$ and zero ordinate, as shown in Fig. 12. Point A represents the facet with a semi-normal, whose orientation is defined by an angle α measured from the principal direction 1, positive in the counterclockwise direction. (Fig. 11, with $\theta = \alpha$). Orthogonal facets are represented by opposite points in the Mohr's circle, since an α-rotation of the facet corresponds to a 2α-rotation of its representation.

From a quick glance at Fig. 12, the following conclusions may immediately be drawn:

- the maximum value of the shearing stress is $\tau_{\max} = \frac{\sigma_1-\sigma_2}{2}$ (radius of the Mohr's circle);
- the maximum shearing stress occurs in facets with a $45°$-orientation, in relation to the principal directions ($2\alpha = 90°$ – point B – and $2\alpha = 270°$ – point C);
- in the facets where the normal stress attains its extreme values the shearing stress takes a zero value (points on the axis of abscissas).

Irradiation Poles

The *irradiation poles* enable a graphic relation to be established between the Mohr's circle and the facet representation (Fig. 11). Irradiation poles for the facets and for the normals to the facets may be defined. Figure 13 presents the graphical construction leading to the facet irradiation pole.

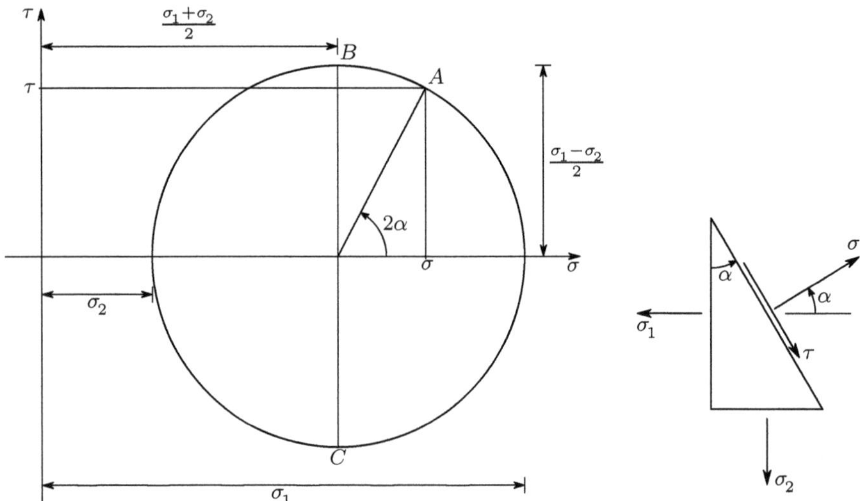

Fig. 12. Mohr's representation for the two-dimensional stress state

In this figure, the position of the facet irradiation pole I_f is first obtained by drawing a parallel line to facet a, which contains the point representing this facet in the Mohr's circle (obviously, facet b could also be used). The point representing the generic facet c on the Mohr's circle may then be obtained by drawing a line passing by the irradiation pole I_f, which is parallel to facet c. This line intersects the Mohr's circle in the point which represents facet c: σ_c and τ_c are the normal and shearing stresses acting on facet c. The rightness of this procedure is easily demonstrated: as the angle between facets a and c is β, their representations on the Mohr's circle (a and c on the circle, Fig. 13) are at the distance defined by the central angle 2β. As a consequence, the circumferential angle $(a,\ I_f,\ c)$ measures β, since it must take half the value of 2β. Thus, if the line $\overline{aI_f}$ is parallel to the facet a, then the line $\overline{cI_f}$ is parallel to facet c.

If the direction of the normals to the facets is used instead of the facet direction, the irradiation pole of the normals, I_f, is obtained (Fig. 13). Most times, the irradiation pole of the facets is used. For simplicity, it is usually denoted by I.

The principal directions may be obtained directly from the irradiation poles:

– if the irradiation pole of the facets is used, the line joining this point with the one representing the facet where the principal stress σ_1 acts (point B), is parallel to this facet; thus, it corresponds to principal direction 2; in the same way the line $\overline{I_f\,A}$ gives the principal direction 1;

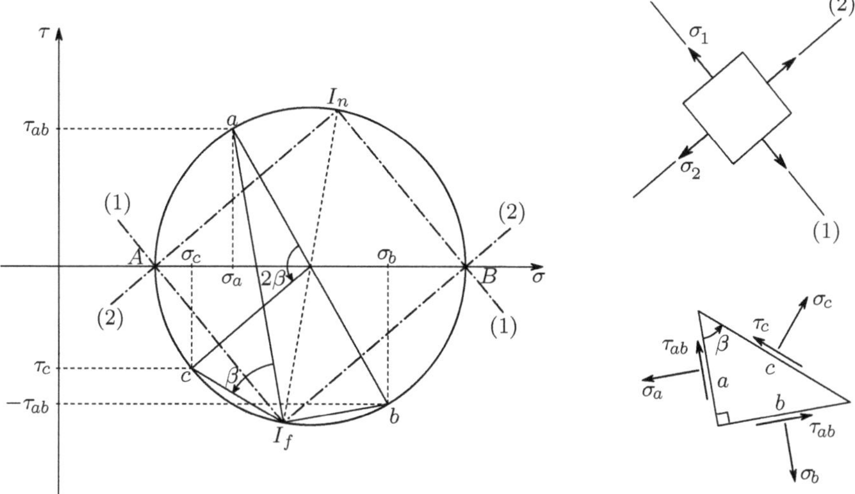

Fig. 13. Irradiation poles

– if the irradiation pole of the normals is used, as principal direction 1 is the
normal to the facet, where σ_1 acts (point B), then the line $\overline{I_n\,B}$ is parallel
to the principal direction 1.

II.10 Three-Dimensional Mohr's Circles

If a two-dimensional analysis is performed in each of the principal planes
(planes defined by the principal directions), it is easily concluded that the
stresses in the three families of facets that are parallel to each of the three
principal directions may be represented in the Mohr's plane by the three circles
defined by the three pairs of principal stresses, as shown in Fig. 14.

The facets which are not parallel to any of the principal directions are
represented by points contained in the shaded area of Fig. 14.[4] The demonstration of this statement is based on the solution of the system of equations
(cf. (2), (11) and (12))

$$\begin{cases} l^2 + m^2 + n^2 = 1 \\ l^2\sigma_1 + m^2\sigma_2 + n^2\sigma_3 = \sigma \\ l^2\sigma_1^2 + m^2\sigma_2^2 + n^2\sigma_3^2 = \sigma^2 + \tau^2 \end{cases} \Leftrightarrow \begin{bmatrix} 1 & 1 & 1 \\ \sigma_1 & \sigma_2 & \sigma_3 \\ \sigma_1^2 & \sigma_2^2 & \sigma_3^2 \end{bmatrix} \left\{ \begin{array}{c} l^2 \\ m^2 \\ n^2 \end{array} \right\} = \left\{ \begin{array}{c} 1 \\ \sigma \\ \sigma^2 + \tau^2 \end{array} \right\}.$$

$$(40)$$

---

[4]Only the upper half is considered, since it is not possible to make a general
distinction between positive and negative shearing stresses in an inclined facet in a
three-dimensional space.

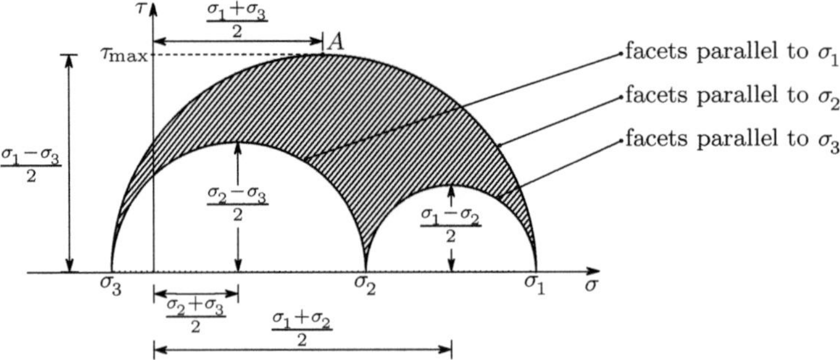

Fig. 14. Mohr's representation of the stress state in the three-dimensional case

The solution of this system may be obtained by means of determinants, yielding

$$l^2 = \frac{1}{D} \begin{vmatrix} 1 & 1 & 1 \\ \sigma & \sigma_2 & \sigma_3 \\ \sigma^2 + \tau^2 & \sigma_2^2 & \sigma_3^2 \end{vmatrix} = \frac{\tau^2 + (\sigma - \sigma_2)(\sigma - \sigma_3)}{(\sigma_1 - \sigma_2)(\sigma_1 - \sigma_3)}$$

$$m^2 = \frac{1}{D} \begin{vmatrix} 1 & 1 & 1 \\ \sigma_1 & \sigma & \sigma_3 \\ \sigma_1^2 & \sigma^2 + \tau^2 & \sigma_3^2 \end{vmatrix} = \frac{\tau^2 + (\sigma - \sigma_1)(\sigma - \sigma_3)}{(\sigma_2 - \sigma_1)(\sigma_2 - \sigma_3)}$$

$$n^2 = \frac{1}{D} \begin{vmatrix} 1 & 1 & 1 \\ \sigma_1 & \sigma_2 & \sigma \\ \sigma_1^2 & \sigma_2^2 & \sigma^2 + \tau^2 \end{vmatrix} = \frac{\tau^2 + (\sigma - \sigma_1)(\sigma - \sigma_2)}{(\sigma_3 - \sigma_1)(\sigma_3 - \sigma_2)} \, ,$$

where $D = (\sigma_1 - \sigma_2)(\sigma_2 - \sigma_3)(\sigma_3 - \sigma_1)$ is the system's determinant (Expr. 40). After some algebraic manipulations, these expressions may be given the forms (cf. e.g. [1])

$$\tau^2 + \left(\sigma - \frac{\sigma_2 + \sigma_3}{2} \right)^2 = \left(\sigma_1 - \frac{\sigma_2 + \sigma_3}{2} \right)^2 l^2 + \left(\frac{\sigma_2 - \sigma_3}{2} \right)^2 (1 - l^2)$$

$$\tau^2 + \left(\sigma - \frac{\sigma_1 + \sigma_3}{2} \right)^2 = \left(\sigma_2 - \frac{\sigma_1 + \sigma_3}{2} \right)^2 m^2 + \left(\frac{\sigma_1 - \sigma_3}{2} \right)^2 (1 - m^2)$$

$$\tau^2 + \left(\sigma - \frac{\sigma_1 + \sigma_2}{2} \right)^2 = \left(\sigma_3 - \frac{\sigma_1 + \sigma_2}{2} \right)^2 n^2 + \left(\frac{\sigma_1 - \sigma_2}{2} \right)^2 (1 - n^2) \, .$$

$$(41)$$

Considering the first of these equations, for example, we easily confirm that it represents, in the (σ, τ)-plane, a family of circles with center in the point of coordinates $\sigma = \frac{\sigma_2 + \sigma_3}{2}$ and $\tau = 0$. The radius of each circle depends on the value of l^2, which varies between 0 and 1. As this equation depends linearly on l^2, the extreme values of the radius are $\frac{\sigma_2 - \sigma_3}{2}$ and $\sigma_1 - \frac{\sigma_2 + \sigma_3}{2}$, respectively for $l^2 = 0$ and $l^2 = 1$. The other two equations represent the other two families of circles, as shown in Fig. 15.

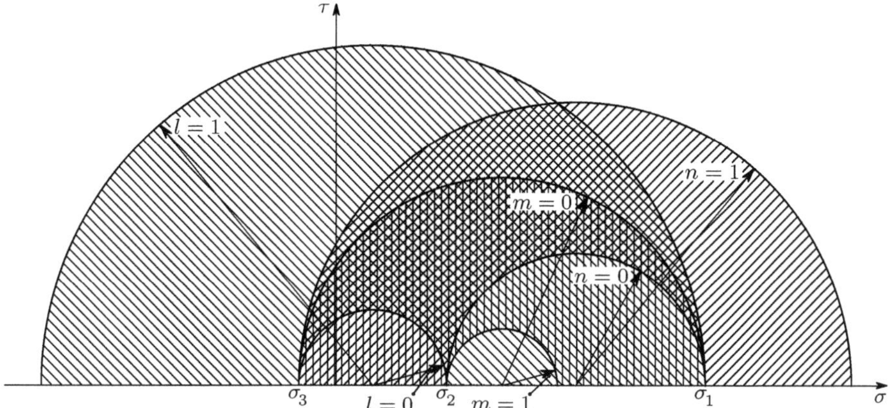

Fig. 15. Families of circles described by Expressions 41 in the Mohr's plane

The normal and shearing stresses, σ and τ, in a facet, whose semi-normal has an orientation defined by the direction cosines l, m and n in a principal reference system (axes parallel to the principal directions o the stress tensor), must obey the three Expressions 41. As we have, simultaneously,

$$0 \leq l^2 \leq 1 \qquad 0 \leq m^2 \leq 1 \qquad 0 \leq n^2 \leq 1 \,, \tag{42}$$

the points representing facets of the stress state defined by σ_1, σ_2 and σ_3 in the Mohr's plane must be on the surface containing the points whose coordinates obey the conditions 41 and 42. They are therefore on the shaded area of Fig. 14, which corresponds to the triple shaded area in Fig. 15.

The point representing a facet defined by a set of direction cosines may be found by the intersection of two of the three circles defined by Equations 41, for the corresponding values of l, m and n. The three circles intersect in this point, since the three Equations 41 must be satisfied simultaneously.

The position of this point can also be obtained graphically. However, as the explanation of the corresponding procedure is relatively lengthy and the importance of the quantitative graphical methods has substantially declined since the appearance of the computer, this method is not described here. Quite a detailed description of this procedure can be found in reference [1].

The actual importance of the Mohr's representation of the stress tensor resides in the fact, that it provides a simple global visualization of the stress state, making some conclusions obvious whose demonstration would be more difficult by other methods. From it we conclude, for example, that the maximum shearing stress occurs in facets which are parallel to the middle principal direction (direction of σ_2) and make a $45°$-angle with the directions of the maximum and minimum principal stresses (point A in Fig. 14). In these facets the normal and shearing stress take the values $\frac{\sigma_1+\sigma_3}{2}$ and $\frac{\sigma_1-\sigma_3}{2}$, respectively, as shown in Fig. 14. We can also confirm the conclusion obtained by means of

Lamé's ellipsoid, that σ_1 and σ_3 are the extreme values attained by the total stress acting in the family of facets, passing through the point, whose stress tensor has the principal stresses σ_1, σ_2 and σ_3.

II.11 Conclusions

In the theory presented in this Chapter, we have mainly analysed the stress state around a point, i.e. in an infinitesimal neighborhood of a point inside or on the surface of a solid body, or of a liquid mass, under the action of forces. This spatial restriction makes it possible to treat the stress state as homogeneous, i.e., as if it would not vary from point to point.

The expressions defining the components of the stress tensor as functions of the coordinates x, y and z were used only to develop the differential equations of equilibrium (5). In the rest of the Chapter, only the elements of the stress tensor at a given point were considered Those functions, however, play an important role in the analytical solutions for the stress distribution inside a body, which are obtained using the Theory of Elasticity (cf. e.g. Reference [4]).

In relation to the sign conventions used for the normal and shearing stresses, it should be mentioned, that, while the same convention could always be retained for the normal stresses, in the case of the shearing stresses it was necessary to abandon the initial convention (the Von-Karman convention), when studying the Mohr's representation. This is a consequence of the fact that the sign convention for the normal stresses is based on the physical concept of the tensile force as a force which causes an increase of the distance between two points, while such a physically grounded convention does not exist for the shearing stresses. In fact, in the Von Karman convention the positive stress corresponds to the positive direction of the reference axes, which have arbitrary directions; in the Mohr's circle, the positive shearing stress is defined by a direction of rotation, which depends on the observer's position. Finally, it is not possible to define a positive direction for the shearing stress in a facet in the three-dimensional case. For these reasons, a physical distinction between positive and negative stresses only makes sense for the normal stresses.

The validity of the theory expounded in this chapter is only limited by the hypothesis of continuity. Thus, it is valid in a solid with small or large deformations, in static equilibrium or in dynamic motion, or in a fluid in steady or unsteady motion. However, in the case of a solid body under finite deformations (deformations which are not small enough to be considered as infinitesimal quantities), it should be noted, that the coordinates x, y, z of the points of the body refer to the deformed configuration and not to the initial geometry of the body.

There are, however, tensors which describe the stress state using the coordinates corresponding to the undeformed geometry of the body, even in the case of large deformations (Lagrange and Piola-Kirchhoff stress tensors, cf., e.g. [2]). However, the deformations of the structures used in the engineering problems, which are solved by means of the Solid Mechanics (Civil, Mechanical, Aeronautical Engineering, etc. structures) are mostly small enough, to be treated as infinitesimal quantities. Furthermore, as the study of these tensors is rather involved and fairly abstract, they are not included in this introduction to the Mechanics of Materials.

II.12 Examples and Exercises

II.1. Using the theory described in Sects. II.6 to II.8, derive expressions for the direct computation of the principal stresses and directions.

Principal Stresses

Considering only the deviatoric component of the stress tensor, the corresponding characteristic equation takes the form

$$\sigma'^3 + I_2'\sigma' - I_3' = 0 ,$$

where $I_2' < 0$, as demonstrated in Sect. II.8. As the three principal stresses always exist (Sect. II.6.a), we can compute the roots of this equation using the algorithm (cf., e.g. [5], Sect. 2.4.2.3, or [2], prob. 3.5)

$$
\begin{aligned}
\sigma_1' &= 2\sqrt[3]{\alpha}\cos\beta \\
\sigma_2' &= 2\sqrt[3]{\alpha}\cos\left(\beta + \frac{4\pi}{3}\right) \\
\sigma_3' &= 2\sqrt[3]{\alpha}\cos\left(\beta + \frac{2\pi}{3}\right)
\end{aligned}
\quad\text{with}\quad
\begin{cases}
\alpha = \sqrt{-\dfrac{I_2'^3}{27}} \\[2ex]
\beta = \dfrac{1}{3}\arccos\left(\dfrac{I_3'}{2\alpha}\right) .
\end{cases}
$$

α is always a real quantity, since $I_2' < 0$. The expression of parameter β shows that it takes values between 0 and $\frac{\pi}{3}$, since $0 \le 3\beta \le \pi$. With these limits, it is easily verified that we always have

$$
\begin{cases}
0.5 \le \cos\beta \le 1 \\
-0.5 \le \cos\left(\beta + \dfrac{4\pi}{3}\right) \le 0.5 \\
-1 \le \cos\left(\beta + \dfrac{2\pi}{3}\right) \le -0.5
\end{cases}
\Rightarrow \cos\beta \ge \cos\left(\beta + \frac{4\pi}{3}\right) \ge \cos\left(\beta + \frac{2\pi}{3}\right) .
$$

From this we conclude that $\sigma_1' \ge \sigma_2' \ge \sigma_3'$.

The principal stresses of the total stress state may then be found by adding the isotropic stress (mean normal stress) to the principal stress of the deviatoric tensor component

$$\sigma_i = \sigma_i' + \frac{\sigma_x + \sigma_y + \sigma_z}{3} \quad\text{with}\quad i = 1, 2, 3 .$$

It is obvious that $\sigma_1' \ge \sigma_2' \ge \sigma_3'$ implies $\sigma_1 \ge \sigma_2 \ge \sigma_3$.

Principal Directions

An expression for the computation of the direction cosines of principal direction i may be obtained by ascertaining that the values of l, m and n given by

$$l = KL \qquad m = KM \qquad n = KN \,,$$

where K is an arbitrary constant and L, M and N take the values

$$L = \begin{vmatrix} \sigma_y - \sigma_i & \tau_{yz} \\ \tau_{yz} & \sigma_z - \sigma_i \end{vmatrix} \qquad M = - \begin{vmatrix} \tau_{xy} & \tau_{yz} \\ \tau_{xz} & \sigma_z - \sigma_i \end{vmatrix} \qquad N = \begin{vmatrix} \tau_{xy} & \sigma_y - \sigma_i \\ \tau_{xz} & \tau_{yz} \end{vmatrix} \,,$$

satisfy (17). This is easily confirmed, since the product of the first line of matrix $[C]$ with this vector of direction cosines $\{l, m, n\}$ corresponds to the product of K with the determinant of matrix $[C]$ computed by decomposition, using the elements of the first line and their complementary minors. This determinant vanishes, when σ takes the value of a principal stress, as is the case. The products of the second and third lines of matrix $[C]$ by the same vector are zero as well, since they represent determinants of matrices with two equal lines.

The value of K may then be computed by means of Expression 2, yielding

$$L^2 K^2 + M^2 K^2 + N^2 K^2 = 1 \;\Rightarrow\; K = \pm \frac{1}{\sqrt{L^2 + M^2 + N^2}} \,.$$

The two vectors obtained in this way, corresponding to the two possible signs of K, represent the two opposite senses of principal direction i.

As an example, consider the stress state defined by the tensor

$$\sigma_x = 30 \quad \sigma_y = -40 \quad \sigma_z = 60 \quad \tau_{xy} = -20 \quad \tau_{xz} = 25 \quad \tau_{yz} = 50 \,.$$

Using the previously developed expressions on this tensor, the principal stresses and directions are obtained

$$\begin{aligned}
\sigma_1 &= 85.4719 & l_1 &= 0.294148 & m_1 &= 0.312980 & n_1 &= 0.903062 \\
\sigma_2 &= 33.3299 & l_2 &= -0.914713 & m_2 &= 0.366114 & n_2 &= 0.171057 \\
\sigma_3 &= -68.8017 & l_3 &= 0.277086 & m_3 &= 0.876358 & n_3 &= -0.393978 \,.
\end{aligned}$$

The exactness of these values is easily verified by using (15). By transposing the reference axes to the principal directions of the stress state, we get a diagonal tensor with the principal stresses

$$\begin{bmatrix} l_1 & m_1 & n_1 \\ l_2 & m_2 & n_2 \\ l_3 & m_3 & n_3 \end{bmatrix} \times \begin{bmatrix} 30 & -20 & 25 \\ -20 & -40 & 50 \\ 25 & 50 & 60 \end{bmatrix} \times \begin{bmatrix} l_1 & l_2 & l_3 \\ m_1 & m_2 & m_3 \\ n_1 & n_2 & n_3 \end{bmatrix} = \begin{bmatrix} 85.47 & 0 & 0 \\ 0 & 33.33 & 0 \\ 0 & 0 & -68.80 \end{bmatrix} \,.$$

II.2. Verify the differential equations of equilibrium in the stress field installed in a still liquid under its own weight.

Resolution

Considering a reference system with the origin on the free surface of the liquid, whose axis y is vertical and points downwards, the stress field has the components

$$\sigma_x = \sigma_y = \sigma_z = \rho y \qquad \text{and} \qquad \tau_{xy} = \tau_{xz} = \tau_{yz} = 0 \,,$$

where ρ represents the mass density of the liquid. Since the only body force is the gravity force, we have

$$X = Z = 0 \qquad \text{and} \qquad Y = -\rho \,.$$

Substituting these values in Expressions 5, we immediately see that they are satisfied.

II.3. In a body under a plane stress state the body forces are zero and the stresses have been approximated by the expressions (η, ρ, H and λ are constants)

$$\sigma_x = \sigma_y = \eta \rho \left(H - y - \frac{x}{\lambda} \right) \qquad \tau_{xy} = 0 \,.$$

(a) Verify that these functions cannot represent the stress distribution in the body.
(b) Determine the conditions which the body forces have to obey so that these expressions can represent a possible stress distribution.

III

The Strain Tensor

III.1 Introduction

When the material points inside a solid body or a liquid mass suffer a displacement, this may be a consequence of a rigid body motion or of a deformation. Forces are not necessarily involved in a rigid body motion, unless the displacement is accompanied by acceleration. On the contrary, the deformation is almost always a consequence of internal forces. Other causes may be a temperature variation or similar phenomena, like the retraction of a concrete mass during the curing process.

In Solid Mechanics consideration of the deformation associated with the displacement field is generally unavoidable, since, unless the case under consideration fits into the rare category of fully statically determinate problems, the way the material deforms influences the way the internal forces are distributed inside the body.

When the validity of the continuum hypothesis is accepted, the internal forces may be defined by the stress tensor, as explained in the previous chapter. In the same way, as mentioned in Chap. I, the deformations of a continuous material may be defined independently of the geometrical dimensions of the continuum, by means of the strain definition. If these strains are defined appropriately, they define a symmetrical second order tensor, with exactly the same mathematical characteristics as the stress tensor. This chapter analyses both the properties of this tensor and the relations of its components with the functions that represent the displacement field describing the motion of the material points.

III.2 General Considerations

The deformation caused in a body by external forces or other actions generally varies from one point to another, i.e., it is not homogeneous. In fact, a homogeneous deformation is rare. It occurs, for example, in a body with iso-static supports under a uniform temperature variation or in a slender member under constant axial force.

A non-homogeneous deformation may by more clearly understood by imagining small line segments in several places of the body, which, before the deformation, have the same infinitesimal length, ds, and are parallel. As a rule, the deformation causes various rotations and elongations in the different line segments, as represented schematically in Fig. 16.

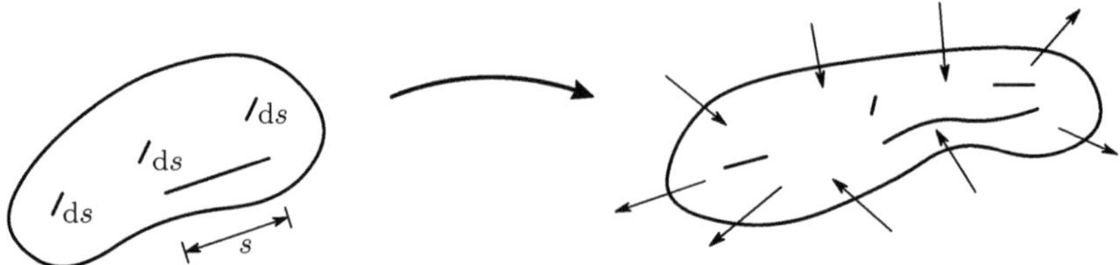

Fig. 16. Non-homogeneous deformation of a body

If the deformation is homogeneous, however, i.e., if it does not vary from point to point, the elongation and the rotation are the same all along the line segments, which means that two parallel straight lines of equal length (now not necessarily infinitesimal) remain straight, parallel and with the same length after the deformation. As a consequence, a homogeneous deformation transforms triangles into triangles, rectangles into parallelograms, tetrahedra into tetrahedra and rectangular parallelepipeds into, generally non-rectangular, parallelepipeds (Fig. 17).

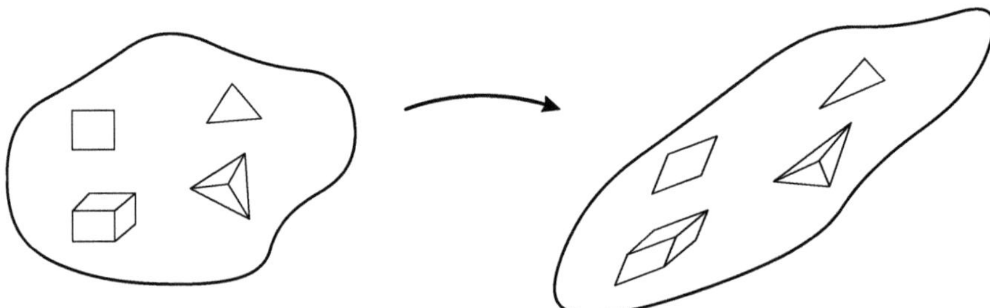

Fig. 17. Homogeneous deformation of a body

From these considerations we conclude that a homogeneous deformation may be fully defined by the six quantities that are required to define the shape and dimensions of a non-rectangular parallelepiped (e.g. the length of the three sides and the three independent angles between non-parallel sides) or of a tetrahedron (e.g. the length of its six sides). In the two-dimensional case the three quantities needed to define a parallelogram or a triangle, are enough.

As mentioned above, the deformation is not usually homogeneous, but varies from point to point. But it may be treated as such *if we limit the deformation analysis to an infinitesimal neighbourhood of a point*. This statement may be easily proved by developing in series the functions that define the coordinates of the material points after the deformation [1].

A physical visualization is, however, more indicative. To this end, let us consider the shapes which result from the non-homogeneous deformation of a rectangle, a triangle and a rectangular parallelepiped (Fig. 18).

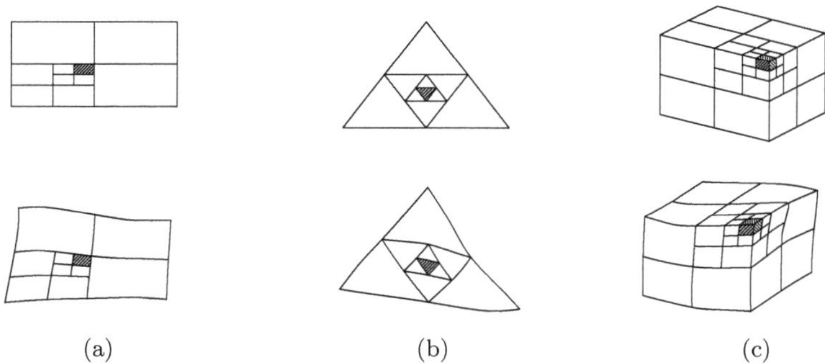

(a)   (b)   (c)

Fig. 18. Homogeneous deformation of an infinitesimal region

The non-homogeneous deformation results in the initially simple geometrical shapes transforming into complex shapes, which cannot be described by means of a reduced number of parameters. However, by subdividing the initial geometrical shapes into others of the same type, we find that the finer the subdivision, the closer the deformation gets to a homogeneous deformation. Ultimately, when the dimensions of the smallest shape go to zero, the deformation is homogeneous, since we can accept that the shaded rectangle in Fig. 18-a has become a parallelogram, the shaded triangle in Fig. 18-b remains a triangle and the rectangular parallelepiped with shaded faces in Fig. 18-c changes into a (non-rectangular) parallelepiped. From these considerations, we conclude that the definition of the *state of deformation* in an infinitesimal neighbourhood around a point needs six parameters, and these are the six quantities necessary to define a homogeneous deformation in the three-dimensional case (or three, in the two-dimensional case).

The above considerations are valid irrespective of the size of the deformation. However, as will be seen later, the expressions which relate the functions describing the displacement of the material points with the strain may be greatly simplified if the deformations and rotations are sufficiently small to be considered as infinitesimal quantities. Furthermore, the restriction on infinitesimal deformations and rotations allows the superposition of the strains associated with different displacement fields.

III.3 Components of the Strain Tensor

The general considerations discussed in the previous section will now be quantified by using a rectangular Cartesian reference frame, xyz. It will be seen later that, in this reference system, the strain tensor has components, which, for infinitesimal deformations, correspond to the elongation per unit length of line segments having the direction of the reference axes, and to half the angular variation of what were initially right-angles between these line segments (three pairs). The three elongations – the *longitudinal strains* – and the three angular variations – the *shearing strains* – are the six quantities necessary (and sufficient) to define the state of deformation around a point.

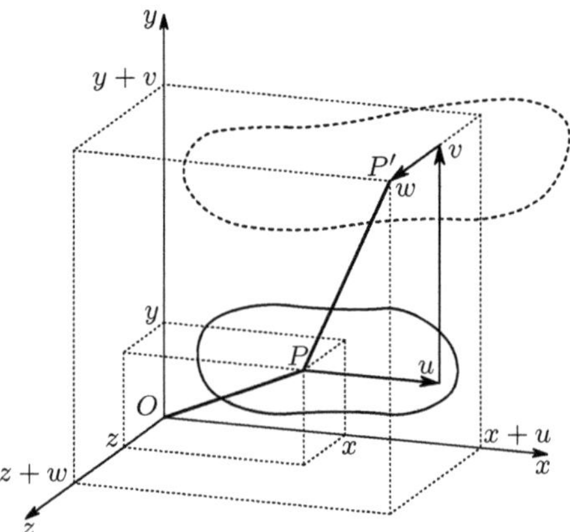

Fig. 19. Displacement of a material point P inside a body: —— before the deformation; ------ after the deformation

Then the initial position of the material points of the body is described by the coordinates x, y, z of the generic point P and its displacement is defined by vector $\overrightarrow{PP'}$ with components u, v, w in the reference directions x, y, z, respectively (Fig. 19). The position of the point after the deformation is therefore given by the coordinates $x + u$, $y + v$, $z + w$. If the material is continuous before and after the deformation, the functions $u(x, y, x)$, $v(x, y, z)$, $w(x, y, z)$ are continuous functions of the position coordinates of the body before the deformation, x, y, z.

Now, let us consider a straight line of infinitesimal length $\mathrm{d}x$, which is parallel to axis x in the undeformed configuration and is defined by the two close points $P_0(x_0, y_0, z_0)$ and $P_1(x_0 + \mathrm{d}x, y_0, z_0)$, as represented in Fig. 20.

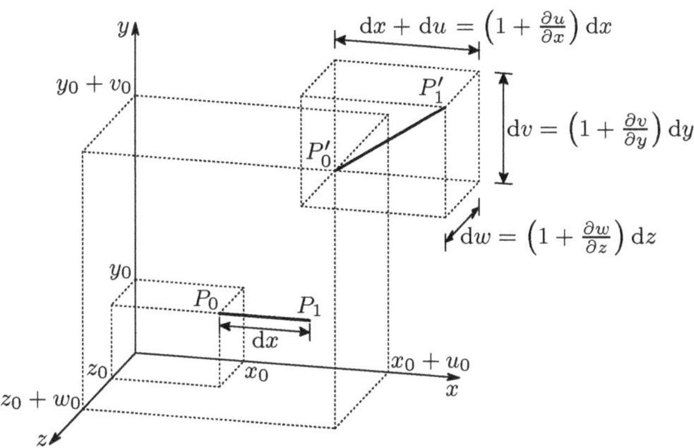

Fig. 20. Computation of the strain ε_x

After the deformation, these points will occupy the positions defined by the coordinates $P_0'(x_0 + u_0, y_0 + v_0, z_0 + w_0)$ and $P_1'(x_0 + \mathrm{d}x + u_1, y_0 + v_1, z_0 + w_1)$. As point P_1 is at an infinitesimal distance from P_0 and, in going from P_0 to P_1 only the coordinate x suffers an increment $\mathrm{d}x$ (undeformed configuration), we have

$$
\begin{cases}
u_1 = u_0 + \mathrm{d}u = u_0 + \dfrac{\partial u}{\partial x}\,\mathrm{d}x \\[2mm]
v_1 = v_0 + \mathrm{d}v = v_0 + \dfrac{\partial v}{\partial x}\,\mathrm{d}x \\[2mm]
w_1 = w_0 + \mathrm{d}w = w_0 + \dfrac{\partial w}{\partial x}\,\mathrm{d}x\ .
\end{cases}
$$

The deformation transforms the line segment $\overline{P_0 P_1}$ into the line segment $\overline{P_0' P_1'}$, which is generally no longer parallel to axis x and suffers an elongation. The projections of $\overline{P_0' P_1'}$ in the reference directions are then

$$
\begin{cases}
\underbrace{x_0 + \mathrm{d}x}_{x_1} + \underbrace{u_0 + \dfrac{\partial u}{\partial x}\,\mathrm{d}x}_{u_1} - (x_0 + u_0) = \left(1 + \dfrac{\partial u}{\partial x}\right)\mathrm{d}x \\[4mm]
y_0 + \underbrace{v_0 + \dfrac{\partial v}{\partial x}\,\mathrm{d}x}_{v_1} - (y_0 + v_0) = \dfrac{\partial v}{\partial x}\,\mathrm{d}x \\[4mm]
z_0 + \underbrace{w_0 + \dfrac{\partial w}{\partial x}\,\mathrm{d}x}_{w_1} - (z_0 + w_0) = \dfrac{\partial w}{\partial x}\,\mathrm{d}x\ .
\end{cases}
\tag{43}
$$

Defining longitudinal strain (or simply strain) as the elongation per unit length, the strain in direction x (strain of the line segment $\mathrm{d}x$) takes the value

$$
\varepsilon_x = \frac{\overline{P_0' P_1'} - \overline{P_0 P_1}}{\overline{P_0 P_1}} = \frac{\overline{P_0' P_1'} - \mathrm{d}x}{\mathrm{d}x} \Rightarrow \overline{P_0' P_1'} = (1 + \varepsilon_x)\,\mathrm{d}x\ .
\tag{44}
$$

Using the Pythagorean theorem, the length $\overline{P'_0 P'_1}$ may be computed from its projections in the coordinate axes, yielding, from expressions 43 and 44

$$\left(\overline{P'_0 P'_1}\right)^2 = (1 + \varepsilon_x)^2 \, dx^2$$
$$= \left[\left(1 + \frac{\partial u}{\partial x}\right)^2 + \left(\frac{\partial v}{\partial x}\right)^2 + \left(\frac{\partial w}{\partial x}\right)^2\right] dx^2 \tag{45}$$
$$\Rightarrow \varepsilon_x + \frac{\varepsilon_x^2}{2} = \frac{\partial u}{\partial x} + \frac{1}{2}\left[\left(\frac{\partial u}{\partial x}\right)^2 + \left(\frac{\partial v}{\partial x}\right)^2 + \left(\frac{\partial w}{\partial x}\right)^2\right].$$

In the same way, the expressions relating strains in the directions y and z to the displacement functions may be established, yielding

$$\varepsilon_y + \frac{\varepsilon_y^2}{2} = \frac{\partial v}{\partial y} + \frac{1}{2}\left[\left(\frac{\partial u}{\partial y}\right)^2 + \left(\frac{\partial v}{\partial y}\right)^2 + \left(\frac{\partial w}{\partial y}\right)^2\right]$$
$$\varepsilon_z + \frac{\varepsilon_z^2}{2} = \frac{\partial w}{\partial z} + \frac{1}{2}\left[\left(\frac{\partial u}{\partial z}\right)^2 + \left(\frac{\partial v}{\partial z}\right)^2 + \left(\frac{\partial w}{\partial z}\right)^2\right]. \tag{46}$$

Let us consider now two straight lines of infinitesimal lengths, dx and dy, which, in the undeformed configuration, are parallel to axes x and y and are defined by points P_0, P_1 and P_2 (Fig. 21). The deformation transforms these straight lines into $\overline{P'_0 P'_1}$ and $\overline{P'_0 P'_2}$. Following the same line of reasoning as above, these line segments have the components

$$\overline{P'_0 P'_1} \longrightarrow \left\{\begin{array}{c} \left(1 + \dfrac{\partial u}{\partial x}\right) dx \\[2mm] \dfrac{\partial v}{\partial x} dx \\[2mm] \dfrac{\partial w}{\partial x} dx \end{array}\right\} \quad \text{and} \quad \overline{P'_0 P'_2} \longrightarrow \left\{\begin{array}{c} \dfrac{\partial u}{\partial y} dy \\[2mm] \left(1 + \dfrac{\partial v}{\partial y}\right) dy \\[2mm] \dfrac{\partial w}{\partial y} dy \end{array}\right\}.$$

In accordance with the definition of strain given earlier, the line segments $\overline{P'_0 P'_1}$ and $\overline{P'_0 P'_2}$ have the lengths $(1+\varepsilon_x)\, dx$ and $(1+\varepsilon_y)\, dy$, respectively. The scalar product of vectors $\overrightarrow{P'_0 P'_1}$ and $\overrightarrow{P'_0 P'_2}$ may be expressed by $(\cos\left(\frac{\pi}{2} - \theta\right) = \sin\theta)$

$$(1 + \varepsilon_x)\, dx\, (1 + \varepsilon_y)\, dy \cos\left(\frac{\pi}{2} - \theta_{xy}\right)$$
$$= \left(1 + \frac{\partial u}{\partial x}\right) dx \frac{\partial u}{\partial y} dy + \frac{\partial v}{\partial x} dx \left(1 + \frac{\partial v}{\partial y}\right) dy + \frac{\partial w}{\partial x} dx \frac{\partial w}{\partial y} dy \tag{47}$$
$$\Rightarrow \sin\theta_{xy} = \frac{\frac{\partial u}{\partial y} + \frac{\partial v}{\partial x} + \frac{\partial u}{\partial x}\frac{\partial u}{\partial y} + \frac{\partial v}{\partial x}\frac{\partial v}{\partial y} + \frac{\partial w}{\partial x}\frac{\partial w}{\partial y}}{(1 + \varepsilon_x)(1 + \varepsilon_y)}.$$

Angle θ_{xy} represents the decrease of the initially right-angle between the line segments dx and dy. It therefore defines the *distortion* (double shearing

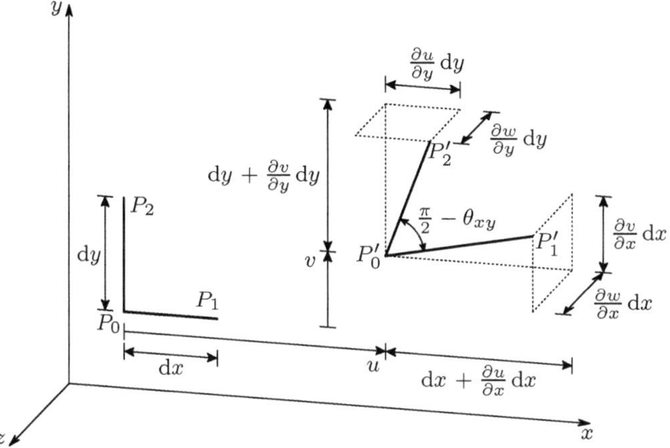

Fig. 21. Motion of the line segments $\mathrm{d}x = \overline{P_0P_1}$ and $\mathrm{d}y = \overline{P_0P_2}$ during the deformation (not considering the displacement of P_0 in direction z)

strain) of directions x and y after the deformation. In the same way, the distortions θ_{xz} and θ_{yz} may be related to the derivatives of the displacement functions, yielding

$$
\sin\theta_{xz} = \frac{\dfrac{\partial u}{\partial z} + \dfrac{\partial w}{\partial x} + \dfrac{\partial u}{\partial x}\dfrac{\partial u}{\partial z} + \dfrac{\partial v}{\partial x}\dfrac{\partial v}{\partial z} + \dfrac{\partial w}{\partial x}\dfrac{\partial w}{\partial z}}{(1+\varepsilon_x)(1+\varepsilon_z)}
$$

$$
\sin\theta_{yz} = \frac{\dfrac{\partial v}{\partial z} + \dfrac{\partial w}{\partial y} + \dfrac{\partial u}{\partial y}\dfrac{\partial u}{\partial z} + \dfrac{\partial v}{\partial y}\dfrac{\partial v}{\partial z} + \dfrac{\partial w}{\partial y}\dfrac{\partial w}{\partial z}}{(1+\varepsilon_y)(1+\varepsilon_z)}\, .
$$

$$(48)$$

The quantities ε_x, ε_y, ε_z, θ_{xy}, θ_{xz} and θ_{yz}, described by expressions 45, 46, 47 and 48, fully define the state of deformation around point P_0, since they are enough to define the shape and dimensions of the generally non-rectangular parallelepiped resulting from the homogeneous deformation of the rectangular parallelepiped, defined by the line segments $\mathrm{d}x$, $\mathrm{d}y$ and $\mathrm{d}z$, which are parallel to the coordinate axes in the undeformed configuration. However, the analytical treatment of these expressions is not simple, since they contain those quantities implicitly.[1] Moreover, they are not linear.

In most problems of Solid Mechanics that arise in structural Engineering the longitudinal and shearing strains are small enough to be considered as infinitesimal quantities, which allows the simplifications $\varepsilon^2 + 2\varepsilon \approx 2\varepsilon$ and

---

[1]For this reason, when the deformations are too large to be considered as infinitesimal, the strains are defined in a different way. Instead of considering the elongation per unit length $\varepsilon = \frac{l-l_0}{l_0}$, half the relative variation of the square of the length $E = \frac{1}{2}\frac{l^2-l_0^2}{l_0^2}$ is considered, which considerably simplifies the expressions, since we have $E = \varepsilon + \frac{\varepsilon^2}{2}$.

$\sin\theta \approx \theta$ and makes it possible to disregard the strains in the denominators of expressions 47 and 48. The strains may therefore be defined by explicit expressions of the type (γ_{xy} is the infinitesimal distortion of directions x and y)

$$\varepsilon_x = \frac{\partial u}{\partial x} + \frac{1}{2}\left[\left(\frac{\partial u}{\partial x}\right)^2 + \left(\frac{\partial v}{\partial x}\right)^2 + \left(\frac{\partial w}{\partial x}\right)^2\right]$$

$$\frac{\theta_{xy}}{2} = \frac{\gamma_{xy}}{2} = \varepsilon_{xy} = \frac{1}{2}\left[\frac{\partial u}{\partial y} + \frac{\partial v}{\partial x} + \frac{\partial u}{\partial x}\frac{\partial u}{\partial y} + \frac{\partial v}{\partial x}\frac{\partial v}{\partial y} + \frac{\partial w}{\partial x}\frac{\partial w}{\partial y}\right].$$

$$(49)$$

Furthermore, if the rotations are sufficiently small to be considered as infinitesimal quantities, the squares and the products of the derivatives contained in expressions 45 to 49 may be disregarded, since, with infinitesimal strains and rotations, these derivatives will also be infinitesimal, as may easily be concluded from Fig. 21. In this case, the strains may be defined by the linear expressions[2]

$$\begin{cases}\varepsilon_x = \dfrac{\partial u}{\partial x} \\[2ex] \varepsilon_y = \dfrac{\partial v}{\partial y} \quad\text{and} \\[2ex] \varepsilon_z = \dfrac{\partial w}{\partial z}\end{cases} \qquad \begin{cases}\dfrac{\gamma_{xy}}{2} = \varepsilon_{xy} = \dfrac{1}{2}\left(\dfrac{\partial u}{\partial y} + \dfrac{\partial v}{\partial x}\right) \\[2ex] \dfrac{\gamma_{xz}}{2} = \varepsilon_{xz} = \dfrac{1}{2}\left(\dfrac{\partial u}{\partial z} + \dfrac{\partial w}{\partial x}\right) \\[2ex] \dfrac{\gamma_{yz}}{2} = \varepsilon_{yz} = \dfrac{1}{2}\left(\dfrac{\partial v}{\partial z} + \dfrac{\partial w}{\partial y}\right).\end{cases} \qquad (50)$$

We shall now consider that the strain-displacement relations are defined by these simple expressions. One of the most useful consequences of this simplification is that it makes it possible to superpose the deformations associated with distinct displacement fields. This is quite evident if we consider the displacements u_1, v_1, w_1 and u_2, v_2, w_2, with which the strains $^1\varepsilon$ and $^2\varepsilon$ are respectively associated. From expressions 50 we immediately conclude

$$\begin{cases}\varepsilon_x = \dfrac{\partial(u_1 + u_2)}{\partial x} = \dfrac{\partial u_1}{\partial x} + \dfrac{\partial u_2}{\partial x} = {}^1\varepsilon_x + {}^2\varepsilon_x \\[2ex] \quad\vdots \\[2ex] \varepsilon_{yz} = \dfrac{1}{2}\left[\dfrac{\partial(v_1 + v_2)}{\partial z} + \dfrac{\partial(w_1 + w_2)}{\partial y}\right] \\[3ex] \qquad = \dfrac{1}{2}\left(\dfrac{\partial v_1}{\partial z} + \dfrac{\partial w_1}{\partial y}\right) + \dfrac{1}{2}\left(\dfrac{\partial v_2}{\partial z} + \dfrac{\partial w_2}{\partial y}\right) = {}^1\varepsilon_{yz} + {}^2\varepsilon_{yz}.\end{cases} \qquad (51)$$

In the case of a flowing liquid (Fluid Mechanics), the deformations and rotations in relation to the original configuration obviously cannot be considered

---

[2]It must, however, be pointed out that, as a consequence of the last simplification (infinitesimal rotations), the force-displacement relations obtained from these linearized strain-displacement relations (cf. Sect. I.3) cannot capture instability phenomena, in which sudden rotations of parts of the body occur.

as infinitesimal. However, taking the position of the points constituting the liquid mass in the instant t_0 as reference configuration, the rotations and deformations in instant $t_0 + dt$ (dt is an infinitesimal time step) may be considered as infinitesimal and therefore expressions 50 may be used. Obviously, this also holds in the case of large deformations of solid bodies.

Expressions 50 furnish the components of the strain tensor in a rectangular Cartesian reference system for infinitesimal deformations and rotations. As we shall see below (Sect. III.6), the factor $\frac{1}{2}$ in the expressions concerning the shearing strains is necessary so that the quantities defining the state of deformation, ε_x, ε_y, ε_z, ε_{xy}, ε_{xz} and ε_{yz}, can form a tensor in the Cartesian space xyz, in the mathematical sense of the term.[3] The main advantage of having the deformation state defined as a tensor is that tensor mathematics can be used in its analytical treatment, which is obviously the same, regardless of the particular tensor under consideration: the stress tensor, the strain tensor or any other symmetric second order tensor. This fact allows conclusions to be drawn about the properties of the strain tensor, which are taken by analogy with the stress tensor, as will be seen later.

III.4 Pure Deformation and Rigid Body Motion

In Sect. III.3 we developed expressions to compute the elements of the strain tensor from the displacement functions u, v, w. The infinitesimal neighbourhood around a point suffers not only pure deformation but a rigid body motion, too, due to the deformation of other regions of the body. In the case of infinitesimal deformations and rotations, the motions associated with different displacement fields may be superposed, which allows the rigid body motion of the infinitesimal region around a point to be identified.

The motion of the infinitesimal region is fully defined by the quantities u, v, w, $\frac{\partial u}{\partial x}$, $\frac{\partial u}{\partial y}$, $\frac{\partial u}{\partial z}$, $\frac{\partial v}{\partial x}$, $\frac{\partial v}{\partial y}$, $\frac{\partial v}{\partial z}$, $\frac{\partial w}{\partial x}$, $\frac{\partial w}{\partial y}$ and $\frac{\partial w}{\partial z}$. Quantities u, v and w obviously represent the translation motion. The remaining nine quantities contain the deformation and the rigid body rotation. In order to identify the latter, we should point out that, in a rigid body motion, longitudinal and shearing strains vanish, which means (cf. (50))

---

[3]Following the mathematical definition of a second order tensor, its components transform as defined by expression 15 (stress tensor), when the reference system rotates. As we have already seen, the state of deformation may be defined by any set of six quantities, enabling the quantification of the deformation of the elementary parallelepiped. However, only the six quantities defined as represented in expression 50 define the components of a symmetric second order tensor in the Cartesian reference frame xyz.

$$
\begin{cases}
\varepsilon_x = \dfrac{\partial u}{\partial x} = 0 \\[2ex]
\varepsilon_y = \dfrac{\partial v}{\partial y} = 0 \\[2ex]
\varepsilon_z = \dfrac{\partial w}{\partial z} = 0
\end{cases}
\quad \text{and} \quad
\begin{cases}
\gamma_{xy} = 0 \Rightarrow \dfrac{\partial u}{\partial y} = -\dfrac{\partial v}{\partial x} \\[2ex]
\gamma_{xz} = 0 \Rightarrow \dfrac{\partial u}{\partial z} = -\dfrac{\partial w}{\partial x} \\[2ex]
\gamma_{yz} = 0 \Rightarrow \dfrac{\partial v}{\partial z} = -\dfrac{\partial w}{\partial y} \; .
\end{cases}
$$

A displacement field u_r, v_r, w_r, where we have $\frac{\partial v_r}{\partial x} = -\frac{\partial u_r}{\partial y} = \omega_{xy}$, $\frac{\partial u_r}{\partial z} = -\frac{\partial w_r}{\partial x} = \omega_{xz}$ and $\frac{\partial w_r}{\partial y} = -\frac{\partial v_r}{\partial z} = \omega_{yz}$ thus describes a rigid body rotation of the infinitesimal region, with ω_{yz}, ω_{xz} and ω_{xy} representing positive rotations around the reference axes x, y and z, respectively. In this motion the pairs of line segments (dy,dz), (dz,dx) and (dx,dy) rotate around axes x, y and z, respectively, and remain perpendicular to each other.

If, in addition to the rigid body motion, the infinitesimal region suffers a deformations, this does not take place anymore, as seen in Sect. III.3. However, we can define the rigid body rotations around the reference axes, under the action of the deformation field u, v, w, as the *mean rotations of the line segments* dx, dy, *and* dz *around those axes*

$$
\begin{cases}
\omega_x = \dfrac{1}{2}\left(\dfrac{\partial w}{\partial y} - \dfrac{\partial v}{\partial z} \right) = \omega_{yz} \\[2ex]
\omega_y = \dfrac{1}{2}\left(\dfrac{\partial u}{\partial z} - \dfrac{\partial w}{\partial x} \right) = \omega_{xz} \\[2ex]
\omega_z = \dfrac{1}{2}\left(\dfrac{\partial v}{\partial x} - \dfrac{\partial u}{\partial y} \right) = \omega_{xy} \; .
\end{cases}
\tag{52}
$$

By subtracting the rigid body rotation from the total motion of the infinitesimal region, the pure deformation is obtained, which is defined by a motion where the mean rotations of the three pairs of line segments vanish. Figure 22 illustrates these situations, with the example of the rotation around axis z. Obviously this decomposition eliminates the rigid body rotation only in the infinitesimal region under consideration, since it generally varies from point to point.

We will see later (Sect. III.6) that this definition of rigid body motion is independent of the spatial orientation of the reference frame.

The additive decomposition of the strain tensor presented in this section is only valid in the case of infinitesimal deformation, where the simplified linear form of the strain-displacement relations (50) can be used. However, it is also possible to define a rigid body rotation in the case of large deformations and rotations, by using the *polar decomposition theorem*, which is based on a multiplicative decomposition of the deformation gradient (cf., e.g. [15]). Nevertheless, in this case, the mean rotation of two orthogonal line segments (in the initial configuration) does not define the rigid body rotation anymore. This

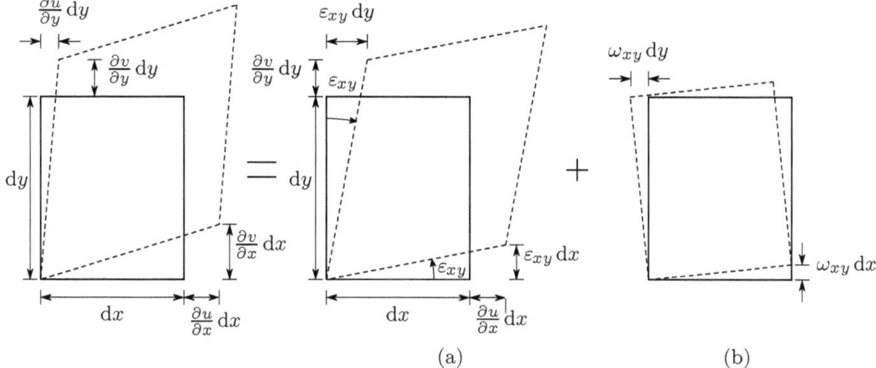

Fig. 22. Decomposition of the motion of an infinitesimal region in pure deformation and rigid body motion (infinitesimal deformations and rotations): $\varepsilon_{xy}\,\mathrm{d}x + \omega_{xy}\,\mathrm{d}x = \frac{1}{2}\left(\frac{\partial u}{\partial y} + \frac{\partial v}{\partial x}\right)\mathrm{d}x + \frac{1}{2}\left(\frac{\partial v}{\partial x} - \frac{\partial u}{\partial y}\right)\mathrm{d}x = \frac{\partial v}{\partial x}\,\mathrm{d}x$ $\varepsilon_{xy}\,\mathrm{d}y - \omega_{xy}\,\mathrm{d}y = \frac{1}{2}\left(\frac{\partial u}{\partial y} + \frac{\partial v}{\partial x}\right)\mathrm{d}y - \frac{1}{2}\left(\frac{\partial v}{\partial x} - \frac{\partial u}{\partial y}\right)\mathrm{d}y = \frac{\partial u}{\partial y}\,\mathrm{d}y$

theorem is not given here, since a deeper insight into the finite deformation theory is beyond the scope of this introductory text.

III.5 Equations of Compatibility

As mentioned in the first chapter, we accept that the material is continuous before the deformation and remains continuous after it. This continuity condition will be satisfied if the displacement functions u, v and w are continuous, since the coordinates of the material points after the deformation are given by the expressions $x' = x + u$, $y' = y + v$ and $z' = z + w$. Therefore, if u, v and w are continuous, two points, which lie at an infinitesimal distance from each other before the deformation, will remain at an infinitesimal distance after it.

The question of the deformation's compatibility arises when, given six strain functions $\varepsilon_x(x, y, z), \ldots, \gamma_{yz}(x, y, z)$, we want to know if they represent a *compatible deformation*, i.e., a deformation where the material remains continuous. It is expected that some conditions exist between the six strain functions in a compatible deformation, as this is completely defined by the three displacement functions, u, v and w, which means that the system of equations formed by Expressions 50 has only three unknowns.

The existence of compatibility conditions may also be understood by means of geometrical considerations. For this purpose, let us imagine the continuum divided into very small parallelepipeds, so that the deformation of each one may be considered as homogeneous. A compatible deformation will be a deformation, in which the deformed parallelepipeds fit perfectly together. An incompatible deformation, however, will lead either to gaps between the parallelepipeds, or to other material discontinuities.

The compatibility conditions are obtained by eliminating the displacements u, v and w from the system formed by (50).

A first group is obtained from the relations between the longitudinal strains ε_x, ε_y, ε_z and the displacement functions. Taking, for example, the longitudinal strains in the xy-plane, we get

$$
\begin{cases}
\dfrac{\partial^2 \varepsilon_x}{\partial y^2} = \dfrac{\partial^3 u}{\partial x \partial y^2} \\[2ex]
\dfrac{\partial^2 \varepsilon_y}{\partial x^2} = \dfrac{\partial^3 v}{\partial x^2 \partial y} \\[2ex]
\dfrac{\partial^2 \gamma_{xy}}{\partial x \partial y} = \dfrac{\partial^3 u}{\partial x \partial y^2} + \dfrac{\partial^3 v}{\partial x^2 \partial y}
\end{cases}
\quad \Rightarrow \quad
\dfrac{\partial^2 \varepsilon_x}{\partial y^2} + \dfrac{\partial^2 \varepsilon_y}{\partial x^2} = \dfrac{\partial^2 \gamma_{xy}}{\partial x \partial y} \; .
$$

Proceeding in the same way with the other two pairs of reference directions, x, z and y, z, two other similar equations are obtained.

Another group of three equations may be obtained by derivation of the shearing strains with respect to the coordinate, which is absent from its indexes, and combining the obtained relations as follows

$$
\begin{cases}
\dfrac{\partial \gamma_{xy}}{\partial z} = \dfrac{\partial^2 u}{\partial y \partial z} + \dfrac{\partial^2 v}{\partial x \partial z} \\[2ex]
\dfrac{\partial \gamma_{xz}}{\partial y} = \dfrac{\partial^2 u}{\partial y \partial z} + \dfrac{\partial^2 w}{\partial x \partial y} \\[2ex]
\dfrac{\partial \gamma_{yz}}{\partial x} = \dfrac{\partial^2 v}{\partial x \partial z} + \dfrac{\partial^2 w}{\partial x \partial y}
\end{cases}
\;\Rightarrow\;
\begin{aligned}
&\dfrac{\partial}{\partial x}\left(\dfrac{\partial \gamma_{xy}}{\partial z} + \dfrac{\partial \gamma_{xz}}{\partial y} - \dfrac{\partial \gamma_{yz}}{\partial x} \right) \\[1.5ex]
&= 2\dfrac{\partial^3 u}{\partial x \partial y \partial z} = 2\dfrac{\partial^2 \varepsilon_x}{\partial y \partial z} \; .
\end{aligned}
$$

Two other relations of this type are obtained by a similar process.

The complete set of conditions, to which the strain functions $\varepsilon_x(x, y, z), \dots,$ $\gamma_{yz}(x, y, z)$ must obey, in order to define a compatible deformation, in the sense that the deformed infinitesimal parallelepipeds fit perfectly together are then

$$
\begin{aligned}
&\dfrac{\partial^2 \varepsilon_x}{\partial y^2} + \dfrac{\partial^2 \varepsilon_y}{\partial x^2} = \dfrac{\partial^2 \gamma_{xy}}{\partial x \partial y}
&& 2\dfrac{\partial^2 \varepsilon_x}{\partial y \partial z} = \dfrac{\partial}{\partial x}\left(\dfrac{\partial \gamma_{xy}}{\partial z} + \dfrac{\partial \gamma_{xz}}{\partial y} - \dfrac{\partial \gamma_{yz}}{\partial x} \right) \\[2ex]
&\dfrac{\partial^2 \varepsilon_x}{\partial z^2} + \dfrac{\partial^2 \varepsilon_z}{\partial x^2} = \dfrac{\partial^2 \gamma_{xz}}{\partial x \partial z}
&& 2\dfrac{\partial^2 \varepsilon_y}{\partial x \partial z} = \dfrac{\partial}{\partial y}\left(\dfrac{\partial \gamma_{xy}}{\partial z} - \dfrac{\partial \gamma_{xz}}{\partial y} + \dfrac{\partial \gamma_{yz}}{\partial x} \right) \\[2ex]
&\dfrac{\partial^2 \varepsilon_y}{\partial z^2} + \dfrac{\partial^2 \varepsilon_z}{\partial y^2} = \dfrac{\partial^2 \gamma_{yz}}{\partial y \partial z}
&& 2\dfrac{\partial^2 \varepsilon_z}{\partial x \partial y} = \dfrac{\partial}{\partial z}\left(-\dfrac{\partial \gamma_{xy}}{\partial z} + \dfrac{\partial \gamma_{xz}}{\partial y} + \dfrac{\partial \gamma_{yz}}{\partial x} \right) \; .
\end{aligned}
\tag{53}
$$

These conditions are necessary to ensure deformation compatibility *at the local level*, i.e., at the level of the infinitesimal neighbourhood of a point, since they contain only derivatives of the strain functions.[4] However, these conditions are

---

[4] Using the model of the infinitesimal parallelepipeds, local compatibility means that each parallelepiped fits perfectly with those in contact with it.

sufficient only in the case of *simply-connected bodies*. In a *multiply-connected body* supplementary conditions are needed to ensure compatibility. These are the *integral conditions of compatibility.*

The mathematical demonstration of these considerations is rather long and time-consuming, so it is not presented here (cf. e.g. [2]). However, a physical explanation on the basis of geometrical considerations is substantially simpler and more indicative.

A simply-connected body is a body where any closed line, fully contained in the body, can be shrunk to a point without leaving the body. Thus a two-dimensional region will be simply-connected if its boundary is defined by only one closed line, i.e., if it has no holes. A three-dimensional body may have holes and be simply-connected: for example a body defined by the space between two concentric spheres (a hollow sphere) is simply-connected, since any closed lined defined in it can shrink to a point without touching the boundaries of the body. An o-ring (torus), on the contrary, is not simply-connected, since a closed line around the hole cannot shrink to a point, without leaving the body.

The degree of connection may be defined as the number of cuts required to render the body simply-connected plus one (the intersection of the cut with the boundary of the body must be a closed line). It can also be defined as the maximum number of cuts which can be made without dividing the body into two, plus one. Some examples of the determination of the degree of connection are presented in Fig. 23.

The fact that the local compatibility conditions are sufficient to ensure the continuity of a simply-connected body after the deformation may be easily understood with the help of the two-dimensional example presented in Fig. 24-a. Clearly, if the deformed parallelepipeds fit perfectly with the neighbouring ones, i.e., if the local compatibility conditions are satisfied, the deformed body will be continuous.

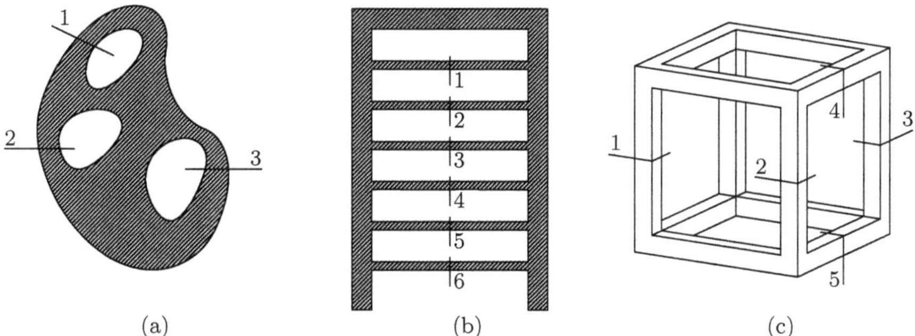

Fig. 23. Cuts required to render a body simply-connected Degrees of connection: (a) 4, (b) 7, (c) 6

On the other side, in the doubly-connected example presented in Fig. 24-b, despite the fact that in every point the local compatibility conditions are satisfied (the line $a'b'$ fits perfectly with line $a''b''$, so that every infinitesimal parallelepiped fits with the neighbouring ones), the deformation is not compatible, since the deformed body displays a discontinuity in the points belonging to line $\overline{ab}$. This is only possible, because the hole exists, i.e., because the degree of connection is superior to one.

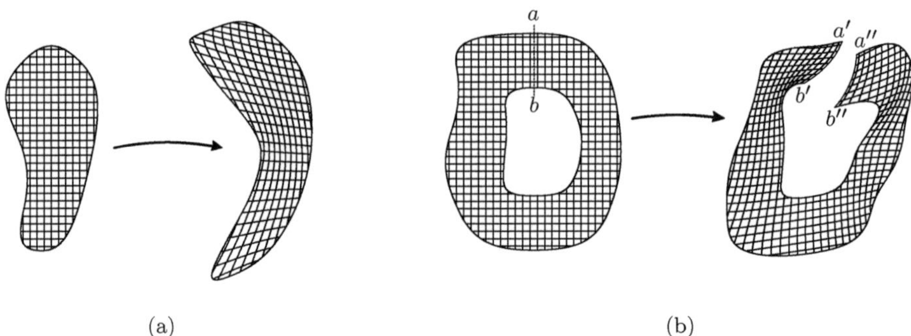

Fig. 24. Local compatibility conditions: (**a**) simply-connected body: necessary and sufficient condition; (**b**) doubly-connected body: necessary, but not sufficient condition

The expressions corresponding to the integral conditions of compatibility are not shown here. They will be studied in the second part (Strength of Materials), in the particular case of the computation of internal forces in hyperstatic (statically indeterminate) frames.

III.6 Deformation in an Arbitrary Direction

In the preceding sections the deformations of line segments, which are parallel to the reference axes in the undeformed configurations have been analysed. These deformations define the elements of the strain tensor, so they therefore allow the computation of the longitudinal and shearing deformations in arbitrary directions.

To this end, let us consider a line segment with infinitesimal unit length, whose orientation in relation to the coordinate axes is defined by the direction cosines l, m, n, which simultaneously define the components of vector $\overrightarrow{OP}$. Figure 25 illustrates the motion of the infinitesimal region around the line segment. Excluding the translations, this motion may be defined by the displacement $\overrightarrow{PP'}$ of the tip P of the unit vector $\overrightarrow{OP}$ (Fig. 25-a). The projection in the xy-plane of the non rectangular parallelepiped, which resulted from

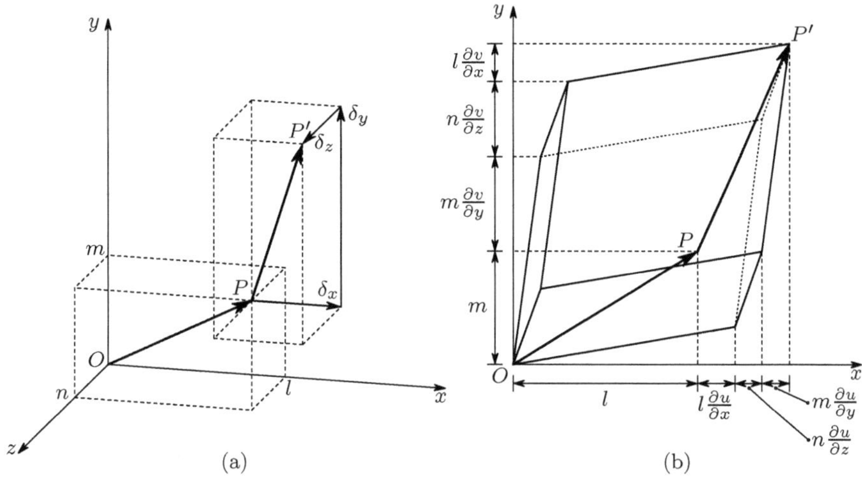

Fig. 25. Motion of the infinitesimal region around a point

the initial rectangular parallelepiped defined by the vector $\overrightarrow{OP}$, is depicted in Fig. 25-b.

The components in directions x and y of the displacement vector $\overrightarrow{PP'}$ may be obtained directly from Fig. 25-b. The projection of the deformed parallelepiped in the yz or in the zx plane would show the three contributions of the displacement in direction z. The components of vector $\overrightarrow{PP'}$ are then.

$$\begin{cases} \delta_x = l\dfrac{\partial u}{\partial x} + m\dfrac{\partial u}{\partial y} + n\dfrac{\partial u}{\partial z} \\[2ex] \delta_y = l\dfrac{\partial v}{\partial x} + m\dfrac{\partial v}{\partial y} + n\dfrac{\partial v}{\partial z} \\[2ex] \delta_z = l\dfrac{\partial w}{\partial x} + m\dfrac{\partial w}{\partial y} + n\dfrac{\partial w}{\partial z} \end{cases} \quad\Leftrightarrow\quad \begin{Bmatrix} \delta_x \\ \delta_y \\ \delta_z \end{Bmatrix} = \begin{bmatrix} \dfrac{\partial u}{\partial x} & \dfrac{\partial u}{\partial y} & \dfrac{\partial u}{\partial z} \\[1.5ex] \dfrac{\partial v}{\partial x} & \dfrac{\partial v}{\partial y} & \dfrac{\partial v}{\partial z} \\[1.5ex] \dfrac{\partial w}{\partial x} & \dfrac{\partial w}{\partial y} & \dfrac{\partial w}{\partial z} \end{bmatrix} \begin{Bmatrix} l \\ m \\ n \end{Bmatrix}.$$

$$(54)$$

These expressions could also be obtained by simple differentiation of the displacement functions u, v and w, since the displacements δ_x, δ_y and δ_z, represent the difference between the displacements of points O and P in the reference directions, and l, m and n are the increments of coordinates x, y and z from point O to point P.

Since vector $\overrightarrow{OP}$ has unit length and only infinitesimal deformations and rotations are considered, the longitudinal strain in its direction may by obtained by the projection of vector $\overrightarrow{PP'}$ in the direction $\overline{OP}$, yielding

$$\varepsilon = l\delta_x + m\delta_y + n\delta_z$$

$$= l^2\frac{\partial u}{\partial x} + m^2\frac{\partial v}{\partial y} + n^2\frac{\partial w}{\partial z}$$

$$+ lm\underbrace{\left(\frac{\partial u}{\partial y} + \frac{\partial v}{\partial x}\right)}_{\gamma_{xy}=2\varepsilon_{xy}} + ln\underbrace{\left(\frac{\partial u}{\partial z} + \frac{\partial w}{\partial x}\right)}_{\gamma_{xz}=2\varepsilon_{xz}} + mn\underbrace{\left(\frac{\partial v}{\partial z} + \frac{\partial w}{\partial y}\right)}_{\gamma_{yz}=2\varepsilon_{yz}}$$

$$= l^2\varepsilon_x + m^2\varepsilon_y + n^2\varepsilon_z + 2lm\varepsilon_{xy} + 2ln\varepsilon_{xz} + 2mn\varepsilon_{yz} \ . \tag{55}$$

This expression is perfectly analogous to Expr. 11, which, in the stress state around a point, gives the normal stress in a facet whose semi-normal has the direction cosines l m and n. This analogy arises, because the elements of the strain tensor have been used to define the homogeneous deformation of the infinitesimal region around point O.

The transversal component δ_t of vector $\overrightarrow{PP'}$ gives the rotation of vector $\overrightarrow{OP}$. This rotation generally has a rigid body rotation and a shearing strain component. The rigid body rotation may be eliminated by considering, instead of the total displacements u, v and w, the displacements associated with the pure deformation of the infinitesimal region under consideration u', v' and w'. In this case we have (cf. Sect. III.4, Expr. 52 and Fig. 22)

$$\omega_x = \omega_y = \omega_z = 0 \Rightarrow \begin{cases} \dfrac{\partial u'}{\partial y} = \dfrac{\partial v'}{\partial x} = \varepsilon_{xy} \\[2mm] \dfrac{\partial u'}{\partial z} = \dfrac{\partial w'}{\partial x} = \varepsilon_{xz} \\[2mm] \dfrac{\partial v'}{\partial z} = \dfrac{\partial w'}{\partial y} = \varepsilon_{yz} \ . \end{cases}$$

With this modification Expr. 55 does not change, since $\frac{\partial u'}{\partial x} = \varepsilon_x, \ldots, \frac{\partial v'}{\partial z} + \frac{\partial w'}{\partial y} = \gamma_{yz}$ and Expr. 54 takes the symmetrical form

$$\underbrace{\begin{Bmatrix} \delta'_x \\ \delta'_y \\ \delta'_z \end{Bmatrix}}_{\{\delta'\}} = \underbrace{\begin{bmatrix} \varepsilon_x & \varepsilon_{xy} & \varepsilon_{xz} \\ \varepsilon_{xy} & \varepsilon_y & \varepsilon_{yz} \\ \varepsilon_{xz} & \varepsilon_{yz} & \varepsilon_z \end{bmatrix}}_{[\varepsilon]} \underbrace{\begin{Bmatrix} l \\ m \\ n \end{Bmatrix}}_{\{l\}} \ . \tag{56}$$

This expression is perfectly analogous to Expr. 10 of the strain tensor, with the difference that it contains the elements of the strain tensor instead of the elements of the stress tensor. As a matter of fact, the operations performed in the analysis of the stress tensor after Expr. 10 are solely tensorial operations on a second order symmetric tensor (note that no equilibrium conditions were used in that development). These operations are therefore also valid in the case of the deformation state, since it is also described by a symmetrical second order tensor, although with different physical quantities. In this sense,

Expr. 12, which furnishes the shearing stress in an arbitrary oriented facet, is analogous to the expression of the transversal displacement in the pure deformation displacement field δ'_t, which is given by $(\vec{\delta'} = \varepsilon + \vec{\delta'})$

$$\begin{cases} \delta'_{tx} = \delta'_x - l\varepsilon \\[2mm] \delta'_{ty} = \delta'_y - m\varepsilon \\[2mm] \delta'_{tz} = \delta'_z - n\varepsilon \ . \end{cases}$$

This analogy and the reciprocity of the shearing stresses allow the conclusion, that the rotation of a line segment $\overrightarrow{OQ}$, which has the same direction as δ'_t, in the plane defined by vectors $\overrightarrow{OP}$ and $\vec{\delta'}$, is equal and has the opposite direction to the rotation of vector $\overrightarrow{OP}$. In fact, if $\overrightarrow{OQ}$ is a unit vector, then $|\overrightarrow{OQ}|=|\overrightarrow{OP}|$ and, as a consequence of the analogy, $\delta'_{tP} = \delta''_{tQ}$ (cf. Fig. 26). Therefore, δ'_t actually represents the maximum shearing strain between direction $\overrightarrow{OP}$ and orthogonal directions, i.e., $\delta'_t = \frac{\gamma}{2} = \sqrt{\delta'^2_x + \delta'^2_y + \delta'^2_z - \varepsilon^2}$.

This fact leads to the conclusion that the definition of pure deformation, which was stated in Sect. III.4 for a reference system x, y, z, is independent of the coordinate axes, since, once the rigid body rotation is eliminated for those directions, it is also eliminated for all other directions.

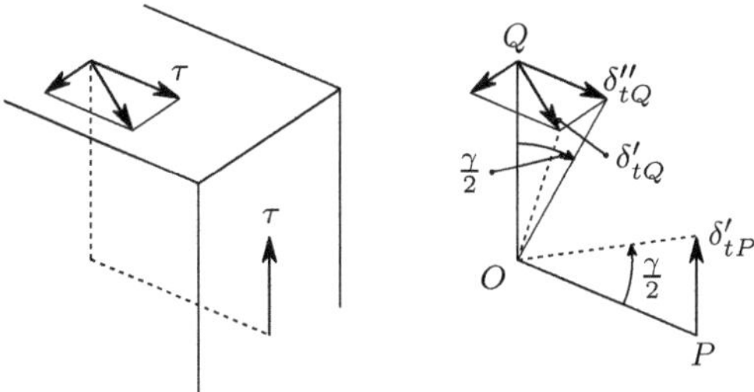

Fig. 26. Analogy between the reciprocity of the shearing stresses and the rotation of two orthogonal directions in the pure deformation

The complete analogy between the tensor operations on the stress and strain tensors, represented by the analogy between Expressions 10 and 56, allows some immediate conclusions to be drawn about the strain tensor representing a deformation state, as follows:

– The reference axes of the strain tensor may be transposed by means of the matrix operation

$$[\varepsilon'] = [l]^t [\varepsilon][l], \tag{57}$$

– where $[\varepsilon']$ contains the tensor components in the new reference axes and the orthogonal matrix $[l]$ contains the same direction cosines as in Expr. 15.

- In the deformation state around a point there are at least three orthogonal directions which do not undergo shearing strain, i.e., where $\delta'_t = 0$. These are the *principal directions* of the deformation state. The (longitudinal) strains in these directions are the *principal strains*, ε_1, ε_2 and ε_3; as a rule, a descending order is adopted, $\varepsilon_1 > \varepsilon_2 > \varepsilon_3$.
- The *characteristic equation of the strain tensor* is given by the expression

$$-\varepsilon^3 + I_1\varepsilon^2 - I_2\varepsilon + I_3 = 0 \ ,$$

- where I_1, I_2 and I_3 are the invariants of the strain tensor and take the values given by

$$I_1 = \varepsilon_x + \varepsilon_y + \varepsilon_z$$

$$I_2 = \begin{vmatrix} \varepsilon_x & \varepsilon_{xy} \\ \varepsilon_{xy} & \varepsilon_y \end{vmatrix} + \begin{vmatrix} \varepsilon_x & \varepsilon_{xz} \\ \varepsilon_{xz} & \varepsilon_z \end{vmatrix} + \begin{vmatrix} \varepsilon_y & \varepsilon_{yz} \\ \varepsilon_{yz} & \varepsilon_z \end{vmatrix}$$

$$I_3 = \begin{vmatrix} \varepsilon_x & \varepsilon_{xy} & \varepsilon_{xz} \\ \varepsilon_{xy} & \varepsilon_y & \varepsilon_{yz} \\ \varepsilon_{xz} & \varepsilon_{yz} & \varepsilon_z \end{vmatrix} \ .$$

- A Lamé's ellipsoid may be drawn for the strain tensor, in the same way as for the stress tensor (Fig. 9, with principal semi-axes $\overline{OA} = \varepsilon_1$, $\overline{OB} = \varepsilon_2$ and $\overline{OC} = \varepsilon_3$).
- The strain tensor may be decomposed into isotropic and deviatoric (distortional) components ($\varepsilon_m = \frac{\varepsilon_x + \varepsilon_y + \varepsilon_z}{3}$)

$$\begin{bmatrix} \varepsilon_x & \varepsilon_{xy} & \varepsilon_{xz} \\ \varepsilon_{xy} & \varepsilon_y & \varepsilon_{yz} \\ \varepsilon_{xz} & \varepsilon_{yz} & \varepsilon_z \end{bmatrix} = \underbrace{\begin{bmatrix} \varepsilon_m & 0 & 0 \\ 0 & \varepsilon_m & 0 \\ 0 & 0 & \varepsilon_m \end{bmatrix}}_{\text{isotropic tensor component}} + \underbrace{\begin{bmatrix} \varepsilon_x - \varepsilon_m & \varepsilon_{xy} & \varepsilon_{xz} \\ \varepsilon_{xy} & \varepsilon_y - \varepsilon_m & \varepsilon_{yz} \\ \varepsilon_{xz} & \varepsilon_{yz} & \varepsilon_z - \varepsilon_m \end{bmatrix}}_{\text{distortional tensor component}} \ .$$

- The octahedral longitudinal and shearing strains are defined by the expressions (cf. (29) and (31))

$$\varepsilon_{oct} = \frac{\varepsilon_x + \varepsilon_y + \varepsilon_z}{3}$$

$$\frac{\gamma_{oct}}{2} = \frac{1}{3}\sqrt{(\varepsilon_x - \varepsilon_y)^2 + (\varepsilon_x - \varepsilon_z)^2 + (\varepsilon_y - \varepsilon_z)^2 + 6\left(\varepsilon_{xy}^2 + \varepsilon_{xz}^2 + \varepsilon_{yz}^2\right)} \ .$$

- A Mohr's representation of the strain tensor, similar to that displayed in Fig. 14 for the stress tensor, may be made, with the longitudinal strains ε in the axis of abscissas and the shearing strain $\frac{\gamma}{2}$ in the axis of ordinates.

III.7 Volumetric Strain

The deformation usually causes a volume change. We define *volumetric strain* as the volume change per unit of initial volume. Since, at this point, only small deformations are considered, the changes to the initially right angles of the infinitesimal parallelepiped, $\gamma_{xy} = 2\varepsilon_{xy}$, $\gamma_{yz} = 2\varepsilon_{yz}$ and $\gamma_{xz} = 2\varepsilon_{xz}$, may be considered as infinitesimal quantities and, therefore, they do not cause volume change. Thus, the volume V of the generally non-rectangular parallelepiped, which results from the initial rectangular parallelepiped defined by the infinitesimal distances $\mathrm{d}x$, $\mathrm{d}y$ and $\mathrm{d}z$, may be computed as though it were rectangular, yielding ($V_0 = \mathrm{d}x\,\mathrm{d}y\,\mathrm{d}z$ is the initial volume of the infinitesimal parallelepiped)

$$V = (1 + \varepsilon_x)\,\mathrm{d}x\,(1 + \varepsilon_y)\,\mathrm{d}y\,(1 + \varepsilon_z)\,\mathrm{d}z$$
$$= (1 + \varepsilon_x)\,(1 + \varepsilon_y)\,(1 + \varepsilon_z)\,V_0\;.$$

Bearing in mind, that the longitudinal strains are also infinitesimal quantities, the products of these strains are infinitesimal quantities of higher order, so they may be disregarded. Thus, the volumetric strain is given by

$$\varepsilon_v = \frac{V - V_0}{V_0} = \varepsilon_x + \varepsilon_y + \varepsilon_z + \varepsilon_x\varepsilon_y + \varepsilon_x\varepsilon_z + \varepsilon_y\varepsilon_z + \varepsilon_x\varepsilon_y\varepsilon_z$$
$$\approx \varepsilon_x + \varepsilon_y + \varepsilon_z = I_1\;. \tag{58}$$

We may conclude that, in the case of small deformations, the first invariant of the strain tensor takes the value of the volumetric strain.

III.8 Two-Dimensional Analysis of the Strain Tensor

III.8.a Introduction

In the same way as in the case of the stress tensor, a two-dimensional analysis of the strain tensor can also be performed, if one of the principal directions is known. If the principal strain associated with this direction is zero, we have a state of *plane strain.*

As noted in Sect. II.9, for the plane state of stress, the two-dimensional analysis of the strain tensor could be performed by particularizing the expressions developed for the general three-dimensional case to the stresses contained in the plane defined by two principal directions. However, for the same reasons as explained in that section, an independent development of the two-dimensional expressions is preferable.

Only the general considerations presented in Sects. III.1 and III.2 are needed to understand the following explanation. The expressions developed here are only valid in the linear case, where both the deformations and the rotations take infinitesimal values. For simplicity, we consider that the known principal direction is direction z, so that the two-dimensional analysis is performed in the xy plane.

III.8.b Components of the Strain Tensor

As discussed in Sect. III.2, the homogeneous deformation of a rectangle may be defined by the elongation of its sides and by the variation of the initially right-angle between two sides. These three quantities (and the initial dimensions) define the parallelogram, which results from the rectangle.

Let us now consider an infinitesimal rectangle, whose sides are parallel to the Cartesian reference axes x, y and have the infinitesimal lengths dx and dy. The elongation of its sides, $\Delta\,dx$ and $\Delta\,dy$, divided by the initial lengths, gives the longitudinal strains $\varepsilon_x = \frac{\Delta\,dx}{dx}$ and $\varepsilon_y = \frac{\Delta\,dy}{dy}$. The variation of the initially right-angle between dx and dy, $\gamma_{xy} = 2\varepsilon_{xy}$, defines the double shearing strain or distortion. These three dimensionless quantities, ε_x, ε_y and γ_{xy}, fully define the state of deformation around a point in the two-dimensional case, since they allow the computation of the strain in any arbitrary direction of plane xy, as will be seen in the following Sub-section.

As in the three-dimensional case, a sign convention is used, in which a positive longitudinal strain corresponds to an increase in length and a positive shearing strain corresponds to a decrease in the angle defined by the positive directions of the reference axes (cf. Fig. 27).

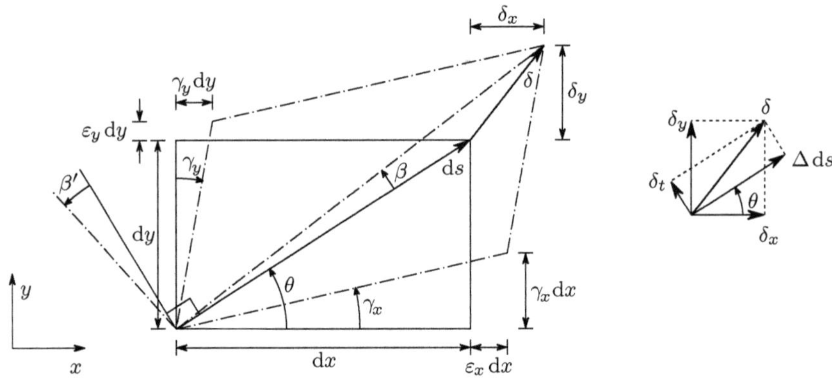

Fig. 27. Components of the deformation of a line segment with arbitrary direction

III.8.c Strain in an Arbitrary Direction

Let us consider an infinitesimal line segment with infinitesimal length ds and orientation defined by the angle θ, measured from axis x in the positive direction (from x to y), as represented in Fig. 27. As a consequence of the longitudinal and shearing strains ε_x, ε_y and γ_{xy} this line segment undergoes a longitudinal strain and a rotation. Denoting the rotations of the line segments $dx = ds\cos\theta$ and $dy = ds\sin\theta$ by γ_x and γ_y, respectively, positive if they lead to a decrease in the angle between the positive semi-axes x and y, the geometrical considerations depicted in Fig. 27 may be established.

In the displacements represented in this Figure the products of longitudinal and shearing strains have been disregarded, since they are infinitesimal quantities of higher order, because only infinitesimal deformations and rotations are considered (for example, we have considered $\gamma_x \left(dx + \varepsilon_x\, dx \right) \approx \gamma_x\, dx$). Furthermore, as all the rotations are infinitesimal the simplifications $\cos\gamma \approx 1$ and $\sin\gamma \approx \tan\gamma \approx \gamma$ have been made.

The displacement δ of the tip of vector ds may be defined by its components (cf. Fig. 27)

$$\begin{cases} \delta_x = \varepsilon_x\, dx + \gamma_y\, dy \\ \delta_y = \varepsilon_y\, dy + \gamma_x\, dx \ . \end{cases} \tag{59}$$

The projection of δ_x and δ_y in the direction of ds gives the elongation of this line segment

$$\Delta\, ds = \delta_x \cos\theta + \delta_y \sin\theta \ , \tag{60}$$

or, substituting (59) in (60) and dividing by the initial length ds

$$\varepsilon_\theta = \frac{\Delta\, ds}{ds} = \varepsilon_x \frac{dx}{ds}\cos\theta + \gamma_y \frac{dy}{ds}\cos\theta + \varepsilon_y \frac{dy}{ds}\sin\theta + \gamma_x \frac{dx}{ds}\sin\theta \ .$$

Taking into consideration that $\frac{dx}{ds} = \cos\theta$ and $\frac{dy}{ds} = \sin\theta$, we get

$$\varepsilon_\theta = \varepsilon_x \cos^2\theta + \varepsilon_y \sin^2\theta + 2\frac{\gamma_{xy}}{2}\sin\theta\cos\theta \ , \tag{61}$$

since $\gamma_x + \gamma_y = \gamma_{xy}$. This expression is analogous to Expr. 33, which furnishes the stress in a facet, whose orientation is defined by angle θ (cf. Fig.11). By projecting δ_x and δ_y on the normal direction to ds, the transversal displacement δ_t of the tip of ds is obtained

$$\delta_t = -\delta_x \sin\theta + \delta_y \cos\theta \ .$$

In this expression, the displacement δ_t is considered positive if it corresponds to a rotation of ds in the counterclockwise direction. By dividing by ds and taking (59) into consideration, as well as the relations $dx = ds \cos\theta$ and $dy = ds \sin\theta$, the rotation β (cf. Fig. 27) is obtained

$$\beta = \frac{\delta_t}{ds} = (\varepsilon_y - \varepsilon_x)\sin\theta\cos\theta + \gamma_x \cos^2\theta - \gamma_y \sin^2\theta \ . \tag{62}$$

The rotation β' of a line segment ds', which makes a right-angle, in the positive (counterclockwise) direction, with ds (cf. Fig. 27) may be computed by substituting θ by $\theta + \frac{\pi}{2}$ in (62), yielding

$$\beta' = -(\varepsilon_y - \varepsilon_x)\sin\theta\cos\theta + \gamma_x \sin^2\theta - \gamma_y \cos^2\theta,$$

since $\sin\left(\theta + \frac{\pi}{2}\right) = \cos\theta$ and $\cos(\theta + \frac{\pi}{2}) = -\sin\theta$. The double shearing strain γ_θ between the directions defined by the angles θ and $\theta + \frac{\pi}{2}$ is then given by (cf. Fig. 27).

$$\gamma_\theta = \beta - \beta' = (\varepsilon_y - \varepsilon_x)\, 2\sin\theta\cos\theta + \underbrace{\left(\gamma_x + \gamma_y\right)}_{=\gamma_{xy}} \left(\cos^2\theta - \sin^2\theta\right). \tag{63}$$

Taking into consideration that an infinitesimal rotation of the reference axes causes only infinitesimal changes in the components of the strain tensor (this may be easily verified by substituting in (61) θ by $d\theta \Rightarrow \varepsilon_\theta \approx \varepsilon_x$, or by $\frac{\pi}{2} + d\theta \Rightarrow \varepsilon_\theta \approx \varepsilon_y$ and in (63) θ by $d\theta \Rightarrow \gamma_\theta \approx \gamma_{xy}$). Therefore, we may consider in Fig. 27 and in (62) that $\gamma_x = \gamma_y = \frac{\gamma_{xy}}{2}$. In this case, (62) immediately gives the shearing strain $\frac{\gamma_\theta}{2}$ in the orthogonal directions θ and $\theta + \frac{\pi}{2}$, yielding[5]

$$\gamma_x = \gamma_y = \frac{\gamma_{xy}}{2}$$
$$\Rightarrow \beta = -\beta' = \frac{\gamma_\theta}{2} = (\varepsilon_y - \varepsilon_x)\sin\theta\cos\theta + \frac{\gamma_{xy}}{2}\left(\cos^2\theta - \sin^2\theta\right). \tag{64}$$

Equation (64) is formally analogous to (34). By transforming (61) and (64) in the same way as (33) and (34) were transformed in Subsect. II.9.b, we get, from (61)

$$\varepsilon_\theta = \frac{\varepsilon_x + \varepsilon_y}{2} + \frac{\varepsilon_x - \varepsilon_y}{2}\cos 2\theta + \frac{\gamma_{xy}}{2}\sin 2\theta. \tag{65}$$

In the same way, we get from (64)

$$\frac{\gamma_\theta}{2} = -\frac{\varepsilon_x - \varepsilon_y}{2}\sin 2\theta + \frac{\gamma_{xy}}{2}\cos 2\theta. \tag{66}$$

These two expressions (65 and 66) are formally analogous to those obtained in the two-dimensional analysis of the stress tensor for the normal and shearing stresses in an arbitrary oriented facet ((35) and (36), respectively). As the further developments based on these expressions were based solely on mathematical considerations, they are also valid for the strain tensor, if we substitute σ_x, σ_y and τ_{xy} by ε_x, ε_y and $\varepsilon_{xy} = \frac{\gamma_{xy}}{2}$, respectively. The following conclusions may therefore be drawn:

- There are two orthogonal directions, which do not suffer shearing strain during the deformation. These are the principal directions of the strain tensor and may be computed by an expression analogous to (37)

$$\theta = \frac{1}{2}\arctan\frac{\gamma_{xy}}{\varepsilon_x - \varepsilon_y}. \tag{67}$$

---

[5]This conclusion confirms the considerations established in Sects. III.4 and III.6 about the decomposition of the motion of the material points in an infinitesimal region in pure deformation and rigid body rotation. The infinitesimal rotation of the reference axes, so that $\gamma_x = \gamma_y$, is equivalent to the elimination of the rigid body motion.

– The longitudinal strain ε reaches extreme values (maximum and minimum) in the principal directions. These are the principal strains, which may be computed by the expression

$$\left\{ \begin{array}{c} \varepsilon_1 \\ \varepsilon_2 \end{array} \right\} = \frac{\varepsilon_x + \varepsilon_y}{2} \left\{ \begin{array}{c} + \\ - \end{array} \right\} \sqrt{\left(\frac{\varepsilon_x - \varepsilon_y}{2} \right)^2 + \left(\frac{\gamma_{xy}}{2} \right)^2} \, .$$

– A Mohr's circle may be drawn for the two-dimensional strain tensor, where the axis of abscissas contains the longitudinal strains and the axis of ordinates the shearing strains.

III.9 Conclusions

In this chapter we have mainly analysed the physical aspects of the deformation. This has been possible because the conclusion that the state of deformation, as the state of stress, may be described by a symmetric second order tensor, allows a full analogy between the purely mathematical tensor transformations in the two cases. Thus, we have concluded that the tensorial operations described in Sects. II.5 to II.8 for the stress tensor are also valid in the case of the strain tensor.

As in Chap. II for the stress theory, the analysis is mainly performed in an infinitesimal neighbourhood around a point. The functions describing the evolution of the elements of the strain tensor in the continuum were taken into consideration only for developing the equations of compatibility. Here it should be noted that, while the six elements of the strain tensor are completely independent of each other, the six functions defining the elements of the strain tensor must obey the compatibility conditions.[6]

The two restrictions used in the development of the mathematical expressions for the deformation state are of completely different nature.

The first – restriction of the analysis to an infinitesimal neighbourhood around a point, so that the deformation may be considered as homogeneous – has consequences on the level of the mathematical tools used: the simplifications made possible by the consideration of a a homogeneous deformation impose the use of integral and differential calculus.[7]

The second – consideration of infinitesimal deformations and rotations – has consequences on the level of the problem's physics. As a consequence, no matter how good the mathematical or numerical tools used are, an error

---

[6]The same conclusion may be drawn in relation to the functions defining the stresses in the continuum: the six elements of the stress tensor are independent of each other, but the six functions, which define the same stresses as functions of the coordinates x, y, z must obey the differential equations of equilibrium (5).

[7]The corresponding restriction in the stress state is the consideration of infinitesimal facets. The consequences of this restriction are, as in the deformation state, only on the level of the mathematical formulation of the problem.

is always present, and this becomes larger when deformations and rotations grow. As mentioned in Footnote 6, the restriction to small rotations even excludes the capacity to consider structural instability phenomena. For this reason, the analysis of the buckling of slender members (Chap. XI) is based on the bending theory, where, as Chap. VII will show, the validity of the relation between the motion of cross sections and strains is not limited to small rotations.

III.10 Examples and Exercises

III.1. Displacements were measured in a deformed body, which may be approximated by the expressions

$$\begin{cases} u = 5x^2 + 3xy + 4 + 4y^2 + 3yz \\ v = 6xy + 4y^2 + 5 + 2z^2 \\ w = 4xz + 2y^2 + 3y + 6z^2 \ . \end{cases}$$

Knowing that both deformations and rotations are sufficiently small to be considered as infinitesimal, determine the functions describing the strains and the rotations in the body.

III.2. Displacements were measured in the deformation of a body, which may be approximated by the expressions ($A, B, \dots, H$ are constants)

$$\begin{cases} u = Ax^3 + By^2 + Cyz \\ v = Dx^2y + Ey^3 + Fy^2z \\ w = Gxz^2 + Hyz^2 \ . \end{cases}$$

Knowing that, although the rotations are of considerable magnitude, the deformations are sufficiently small to be considered as infinitesimal, compute the longitudinal strain of an infinitesimal line segment which is parallel to axis x and located in an infinitesimal neighbourhood of the point of coordinates (2,-3,5).

III.3. What are the degrees of connection of the following bodies:
   (a) a body composed by six bars linked like the edges of a tetrahedron;
   (b) a prism with a square base and an interior cavity, which intersects the four side faces and does not intersect the top and bottom faces;
   (c) a ring with a tubular cross section.

III.4. Write a computation sequence to verify the reciprocity of the rotations in a pure deformation (cf. Fig. 26).

Resolution

Given data: elements of the strain tensor, ε_x, ε_y, ε_z, ε_{xy}, ε_{xz} and ε_{yz}; direction cosines of direction OP, l, m and n.

Computation Sequence

1. components of the displacement of the tip of vector $\overrightarrow{OP}$:

$$\begin{cases} \delta'_{px} = l\varepsilon_x + m\varepsilon_{xy} + n\varepsilon_{xz} \\ \delta'_{py} = l\varepsilon_{xy} + m\varepsilon_y + n\varepsilon_{yz} \\ \delta'_{pz} = l\varepsilon_{xz} + m\varepsilon_{yz} + n\varepsilon_z \ ; \end{cases}$$

2. longitudinal strain in direction OP:

$$\varepsilon_p = l\delta'_{px} + m\delta'_{py} + n\delta'_{pz};$$

3. transversal displacement of the tip of vector $\overrightarrow{OP}$:

$$\delta'_{tp} = \sqrt{\delta'^2_{px} + \delta'^2_{py} + \delta'^2_{pz} - \varepsilon^2_p};$$

4. direction cosines of direction OQ $(\overrightarrow{OQ}\|)\delta'_{tp}$:

$$l_q = \frac{\delta'_{px} - l\varepsilon_p}{\delta'_{tp}} \qquad m_q = \frac{\delta'_{py} - m\varepsilon_p}{\delta'_{tp}} \qquad n_q = \frac{\delta'_{pz} - n\varepsilon_p}{\delta'_{tp}};$$

5. components of the displacement of the tip of vector $\overrightarrow{OQ}$:

$$\begin{cases} \delta'_{qx} = l_q\varepsilon_x + m_q\varepsilon_{xy} + n_q\varepsilon_{xz} \\ \delta'_{qy} = l_q\varepsilon_{xy} + m_q\varepsilon_y + n_q\varepsilon_{yz} \\ \delta'_{qz} = l_q\varepsilon_{xz} + m_q\varepsilon_{yz} + n_q\varepsilon_z; \end{cases}$$

6. longitudinal strain in direction OQ:

$$\varepsilon_q = l_q\delta'_{qx} + m_q\delta'_{qy} + n_q\delta'_{qz};$$

7. components of the transversal displacement of the tip of vector $\overrightarrow{OQ}$:

$$\delta'_{tqx} = \delta'_{qx} - l_q\varepsilon_q \qquad \delta'_{tqy} = \delta'_{qy} - m_q\varepsilon_q \qquad \delta'_{tqz} = \delta'_{qz} - n_q\varepsilon_q;$$

8. projection of δ'_{tq} in direction OP:

$$\delta''_{tq} = l\delta'_{tqx} + m\delta'_{tqy} + n\delta'_{tqz};$$

9. verification:

$$\delta''_{tq} = \delta'_{tp}.$$

IV

Constitutive Law

IV.1 Introduction

In the last two chapters the relations between the external forces and the stresses and between the displacements and the strains have been analysed. The constitutive law is the third element of the chain which links the external forces with the displacements they cause in a continuous body, as described in Sect. I.3.

In this element of the chain the rheological behaviour of the material plays the central role, since the relations between the internal forces and deformations (described by the stress and strain tensors, respectively) must be established.

The constitutive law may be described by a tensor, as the internal forces and deformations. This is actually a fourth order tensor, since it describes the relations between two second order tensors. However, in the explanation that follows, the tensor mathematics will not be used to define the constitutive law. In fact, although the mathematical approach offers more systematization possibilities, the other approach – the physical evidence, with adequate justification – allows a better physical understanding of the conclusions arrived at, which is of crucial importance in the first stages of the study of Mechanics of Materials.

IV.2 General Considerations

As was made clear in the first chapter (Sect. I.1), a phenomenological approach must be used in order to quantify the rheological behaviour of materials. In this approach the results of experimental tests are used as a basis for the typification and quantification of the rheological behaviour. This process always uses *behaviour models*. The oldest and simplest of these models is the one defined by Robert Hooke in 1678 (Hooke's law) [2], which states that strain is proportional to stress. In this case the *mathematical model* of proportionality is used in the definition of the material behaviour. *Physical models* may also be used: for example, the influence of viscosity on the material's behaviour may be understood by analogy with the deformation of the dashpot in a shock absorber device, such as a car damper. The dashpot is in this case the physical model. Some of these models are analysed in Sect. IV.3. Some commonly used concepts are now defined:

- *Isotropic material* – material with the same rheological characteristics in every direction, or, in other words, material with rheological properties, which are symmetrical in relation to any plane containing the material point under consideration. Steel and concrete – the materials most used in civil engineering structures – are examples of materials which may be considered as isotropic.
- *Monotropic material* – material with symmetrical properties in relation to planes, which are parallel or perpendicular to a particular direction: the direction of monotropy of the material. Wood is an example of a material, which may be considered as monotropic, with the fibres defining direction of monotropy, since it displays symmetrical properties in relation to planes parallel or perpendicular to the fibres' direction. Other examples include all fibre-reinforced isotropic materials whose fibres are in one direction, and materials composed by alternate layers of two or more isotropic materials, if the thickness of the layers is sufficiently small to consider the material as continuous in the direction perpendicular to the layers. This direction describes the direction of monotropy. The monotropic material is also called *transversal isotropic material,* since it displays isotropic properties in the plane perpendicular to the monotropy direction.
- *Orthotropic material* – material with symmetrical properties in relation to three orthogonal planes. An example would be an isotropic material reinforced by fibres placed in orthogonal directions, or in such a way that three orthogonal planes of symmetry are obtained. Composites made of carbon or glass fibre reinforced plastics are mostly orthotropic (cf. example IV.12).
- *Elastic deformation* – recoverable deformation which occurs at the same time as the loading which causes it. Furthermore, the stress-strain relation is the same in the loading and unloading phases. If this relation is linear, a linear-elastic behaviour is obtained.
- *Plastic deformation* – non-recoverable deformation, which occurs at the same time as the loading. Usually, the plastic deformation takes place only if the stress reaches a certain threshold value. In some metals, such as mild steel, a significant part of the plastic deformation occurs suddenly, when the stress reaches a value called the *yielding stress* of the material.
- *Viscous deformation* – time-dependent deformation, in which the rate of deformation is a function of the stresses. This deformation takes place while the material is under loading, i.e., it increases with the time, even with constant internal forces. In the special case of *linear viscosity* the deformation rate is proportional to the stress.

– *Ductile material* – material, which is able to undergo a substantial amount of plastic deformation, generally much larger than the elastic deformation, before rupture.
– *Brittle material* – material, which attains rupture with very little non-elastic deformation, i.e., with elastic behaviour practically until rupture.
– *Creep* – increase of deformation with the time without loading change. This rheological behaviour may be explained as a combination of elastic and viscous behaviour (visco-elastic deformation, recoverable), plastic and viscous behaviour (visco-plastic deformation, non-recoverable) or even a combination of the three fundamental deformation types: elastic, plastic and viscous (elasto-visco-plastic behaviour). Usually the deformation rate decreases with time.
– *Relaxation* – Decrease of the internal forces with time, without variation of the deformation. This phenomenon may also be explained as a consequence of visco-elastic, visco-plastic or elasto-visco-plastic rheological behaviour. Usually the rate of decay of the internal forces decreases with time.
– *Liquid* – from a rheological point of view, a liquid is a material, in which only isotropic stress states are possible, unless the deformation varies with time and the material displays viscosity. As a consequence, in a non-flowing liquid there are no shearing stresses at all.
– *Solid* – in contrast to a liquid, from a rheological point of view, a solid is a material, in which non-isotropic stress states are possible, even when the deformation rate vanishes in every region of the body.

IV.3 Ideal Rheological Behaviour – Physical Models

The formerly defined basic types of material behaviour may be understood with the help physical models. Physical modelling, although not absolutely necessary, does facilitate the understanding of the possible mechanisms of the material structure, which determine its rheology and, if consistently followed, it avoids the development of rheological models, which do not obey fundamental laws, such as the second law of thermodynamics (law of increasing entropy, or of energy dissipation).

Elastic behaviour may be easily understood by imagining a spring under the action of a force. When the force is applied, the spring deforms, if the force is increased the deformation increases, if the force remains unchanged, the deformation does not change and if the force is withdrawn, the spring recovers its initial dimensions. The relation between the force and the deformation is the same, either when the force is increasing (loading) or when the force is decreasing (unloading), i.e., the rheological behaviour is the same in the loading and unloading phases. If the force is proportional to the deformation the constitutive law of the model reads (K is a constant)

$$ F = K\delta \qquad \Longleftrightarrow \qquad \sigma = E\varepsilon \quad \text{(Hooke's law)} . $$

These considerations are summarized in Fig. 28.

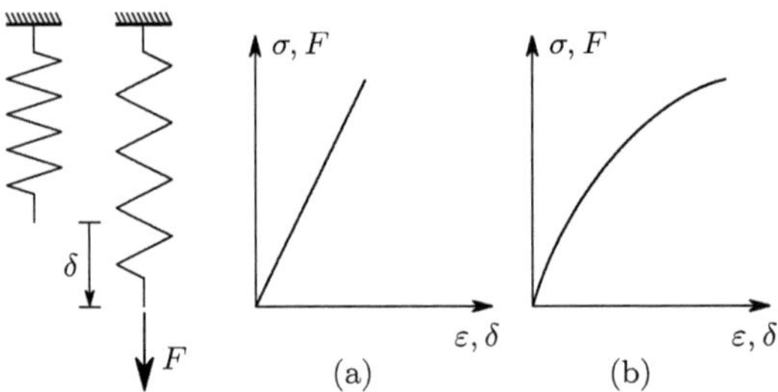

Fig. 28. Physical model for the elastic behaviour: (**a**) linear; (**b**) non-linear

The plastic deformation has force-displacement and energy dissipation features, which may be understood by analogy with the displacement of a body under its own weight, lying on a horizontal surface, when a horizontal force is applied, as represented in Fig. 29: the force causes a displacement of the body only if its intensity reaches the value corresponding to the friction force between the body and the surface. After this, the displacement may be increased without variation in the force, and the body does not go back to the initial position when the horizontal force disappears. The deformation is therefore not recoverable (unless a force with the opposite direction is applied), which means that the work done by the horizontal force is dissipated: it is transformed in heat in the contact surface.

The viscous deformation may be modelled by a cylinder with a liquid, where a piston with a small perforation moves, as represented in Fig. 30. When a force P is applied to the piston, the liquid flows through the perforation with a flow volume which depends on the pressure under the piston. The higher the intensity of the force P, the higher this pressure will be. Thus the rate of displacement of the piston $\dot{\delta} = \frac{\mathrm{d}\delta}{\mathrm{d}t}$ may be directly related to the force P. In terms of stresses and strains linear viscosity may be expressed by the constitutive law (η is the viscosity modulus)

$$\dot{\varepsilon} = \frac{\mathrm{d}\varepsilon}{\mathrm{d}t} = \frac{\sigma}{\eta}. \tag{68}$$

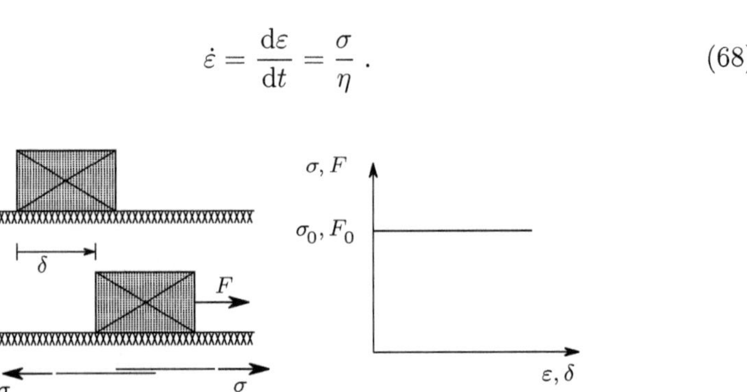

Fig. 29. Physical model for the plastic behaviour: $F < F_0 \Rightarrow \delta = 0 \; \sigma < \sigma_0 \Rightarrow \varepsilon = 0$

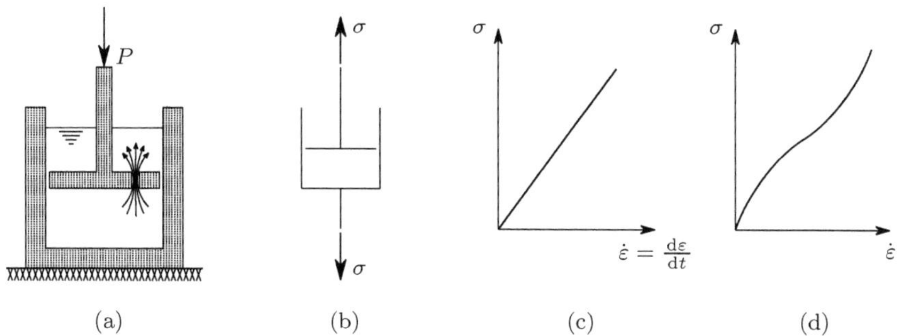

(a)   (b)   (c)   (d)

Fig. 30. Physical model for the viscous deformation: (**a**) physical model: dashpot (**b**) schematic representation (**c**) linear viscosity (**d**) non-linear viscosity

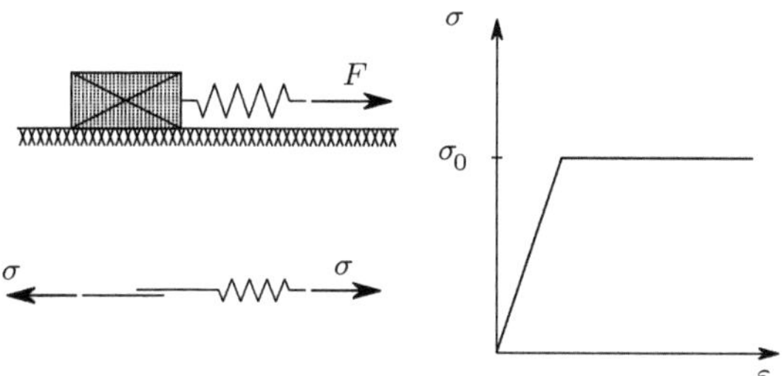

Fig. 31. Rheological model for the elastic, perfectly plastic behaviour

By connecting together the base models presented in Figs. 28, 29 and 30, more elaborate models may be constructed. For example, connecting the elastic and the plastic model in a series, an elastic, perfectly plastic behaviour is modelled, as depicted in Fig. 31.

The visco-elastic behaviour of a solid material may be modelled by the *Kelvin solid*, which consists of a parallel association of a spring and a dashpot.[1] The constitutive law of this model may be obtained by taking into consideration that the total stress is the sum of the stresses in the spring and the dashpot, yielding

$$\sigma = E\varepsilon + \eta\dot{\varepsilon} \ . \tag{69}$$

If the strain is prescribed, this expression furnishes the corresponding stress directly. However, if the stress is prescribed, it is necessary to solve the differential equation 69, to compute the corresponding strain. For example, if a stress σ_0 is applied to the model at the time t_0 and if the stress remains unchanged, the corresponding deformation is given by (a particular integral of the complete equation is $\varepsilon = \frac{\sigma_0}{E}$ and C is an integration constant)

---

[1] In the parallel associations it is assumed that all elements suffer the same deformation.

$$\begin{cases} \varepsilon = \dfrac{\sigma_0}{E} + C\,\mathrm{e}^{-\frac{E}{\eta}t} \\[2mm] t = t_0 \Rightarrow \varepsilon = 0 \end{cases} \Rightarrow C = -\dfrac{\sigma_0}{E}\,\mathrm{e}^{\frac{E}{\eta}t_0} \Rightarrow \varepsilon = \dfrac{\sigma_0}{E}\left[1 - \mathrm{e}^{-\frac{E}{\eta}(t-t_0)}\right]. \qquad (70)$$

This expression represents a strain which tends asymptotically to the value $\frac{\sigma_0}{E}$. This deformation has the characteristics of a creep deformation, since it increases without variation of stress and the deformation rate decays with time. At time instant t_1 the deformation takes the value

$$t = t_1 \Rightarrow \varepsilon = \varepsilon_1 = \frac{\sigma_0}{E}\left[1 - \mathrm{e}^{-\frac{E}{\eta}(t_1-t_0)}\right].$$

If the model is unloaded at time t_1, the rheological behaviour may be computed by solving (69) again, yielding

$$\begin{cases} \sigma = 0 \Rightarrow E\varepsilon + \eta\dot{\varepsilon} = 0 \\[2mm] t = t_1 \Rightarrow \varepsilon = \varepsilon_1 \end{cases} \Rightarrow C = \varepsilon_1\,\mathrm{e}^{\frac{E}{\eta}t_1} \Rightarrow \varepsilon = \varepsilon_1\,\mathrm{e}^{-\frac{E}{\eta}(t-t_1)}.$$

We can see that, although the dashpot dissipates energy, the deformation is recoverable, since it tends asymptotically to zero.

Figure 32 illustrates the considerations about the Kelvin solid established here.

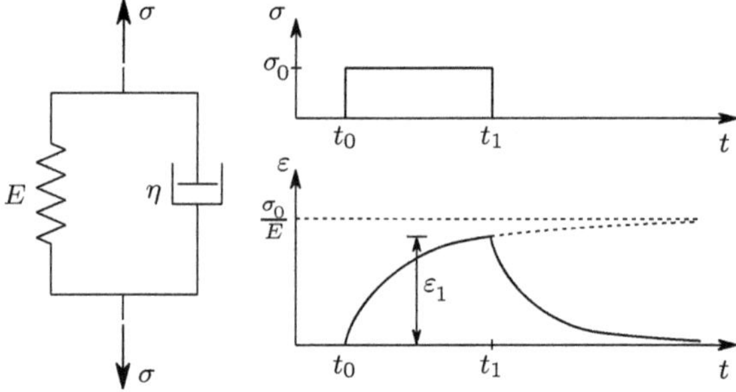

Fig. 32. Deformation and recovery of the Kelvin solid under constant stress

The shape of the creep curves may be easily explained, taking into account that the dashpot does not deform instantaneously, since an infinite strain rate would demand an infinite stress (cf. (68)). Thus, at the moment the stress is applied (instant t_0) the load is totally supported by the dashpot, since the spring is undeformed and therefore its stress is zero. As the deformation increases with time, the stress in the spring increases and, as a consequence, it decreases in the dashpot, which causes a fall in its (and in the model's) deformation rate. This rate vanishes when all the stress is transferred to the spring, which theoretically only happens for $t = \infty$, when $\varepsilon = \frac{\sigma_0}{E}$.

The deformation recovery after $t = t_1$ proceeds in a similar way. Immediately after the stress is released, the installed deformation causes a stress in the spring, which must now be balanced by an opposite stress in the dashpot, since the total stress is zero. This stress, which causes the deformation recovery, decreases as the strain in the model decreases, which causes a reduction in the recovery rate. Theoretically, the deformation in the Kelvin model disappears completely ($\varepsilon = 0$) only for $t = \infty$.

The quantity $\frac{E}{\eta}$ defines the reaction time of the model and is usually called the *creep modulus* or *retardation coefficient* of the model. Its inverse $\frac{\eta}{E}$ has time dimensions and is known as the *retardation time*.

A serial association of a spring and a dashpot defines the *Maxwell model*.[2] The rate of deformation of this model is the sum of the rates of deformation in the spring and in the dashpot. Its constitutive law may therefore be represented by the expression (cf. (68))

$$\dot{\varepsilon} = \frac{\dot{\sigma}}{E} + \frac{\sigma}{\eta} \; . \tag{71}$$

Contrary to the Kelvin model, in the Maxwell model the computation of the strain for given stress is immediate (cf. Fig. 33-a), while the computation of the stress for given strains demands the solution of the differential equation represented by (71), when the stress is unknown. Imposing to the rheological model a strain ε_0 at the time t_0 and keeping it constant, the stress at once takes the value $E\varepsilon_0$ and relaxes as the dashpot deforms. The interaction between the deformation in the dashpot and the stress in the model leads to a stress-time relation represented by an exponential law which tends asymptotically to zero, as represented in Fig. 33-b. This relation takes the form

$$\begin{cases} \dot{\varepsilon} = 0 \Rightarrow \dfrac{\dot{\sigma}}{E} + \dfrac{\sigma}{\eta} = 0 \\ t = t_0 \Rightarrow \sigma = E\varepsilon_0 \end{cases} \Rightarrow \sigma = E\varepsilon_0\, \mathrm{e}^{-\frac{E}{\eta}(t-t_0)} \; . \tag{72}$$

This expression is obtained by solving (71) for constant ε, with an initial stress $\sigma = E\varepsilon_0$ for $t = t_0$.

Figure 33 summarizes the considerations about the Maxwell model established here.

The shape of the relaxation curve may be physically explained by similar considerations to those for the creep deformation of the Kelvin model under constant stress. As mentioned previously, since the dashpot does not deform instantaneously, the imposed deformation is at first completely transmitted to the spring, which causes the initial stress $\sigma = E\varepsilon_0$. This stress decays with time, as the dashpot deforms, since this deformation causes a reduction of the strain in the spring (the total strain remains constant). The rate of decay

---

[2] If the stress acting on the model belongs to the deviatoric component of the stress tensor, the model represents the rheological behaviour of a liquid, since the stress is different from zero, only if the rate of deformation of the dashpot does not vanish, which is in accordance with the definition of liquid given in Sect. IV.2.

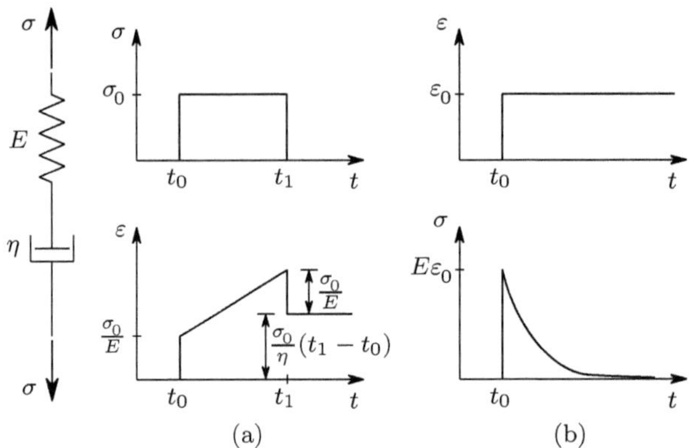

Fig. 33. Maxwell liquid: (**a**) creep; (**b**) relaxation

decreases with time, since the stress relaxation causes a fall in the deformation rate on the dashpot.

The reaction time of the model is determined by the quantity $\frac{E}{\eta}$ (cf. (72)). This relation is usually called *relaxation modulus* and its inverse $\frac{\eta}{E}$ is the *relaxation time*.

Although the rheological behaviour observed experimentally in actual materials may be qualitatively described by the simple rheological models of Kelvin and Maxwell, a quantitative approximation with a sufficient degree of precision demands generally more sophisticated models. These usually have a larger number of base elements. The development and description of these models is beyond the scope of this text. However, it may be pointed out that, for visco-elastic materials, two main families of models are generally used: the *Kelvin chain*, composed of a serial association of Kelvin elements (Fig. 32), where the computation of the strain for a given stress history is relatively simple, since the stress is the same in all elements, and the *generalized Maxwell model*, composed of a parallel association of Maxwell elements (Fig. 33), where the computation of the stress for a given strain history is simple, since the deformation is the same in all elements.

If the viscosity and elasticity moduli, η and E, are constant, we have linear visco-elasticity, since the effects of different stress or strain histories imposed in visco-elastic materials may be superposed, as is easily confirmed in the example of the equations representing the constitutive laws of the models of Kelvin and Maxwell analysed above. Thus, we get from (69)

$$\begin{cases} E\varepsilon_1 + \eta\dot{\varepsilon}_1 = \sigma_1 \\ E\varepsilon_2 + \eta\dot{\varepsilon}_2 = \sigma_2 \\ \varepsilon_1 + \varepsilon_2 = \varepsilon_3 \end{cases} \Rightarrow \quad \begin{aligned} E\varepsilon_3 + \eta\dot{\varepsilon}_3 &= E\left(\varepsilon_1 + \varepsilon_2\right) + \eta\left(\dot{\varepsilon}_1 + \dot{\varepsilon}_2\right) \\ &= \underbrace{E\varepsilon_1 + \eta\dot{\varepsilon}_1}_{\sigma_1} + \underbrace{E\varepsilon_2 + \eta\dot{\varepsilon}_2}_{\sigma_2} = \sigma_3 \,. \end{aligned} \qquad (73)$$

From this expression we verify that the effects of the strains ε_1 and ε_2, which correspond to the stresses σ_1 and σ_2 may be superposed. This proof also works in the inverse problem, i.e., starting from a stress $\sigma_3 = \sigma_1 + \sigma_2$, (73) shows that the corresponding strain is $\varepsilon_3 = \varepsilon_1 + \varepsilon_2$. The same line of reasoning could also be applied to the constitutive law of the Maxwell model (71).

In the case of the plastic model presented in Fig. 29, the effects of stresses and strains can not be superposed. This can easily be verified by considering two values for the stress, whose sum exceeds the yielding stress σ_0, with each stress being inferior to this value. If only one of these stresses is acting, there is no deformation, while the two stresses acting simultaneously cause an infinite deformation.

IV.4 Generalized Hooke's Law

IV.4.a Introduction

In the previous section some considerations about the fundamental types of rheological behaviour were established and a physical interpretation of these phenomena has been made with the help of one-dimensional models.

However, in the problems arising in the practical application of the Mechanics of Materials the stress and strain states are frequently two- or three-dimensional, so that the generalization of the one-dimensional models to two or three dimensions is necessary. In this section, we will study three-dimensional constitutive laws in the linear case (i.e., when the stresses and strains are related by linear functions), considering isotropic, monotropic and orthotropic materials.

IV.4.b Isotropic Materials

As defined in Sect. IV.2, isotropic materials display symmetrical features in relation to any plane. Therefore, the three planes define by the reference axes in any rectangular Cartesian reference system are symmetry planes in relation to the rheological behaviour of the material.

Let us first consider the isolated action of the normal stress σ_x, as represented in Fig. 34. Since this stress pair describes a symmetrical system of forces in relation to the symmetry planes of the elementary parallelepiped, which are also symmetry planes of the rheological properties of the material, the deformation must also be symmetric in relation to those planes. As a consequence, after the deformation the parallelepiped remains rectangular and only changes in the lengths of its sides take place. Besides, as the directions y and z are in the same conditions in relation to axis x, its strain must be the same. Consequently, the parallelepiped's deformation is completely defined by the longitudinal strains in directions x and y or z. The transversal strains (y and z) are usually negative (length reduction), when the longitudinal one (x in this case) is positive (length increase). Furthermore, as only linear elastic behaviour is considered, all the deformations are proportional to the stress

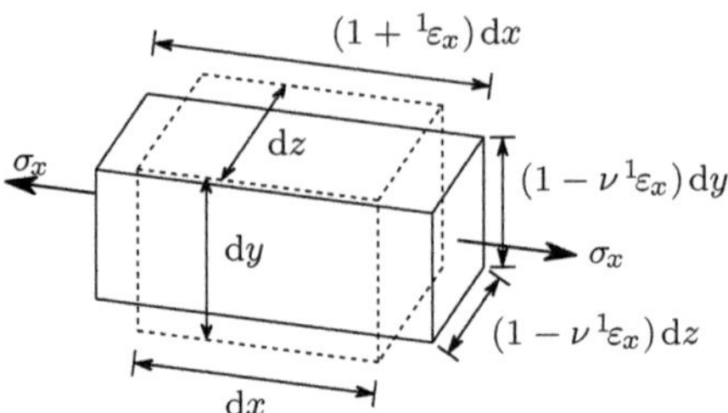

Fig. 34. Deformation caused by the isolated actuation of σ_x: ⋯⋯ original configuration; —— deformed configuration

σ_x, which means that the transversal strains are proportional to the strain in longitudinal direction x.

The relation between the normal stress (in this case σ_x) and the longitudinal strain in the same direction (in this case ε_x) is called the *longitudinal modulus of elasticity* or *Young's modulus* of the material. The relation between the transversal and longitudinal strains, multiplied by -1, is known as the *Poisson's coefficient* of the material (ν). The strains caused by the stress σ_x are then (cf. Fig. 34)

$$\sigma_x \rightarrow \begin{cases} {}^1\varepsilon_x = \dfrac{\sigma_x}{E} \\[2mm] {}^1\varepsilon_y = -\nu\,{}^1\varepsilon_x = -\dfrac{\nu}{E}\sigma_x \\[2mm] {}^1\varepsilon_z = -\nu\,{}^1\varepsilon_x = -\dfrac{\nu}{E}\sigma_x \,. \end{cases}$$

If only infinitesimal deformations are considered, we can accept that the parallelepiped's geometry remains unchanged, when the effects of σ_y and σ_z are considered. Since the stress-strain relation is linear, it does not change with the actuation of σ_x. Thus, if the stresses σ_y or σ_z are applied after σ_x, they cause the same deformations that would occur under the isolated action of each of them. The total strains may therefore be computed by adding the strains, which would be produced by the isolated action of each stress. This conclusion describes the so-called *superposition principle*, which is valid for all solid bodies if the deformations are small and if the material has a linear constitutive law. This principle is described in more detail in Sect. V.8.

The isolated actions of σ_y and σ_z would cause the strains

$$\sigma_y \rightarrow \begin{cases} {}^2\varepsilon_x = -\dfrac{\nu}{E}\sigma_y \\[2mm] {}^2\varepsilon_y = \dfrac{1}{E}\sigma_y \\[2mm] {}^2\varepsilon_z = -\dfrac{\nu}{E}\sigma_y \end{cases} \qquad \sigma_z \rightarrow \begin{cases} {}^3\varepsilon_x = -\dfrac{\nu}{E}\sigma_z \\[2mm] {}^3\varepsilon_y = -\dfrac{\nu}{E}\sigma_z \\[2mm] {}^3\varepsilon_z = \dfrac{1}{E}\sigma_z \,. \end{cases}$$

The superposition principle allows the computation of the total strain by adding the strains caused by the isolated actions of the stresses σ_x, σ_y and σ_z,

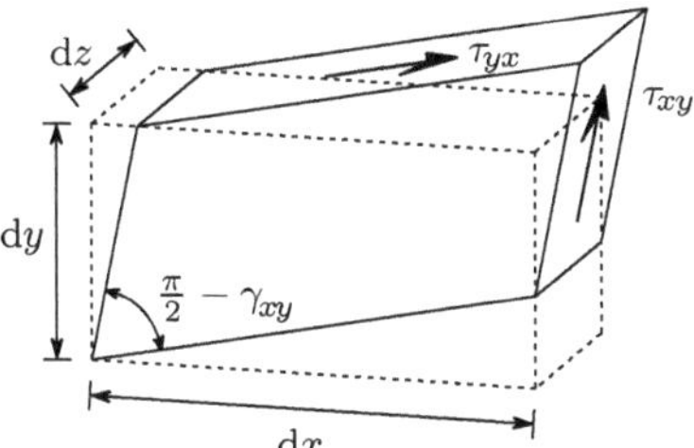

Fig. 35. Deformation caused by the shearing stress τ_{xy}: ------ original configuration;
—— deformed configuration

yielding

$$\left\{ \begin{aligned} \varepsilon_x &= {}^1\varepsilon_x + {}^2\varepsilon_x + {}^3\varepsilon_x = \frac{1}{E}\left[\sigma_x - \nu\left(\sigma_y + \sigma_z\right)\right] \\[2mm] \varepsilon_y &= {}^1\varepsilon_y + {}^2\varepsilon_y + {}^3\varepsilon_y = \frac{1}{E}\left[\sigma_y - \nu\left(\sigma_x + \sigma_z\right)\right] \\[2mm] \varepsilon_z &= {}^1\varepsilon_z + {}^2\varepsilon_z + {}^3\varepsilon_z = \frac{1}{E}\left[\sigma_z - \nu\left(\sigma_x + \sigma_y\right)\right] \ . \end{aligned} \right. \tag{74}$$

These expressions were obtained from symmetry considerations and relate the
normal stresses with the longitudinal strains. The same symmetry considerations lead to the conclusion that the shearing strains cause distortions only
in their plane, since the deformed parallelepiped must remain symmetrical
in relation to the plane containing the shearing stresses, as represented in
Fig. 35. The constant of proportionality between the shearing stress and the
shearing strain is known as the *shear modulus* of the material (G), also called
the *transversal modulus of elasticity*. The constitutive law of an isotropic material, defined in terms of normal stresses and longitudinal strains by (74), is
completed by the relations between shearing stresses and shearing strains

$$\left\{ \begin{aligned} \gamma_{xy} &= 2\varepsilon_{xy} = \frac{1}{G}\tau_{xy} \\[2mm] \gamma_{xz} &= 2\varepsilon_{xz} = \frac{1}{G}\tau_{xz} \\[2mm] \gamma_{yz} &= 2\varepsilon_{yz} = \frac{1}{G}\tau_{yz} \ . \end{aligned} \right. \tag{75}$$

These expressions show that the shearing strain vanishes if the shearing
stresses have zero values. Taking as reference system axes which are parallel
to the principal directions of the stress tensor, a strain tensor with non-zero
elements only in the diagonal is obtained, which means that *in an isotropic
material the principal directions of the stress and strain tensors coincide.*

Among the three rheological parameters defined until now (E, ν, G), only
two are independent, since a relation between them can be established. This
relation may be obtained by considering a two-dimensional stress state with

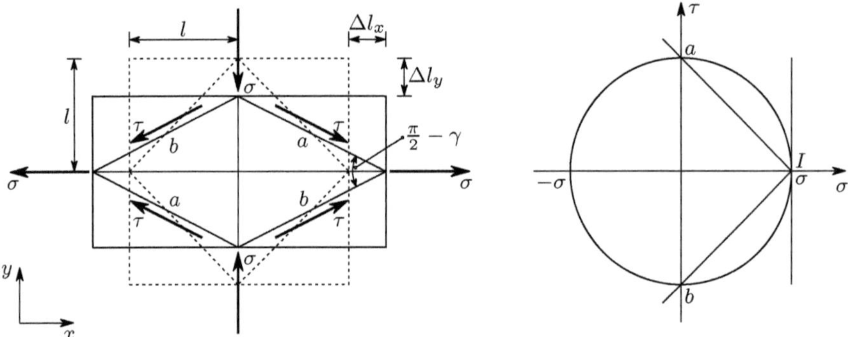

Fig. 36. Purely deviatoric plane stress state and corresponding Mohr's circle

$\sigma_x = \sigma$, $\sigma_y = -\sigma$ and $\tau_{xy} = 0$. This stress state and the corresponding Mohr's circle are represented in Fig. 36.[3] From the Mohr's circle, we verify that, in facets at an angle of 45°, in relation to the principal directions x and y, the normal stress vanishes and the shearing stress takes the value $\tau = \sigma$. Considering the deformation of the square defined by the two pairs of facets inclined 45°, a and b (cf. Fig. 36), we get for the deformation of its semi-diagonals, which have the initial length l,

$$\begin{cases} \Delta l_x = l\varepsilon_x = l\dfrac{1}{E}\left(\sigma_x - \nu\sigma_y\right) = l\dfrac{1+\nu}{E}\sigma \\[2mm] \Delta l_y = l\varepsilon_y = l\dfrac{1}{E}\left(\sigma_y - \nu\sigma_x\right) = -l\dfrac{1+\nu}{E}\sigma \ . \end{cases}$$

As a consequence of these deformations, the angle between facet a and axis x suffers a decrease $\frac{\gamma}{2}$, which represents the shearing strain between directions a and b. After the deformation we have, therefore

$$\tan\left(\frac{\pi}{4} - \frac{\gamma}{2}\right) = \frac{\tan\frac{\pi}{4} - \tan\frac{\gamma}{2}}{1 + \tan\frac{\pi}{4}\tan\frac{\gamma}{2}} = \frac{l + \Delta l_y}{l + \Delta l_x} = \frac{l - l\frac{1+\nu}{E}\sigma}{l + l\frac{1+\nu}{E}\sigma} \ . \tag{76}$$

As only small deformations are considered here, angle γ may be considered an infinitesimal quantity, which allows the simplification $\tan\frac{\gamma}{2} \approx \frac{\gamma}{2}$. Furthermore, as $\sigma = \tau$, we conclude that the shear modulus is related to the Young's modulus and to the Poisson's coefficient by the expression

$$\frac{1 - \frac{\gamma}{2}}{1 + \frac{\gamma}{2}} = \frac{1 - \frac{1+\nu}{E}\tau}{1 + \frac{1+\nu}{E}\tau} \Rightarrow \frac{\gamma}{2} = \frac{1+\nu}{E}\tau \Rightarrow \tau = G\gamma$$

$$\text{with} \quad G = \frac{E}{2\left(1 + \nu\right)} \ . \tag{77}$$

By drawing the Mohr circle of the state of deformation corresponding to the purely deviatoric plane state of stress ($\varepsilon_x = \frac{1+\nu}{E}\sigma$, $\varepsilon_y = -\frac{1+\nu}{E}\sigma$ and

---

[3]This is the only purely deviatoric stress state possible in the plane stress case.

$\gamma_{xy} = 0$), we can immediately confirm that in the direction of facets a and b the longitudinal strain is zero. This confirms that the shearing stress causes shearing strain alone, in the directions in which it acts. The relation between G, E and ν, expressed by (77), could also be obtained without using the trigonometric relation contained in (76), by considering the shearing strain in a direction at 45° to the principal axes, using the Mohr's circle of the strain tensor or (66) (cf. example IV.1).

The volumetric strain is caused only by the normal stresses, since, in isotropic materials, the longitudinal strains in the reference directions depend only on these stresses, as stated by (74). By substituting these relations in (58), we get

$$\varepsilon_v = \varepsilon_x + \varepsilon_y + \varepsilon_z = \frac{1 - 2\nu}{E}\left(\sigma_x + \sigma_y + \sigma_z\right) = \frac{3\left(1 - 2\nu\right)}{E}\sigma_m = \frac{\sigma_m}{K}$$

$$\text{with} \quad K = \frac{E}{3\left(1 - 2\nu\right)} \, . \tag{78}$$

The constant of proportionality K between the mean normal stress σ_m and the volumetric strain ε_v is called the *volumetric modulus of elasticity* or *bulk modulus* of the material.

Expressions to compute the stress for given strains may be obtained by solving (74) and (75), yielding

$$\begin{cases} \sigma_x = \dfrac{E}{(1+\nu)(1-2\nu)}\left[(1-\nu)\varepsilon_x + \nu\varepsilon_y + \nu\varepsilon_z\right] \\ \sigma_y = \dfrac{E}{(1+\nu)(1-2\nu)}\left[\nu\varepsilon_x + (1-\nu)\varepsilon_y + \nu\varepsilon_z\right] \\ \sigma_z = \dfrac{E}{(1+\nu)(1-2\nu)}\left[\nu\varepsilon_x + \nu\varepsilon_y + (1-\nu)\varepsilon_z\right] \end{cases} \text{and} \quad \begin{cases} \tau_{xy} = 2G\varepsilon_{xy} \\ \tau_{xz} = 2G\varepsilon_{xz} \\ \tau_{yz} = 2G\varepsilon_{yz} \, . \end{cases}$$

$$\tag{79}$$

Expressions (74) and (79) may be given another form by defining the constant quantities

$$e = \varepsilon_x + \varepsilon_y + \varepsilon_z = I_{1e} \text{ (first invariant of the strain tensor)}$$
$$\theta = \sigma_x + \sigma_y + \sigma_z = I_{1t} \text{ (first invariant of the stress tensor)} \tag{80}$$
$$\lambda = \frac{\nu E}{(1+\nu)(1-2\nu)} \quad \text{(Lamé's constant)} \, ,$$

which leads to the more simple expressions

$$\begin{cases} \varepsilon_x = \dfrac{1}{E}\left[(1+\nu)\sigma_x - \nu\theta\right] \\ \varepsilon_y = \dfrac{1}{E}\left[(1+\nu)\sigma_y - \nu\theta\right] \\ \varepsilon_z = \dfrac{1}{E}\left[(1+\nu)\sigma_z - \nu\theta\right] \end{cases} \quad \begin{cases} \sigma_x = \lambda e + 2G\varepsilon_x \\ \sigma_y = \lambda e + 2G\varepsilon_y \\ \sigma_z = \lambda e + 2G\varepsilon_z \, . \end{cases} \tag{81}$$

From these expressions we conclude immediately that, if $\sigma_x > \sigma_y > \sigma_z$, then $\varepsilon_x > \varepsilon_y > \varepsilon_z$. Furthermore, in a purely deviatoric stress state ($e = 0$), each element of the stress tensor is proportional to the correspondent element of the strain tensor, with the quantity $2G$ as proportionality constant (last expressions in (79) and (81)).

xsxs The quantities E, G and K must take positive values. In fact the work done by one force applied to a body must be positive, which requires that a displacement of the point of application of the force has a component with the same direction as the force itself.[4] This condition leads to the limiting values for the Poisson's coefficient

$$\begin{cases} G > 0 \;\Rightarrow\; 1 + \nu > 0 \Rightarrow \nu > -1 \\ K > 0 \;\Rightarrow\; 1 - 2\nu > 0 \Rightarrow \nu < 0.5 \end{cases} \;\Rightarrow\; -1 \le \nu \le 0.5 \,.$$

Usually this coefficient takes positive values, so that the limits $0 \le \nu \le 0.5$ are generally accepted. The maximum value $\nu = 0.5$ corresponds to an incompressible material, since it leads to an infinite value of the bulk modulus ($1 - 2\nu = 0 \Rightarrow K = \infty$).

IV.4.c · Monotropic Materials

While in isotropic materials any rectangular reference system defines symmetry planes in relation to the rheological properties of the material, in a monotropic material this happens only if one of the reference axes is parallel to the direction of monotropy of the material (cf. Fig. 37).

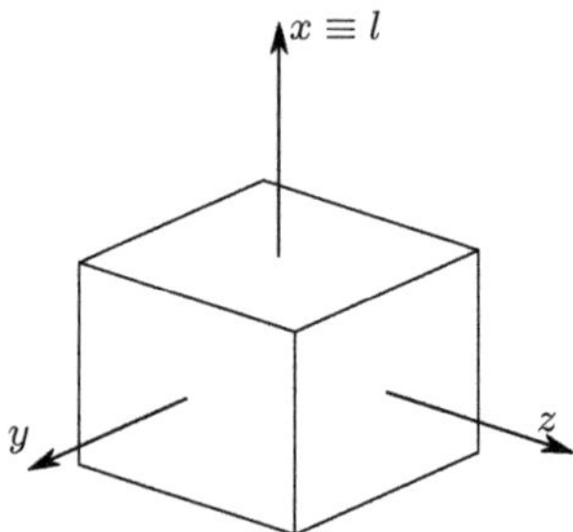

Fig. 37. Reference direction for the definition of the constitutive law of a monotropic material: l – monotropy direction

The constitutive law presented here is developed from symmetry considerations. Therefore, it is only valid for a reference frame, which obeys the

---

[4]If the force and the displacement were to have opposite directions, the principle of energy conservation would not be satisfied, since both the potential energy of the force and the elastic potential energy stored by the body would increase. These topics are expounded in Sect. IV.6 and in Chap. XII.

above referred symmetry conditions. Thus, it may only be used for stress and strain tensors represented by its components in a Cartesian rectangular reference frame with one of its axes parallel to the direction of monotropy of the material.

In order to develop the constitutive law of monotropic materials with linear elastic behaviour, let us consider a Cartesian rectangular reference frame x, y, z with the x-axis parallel to the direction of monotropy. Axes y and z therefore define the transversal isotropy plane (cf. Fig. 37). Defining the rheological parameters (ν_{ij} – Poisson's coefficient, when the stress acts in direction i and the deformation is measured in direction j):

1. E_l – longitudinal modulus of elasticity in the monotropy direction;
2. E_t – longitudinal modulus of elasticity in one direction of the transversal isotropy plane (y, z-plane);
3. $\nu_{xy} = \nu_{xz} = \nu_l$ – Poisson's coefficient for a normal stress acting in the monotropy direction $x \equiv l$ (σ_x);
4. $\nu_{yx} = \nu_{zx}$ – Poisson's coefficient for the longitudinal strain in direction x, when the stress acts in a transversal direction (y, z-plane);
5. $\nu_{yz} = \nu_{zy} = \nu_t$ – Poisson's coefficient in the isotropy plane yz;
6. $G_{xy} = G_{xz} = G_t$ – shear modulus for shearing stresses, which are parallel to the monotropy direction $x \equiv l$ (τ_{xy} and τ_{xz});
7. $G_{yz} = \dfrac{E_t}{2(1+\nu_t)}$ – shear modulus for shearing stresses contained in the transversal isotropy plane yz (τ_{yz}).

The seventh rheological parameter is defined in terms of the second and fifth ones. The remaining six parameters are not completely independent either, since, as consequence of a theorem on the energy of deformation in materials with linear elastic behaviour – the theorem of Maxwell, a relation may be established between parameters E_l, E_t and ν_{yz}. This theorem, whose demonstration is presented in Chap. XII, allows the following conclusion: a stress σ applied in direction x causes a strain ($\varepsilon_y = -\frac{\nu_l}{E_l}\sigma$) in direction y, which is equal to the strain caused in direction x by a stress with the same magnitude σ, applied in direction y ($\varepsilon_x = -\frac{\nu_{yx}}{E_t}\sigma$), which means[5]

$$\frac{\nu_{yx}}{E_t} = \frac{\nu_{zx}}{E_t} = \frac{\nu_l}{E_l} \,. \tag{82}$$

The number of independent parameters needed to describe the rheological behaviour of a monotropic material is therefore reduced to five. This constitutive law may then be expressed by the relation (x – direction of monotropy)

---

[5]This relation is also demonstrated in a particular form in Subsect. IV.6.b. This demonstration is also valid for Expressions 84.

$$\left\{ \begin{array}{c} \varepsilon_x \\ \varepsilon_y \\ \varepsilon_z \end{array} \right\} = \left[\begin{array}{ccc} \dfrac{1}{E_l} & -\dfrac{\nu_{yx}}{E_t} = -\dfrac{\nu_l}{E_l} & -\dfrac{\nu_{zx}}{E_t} = -\dfrac{\nu_l}{E_l} \\[2ex] -\dfrac{\nu_l}{E_l} & \dfrac{1}{E_t} & -\dfrac{\nu_t}{E_t} \\[2ex] -\dfrac{\nu_l}{E_l} & -\dfrac{\nu_t}{E_t} & \dfrac{1}{E_t} \end{array} \right] \left\{ \begin{array}{c} \sigma_x \\ \sigma_y \\ \sigma_z \end{array} \right\} \tag{83}$$

$$\gamma_{xy} = \frac{\tau_{xy}}{G_t} \qquad \gamma_{xz} = \frac{\tau_{xz}}{G_t} \qquad \gamma_{yz} = \frac{2\,(1+\nu_t)}{E_t}\tau_{yz}\;.$$

In this case the five rheological parameters used are E_l, E_t, ν_l, ν_t and G_t.

IV.4.d Orthotropic Materials

In orthotropic materials the symmetry conditions of the rheological properties of the material are satisfied only in relation to the planes defined by the orthotropy directions of the material. The axes of the rectangular reference frame must therefore be parallel to the material orthotropy directions, if the symmetry conditions are to be used in the development of the constitutive law. Thus, the obtained expressions are valid only for this reference frame.

As the reference directions are symmetry directions, the normal stresses do not cause shearing strain and the shearing stresses cause distortions only in the plane in which they are contained. Twelve elastic material parameters may thus be defined: three longitudinal moduli of elasticity (E_x, E_y and E_z), three shear moduli (G_{xy}, G_{xz} and G_{yz}) and six Poisson's coefficients (ν_{xy}, ν_{yx}, ν_{xz}, ν_{zx}, ν_{yz} and ν_{zy}). Of these twelve constants only nine are independent, since the theorem of Maxwell, applied to the three pairs of orthotropy directions, yields the relations (cf. Footnote 16)

$$\frac{\nu_{xy}}{E_x} = \frac{\nu_{yx}}{E_y} \qquad \frac{\nu_{xz}}{E_x} = \frac{\nu_{zx}}{E_z} \qquad \frac{\nu_{yz}}{E_y} = \frac{\nu_{zy}}{E_z}\;. \tag{84}$$

The constitutive law of an orthotropic material with linear elastic behaviour, which is valid only for the reference axes parallel to the orthotropy directions, is then

$$\left\{ \begin{array}{c} \varepsilon_x \\ \varepsilon_y \\ \varepsilon_z \end{array} \right\} = \left[\begin{array}{ccc} \dfrac{1}{E_x} & -\dfrac{\nu_{yx}}{E_y} = -\dfrac{\nu_{xy}}{E_x} & -\dfrac{\nu_{zx}}{E_z} = -\dfrac{\nu_{xz}}{E_x} \\[2ex] -\dfrac{\nu_{xy}}{E_x} & \dfrac{1}{E_y} & -\dfrac{\nu_{zy}}{E_z} = -\dfrac{\nu_{yz}}{E_y} \\[2ex] -\dfrac{\nu_{xz}}{E_x} & -\dfrac{\nu_{yz}}{E_y} & \dfrac{1}{E_z} \end{array} \right] \left\{ \begin{array}{c} \sigma_x \\ \sigma_y \\ \sigma_z \end{array} \right\} \tag{85}$$

$$\gamma_{xy} = \frac{\tau_{xy}}{G_{xy}} \qquad \gamma_{xz} = \frac{\tau_{xz}}{G_{xz}} \qquad \gamma_{yz} = \frac{\tau_{yz}}{G_{yz}}\;.$$

Here it should be noted that in orthotropic as in monotropic materials (and generally in anisotropic materials), the principal directions of the stress and strain tensors are generally not the same, unless they are parallel to the monotropy or orthotropy directions of the material, or in the case of other particular stress tensors, as, for example, a purely deviatoric stress tensor where $\sigma_x = \sigma_y = \sigma_z = \tau_{xz} = \tau_{yz} = 0$ and $\tau_{xy} \neq 0$ (see example IV.4).

IV.4.e Isotropic Material with Linear Visco-Elastic Behaviour

The one-dimensional stress-strain relations for materials with constant elastic and viscous rheological parameters (linear visco-elastic behaviour), as, for example, (69) and (71), may easily be generalized to the isotropic three-dimensional case, if the Poisson's coefficient is considered constant. For this purpose, let us note that the three-dimensional Hooke's law of an isotropic material may be put in a form, in which the Young's modulus (or its inverse) appears as a factor applied to all relations, as is easily verified by writing (74) and (75) as follows

$$
\begin{Bmatrix} \varepsilon_x \\ \varepsilon_y \\ \varepsilon_z \\ \varepsilon_{xy} \\ \varepsilon_{xz} \\ \varepsilon_{yz} \end{Bmatrix}
= \frac{1}{E}
\begin{bmatrix}
1 & -\nu & -\nu & 0 & 0 & 0 \\
-\nu & 1 & -\nu & 0 & 0 & 0 \\
-\nu & -\nu & 1 & 0 & 0 & 0 \\
0 & 0 & 0 & 1+\nu & 0 & 0 \\
0 & 0 & 0 & 0 & 1+\nu & 0 \\
0 & 0 & 0 & 0 & 0 & 1+\nu
\end{bmatrix}
\begin{Bmatrix} \sigma_x \\ \sigma_y \\ \sigma_z \\ \tau_{xy} \\ \tau_{xz} \\ \tau_{yz} \end{Bmatrix} .
\tag{86}
$$

As seen in Sect. IV.3, in a linear visco-elastic material stresses and strains may be superposed. Thus, the loading history may be defined by small stress or strain impulses, which are applied at given time instants and kept indefinitely constant. Unloading is represented by negative impulses, as presented schematically in Fig. 38.

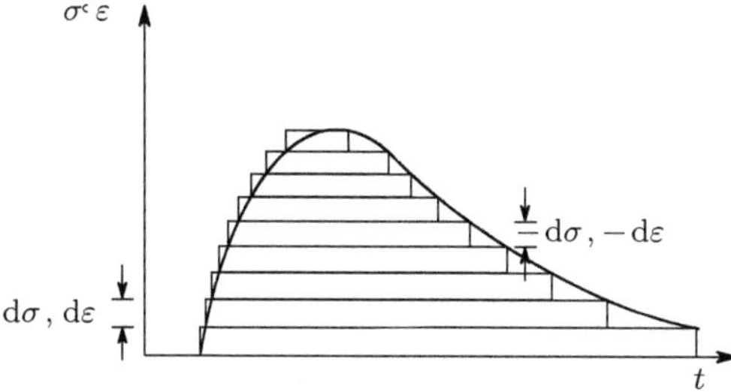

Fig. 38. Loading history of a visco-elastic material represented by impulses

The strain corresponding to a stress impulse or the stress corresponding to a strain impulse depends on the time elapsed since the onset of the impulse. These relations are of the type described by (70) and (72) for the Kelvin solid and the Maxwell liquid, respectively. Since the elasticity modulus is the relation between the stress and the corresponding strain, equivalent elasticity moduli may be defined from those equations. These moduli therefore depend on the time elapsed between the instant of application of the stress impulse and the time at which the strain is computed, or vice versa

$$\frac{\mathrm{d}\varepsilon}{\mathrm{d}\sigma} = \frac{1}{E}\left[1 - \mathrm{e}^{-\frac{E}{\eta}(t-t_0)}\right] = \frac{1}{E'(t)} \qquad \text{(Kelvin solid)}$$

$$\frac{\mathrm{d}\sigma}{\mathrm{d}\varepsilon} = E\,\mathrm{e}^{-\frac{E}{\eta}(t-t_0)} = E'(t) \qquad \text{(Maxwell's liquid)} .$$

(87)

By substituting in (86) E by $E'(t)$, the one-dimensional linear visco-elastic model under consideration is generalized to the three-dimensional case. If more elaborate rheological models, like Kelvin chains or generalized Maxwell models, are used, the procedure is similar to the one represented by (87). In other words, the equivalent elasticity modulus is computed by means of the creep relation between a constant stress and the corresponding strain in the Kelvin chain, or, in the generalized Maxwell model, by means of the relaxation relation between a constant strain and the corresponding stress decay.

IV.5 Newtonian Liquid

Liquids are isotropic materials, since their rheological properties are independent of the direction considered. When a stress field is applied to a liquid, its effect has two distinct components: the isotropic component of the stress tensor – the pressure – causes a volumetric strain, if the liquid is compressible, or has no effect vis-à-vis the deformation, if the liquid may be considered as incompressible; the deviatoric component, to which shearing stresses are always associated, causes distortional deformation rates. These latter depend on the viscosity, since, by definition, a non-flowing liquid does not resist shearing stresses. The volumetric deformation has the characteristics of an elastic deformation, since it does not depend on the time and since the plasticity is mostly associated with the deviatoric (distortional) component of the deformation, even in solid materials. The deformation associated with the deviatoric stress component is a viscous deformation, as concluded above.

A *perfect liquid* is defined as an incompressible liquid without viscosity. Thus, in this material there are no shearing stresses at all, since even when the liquid is flowing, those stresses cannot be resisted. For this reason, in a perfect liquid, the body forces (including inertial forces) are balanced only by the variation of the isotropic stress, as required by the differential equations of equilibrium (cf. (5) and (6)).

Actual liquids are compressible and have non-zero viscosity, which means that shearing stresses are generally present when the liquid is flowing. The relation between the stresses in the deviatoric component of the stress tensor and the viscous deformation rates associated with the distortional deformation, and the relation between the isotropic stress and the volumetric strain define the constitutive law of the liquid. In most liquids these relations may be accepted as linear. In that case we have a *Newtonian liquid.*

The constitutive law of a Newtonian liquid could be established by means of symmetry considerations, similar to those described in Sect. IV.4.b for solid isotropic materials. However, precisely because the symmetry considerations are the same for isotropic solids and isotropic liquids, an analogy with Hooke's law may be used to find the constitutive equations of the Newtonian liquid more quickly. Thus, from Expressions 77, 78 and 80 we can easily conclude that the Lamé's constant λ may be related with the shear and bulk moduli, G and K, respectively. This relation is

$$ K - \frac{2}{3}G = \frac{E}{3(1-2\nu)} - \frac{2}{3}\frac{E}{2(1+\nu)} = \frac{\nu E}{(1+\nu)(1-2\nu)} = \lambda \ . $$

By substituting this expression in the second of (81), we get

$$ \lambda e = \underbrace{Ke}_{\sigma_m} - \frac{2}{3}Ge \Rightarrow \begin{cases} \sigma_x - \sigma_m = -\frac{2}{3}Ge + 2G\varepsilon_x \\[2mm] \sigma_y - \sigma_m = -\frac{2}{3}Ge + 2G\varepsilon_y \\[2mm] \sigma_z - \sigma_m = -\frac{2}{3}Ge + 2G\varepsilon_z \ . \end{cases} \tag{88} $$

This expression relates the normal components of the deviatoric stress tensor with the components of the strain tensor. The second of (79) contains the remaining (shearing) components of the deviatoric stress tensor. The only rheological parameter in these relations is the shear modulus G.

Now, it should be noted that, in a viscous flow, the viscosity modulus, usually denoted as μ in Fluid Mechanics, plays the same role in the relation between the rate of shear deformation and the shearing stresses as the shear modulus G in the relation between shearing stress and shearing strain in an elastic solid, since we have

$$ \tau = \mu\dot{\gamma} \qquad \text{(Newtonian viscous flow)} $$
$$ \tau = G\gamma \qquad \text{(linear elastic solid)} \ , $$

where $\dot{\gamma} = \frac{\partial \gamma}{\partial t}$ represents the rate of distortion (shear deformation). These considerations are summarized in Fig. 39 (v_x is the velocity of the liquid particles in direction x).

Since the symmetry considerations used in the development of (88) are also valid in the case of the Newtonian liquid, and since the deformation rate and the viscosity modulus play the same roles as the deformation and the shear modulus in the solid material, we can substitute in (88) and in the second of (79) the strains by the strain rates and the shear modulus by the viscosity modulus, which yields the constitutive law of the Newtonian liquid, in relation to viscous deformations[6]

---

[6]Example IV.2 gives the development of these expressions directly from the relation $\tau = \mu\dot{\gamma}$.

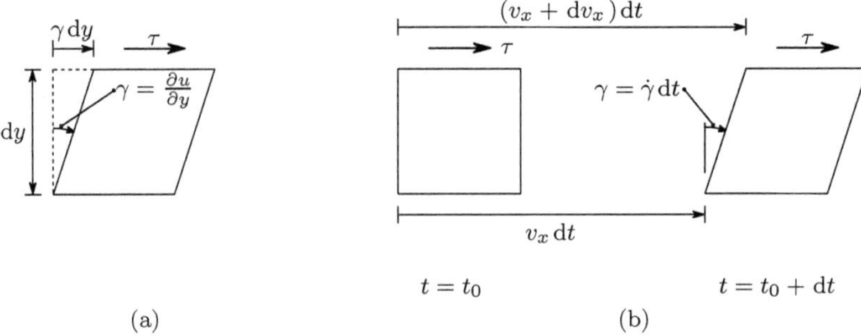

Fig. 39. Analogy between the shear modulus in the linear elastic deformation of a solid and the viscosity modulus in the linear viscous rate of deformation of a liquid: (**a**) istortion caused in the solid by the shearing stress τ; (**b**) distortion caused in the liquid by τ in the infinitesimal time interval $\mathrm{d}t$

$$
\begin{cases}
\sigma_x = \sigma_m - \dfrac{2}{3}\mu\left(\dot{\varepsilon}_x + \dot{\varepsilon}_y + \dot{\varepsilon}_z\right) + 2\mu\dot{\varepsilon}_x \\[2ex]
\sigma_y = \sigma_m - \dfrac{2}{3}\mu\left(\dot{\varepsilon}_x + \dot{\varepsilon}_y + \dot{\varepsilon}_z\right) + 2\mu\dot{\varepsilon}_y \\[2ex]
\sigma_z = \sigma_m - \dfrac{2}{3}\mu\left(\dot{\varepsilon}_x + \dot{\varepsilon}_y + \dot{\varepsilon}_z\right) + 2\mu\dot{\varepsilon}_z
\end{cases}
\qquad
\begin{cases}
\tau_{xy} = \mu\dot{\gamma}_{xy} \\[2ex]
\tau_{xz} = \mu\dot{\gamma}_{xz} \\[2ex]
\tau_{yz} = \mu\dot{\gamma}_{yz}\,.
\end{cases}
\qquad (89)
$$

The only rheological parameter present in these expressions is the viscosity modulus μ. The linear elastic relation between the isotropic stress σ_m and the volumetric strain ε_v, defined by the bulk modulus (78), completes the constitutive law of the Newtonian liquid. We conclude that its rheological behaviour, as in the case of linear elastic isotropic solids, is completely described by only two rheological parameters: μ and K.

IV.6 Deformation Energy

IV.6.a General Considerations

When a material deforms under the action of internal stresses, these lose potential energy and do work, which is either transformed into heat and dissipated, as a consequence of certain phenomena which may be interpreted as internal friction, or else it is stored as elastic potential energy. This energy corresponds to the work done by the stresses in the unloading, i.e., when the stresses are progressively reduced until they vanish.[7]

---

[7]The work done by the external forces may also be stored as kinetic energy, which is associated with the speed of a moving mass. This energy is not taken in account, since only the energy associated with the deformation is analysed here. Furthermore, dynamic effects, i.e., inertial forces in accelerated motions, do not need to be considered in an infinitesimal region around a point, since the work done by

In a slow elastic deformation (i.e., without dynamic effects) the work done by the external forces is totally stored as elastic potential energy, since the stress-strain relation is the same in the loading and unloading processes. Which means that the work done in the loading phase is completely recovered in the unloading.

These considerations may be more easily understood, by considering the relation between the stress σ_x and the strain ε_x caused by it in an infinitesimal parallelepiped made of an elastic material (Fig. 40).

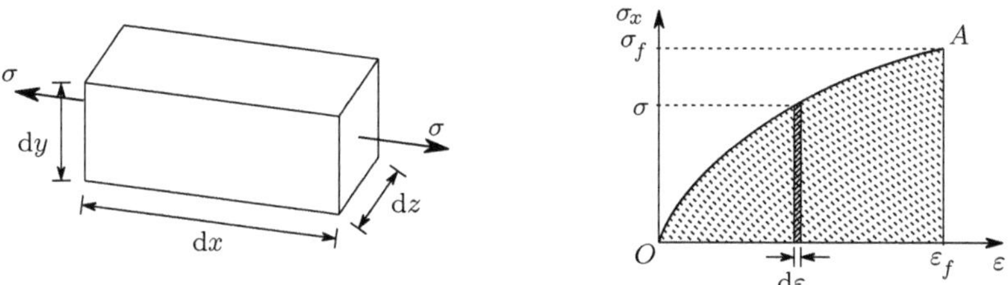

Fig. 40. Work done in the deformation ε_x caused by the stress σ_x

The work done by the force resulting from the stress $\sigma_x = \sigma$ during the displacement which corresponds to the strain increment $\mathrm{d}\varepsilon$ is given by the area of rectangle $\sigma\,\mathrm{d}\varepsilon$ multiplied by the volume of the infinitesimal parallelepiped, as it is easily confirmed by the expression

$$\mathrm{d}\,\mathrm{d}W \;=\; \underbrace{\sigma\,\mathrm{d}y\,\mathrm{d}z}_{\text{force}}\,\overbrace{\mathrm{d}x\,\mathrm{d}\varepsilon}^{\text{displacement}} \;=\; \sigma\,\mathrm{d}\varepsilon\,\mathrm{d}V \;. \tag{90}$$

The total work done, when the strain ε_x varies from zero until the final value ε_f is given by the shaded area in Fig. 40, since we have

$$\mathrm{d}W \;=\; \int_0^{\varepsilon_f} \sigma\,\mathrm{d}\varepsilon\,\mathrm{d}V \;. \tag{91}$$

If the stress is proportional to the strain (linear behaviour), the curve OA becomes a straight line and the shaded area takes the value

$$\mathrm{d}W \;=\; \frac{1}{2}\sigma_f \varepsilon_f\,\mathrm{d}V \;.$$

The work done by this stress is stored as elastic potential energy. The amount of this energy stored by volume unit, U, takes the value

---

these body forces is an infinitesimal quantity of higher order, when compared with the work done by the stresses.

$$dW = dU_0 \;\Rightarrow\; U = \frac{dU_0}{dV} = \frac{1}{2}\sigma_f \varepsilon_f \;. \tag{92}$$

In the plastic and viscous deformations the potential energy lost by the stresses, when they do the deformation work as the body deforms, is transformed into heat by internal friction and dissipated, since in both cases the deformation remains after the force which caused it is removed. Which means that the internal forces do not do work in the unloading process.

In the deformation of actual materials, elastic, plastic and viscous deformations occur simultaneously. As a consequence, one part of the work done by the external forces is dissipated, while other part is stored as elastic potential energy which is recovered in the unloading. Figure 41 illustrates the stored and dissipated parts of the deformation energy in a one-dimensional stress state.

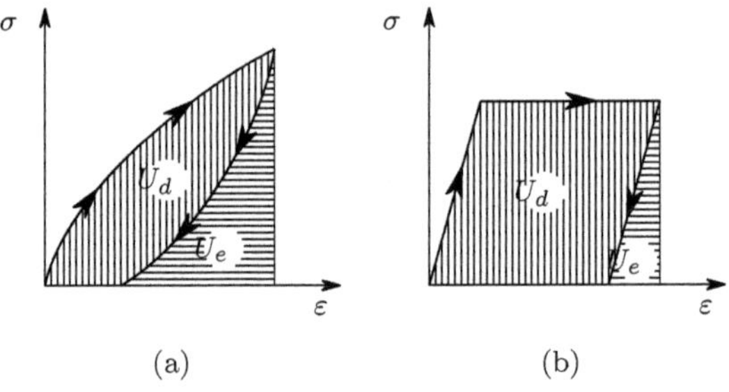

Fig. 41. Elastic potential energy (U_e) and dissipated energy (U_d): (**a**) general case; (**b**) elastic perfectly plastic material

The above considerations are also completely valid for shearing stresses and the corresponding shearing strains. This fact may easily be understood by considering the deformation represented in Fig. 35. By means of a rigid body motion, in which no work is done, as the stresses are in balance, we may consider that the bottom face of the parallelepiped remains horizontal. Under these conditions only the resultant of the shearing stress acting in the upper face, τ_{yx}, ($\tau_{yx}\,dx\,dz$) does work when it suffers the displacement $\gamma_{xy}\,dy$. The equations expressing the work done by the shearing stress therefore take the following forms, which are perfectly similar to the corresponding forms for the normal stresses ((90) and (91))

$$d\,dW = \underbrace{\tau\,dx\,dz}_{\text{force}}\,\overbrace{d\gamma\,dy}^{\text{displacement}} \;\Rightarrow\; dW = \int_0^{\gamma_f}\tau\,d\gamma\,dV \;.$$

IV.6.b Superposition of Deformation Energy in the Linear Elastic Case

Deformation energy does not depend linearly on the stress, even if the stress-strain relation is linear, as is easily confirmed by substituting ε_f by $\frac{\sigma_f}{E}$ in (92). For this reason, it is not generally possible to add the energies of deformation associated with two different states of stress, unless the second state does not have stresses in the directions where the deformations caused by the first occur, and vice versa.

However, if the material has linear elastic behaviour, the deformation caused by a given stress state is always the same, i.e., it does not depend on other stress states previously applied to the material. This is because the deformation is proportional to the stress and the deformations are small enough to be considered as infinitesimal, so that the deformed geometry may be considered the same as the undeformed one. For this reason, the deformations corresponding to different stress states may be computed separately and superposed (added), as described earlier in Sect. IV.4.b. In addition, the elastic potential energy stored by a linear elastic material only depends on the stresses and deformations installed inside it. Thus, elastic potential energy may always be computed as if the forces were applied simultaneously.[8]

With the help of these considerations the symmetry of the constitutive law of monotropic and isotropic materials may easily be demonstrated. For this purpose, let us consider that in a monotropic material (Fig. 37), first a stress is applied in the monotropy direction $\sigma_x = \sigma_l = \sigma$, followed by a second stress in a transversal direction $\sigma_y = \sigma_t = \sigma$. The evolution of the stored elastic potential energy in these two loading phases is (cf. 92):

1. application of $\sigma_l = \sigma$

$$U_1 = \frac{1}{2}\sigma_l \varepsilon_l = \frac{1}{2}\frac{\sigma^2}{E_l}\,;$$

2. application of $\sigma_t = \sigma$

$$U = U_1 + \frac{1}{2}\sigma_t \varepsilon_t + \sigma_l \varepsilon_l'$$
$$= \frac{1}{2}\frac{\sigma^2}{E_l} + \frac{1}{2}\frac{\sigma^2}{E_t} - \frac{\nu_{yx}}{E_t}\sigma^2\,. \tag{93}$$

In the second of these expressions the last element $\left(\sigma_l \varepsilon_l' = -\frac{\nu_{yx}}{E_t}\sigma^2\right)$ represents the work done by the stress in the monotropy direction $\sigma_x = \sigma_l = \sigma$, during the deformation $\varepsilon_l' = -\frac{\nu_{yx}}{E_t}\sigma$ caused in this direction by the stress in the transversal direction y, $\sigma_y = \sigma_t = \sigma$. During this deformation the stress σ_l remains constant. For this reason, this element does not contain the factor $\frac{1}{2}$.

Inverting the loading order (first the stress σ_t), the evolution of the elastic potential energy is given by the expressions:

---

[8]This conclusion is a particular form of Clapeyron's theorem, which will be studied in Chap. XII.

1. application of $\sigma_t = \sigma$

$$U_2 = \frac{1}{2}\sigma_t\varepsilon_t = \frac{1}{2}\frac{\sigma^2}{E_t} \; ;$$

2. application of $\sigma_l = \sigma$

$$U = U_2 + \frac{1}{2}\sigma_l\varepsilon_l + \sigma_t\varepsilon_t'$$

$$= \frac{1}{2}\frac{\sigma^2}{E_t} + \frac{1}{2}\frac{\sigma^2}{E_l} - \frac{\nu_l}{E_l}\sigma^2 \; . \tag{94}$$

Expressions 93 and 94 must yield the same result, since the final deformations do not depend on the order of application of the stresses and the elastic potential energy depends only on the final stresses and deformations. This means that $\frac{\nu_{yx}}{E_t} = \frac{\nu_l}{E_l}$, which confirms the indication presented without the demonstration in Sect. IV.4.c (82). The same line of reasoning is obviously also applicable to orthotropic materials, and, more generally, to any material with linear elastic behaviour. Thus, these considerations also demonstrate (84).

IV.6.c Deformation Energy in Materials with Linear Elastic Behaviour

General Case

As discussed earlier, in a material with linear elastic behaviour, isotropic or non-isotropic, the elastic potential energy may be computed as if the loads were applied simultaneously. Thus, in a body region under the action of a stress tensor described by its components in a rectangular Cartesian reference frame xyz, the elastic potential energy per volume unit may be expressed by

$$U = \frac{1}{2}\left(\sigma_x\varepsilon_x + \sigma_y\varepsilon_y + \sigma_z\varepsilon_z + \tau_{xy}\gamma_{xy} + \tau_{xz}\gamma_{xz} + \tau_{yz}\gamma_{yz}\right) \; . \tag{95}$$

By using the constitutive law of the material under consideration, the elastic potential energy may be represented either in terms of the elements of the stress tensor, or only in terms of the elements of the strain tensor. For an orthotropic material, for example, with the constitutive law defined by (85), we get

$$U = \frac{1}{2}\left[\frac{\sigma_x^2}{E_x} + \frac{\sigma_y^2}{E_y} + \frac{\sigma_z^2}{E_z} - 2\left(\sigma_x\sigma_y\frac{\nu_{xy}}{E_x} + \sigma_x\sigma_z\frac{\nu_{xz}}{E_x} + \sigma_y\sigma_z\frac{\nu_{yz}}{E_y}\right)\right.$$

$$\left. + \frac{\tau_{xy}^2}{G_{xy}} + \frac{\tau_{xz}^2}{G_{xz}} + \frac{\tau_{yz}^2}{G_{yz}}\right] \; , \tag{96}$$

where x, y and z are the material orthotropy directions.

Isotropic materials

In isotropic materials (96) becomes simplified taking the form

$$U = \frac{1}{2E}\left[\sigma_x^2 + \sigma_y^2 + \sigma_z^2 - 2\nu\left(\sigma_x\sigma_y + \sigma_x\sigma_z + \sigma_y\sigma_z\right)\right] + \frac{1}{2G}\left(\tau_{xy}^2 + \tau_{xz}^2 + \tau_{yz}^2\right) .$$

$$(97)$$

As an alternative, the elastic potential energy may be expressed in terms of the elements of the strain tensor. The corresponding expression may be obtained by substituting the second of (81) in (95), yielding

$$U = \frac{1}{2}\lambda e^2 + G\left(\varepsilon_x^2 + \varepsilon_y^2 + \varepsilon_z^2\right) + \frac{1}{2}G\left(\gamma_{xy}^2 + \gamma_{xz}^2 + \gamma_{yz}^2\right) . \qquad (98)$$

From the last two expressions we conclude immediately that

$$\begin{cases} \dfrac{\partial U}{\partial \sigma_i} = \varepsilon_i \\[2ex] \dfrac{\partial U}{\partial \tau_{ij}} = \gamma_{ij} \end{cases} \quad \text{and} \quad \begin{cases} \dfrac{\partial U}{\partial \varepsilon_i} = \lambda e + 2G\varepsilon_i = \sigma_i \\[2ex] \dfrac{\partial U}{\partial \gamma_{ij}} = \tau_{ij} \end{cases} \quad \text{with} \quad \begin{cases} i = x, y, z \\[2ex] j = x, y, z . \end{cases}$$

The elastic potential energy may also be expressed as functions of the invariants of the stress and strain tensors. The corresponding expressions may be obtained more easily by taking the principal directions of the stress and strain tensors as reference system. In this case, we get from (97) (I_{1t}, I_{2t} are the first and second invariants of the stress tensor)

$$U = \frac{1}{2E}\big[\underbrace{\sigma_1^2 + \sigma_2^2 + \sigma_3^2}_{I_{1t}^2 - 2I_{2t}} - 2\nu\underbrace{\left(\sigma_1\sigma_2 + \sigma_1\sigma_3 + \sigma_2\sigma_3\right)}_{I_{2t}}\big] = \frac{I_{1t}^2}{2E} - \frac{1+\nu}{E}I_{2t}$$

$$= \frac{I_{1t}^2}{2E} - \frac{I_{2t}}{2G} . \qquad (99)$$

Similarly, from (98) we get (I_{1e}, I_{2e} are the first and second invariants of the strain tensor)

$$U = \frac{1}{2}\lambda I_{1e}^2 + G\left(I_{1e}^2 - 2I_{2e}\right) .$$

In an isotropic material the isotropic component of the stress tensor does not cause distortional deformation (deviatoric strain tensor) and the deviatoric stress component does not cause isotropic deformation, i.e., volumetric strain. This fact may be understood more clearly by means of the octahedral stresses (cf. Sect. II.8, Fig. 10).

In fact, since in the octahedral facets the normal stress is the mean stress, if the isotropic stress component vanishes only shearing stresses act in those facets. These stresses do not cause elongation in the direction of the normals to the octahedral facets, because, in an isotropic material with linear elastic behaviour, the octahedral strain $\varepsilon_{oct} = \frac{\varepsilon_v}{3}$ vanishes when the isotropic stress σ_m becomes zero, as can be easily confirmed by looking at (78).

As the isotropic stress does not induce any shearing strains, these considerations allow the conclusion that neither the deviatoric stress does work in the isotropic deformation, nor the isotropic tensor does work in the deformation caused by the deviatoric component of the stress tensor. Under these particular conditions, we may decompose the elastic potential energy into two independent components, one related to the volumetric deformation and the other related to the purely deviatoric deformation.

The expression of the first component may be obtained by particularizing (99) for $\sigma_1 = \sigma_2 = \sigma_3 = \sigma_m \Rightarrow I_{2t} = 3\sigma_m^2 = \frac{I_{1t}^2}{3}$, yielding

$$U_v = \frac{I_{1t}^2}{2E} - \frac{I_{1t}^2}{6G} = \frac{1-2\nu}{6E} I_{1t}^2 = \frac{1-2\nu}{6E} \left(\sigma_x + \sigma_y + \sigma_z \right)^2 \ .$$

By subtracting this value from (99) the energy associated with the purely deviatoric deformation U_f is obtained

$$U_f = U - U_v = \frac{I_{1t}^2}{2E} - \frac{I_{2t}}{2G} - \frac{1-2\nu}{6E} I_{1t}^2 = \frac{1}{6G} \underbrace{\left(I_{1t}^2 - 3I_{2t} \right)}_{\frac{9}{2}\tau_{oct}^2} = \frac{3}{4G} \tau_{oct}^2 = -\frac{I_{2t}'}{2G} \ .$$

$$(100)$$

This expression can also be obtained by analysing of the deformation energy per volume unit in the octahedron under the action of the octahedral shearing stress τ_{oct} (cf. Fig. 10 and Example IV.3).

IV.7 Yielding and Rupture Laws

IV.7.a General Considerations

Many ductile materials, especially metallic materials, display a rheological behavior which, until a certain point, may be considered as elastic perfectly plastic. As seen in Sect. IV.3, this ideal behaviour is characterized by a *yielding stress* which defines the onset of the plastic deformation in a one-dimensional stress state. This stress may easily be obtained by means of an experimental test, by submitting a prismatic specimen of the material under consideration to a tensile axial force and plotting the measured force-elongation relation. In two- or three-dimensional stress states the experimental determination of the full spectrum of stress states corresponding to the beginning of the plastic deformation is not feasible any more, since there are infinite stress states which cause plastic deformation. Furthermore, the execution of two- or three-dimensional experimental tests is much more difficult and expensive than one-dimensional tests.

For these reasons, auxiliary theories had to be developed in order to predict the yielding conditions of ductile materials under two- or three-dimensional stress states, on the basis of the information furnished by one-dimensional experimental tests. From these theories several *yielding criteria* are obtained, some of which are described below.

In brittle materials the problem is posed in a similar way: once the conditions which cause rupture under a tensile or compressive one-dimensional stress are known,[9] it is necessary to foresee the conditions under which rupture occurs in two- or three-dimensional stress states. The corresponding theories lead to the so-called *rupture criteria*, only one of which will be analysed in the present text.

In the following discussion only isotropic materials are considered, whose rheological behaviour may considered as linear until yielding (ductile materials) or rupture (brittle materials). As the behaviour of isotropic materials is the same in any direction, the principal directions of the stress and strain tensors may always be used as Cartesian rectangular reference directions, which considerably simplifies the development of the expressions defining the different yielding or rupture laws.

The considerations set forth in this section complete the constitutive laws of ductile solid isotropic materials with elastic perfectly plastic behaviour and of brittle materials with linear elastic behaviour until rupture, whose constitutive law in the elastic phase is explained in Sect. IV.4.b, because the way the material yields or breaks is an important part of its rheological behaviour.

IV.7.b Yielding Criteria

IV.7.b.i Theory of Maximum Normal Stress

A first attempt to extend the results of one-dimensional experimental tests to the prediction of the yielding point in three-dimensional stress states was made by Rankine. This scientist postulated that yielding takes place when the largest principal stress attains the value corresponding to the yielding stress measured in one-dimensional experimental tests. According to this criterion, plastic strains will not occur while the maximum and minimum principal stresses obey the conditions (σ_Y: yielding stress in a one-dimensional test; $\sigma_1 > \sigma_2 > \sigma_3$)

$$\begin{cases} \sigma_1 < \sigma_Y \\ \sigma_3 > -\sigma_Y \ . \end{cases} \tag{101}$$

This theory is not accordance with the results observed in several cases, such as in isotropic compression, for example, where values of pressure much higher than σ_Y can be applied without any plastic deformation. Also, in the case of

---

[9]We will see later (Chap. V) that, while ductile materials usually display the same behaviour in tensile and compressive experimental tests, in brittle materials this does not happen. Generally, these materials display higher strength and stiffness for compressive loading states than for tensile ones.

pure shear ($\sigma_1 = -\sigma_2 = \tau$, cf. Fig. 36), this theory diverges from the experimental observation, since the theory indicates that yielding occurs for a shearing stress $\tau = \sigma_Y$, while the measured shearing yielding stress is lower ($\tau_Y < \sigma_Y$).

IV.7.b.ii Theory of Maximum Longitudinal Deformation

Rankine's theory, in fact, does not make a distinction between one-dimensional and two- or three-dimensional stress states, since it only takes the maximum normal stress into consideration. However, experimental tests show that a transversal tensile stress increases the longitudinal tensile yielding stress, while a transversal compressive stress causes a decrease on the same longitudinal yielding stress. This experimental observation shows that the interaction between the principal stresses does influence the yielding behaviour.

In order to take this interaction into account, Saint-Venant postulated that the yielding is determined by the maximum longitudinal strain. From Hooke's law for isotropic materials (74), we conclude that, according to this criterion, the material remains in the elastic phase as long as the following conditions are satisfied ($\sigma_1 > \sigma_2 > \sigma_3 \Rightarrow \varepsilon_1 > \varepsilon_2 > \varepsilon_3$)

$$\begin{cases} \varepsilon_1 = \frac{1}{E}\left[\sigma_1 - \nu\left(\sigma_2 + \sigma_3\right)\right] < \frac{\sigma_Y}{E} \\ \varepsilon_3 = \frac{1}{E}\left[\sigma_3 - \nu\left(\sigma_1 + \sigma_2\right)\right] > -\frac{\sigma_Y}{E} \end{cases} \Rightarrow \begin{cases} \sigma_1 - \nu\left(\sigma_2 + \sigma_3\right) < \sigma_Y \\ \sigma_3 - \nu\left(\sigma_1 + \sigma_2\right) > -\sigma_Y \,. \end{cases} \tag{102}$$

This theory yields better results than the Rankine criterion, but it, too, is not confirmed by experimental results in several cases. For example, in the plane state of stress defined by $\sigma_1 = \sigma_2 = \sigma$ and $\sigma_3 = 0$, yielding occurs for $\sigma = \sigma_Y$ and not for the higher value given by (102): $\sigma(1 - \nu) = \sigma_Y \Rightarrow \sigma = \frac{\sigma_Y}{1-\nu}$. This is a consequence of the fact that the transversal stress is not complete, since it acts only in one direction ($\sigma_2 = \sigma$, $\sigma_3 = 0$). In the case of isotropic compression, too, this theory leads to a yielding stress $\sigma = -\frac{\sigma_Y}{1-2\nu}$, but the experimental observation shows that much higher values of the pressure may be applied without plastic deformation.

IV.7.b.iii Theory of Maximum Deformation Energy

Another theory which considers the contributions of the three principal stresses on the onset of plastic deformations has been presented by Beltrami. This theory is based on the postulate that yielding is determined by the amount of deformation energy stored per volume unit in the material. According to this criterion yielding does not take place while the stress tensor obeys the condition (U_{Yt}: elastic potential energy stored per unit volume in the one-dimensional stress state defined by $\sigma_1 = \sigma_Y$ and $\sigma_2 = \sigma_3 = 0$, cf. (97))

$$\overbrace{\frac{1}{2E}\left[\sigma_1^2 + \sigma_2^2 + \sigma_3^2 - 2\nu\left(\sigma_1\sigma_2 + \sigma_1\sigma_3 + \sigma_2\sigma_3\right)\right]}^{U} \; < \; \overbrace{\frac{1}{2E}\sigma_Y^2}^{U_{Yt}} \tag{103}$$

$$\Rightarrow \; \sqrt{\sigma_1^2 + \sigma_2^2 + \sigma_3^2 - 2\nu\left(\sigma_1\sigma_2 + \sigma_1\sigma_3 + \sigma_2\sigma_3\right)} \; < \sigma_Y \,.$$

In isotropic compression $\sigma_1 = \sigma_2 = \sigma_3 = -p$ this criterion indicates that yielding occurs when pressure attains the value $p = \frac{\sigma_Y}{\sqrt{3(1-2\nu)}}$. This value is, however, even smaller than the one given by the Saint-Venant criterion, being exceeded by a large amount in experimental tests without yielding.

IV.7.b.iv Theory of Maximum Shearing Stress

The experimental observations that the isotropic stress component of the stress tensor practically does not influence the yielding point, and that in a prismatic specimen under axial tensile loading the plastic deformation starts in the facets where the shearing stress attains a maximum[10] led Tresca to postulate, in 1865 [3], that yielding is determined by the maximum shearing stress. These stresses, as mentioned in Sect. II.10, occur in facets at a 45° angle to the directions of the maximum (σ_1) and minimum (σ_3) principal stresses and take the value $\tau_{\max} = \frac{\sigma_1-\sigma_3}{2}$. Since in the one-dimensional experimental test we have $\tau_{\max} = \frac{\sigma_Y}{2}$, Tresca's criterion postulates that no yielding will occur, while the following condition is satisfied $(\sigma_1 > \sigma_2 > \sigma_3)$

$$\sigma_1 - \sigma_3 < \sigma_Y \,. \tag{104}$$

Experimental tests carried out later (1903) by Guest have supported this theory, which is also known as the Tresca-Guest criterion.

Equation (104) is very simple and gives good results. For this reason it is still frequently used in practice.

IV.7.b.v Theory of Maximum Distortion Energy

In year 1913 the German scientist von Mises presented another theory stating that yielding is determined by the second invariant of the purely deviatoric component of the stress tensor, which is equivalent to the statement that yielding is attained when the deformation energy associated with the deviatoric (distortional) component of the strain tensor reaches a given value. Both of these quantities are proportional to the square of the octahedral stress, as (32) and (100) show (Hencky 1924).

This theory shares with Tresca's criterion the feature of not considering the isotropic component of the stress tensor and gives results which are close to the ones furnished by that criterion, as will be seen later.

---

[10]The one-dimensional tensile test of a ductile material, such as mild steel, is described in greater detail in Sect. V.2.

The energy associated with the distortional deformation may be expressed as a function of the principal stresses. From (100) and (31) we get

$$U_f = \frac{3}{4G}\tau_{oct}^2 = \frac{1+\nu}{6E}\left[(\sigma_1 - \sigma_2)^2 + (\sigma_1 - \sigma_3)^2 + (\sigma_2 - \sigma_3)^2\right] .$$

In a one-dimensional stress state with $\sigma_1 = \sigma_Y$ this expression takes the value $U_f = \frac{1+\nu}{3E}\sigma_Y^2$. Thus, according to this criterion, no plastic deformations will occur, while the following condition is satisfied

$$U_f < \frac{1+\nu}{3E}\sigma_Y^2$$
$$\Rightarrow \frac{1}{\sqrt{2}}\sqrt{(\sigma_1 - \sigma_2)^2 + (\sigma_1 - \sigma_3)^2 + (\sigma_2 - \sigma_3)^2} = \frac{3}{\sqrt{2}}\tau_{oct} < \sigma_Y . \tag{105}$$

The results given by this criterion are the closest to the experimental observations, so that von Mises's criterion may be considered as the standard theory for the prediction of yielding in ductile materials.

IV.7.b.vi Comparison of Yielding Criteria

The several theories described above for predicting the yielding stress state in ductile materials lead obviously to different results, since they are based on different postulates. In order to get an idea about these differences, a graphical analysis of the yielding conditions is presented below. In a two-dimensional case (plane stress state) the five theories are analysed, while in the three-dimensional case only the two more important ones (Tresca and Von Mises's criteria) are considered.

Plane Stress States

By particularizing the expressions defining the different yielding criteria to the plane stress case ($\sigma_3 = 0$), we get the equations which relate the principal stresses σ_1 and σ_2 to the one-dimensional yielding stress σ_Y. Considering a rectangular Cartesian reference system with the principal stresses plotted on its axes, those equations define curves which delimit a zone containing the points which represent elastic stress states. Figure 42 shows the curves corresponding the five yielding theories studied.

As, by definition, every criterion is based on the yielding stress measured in one-dimensional tensile or compressive experimental tests, all the curves representing the different criteria contain the points corresponding to those stress states (points b, d, f and h). In the biaxial case ($\sigma_1 \neq 0$ and $\sigma_2 \neq 0$) the different criteria are represented by:

– *Rankine*: this criterion is represented by the square $aceg$, which means that, while the point representing the plane stress state (coordinates σ_1, σ_2) is

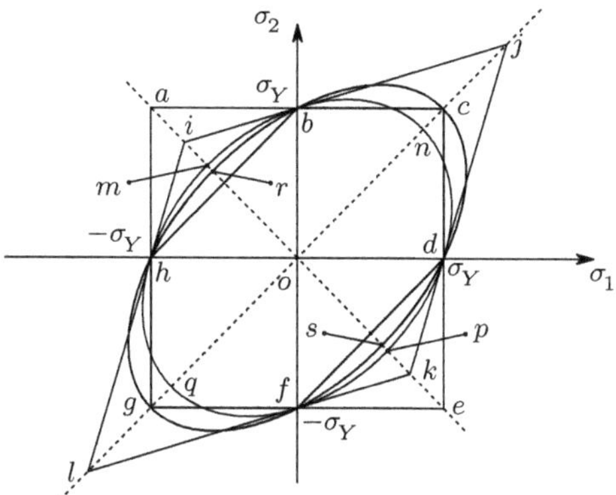

Fig. 42. Comparison of yielding criteria

inside the square ($|\sigma_{\mathrm{max}}| < \sigma_Y$) the material remains in the elastic state. The points outside the square do not represent possible stress states in an elastic perfectly plastic material. After yielding, further deformation causes only a displacement of the point representing the stress state, along the border of the elastic zone.

– *Saint-Venant*: this criterion is represented by the rhombus $ijkl$. In a plane stress state (102) takes the form (note that the convention $\sigma_1 > \sigma_2 > \sigma_3$ is not used here)

$$\begin{cases} -\sigma_Y < \sigma_1 - \nu\sigma_2 < \sigma_Y \\ -\sigma_Y < \sigma_2 - \nu\sigma_1 < \sigma_Y \,. \end{cases}$$

The border of the domain defined by these inequalities is defined by the equations

$$\begin{cases} \sigma_1 - \nu\sigma_2 = -\sigma_Y & \text{(line segment } il) \\ \sigma_1 - \nu\sigma_2 = \sigma_Y & \text{(line segment } jk) \\ \sigma_2 - \nu\sigma_1 = -\sigma_Y & \text{(line segment } kl) \\ \sigma_2 - \nu\sigma_1 = \sigma_Y & \text{(line segment } ij) \,. \end{cases}$$

We may easily verify that the coordinates of the vertexes of this rhombus take the values

$$i \to \left\{ \begin{array}{c} -\dfrac{\sigma_Y}{1+\nu} \\[2mm] \dfrac{\sigma_Y}{1+\nu} \end{array} \right\} \quad j \to \left\{ \begin{array}{c} \dfrac{\sigma_Y}{1-\nu} \\[2mm] \dfrac{\sigma_Y}{1-\nu} \end{array} \right\} \quad k \to \left\{ \begin{array}{c} \dfrac{\sigma_Y}{1+\nu} \\[2mm] -\dfrac{\sigma_Y}{1+\nu} \end{array} \right\} \quad l \to \left\{ \begin{array}{c} -\dfrac{\sigma_Y}{1-\nu} \\[2mm] -\dfrac{\sigma_Y}{1-\nu} \end{array} \right\} \,.$$

In the case of a vanishing Poisson's coefficient ($\nu = 0$), this criterion coincides with the Rankine square.

– *Beltrami*: in the plane stress case the expression defining this criterion (103) takes the form given by the inequality

$$\sigma_1^2 + \sigma_2^2 - 2\nu\sigma_1\sigma_2 \leq \sigma_Y^2 \,.$$

In accordance with this expression, yielding takes place when the point representing the stress state reaches the ellipse defined by the equation

$$\sigma_1^2 + \sigma_2^2 - 2\nu\sigma_1\sigma_2 = \sigma_Y^2 \ .$$

This ellipse ($mnpq$, cf. Fig. 42) has principal axes which make angles of $45°$ with the reference axes and its vertexes have the coordinates

$$m \rightarrow \left\{ \begin{array}{c} -\dfrac{\sigma_Y}{\sqrt{2(1+\nu)}} \\[2ex] \dfrac{\sigma_Y}{\sqrt{2(1+\nu)}} \end{array} \right\} \qquad n \rightarrow \left\{ \begin{array}{c} \dfrac{\sigma_Y}{\sqrt{2(1-\nu)}} \\[2ex] \dfrac{\sigma_Y}{\sqrt{2(1-\nu)}} \end{array} \right\}$$

$$p \rightarrow \left\{ \begin{array}{c} \dfrac{\sigma_Y}{\sqrt{2(1+\nu)}} \\[2ex] -\dfrac{\sigma_Y}{\sqrt{2(1+\nu)}} \end{array} \right\} \qquad q \rightarrow \left\{ \begin{array}{c} -\dfrac{\sigma_Y}{\sqrt{2(1-\nu)}} \\[2ex] -\dfrac{\sigma_Y}{\sqrt{2(1-\nu)}} \end{array} \right\} \ . \tag{106}$$

In the case of a vanishing Poisson's coefficient ($\nu = 0$), the ellipse transforms into a circle with a radius σ_Y.

- *Tresca*: this criterion is represented by the hexagon $bcdfgh$. In fact, when σ_1 and σ_2 have the same sign this criterion coincides with the Rankine's criterion, since the maximum principal stress is determined by the stress $\sigma_3 = 0$ and by the stress with the maximum absolute value in the plane of σ_1 and σ_2. The straight lines $\overline{bh}$ and $\overline{df}$ ($\sigma_1\sigma_2 < 0$) are described by the equations $\sigma_1 - \sigma_2 = -\sigma_Y$ and $\sigma_1 - \sigma_2 = \sigma_Y$, respectively.
- *Von Mises*: in the plane stress case the criterion of the maximum distortion energy may be represented by the inequality

$$\sqrt{\sigma_1^2 + \sigma_2^2 - \sigma_1\sigma_2} \leq \sigma_Y \ .$$

The von Mises criterion states, therefore, that the material remains in the elastic phase if the point representing the plane state of stress is inside the region limited by the ellipse represented by the equation

$$\sigma_1^2 + \sigma_2^2 - \sigma_1\sigma_2 = \sigma_Y^2 \ .$$

This ellipse contains the points c and g, i.e., it gives the same values as the Tresca's criterion if $\sigma_1 = \sigma_2$, as it is easily verified by performing $\sigma_1 = \sigma_2 = \sigma \Rightarrow \sigma = \sigma_Y$ (cf. Fig. 42). In a purely deviatoric state of stress (points r and s, Fig. 42) yielding occurs for a value of the shearing stress given by

$$\begin{array}{l} \sigma_1 \ \ = \tau_Y \\ \sigma_2 \ \ = -\tau_Y \end{array} \ \Rightarrow \ \tau_Y = \dfrac{\sigma_Y}{\sqrt{3}} \approx 0.577\sigma_Y \ .$$

For the same stress state Tresca's criterion gives $\tau_Y = 0.5\sigma_Y$. If the Poisson coefficient takes the value $\nu = 0.5$ (incompressible material), the von Mises and Beltrami criteria coincide, since there is no isotropic deformation and, as a consequence, the distortion energy coincides with the total energy. This fact may be easily verified by making $\nu = 0.5$ in (106).

Three-Dimensional Stress State

In the three-dimensional case the yielding criteria may be represented in the space defined by the principal stresses by means of *yielding surfaces*. These surfaces delimit the region of that space which contains the points whose coordinates are the principal stresses of elastic stress states.[11] The equations which define those surfaces are given by the expressions defining the different criteria (101–105), if we substitute in these expressions "$<$" by "$=$". In the two most important of these criteria the yielding surfaces have the shapes:

- *Tresca*: considering any values for the principal stresses (i.e., abandoning the convention $\sigma_1 > \sigma_2 > \sigma_3$), the criterion of the maximum shearing stress indicates that yielding takes place when the point representing the stress state in the principal stress space is on one of the six planes, whose equations are

$$\begin{array}{lll} \sigma_1 - \sigma_2 = -\sigma_Y & \sigma_1 - \sigma_2 = \sigma_Y & (\parallel \sigma_3) \\ \sigma_1 - \sigma_3 = -\sigma_Y & \sigma_1 - \sigma_3 = \sigma_Y & (\parallel \sigma_2) \\ \sigma_2 - \sigma_3 = -\sigma_Y & \sigma_2 - \sigma_3 = \sigma_Y & (\parallel \sigma_1) \, . \end{array} \qquad (107)$$

These planes are all parallel to the direction whose direction cosines take the values $l = m = n = \frac{1}{\sqrt{3}}$, as can be verified by considering the equation of a straight line with this direction and containing the origin of the reference axes, $\sigma_1 = \sigma_2 = \sigma_3$. It is easily verified that none of the points contained in the planes defined by (107) obeys this condition, which means that this straight line does not intersect any of these planes. The three pairs of planes (each pair is parallel to one of the reference axes and to this straight line) define, therefore, a regular hexagonal prism, whose axis is equally inclined in relation to the three principal directions σ_1, σ_2 and σ_3 (Fig. 43). The points inside the prism represent the possible states of stress in the elastic perfectly plastic material.
The intersection of this yielding surface with the plane σ_1, σ_2 defines the hexagon *bcdfgh* represented in Fig. 42.
- *von Mises*: the equation representing the yielding surface corresponding to the theory of maximum distortion energy may be obtained from Expr. 105, yielding

$$(\sigma_1 - \sigma_2)^2 + (\sigma_1 - \sigma_3)^2 + (\sigma_2 - \sigma_3)^2 = 2\sigma_Y^2 \, .$$

The equation of a cylinder with a radius r, with its axis containing the origin of the rectangular Cartesian reference frame xyz, and being equally

---

[11]In an elastic perfectly plastic material it is not possible to have stresses with a higher value than σ_Y, since, once this value is attained, the material deforms without an increase in the stress. However, many ductile materials resist higher stresses than those causing the first plastic deformations. In these materials the plastic deformation changes the rheological behaviour, since in subsequent loadings (after unloading) the stresses necessary to initiate the plastic deformation are higher than in the first loading. This phenomenon, called *hardening*, is briefly described in Sects. V.2 to V.5.

inclined in relation to those axes, takes the form

$$\frac{1}{3}\left[(x-y)^2+(x-z)^2+(y-z)^2\right]=r^2\;.$$

From these two expressions, we conclude that the yielding surface of the von Mises criterion is a cylinder, which is equally inclined in relation to the principal stresses and has the radius $\sqrt{\frac{2}{3}}\sigma_Y$.

The ellipse corresponding to the two-dimensional case (*csgr*, Fig. 42) results from the intersection of this cylinder with the plane σ_1,σ_2.

In Figure 43 the yielding surfaces corresponding to the criteria of Tresca and von Mises are represented.

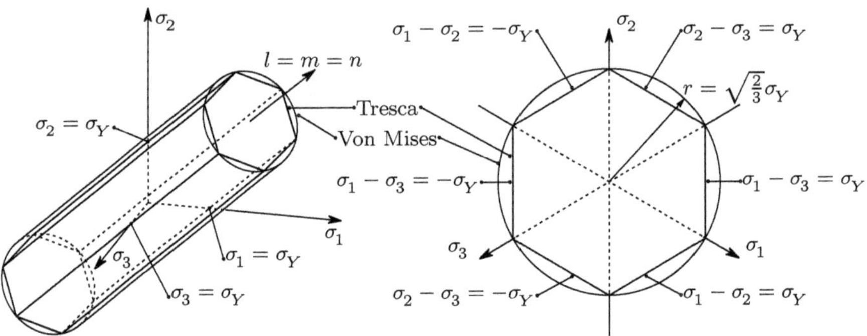

Fig. 43. Yielding surfaces describing the Tresca's and von Mises's criteria

IV.7.b.vii Conclusions About the Yielding Theories

From the considerations described in this subsection in relation to the different yielding theories, we may conclude that plastic deformation is caused mainly by the deviatoric component of the stress tensor. In fact, the two theories, whose predictions are closer to the experimentally observed material behaviour – Tresca and von Mises – postulate that an isotropic tensile or compressive stress do not cause yielding, no matter how high these stresses are.

These conclusions, arrived at from a phenomenological approach, are physically explained by the observation that the macroscopically observed plastic deformation corresponds to shear deformations inside the microscopic crystal structure of the material, caused mainly by the shearing stresses. These stresses do not exist in an isotropic stress state, as seen in Chap. II. Another indication of the importance of the role played by the shearing stresses in the plastic deformation is given by the fact that the experimentally observed tensile and compressive yielding stresses have the same absolute values: although tensile and compressive normal stresses are physically different, the shearing stresses take the same value in the two cases, as can be easily confirmed by drawing the corresponding Mohr circles (see the considerations about the sign conventions in the normal and shearing stresses in Sect. II.11).

IV.7.c Mohr's Rupture Theory for Brittle Materials

As stated earlier (Sect. IV.2) brittle materials rupture, practically without plastic deformation. To predict the stress conditions under which rupture occurs in three-dimensional stress states, the most widely-accepted theory is the Mohr criterion. This theory is based on the *intrinsic strength curve* of the material. This curve is defined as the envelope of the Mohr circles defined by the maximum and minimum principal stress (σ_1 and σ_3) of the stress states which cause rupture. These circles may be obtained by two- or three-dimensional experimental tests, by gradually increasing one or more of the principal stresses, until rupture takes place. The Mohr circle defined by the principal stresses σ_1 and σ_3 at the moment of rupture is tangent to the intrinsic strength curve. By repeating this procedure for several values of the relations between σ_1, σ_2 and σ_3, a sufficient number of rupture circles to define the intrinsic strength curve may be obtained. Figure 44 shows this curve.

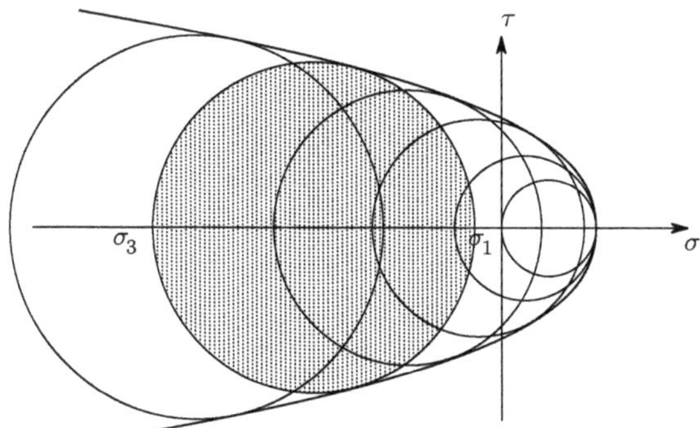

Fig. 44. Intrinsic strength curve of a brittle material

It is obvious that this curve does not fully define the conditions under which rupture takes place, unless the value of the middle principal stress does not play a significant role, since it is not considered in the definition of the intrinsic reference curve. Experimental investigations have shown that this hypothesis may be accepted without introducing a significant error.

Once the material's intrinsic strength curve is obtained, it may easily be verified if a given stress state would cause rupture or not, by drawing the Mohr circle corresponding to the maximum and minimum principal stress and confirming if it intersects that curve or not.

In order to simplify the use of this method, Mohr admitted that the intrinsic curve may be approximated with sufficient precision by two straight lines. With this assumption, that curve can be obtained from one-dimensional tensile and compressive experimental tests, as represented in Fig. 45.

At the rupture, the Mohr's circle corresponding to the maximum and minimum principal stresses, σ_1 and σ_3, is tangent to the intrinsic curve. These stresses may be related to the rupture stresses measured in experimental tensile and compressive tests, σ_t and σ_c (in this analysis the yielding stress σ_c is considered to have a positive value). To this end, the following conditions are considered (cf. Fig. 45)

$$\overline{AB} = \frac{\sigma_c}{2} - \frac{\sigma_t}{2} \qquad\qquad \overline{AE} = \frac{\sigma_c}{2} + \frac{\sigma_t}{2}$$

$$\overline{CD} = \frac{\sigma_1 - \sigma_3}{2} - \frac{\sigma_t}{2} \qquad\qquad \overline{CE} = \frac{\sigma_t}{2} - \frac{\sigma_1 + \sigma_3}{2} \ .$$

By means of triangle similarity considerations in Fig. 45, we easily conclude that the Mohr circle defined by the principal stresses σ_1 and σ_3 does not intersect the intrinsic strength curve, i.e., the material does not rupture, as long as the following condition is satisfied

$$\frac{\overline{AB}}{\overline{AE}} > \frac{\overline{CD}}{\overline{CE}} \ \Rightarrow\ \frac{\sigma_c - \sigma_t}{\sigma_c + \sigma_t} > \frac{\sigma_1 - \sigma_3 - \sigma_t}{\sigma_t - \sigma_1 - \sigma_3} \ . \tag{108}$$

If the three principal stresses are tensile stresses, the Mohr criterion yields results which are no longer confirmed by experimental observations. This is a consequence of the fact that the approximation of the intrinsic curve by two straight lines displays the vertex V (Fig. 45), which does not appear in the real intrinsic strength curve (Fig. 44). In these cases, the Saint-Venant criterion (maximum normal stress) shall be used instead, i.e., the following condition should be satisfied ($\sigma_1 > \sigma_2 > \sigma_3$)

$$\sigma_1 \leq \sigma_t \ .$$

Since the material does not resist tensile stress with a value superior to σ_t ($\sigma_1 \leq \sigma_t$) and (108) only gives good results for $\sigma_3 < 0$, the quantity $\sigma_t - \sigma_1 - \sigma_3$ takes always positive values. Under these conditions, the inequality 108 is equivalent to the condition

$$\frac{\sigma_1}{\sigma_t} - \frac{\sigma_3}{\sigma_c} < 1 \ . \tag{109}$$

This expression represents the Mohr criterion for the prediction of rupture in brittle materials.

In some applications of this criterion, especially in the fields of Soil and Rock Mechanics, the material parameters used to characterize the rupture are the internal friction angle ϕ and the shearing rupture stress c (cohesion), instead of σ_c and σ_t (Fig. 45). Angle ϕ is a measure of the increase of shear

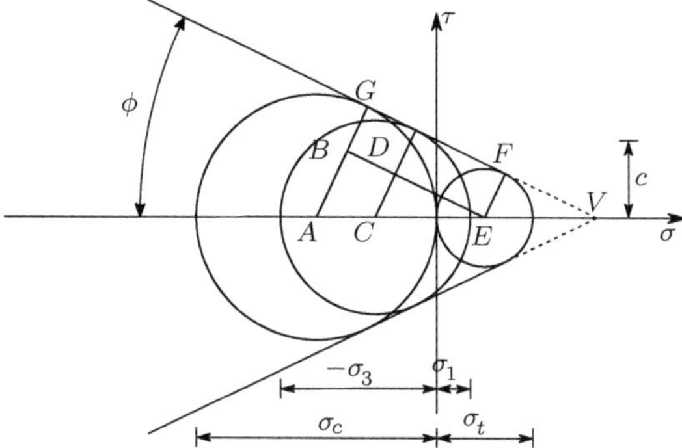

Fig. 45. Mohr's rupture criterion

strength in a facet, when a compressive normal stress acts on it. The cohesion c represents the shear strength in the same facet, when the normal stress is zero. The expression of Mohr's criterion as a function of these rheological parameters may be obtained from the relations between the radii $\overline{EF} = \frac{\sigma_t}{2}$ and $\overline{AG} = \frac{\sigma_c}{2}$ and the parameters c and ϕ. From Fig. 45 we easily obtain the relations

$$\begin{cases} \dfrac{\sigma_t}{2} + \dfrac{\sigma_t}{2}\sin\phi = c\cos\phi \Rightarrow \sigma_t = \dfrac{2c\cos\phi}{1+\sin\phi} \\[2mm] \dfrac{\sigma_c}{2} - \dfrac{\sigma_c}{2}\sin\phi = c\cos\phi \Rightarrow \sigma_c = \dfrac{2c\cos\phi}{1-\sin\phi} \end{cases}.$$

By substituting these expressions of σ_t and σ_c in (109), we get the expression of the Mohr criterion as a function of c and ϕ. Thus, according to this criterion, rupture will not occur while the following condition is satisfied

$$\sigma_1\left(1+\sin\phi\right) - \sigma_3\left(1-\sin\phi\right) < 2c\cos\phi . \tag{110}$$

A graphical representation of this criterion in the two-dimensional space defined by the principal stresses σ_1 and σ_2 with $\sigma_3 = 0$ (plane stress) is presented in Fig. 46-a. The Mohr criterion reduces to the Tresca criterion when the tensile and compressive rupture stresses are equal ($\sigma_c = \sigma_t$).

In the space of the principal stresses σ_1, σ_2 and σ_3 the Mohr criterion is represented by a pyramid with an irregular hexagonal cross-section, whose lateral faces are the six planes defined by the equations (Fig. 46-b)

$$\frac{\sigma_1}{\sigma_t} - \frac{\sigma_2}{\sigma_c} = 1 \quad (\sigma_1 > \sigma_3 > \sigma_2) \qquad \frac{\sigma_2}{\sigma_t} - \frac{\sigma_1}{\sigma_c} = 1 \quad (\sigma_2 > \sigma_3 > \sigma_1)$$

$$\frac{\sigma_1}{\sigma_t} - \frac{\sigma_3}{\sigma_c} = 1 \quad (\sigma_1 > \sigma_2 > \sigma_3) \qquad \frac{\sigma_3}{\sigma_t} - \frac{\sigma_1}{\sigma_c} = 1 \quad (\sigma_3 > \sigma_2 > \sigma_1) \tag{111}$$

$$\frac{\sigma_2}{\sigma_t} - \frac{\sigma_3}{\sigma_c} = 1 \quad (\sigma_2 > \sigma_1 > \sigma_3) \qquad \frac{\sigma_3}{\sigma_t} - \frac{\sigma_2}{\sigma_c} = 1 \quad (\sigma_3 > \sigma_1 > \sigma_2) .$$

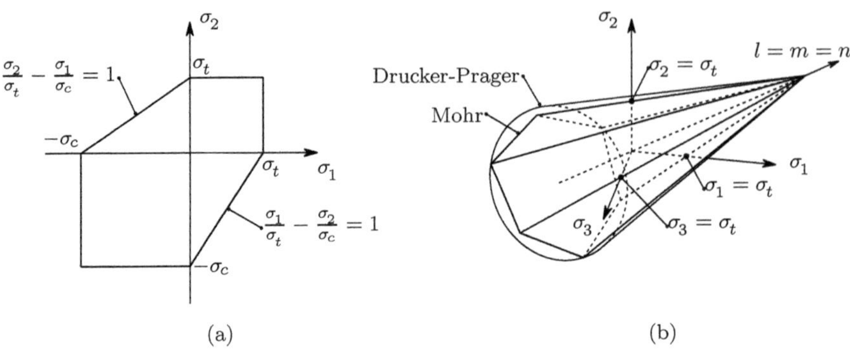

Fig. 46. Graphical representation of Mohr's criterion in the principal stress space: (**a**) two-dimensional case; (**b**) three-dimensional case

The vertex of the pyramid is on the straight line with the equation $\sigma_1 = \sigma_2 = \sigma_3$ and its coordinates are given by the expression

$$\sigma_1 = \sigma_2 = \sigma_3 = \frac{\sigma_t \sigma_c}{\sigma_c - \sigma_t} = \frac{c}{\tan \phi},$$

as it may be easily verified by performing $\sigma_1 = \sigma_2 = \sigma_3 = \sigma$ in (110) and (111).

There is an approximation to Mohr's criterion which was proposed by Drucker and Prager in 1952 [6]. This criterion is represented by a cone with a circular cross-section, which touches the Mohr pyramid in three of its six longitudinal edges, as represented in Fig. 45-b), in the same way as the von Mises cylinder touches the Tresca prism (Fig 43). The Drucker-Prager criterion has the advantage of a simpler computational implementation, since it is described by only one equation (instead of the six equations (111)) [6] and does not have the edges of the Mohr yielding surface.[12] However, the results yielded by the Drucker-Prager criterion are not as close to the experimentally observed failure conditions as the predictions of the Mohr criterion.

It is, however, also possible to obtain single analytical expressions for the Tresca and Mohr criteria, by expressing them as functions of the invariants of the stress tensor. To this end, the maximum and minimum principal stresses, σ_1 and σ_3, in (104) and (109), may be expressed in terms of the invariants, by using, for example, the algorithm presented in Example II.1. Even though, the analytical and numerical treatment of the smooth surfaces (von Mises and Drucker-Prager) is simpler.

If there are no tensile stresses in the material, which happens when the three principal stresses are negative, the rupture criteria may be used as yield-

---

[12]In the modelling of structures where plastic deformations take place (elasto-plastic structural analysis) the normal direction to the yielding surface must be computed. It is obvious that this normal cannot be defined uniquely at an angle point.

ing criteria, even in the case of a brittle material. In fact, the surfaces where material failure has occurred still resist compressive normal stresses and shearing stresses, whose values depend on the friction angle and on the compressive stress. This rupture has the same nature as yielding, since the material loading capacity does not disappear. For this reason, it is possible to perform elastoplastic structural analysis in soil materials, rocks and concrete, although they are brittle materials.

As footnote 20 explains, most brittle materials display larger rupture stresses in compressive than in tensile stress states, since the internal friction angle means that a compressive normal stress acting on a facet increases its shearing strength. In ductile materials, however, the normal stresses do not influence shear strength. This difference between the behaviour of ductile and brittle materials may be explained by the fact that in the latter, it is mainly the cohesion and friction forces between the particles that oppose deformation and rupture, while in plastic deformation the inter-atomic forces in the crystal structure play the most important role in shear deformation. Since those forces are not affected by the normal stresses, the yielding behaviour is the same for tensile and compressive internal forces.

IV.8 Concluding Remarks

From the considerations set forth in this chapter, we conclude that the experimentally observed rheological behaviour of materials may be explained as the combination of three basic types of stress-strain relations; elasticity, plasticity and viscosity. If the relations between these elementary types of deformation and the stresses causing them are linear, the constitutive laws take especially simple forms and, with exception of plastic deformation, the effects of different stresses or strains may be superposed.

When the material behaviour under two- or three-dimensional stress states is analysed, the concept of isotropy plays a fundamental role. In fact, we have concluded that in an isotropic material the principal directions of the stress and strain tensors always coincide. Furthermore, in an isotropic material the constitutive laws described in this chapter are valid for any rectangular Cartesian reference system, while for monotropic and orthotropic materials only the particular cases of reference systems coinciding with material monotropy or orthotropy directions have been considered. By means of tensorial transformations in the fourth order tensor, which defines the constitutive relation, these relations may be generalized to reference systems with arbitrary spacial orientation. These transformations, however, have not been studied in this text, because it is easier (and always possible) to transform the second order tensors describing the stress and strain states to reference systems coinciding with the material symmetry directions. These transformations have been described in Chaps. II and III (15) and (57).

In an anisotropic material with linear elastic rheological behaviour, Hooke's law has a maximum number of 21 independent elastic constants. These constants may be obtained experimentally by measuring the six strains corresponding to each of the six elements of the stress tensor. Of the 36 measured values only 21 are independent, since the constitutive law must be symmetric, as shown in Sect. IV.6.b (see, e.g., [1]).

The yield and rupture theories studied in Sect. IV.7 are valid only for isotropic materials. The first three (Rankine's, Saint-Venant's and Beltrami's yielding postulates) have a merely historical value, since their predictions are not confirmed by experimental tests. The yielding and rupture theories for non-isotropic materials, which have a more involved formulation than the isotropic case, have not been presented here, since they are beyond the scope of this introductory text to the Mechanics of Materials (see, e.g., [2]).

IV.9 Examples and Exercises

IV.1. Find the relation between G, E and ν (77) by means of:
    (a) the Mohr's circle of the stress state;
    (b) Equation (66).

Resolution

Considering the stress state depicted in Fig. 36, $\sigma_x = \sigma$ and $\sigma_y = -\sigma$, the corresponding strains are

$$\varepsilon_x = \frac{1}{E}\left(\sigma_x - \nu\sigma_y\right) = \frac{1+\nu}{E}\sigma$$

$$\varepsilon_y = \frac{1}{E}\left(\sigma_y - \nu\sigma_x\right) = -\frac{1+\nu}{E}\sigma \ .$$

(a) The Mohr's circle of this strain state has the centre at the origin of the $\varepsilon, \frac{\gamma}{2}$ reference frame and its radius takes the value $\frac{1+\nu}{E}\sigma$. Thus, the shearing strain $\frac{\gamma}{2}$ in the pair of orthogonal directions at $45°$ angles to the principal directions x and y takes the value

$$\frac{\gamma}{2} = \frac{1+\nu}{E}\sigma = \frac{1+\nu}{E}\tau \ ,$$

where τ is the shearing stress acting on the facets a and b (Fig. 36). The double shearing strain γ in the pair of orthogonal directions a and b may then be related to the corresponding shearing stress by the expression

$$\gamma = \frac{2\left(1+\nu\right)}{E}\tau = \frac{\tau}{G} \Rightarrow G = \frac{E}{2\left(1+\nu\right)} \ .$$

(b) As an alternative, the double shearing strain γ in the pair of orthogonal directions a and b (Fig. 36) may be obtained from (66), yielding

$$\theta = -45° \text{ (direction } a) \Rightarrow \frac{\gamma}{2} = \frac{\varepsilon_x - \varepsilon_y}{2} = \frac{1+\nu}{E}\sigma = \frac{1+\nu}{E}\tau$$

$$\Rightarrow \gamma = \frac{2\left(1+\nu\right)}{E}\tau = \frac{\tau}{G} \Rightarrow G = \frac{E}{2\left(1+\nu\right)} \;.$$

IV.2. Find the three-dimensional constitutive law of a Newtonian liquid, directly from the relations $\tau = \mu\dot\gamma$ and $\sigma_m = K\varepsilon_v$, i.e., without using the analogy with the constitutive law of the isotropic linear elastic solid.

Resolution

Considering the stress state defined by its components in a rectangular Cartesian reference frame x, y, z, the deformations caused by them may be computed by considering the normal and shearing stresses separately. Thus, the relation between the shearing stresses and the shearing strain rates results directly from the first relation, yielding

$$\tau_{xy} = \mu\dot\gamma_{xy} \qquad \tau_{xz} = \mu\dot\gamma_{xz} \qquad \tau_{yz} = \mu\dot\gamma_{yz} \;.$$

The relation between the normal stresses and the corresponding longitudinal strains may be obtained by considering the stress state described only by the normal stresses σ_x, σ_y and σ_z. As, in this stress tensor, σ_x, σ_y and σ_z are principal stresses, the maximum shearing stresses in the planes x,y, x,z and y,z are, respectively, $\tau_{max-xy} = \frac{\sigma_x - \sigma_y}{2}$, $\tau_{max-xz} = \frac{\sigma_x - \sigma_z}{2}$ and $\tau_{max-yz} = \frac{\sigma_y - \sigma_z}{2}$. These shearing stresses cause the shearing strain rates $\dot\gamma_{max-xy} = \dot\varepsilon_x - \dot\varepsilon_y$, $\dot\gamma_{max-xz} = \dot\varepsilon_x - \dot\varepsilon_z$ and $\dot\gamma_{max-yz} = \dot\varepsilon_y - \dot\varepsilon_z$, respectively, as may be easily verified by a two-dimensional analysis of the strain tensor in each of the three planes defined by the reference axes ((66) with $\gamma_{xy} = 0$ and $\theta = 45°$). Thus, we have

$$\tau_{max-xy} = \mu\dot\gamma_{max-xy} \Rightarrow \dot\varepsilon_x - \dot\varepsilon_y = \frac{\sigma_x - \sigma_y}{2\mu}$$

$$\tau_{max-xz} = \mu\dot\gamma_{max-xz} \Rightarrow \dot\varepsilon_x - \dot\varepsilon_z = \frac{\sigma_x - \sigma_z}{2\mu}$$

$$\tau_{max-yz} = \mu\dot\gamma_{max-yz} \Rightarrow \dot\varepsilon_y - \dot\varepsilon_z = \frac{\sigma_y - \sigma_z}{2\mu} \;.$$

These three equations are not linearly independent. Therefore, a supplementary condition is needed, in order to determine $\dot\varepsilon_x$, $\dot\varepsilon_y$ and $\dot\varepsilon_z$. This condition is provided by the relation between the isotropic stress σ_m and the volumetric strain ε_v, $\sigma_m = K\varepsilon_v$. Thus, we get from the first two expressions and from the condition $\dot\sigma_m = K\dot\varepsilon_v$

$$
\begin{cases}
\dot{\varepsilon}_x - \dot{\varepsilon}_y = \dfrac{1}{2\mu}\left(\sigma_x - \sigma_y\right) \\[2ex]
\dot{\varepsilon}_x - \dot{\varepsilon}_z = \dfrac{1}{2\mu}\left(\sigma_x - \sigma_z\right) \\[2ex]
\dot{\varepsilon}_v = \dot{\varepsilon}_x + \dot{\varepsilon}_y + \dot{\varepsilon}_z = \dfrac{\dot{\sigma}_m}{K}
\end{cases}
\Rightarrow
\begin{aligned}
\dot{\varepsilon}_x &= \dfrac{\dot{\sigma}_m}{3K} + \dfrac{\sigma_x}{2\mu} - \dfrac{1}{6\mu}\underbrace{\left(\sigma_x + \sigma_y + \sigma_z\right)}_{3\sigma_m} \\[2ex]
&= \dfrac{\dot{\sigma}_m}{3K} + \dfrac{\sigma_x - \sigma_m}{2\mu} .
\end{aligned}
$$

Expressions for $\dot{\varepsilon}_y$ and $\dot{\varepsilon}_z$ may be obtained in the same way.

The strain rates induced in the liquid by the stresses σ_x, σ_y, σ_z, τ_{xy}, τ_{xz} and τ_{yz} may, therefore, be computed by the expressions

$$
\begin{cases}
\dot{\varepsilon}_x = \dfrac{\dot{\sigma}_m}{3K} + \dfrac{\sigma_x - \sigma_m}{2\mu} \\[2ex]
\dot{\varepsilon}_y = \dfrac{\dot{\sigma}_m}{3K} + \dfrac{\sigma_y - \sigma_m}{2\mu} \\[2ex]
\dot{\varepsilon}_z = \dfrac{\dot{\sigma}_m}{3K} + \dfrac{\sigma_z - \sigma_m}{2\mu}
\end{cases}
\quad\text{and}\quad
\begin{cases}
\dot{\gamma}_{xy} = \dfrac{\tau_{xy}}{\mu} \\[2ex]
\dot{\gamma}_{xz} = \dfrac{\tau_{xz}}{\mu} \\[2ex]
\dot{\gamma}_{yz} = \dfrac{\tau_{yz}}{\mu} .
\end{cases}
$$

Expressions 89 follow immediately from these, since we have $\dot{\sigma}_m = K(\dot{\varepsilon}_x + \dot{\varepsilon}_y + \dot{\varepsilon}_z)$.

IV.3. Find the expression of the elastic potential energy per volume unit, associated with the purely deviatoric component of the stress tensor, as a function of the octahedral shearing stress (100), directly from the deformation caused by τ_{oct} on the octahedron (Fig. 10).

Resolution

If the isotropic component of the stress tensor is zero, only the octahedral shearing stress acts on the faces of the octahedron. The work done by the forces resulting from these stresses in the octahedron's deformation, divided by its volume, yields that energy. Considering a unit length the distance of the origin of the reference frame to each of the octahedron's faces, measured in the normal direction to the face – line segment $\overline{OP}$ in Fig. IV.3 – we get the dimensions indicated in that figure. The required expression may then be deduced as follows.

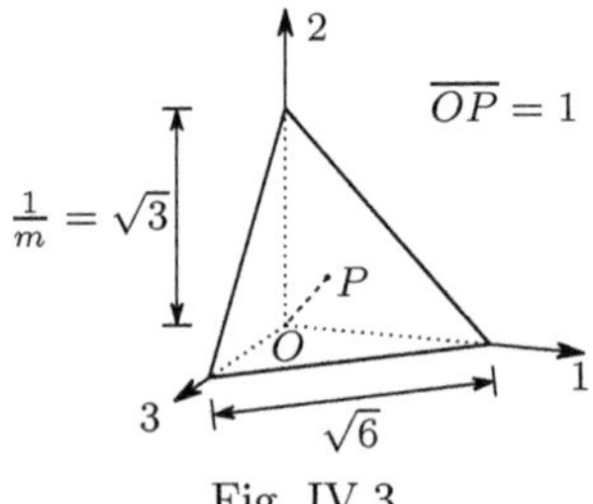

Fig. IV.3

1. Area of each face (equilateral triangle with a $\sqrt{6}$ side length):

$$\Omega = \frac{1}{2}\sqrt{6} \times \sqrt{6}\sin 60° = 3\frac{\sqrt{3}}{2} \ .$$

2. Tangential force resulting from the stress τ_{oct}, in each octahedron's face

$$F = \tau_{oct}\Omega = 3\frac{\sqrt{3}}{2}\tau_{oct} \ .$$

3. Displacement of each of the tangential forces F:

$$\delta = \frac{\gamma_{oct}}{2} \times 1 = \frac{\tau_{oct}}{2G} \ .$$

4. Work done by the eight forces F:

$$W = 8 \times \frac{1}{2}F\delta = 4 \times \frac{3\sqrt{3}}{2}\tau_{oct}\frac{\tau_{oct}}{2G} = 3\sqrt{3}\frac{\tau_{oct}^2}{G} \ .$$

5. Octahedron's volume:

$$V = 2 \times \frac{1}{3} \times \left(\sqrt{6}\right)^2 \times \sqrt{3} = 4\sqrt{3} \ .$$

6. Elastic potential energy stored by volume unit:

$$U_f = \frac{W}{V} = \frac{3}{4}\frac{\tau_{oct}^2}{G} \ .$$

IV.4. Consider a monotropic material with linear elastic behaviour under a stress state, whose principal stresses take the values $\sigma_1 = 2\sigma$, $\sigma_2 = \sigma$ and $\sigma_3 = -4\sigma$. The direction of monotropy is contained in the plane defined by the principal directions 1 and 2 of the stress tensor and makes a 45° angle with each of them, as represented in Fig. IV.4. The constitutive law of the material is defined by the rheological parameters $E_l = 3E$, $E_t = E$, $\nu_l = 0.25$, $\nu_t = 0.4$ and $G_t = 2E$. Find:

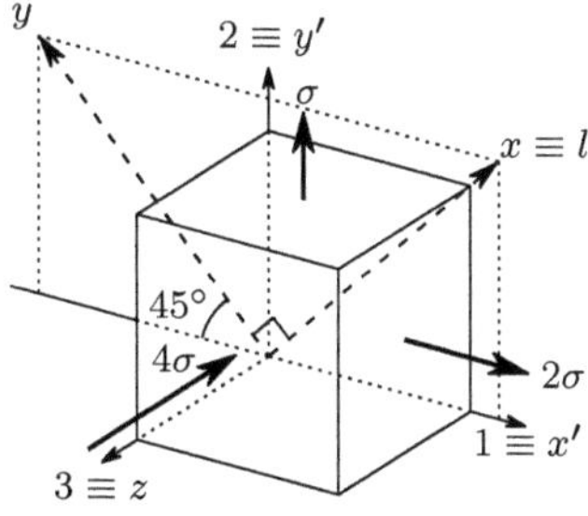

Fig. IV.4

(a) the components of the strain tensor in the reference system defined
by the directions 1, 2 and 3 (Fig. IV.4);
(b) the orientation of the principal directions of the strain tensor.

Resolution

(a) The constitutive law for monotropic materials (83) is only valid if one
of the reference directions is the monotropy direction. Thus it is first
necessary to find the components of the stress tensor in a reference frame
which obeys this condition. The easiest way to express the stress tensor
in such a reference frame is to consider that reference axis x coincides
with the monotropy direction and z coincides with principal direction 3
(Fig. IV.4).

The components of the stress tensor in this reference system may be
obtained by means of a two-dimensional analysis of the stress state in the
plane defined by the principal directions 1 and 2. Thus, we get from (39)

$$\alpha = 45^\circ \Rightarrow \begin{cases} \sigma = \sigma_x = \dfrac{2\sigma + \sigma}{2} + \dfrac{2\sigma - \sigma}{2} \cos 90^\circ = \dfrac{3}{2}\sigma \\[2mm] \tau = \tau_{xy} = -\dfrac{2\sigma - \sigma}{2} \sin 90^\circ = -\dfrac{1}{2}\sigma \end{cases}$$

$$\alpha = -45^\circ \Rightarrow \sigma = \sigma_y = \dfrac{3}{2}\sigma \ .$$

The components of the stress tensor are then

$$\sigma_x = \frac{3}{2}\sigma, \quad \sigma_y = \frac{3}{2}\sigma, \quad \sigma_z = -4\sigma, \quad \tau_{xy} = -\frac{1}{2}\sigma \quad \text{and} \quad \tau_{xz} = \tau_{yz} = 0 \ .$$

By substituting these values in (83), we get the components of the strain
tensor in the x, y, z reference system

$$\begin{Bmatrix} \varepsilon_x \\ \varepsilon_y \\ \varepsilon_z \\ \gamma_{xy} \end{Bmatrix} = \begin{bmatrix} \dfrac{1}{3E} & -\dfrac{0.25}{3E} & -\dfrac{0.25}{3E} & 0 \\[2mm] -\dfrac{0.25}{3E} & \dfrac{1}{E} & -\dfrac{0.4}{E} & 0 \\[2mm] -\dfrac{0.25}{3E} & -\dfrac{0.4}{E} & \dfrac{1}{E} & 0 \\[2mm] 0 & 0 & 0 & \dfrac{1}{2E} \end{bmatrix} \begin{Bmatrix} \dfrac{3}{2}\sigma \\ \dfrac{3}{2}\sigma \\ -4\sigma \\ -\dfrac{1}{2}\sigma \end{Bmatrix} = \dfrac{\sigma}{E} \begin{Bmatrix} 0.7083 \\ 2.975 \\ -4.725 \\ -0.25 \end{Bmatrix} \ .$$

The components of the strain tensor in the reference system defined by
$x' \equiv 1$, $y' \equiv 2$ and $z \equiv 3$ may be computed by means of a two-dimensional
analysis of the strain tensor in plane x, y. Thus, we get from (65) and (66)

$$\theta = -45° \Rightarrow \begin{cases} \varepsilon_\theta = \varepsilon_{x'} = \frac{\sigma}{E}\left[\frac{0.7083+2.975}{2} + \frac{0.7083-2.975}{2}\cos(-90°)\right. \\ \qquad \left. + \frac{-0.25}{2}\sin(-90°)\right] = 1.9667\frac{\sigma}{E} \\[2mm] \frac{\gamma_\theta}{2} = \frac{\gamma_{x'y'}}{2} = \frac{\sigma}{E}\left[-\frac{0.7083-2.975}{2}\sin(-90°)\right. \\ \qquad \left. + \frac{-0.25}{2}\cos(-90)°\right] = -1.1333\frac{\sigma}{E} \end{cases}$$

$$\theta = 45° \Rightarrow \begin{cases} \varepsilon_\theta = \varepsilon_{y'} = \frac{\sigma}{E}\left[\frac{0.7083+2.975}{2} + \frac{0.7083-2.975}{2}\cos(90°)\right. \\ \qquad \left. + \frac{-0.25}{2}\sin(90°)\right] = 1.7167\frac{\sigma}{E}\ . \end{cases}$$

The strain tensor, represented by its components in the original reference system 1,2,3, is then

$$[\varepsilon] = \frac{\sigma}{E}\begin{bmatrix} 1.9667 & -1.1333 & 0 \\ -1.1333 & 1.7167 & 0 \\ 0 & 0 & -4.725 \end{bmatrix}\ .$$

(b) The orientation of the two principal directions of the strain tensor which are in the plane x, y, may be computed directly from (67), yielding

$$\theta_1 = \frac{1}{2}\arctan\frac{\gamma_{x'y'}}{\varepsilon_{x'} - \varepsilon_{y'}} = \frac{1}{2}\arctan\frac{-2.2667}{1.9667 - 1.7167} = -41.853°$$
$$\theta_2 = \theta_1 + 90° = 48.147°\ .$$

θ_1 and θ_2 are the angles between the principal directions of the strain tensor and principal direction 1 of the stress tensor. We confirm, therefore, that the principal direction of the stress tensor which is in the isotropy plane – direction $3 \equiv z$ – is also a principal direction of the strain tensor, while the principal directions which are not in this plane are different in the stress and strain tensors.

IV.5. The stress state in an orthotropic material is defined by its components in a reference system x, y, z. One of the material orthotropy directions makes $60°$ angles with axes x and y and other makes a $45°$ angle with direction z.

    (a) Compute the direction cosines of the orthotropy directions in the reference system x, y, z.

    (b) Write a computation sequence to evaluate the components of the strain tensor in the reference system x, y, z.

Resolution

(a) Denoting by x', y', z' the material orthotropy directions, we get for the first one (x') the direction cosines

$$l_{x'} = \cos(x', x) = \cos 60° = \frac{1}{2}$$

$$m_{x'} = \cos(x', y) = \frac{1}{2}$$

$$n_{x'} = \sqrt{1 - \left(\frac{1}{2}\right)^2 \times 2} = \frac{1}{\sqrt{2}} \ .$$

The direction cosines of the second orthotropy direction (y') may be obtained by solving the system of equations

$$\begin{cases} l_{x'}l_{y'} + m_{x'}m_{y'} + n_{x'}n_{y'} = 0 & (x' \perp y') \\ l_{y'}^2 + m_{y'}^2 + n_{y'}^2 = 1 \\ n_{y'} = \cos 45° = \dfrac{1}{\sqrt{2}} \end{cases} \Rightarrow \begin{cases} l_{y'} = -\dfrac{1}{2} \\ m_{y'} = -\dfrac{1}{2} \\ n_{y'} = \dfrac{1}{\sqrt{2}} \end{cases} .$$

The vector product of the unit vectors defining directions x' and y', yields the components of the unit vector defining the monotropy direction z'

$$l_{z'} = m_{x'}n_{y'} - m_{y'}n_{x'} = \frac{1}{\sqrt{2}}$$

$$m_{z'} = -l_{x'}n_{y'} + l_{y'}n_{x'} = -\frac{1}{\sqrt{2}}$$

$$n_{z'} = l_{x'}m_{y'} - l_{y'}m_{x'} = 0 \ .$$

(b) Since the constitutive law for orthotropic materials described by (85) is only valid for a reference frame, whose axes are parallel to the material orthotropy directions, it is necessary to compute the components of the stress tensor in the reference system x', y', z'. The operation to transpose the reference axes may by performed by means of (15). Matrix $[l]$ contains the direction cosines

$$[l] = \begin{bmatrix} l_{x'} & l_{y'} & l_{z'} \\ m_{x'} & m_{y'} & m_{z'} \\ n_{x'} & n_{y'} & n_{z'} \end{bmatrix} = \begin{bmatrix} \frac{1}{2} & -\frac{1}{2} & \frac{1}{\sqrt{2}} \\ \frac{1}{2} & -\frac{1}{2} & -\frac{1}{\sqrt{2}} \\ \frac{1}{\sqrt{2}} & \frac{1}{\sqrt{2}} & 0 \end{bmatrix} .$$

The computation sequence is therefore:

1. transposition of the stress tensor to the reference axes x', y', z'

$$[\sigma'] = [l]^t [\sigma][l] \ ;$$

2. computation of the elements of the strain tensor in the reference system x', y', z' (85)

$$[\sigma'] \quad \longrightarrow \quad [\varepsilon'] \;;$$

3. transposition of the strain tensor, in order to find its components in the original reference system x, y, z (16, with the strain tensor $[\varepsilon']$ in the place of the stress tensor $[\sigma']$)

$$[\varepsilon] = [l][\varepsilon'][l]^t \;.$$

IV.6. A material has a one-dimensional constitutive law which, in the loading phase, may be described by the expression $\sigma = a\varepsilon - 12a\varepsilon^2$. The unloading follows the linear law described by a straight line which is parallel to the tangent to the loading curve at the origin (E_0, Fig. IV.6).

  (a) Compute the dissipated energy and the elastic potential energy per volume unit, in the longitudinal deformation of the material until a strain value $\varepsilon = 0.04$.

  (b) What kind of deformation corresponds to the dissipated energy?

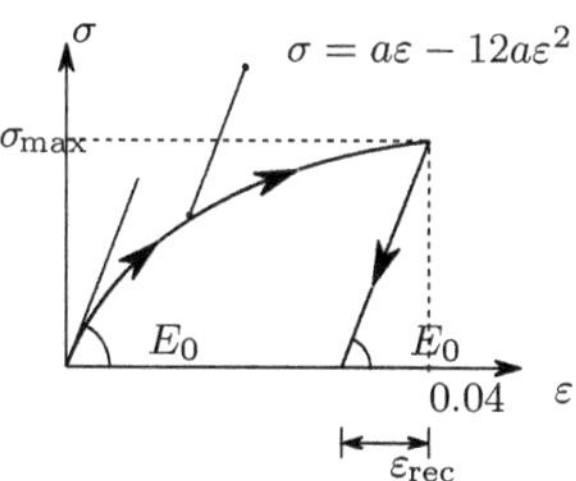

Fig. IV.6

Resolution

(a) From (90) we verify immediately that the work done by the stresses in the deformation of a volume unit of the material – the energy consumed in the loading process – is given by the expression

$$U = \int_0^{0.04} \left(a\varepsilon - 12a\varepsilon^2\right) \, d\varepsilon = 0.000544a \;.$$

In order to compute the part of this energy which is stored as elastic potential energy, the following quantities must be computed (Fig. IV.6)

$$\varepsilon = 0.04 \Rightarrow \sigma = \sigma_{max} = a\left(0.04 - 12 \times 0.04^2\right) = 0.0208a$$

$$E_t = \frac{d\sigma}{d\varepsilon} = a - 24a\varepsilon; \qquad \varepsilon = 0 \Rightarrow E_t = E_0 = a \;.$$

The deformation recovered in the unloading is therefore given by the expression (Fig. IV.6)

$$\varepsilon_{\text{rec}} = \frac{\sigma_{\text{max}}}{E_0} = 0.0208 \ .$$

The elastic potential energy (U_e) and the dissipated energy (U_d) are then

$$U_e = \frac{1}{2}\sigma_{\text{max}}\varepsilon_{\text{rec}} = \frac{1}{2}0.0208a0.0208 = 0.00021632a$$
$$U_d = U - U_e = 0.000544a - 0.00021632a = 0.00032768a \ .$$

(b) The deformation corresponding to the dissipated energy is of the plastic type, since the given constitutive law does not depend on the time (energy is dissipated in plastic deformation and in the viscous deformation; if the stress-strain relation does not depend on the time, there is no viscous deformation).

IV.7. Consider a spherical surface defined in the interior of a undeformed monotropic material. What shape does this surface take, when the material is subjected to an isotropic state of stress?
Answer the same question, supposing that the material is orthotropic.

IV.8. Deduce expressions for the computation of the volumetric strain as a function of the elements of the stress tensor, in
(a) orthotropic materials with linear elastic behaviour;
(b) monotropic materials with linear elastic behaviour.

IV.9. An axisymmetric stress state which is defined by the principal stresses $\sigma_x = \sigma_y = \sigma$ and $\sigma_z = 2\sigma$ acts in a linear elastic monotropic material. Axis x coincides with the material monotropy direction. Imagine a spherical surface with a radius R inside the undeformed material. Define the shape and dimensions of this surface after the deformation. The material's constitutive law is defined by the parameters:

$$E_l = 3E \quad E_t = 2E \quad \nu_l = 0.3 \quad \nu_t = 0.25 \quad G_t = 1.5E \ .$$

IV.10. Compute the energy dissipated in the dashpot of a Maxwell element (Fig. 33) in each of the following loading sequences:
(a) a stress of intensity σ_0 is applied, kept constant and removed after the time interval Δt;
(b) a strain of intensity ε_0 is applied and kept constant until the relaxation reduces to half the stress installed immediately after the application of ε_0; at this instant the stress is removed.

IV.11. Consider an isotropic state of deformation in a monotropic material. What kind of stress state could cause this deformation?

IV.12. A composite material is composed of successive layers of fibres arranged as indicated in the following figure, embedded in an isotropic matrix material.

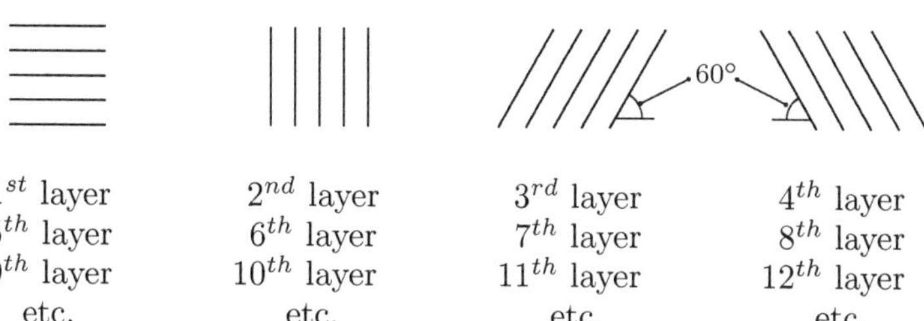

1st layer 2nd layer 3rd layer 4th layer

5th layer 6th layer 7th layer 8th layer

9th layer 10th layer 11th layer 12th layer

etc.   etc.   etc.   etc.

The fibres are all equal and the distance between them is the same in all layers. The composite material has linear elastic behaviour and may be considered as continuous.

How many parameters are necessary to define the constitutive law of this material?

IV.13. Deduce expressions for the volumetric strain caused by an isotropic stress state in

(a) an orthotropic material with linear elastic behaviour;

(b) a monotropic material with linear elastic behaviour.

IV.14. In an orthotropic material a stress state is installed whose principal stresses take the values σ_1, σ_2 and σ_3. The orientations of the principal directions in relation to the material orthotropy directions, x, y and z, are defined by the direction cosines

$$\text{dir. 1} \rightarrow \left\{ \begin{array}{c} l_1 \\ m_1 \\ n_1 \end{array} \right\} \quad \text{dir. 2} \rightarrow \left\{ \begin{array}{c} l_2 \\ m_2 \\ n_2 \end{array} \right\} \quad \text{dir. 3} \rightarrow \left\{ \begin{array}{c} l_3 \\ m_3 \\ n_3 \end{array} \right\} .$$

The material has linear elastic behaviour. Describe computation sequences to obtain:

(a) the volumetric strain;

(b) the strain tensor represented by its components in a reference frame whose axes are parallel to the principal directions of the stress tensor.

IV.15. What is the shape of the relaxation curve of a model composed by a parallel association of a Maxwell-element (Fig. 33) with a spring, when it is instantaneously deformed and this deformation is kept constant indefinitely?

IV.16. What are the principal directions of the strain tensor induced by an isotropic stress tensor in

(a) a monotropic material;

(b) an orthotropic material.

IV.17. Deduce an expression for the elastic potential energy stored per unit volume in a monotropic material with linear elastic behaviour, as a function of the elements of the stress tensor.

IV.18. A material has a one-dimensional constitutive law which, in the loading phase, may be described by the expression $\sigma = \sigma_0 \left(1 - e^{-\alpha\varepsilon}\right)$. The unloading follows a linear law represented by a straight line which is parallel to the tangent to the loading curve at the origin (similar to Fig. IV.6). Find the dissipated and the stored energy components per volume unit in the one-dimensional deformation of a specimen of this material until a strain $\varepsilon = 0.01$.

IV.19. Compute the dissipated energy and the elastic potential energy in the deformation of a Kelvin model (Fig. 32) at a constant strain rate $\dot{\varepsilon} = \frac{d\varepsilon}{dt}$, during a time interval Δt.

Part II

Strength of Materials

V

Fundamental Concepts of Strength of Materials

V.1 Introduction

In Chaps. I to IV expressions were developed for the mathematical description of the relations between external forces, stresses, strains and displacements in a body made of a material which may be considered as continuous. The computation of the stresses and strains induced in the body by a given system of external forces, or by imposed displacements, demands the direct or indirect computation of the solution of systems of equations based on those expressions. Generally, we deal with differential equations (note that many of the main expressions presented in Chaps. I to IV are in differential form) with a degree of complexity which depends on the geometry of the body, on the rheological behaviour of the materials it is made of and on the magnitude of the deformations and rotations. For these reasons, analytical solutions are obtained only for those cases where deformations and rotations are sufficiently small to be considered as infinitesimal, the material is isotropic and has linear elastic behaviour and the geometry of the body has a simple analytical description.

Traditionally the solutions were obtained in two ways:

- *Theory of Elasticity* – this science mainly uses mathematical tools to get analytical solutions for the problems of the Mechanics of Materials. Since the differential equations describing those problems generally have a high degree of complexity, only the simpler problems could be solved. Thus, solutions have been obtained for two-dimensional problems with a simple description for the body geometry and for the loading distribution, in rectangular coordinates (bodies with a rectangular or right triangular border under concentrated and uniformly or linearly distributed external forces), or in polar coordinates (bodies with a circular and/or radial border, problems with some axisymmetry conditions in the stress distribution, etc.). Solutions have also been obtained for some three-dimensional problems, by using rectangular, cylindrical and spherical coordinates [4].

The great advantage of the Theory of Elasticity is that it gives analytical solutions, which allow, for example, a simple investigation about the way a solution changes when the parameters included in it change, which cannot be achieved directly with numerical solutions.

The bigger disadvantage is that it yields solutions only for the simplest cases under ideal conditions. Furthermore, as a consequence of the mathematical approach employed, it is not easy to use physical considerations to get approximate generalizations of its solutions for cases where those conditions do not apply exactly (for example, a slight non-linearity of the material constitutive law).

– *Strength of Materials* – this science favours a more physically grounded, phenomenological and praxis oriented approach. Traditionally, its focus has been on the computation of stresses and deformations in the special case of slender members, although other kinds of structures are also analysed. In fact, these cases belong to the class of problems which can be solved without exaggerated use of mathematical formalism and their solutions were developed prior the appearance of powerful numerical tools. The phenomenological approach and the relatively simple geometry of the problems allow the treatment of a broader spectrum of constitutive laws, as some particular cases of non-linear elasticity, plasticity, etc., the consideration of some material discontinuities, as, for example, slender members made of two or more materials and even non-infinitesimal rotations.

Summarizing, we can make the highly simplified statement that the Theory of Elasticity furnishes mathematical solutions for problems whose geometry is relatively complex (two- or three-dimensional problems), but whose material behaviour is the simplest possible, while the Strength of Materials yields physical solutions for problems with a simpler geometry (slender members), but with some incursions into more complex aspects of material behaviour and non-infinitesimal displacements and rotations. These two sciences are complementary. In fact, the first frequently starts from solutions obtained by means of the second, to develop analytical solutions, and the Strength of Materials often uses solutions obtained by the Theory of Elasticity for particular problems, either for testing a simplifying hypothesis, or to investigate the possibility of generalizing some solutions to problems where the starting conditions are only approximately satisfied.

With the appearance of machines for automatic computation – the computer – it became possible to solve algebraic systems of equations with a large number of unknowns, which made the development of a third method possible: the numerical simulation of structures. This method, by discretizing the continuum and thus allowing the transformation of the differential equations into algebraic equations, took only a few decades to become the most powerful tool for solving problems of Solid Mechanics and, more generally, all continuum problems. Of all the computational tools, the Finite Element Method deserves a special reference, since its flexibility and modularity has allowed it to be successfully applied to practically all kinds of problems of Continuum Mechanics.

This second part of the book introduces structural analysis and the theory of slender members, using the approach which is traditionally called Strength of Materials. In fact, despite of the success of numerical analysis, this subject is still a core part of the Engineering Sciences that deal with Solid Mechanics, since it yields a large number of directly applicable expressions for practical problems. These expressions concern the computation of the effects of the axial and shear forces and of the bending and torsion moments in slender members. Furthermore, as a consequence of the physical approach and of the large number of simple exercises which are solved, it develops in the student a greater capacity for intuitively evaluating the way as a structure behaves.

The notions of stress, strain and rheological behaviour are explained again at the beginning of this chapter. This is because the subjects are introduced differently for the Strength of Materials, compared with Solid Mechanics, and because its is intended that the reader of the second part is able to understand it, without a deep study of the subjects dealt with in the first chapters. However, the reader should already have some knowledge about the two-dimensional analysis of a second order tensor, at least, especially in relation to the transposition of reference axes and Mohr's circle, and the computation of reaction forces in statically determinate structures and internal forces in slender members.

V.2 Ductile and Brittle Material Behaviour

The main characteristics of the rheological behaviour of materials are usually investigated by means of simple experimental tests, in which the relations between forces and deformations in a body with appropriate geometry, made of the material to be studied (test specimen) are measured. The one-dimensional tensile test is the most widely used way to study the behaviour of current structural materials, such as metals. This test determines the relation between an axial force N and the corresponding elongation Δl, as represented schematically in Fig. 47-a.

Consider a prismatic specimen with a doubly symmetric cross-section made of mild steel. If the elongation Δl is gradually increased from zero until the value which causes the rupture of the specimen, and the corresponding axial force N is measured, a relation between these two quantities is obtained. This relation may be represented by a diagram, like that in Fig. 47.[1]

The diagram is typical for a ductile material and is characterized by a zone of purely deformation plastic (irreversible) or *yielding zone*, where deformation

---

[1]The test is carried out with *displacement control*, i.e., by defining a value for the elongation and measuring the corresponding value of the axial force. If *force control* is used instead, the shape of the diagram in the descending zone, where the value of force decreases as the deformation increases (softening), is not correctly captured.

suddenly increases, without a significant increase of the axial force N (line BC). In this diagram distinct zones may be identified. The first corresponds to the straight line OA, where the elongation Δl is completely recoverable and proportional to the axial loading N. Since recoverable deformations are defined as *elastic*, this region is called the linear elastic zone. Usually, the deformations of structural materials are in this zone under service loads.

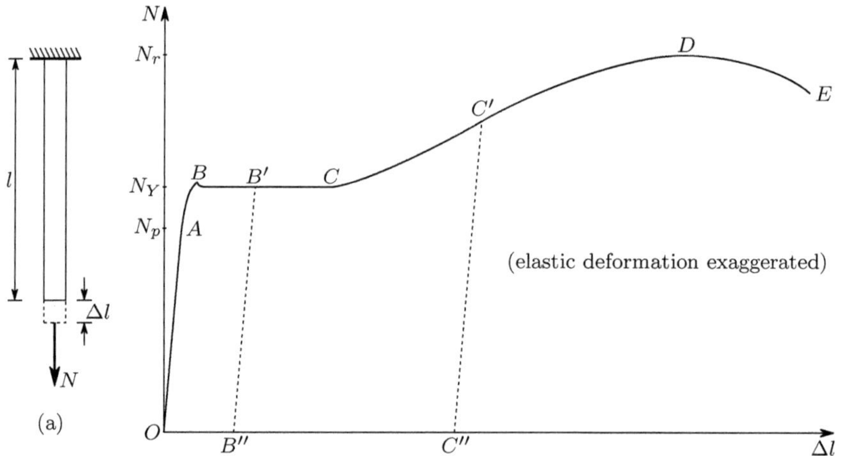

Fig. 47. Force-elongation diagram of mild steel, obtained by means of a one-dimensional tensile test

In the region AB the deformation is still elastic, but there is no proportionality between forces and deformations anymore. Axial force N_p indicates the transition from the linear elastic to the non-linear elastic deformations and is therefore called the *limit of proportionality*. When the loading attains the value N_Y, yielding starts. Region BC is the yielding zone defined above and N_Y is the yielding force. Between these two values (N_p and N_Y) an *elasticity limit* N_e may be defined. This value indicates the maximum value of N, which causes purely elastic deformation, i.e., the maximum value that N can reach, so that the N-Δl diagrams in the loading and unloading phases coincide. In practical terms, the difference between N_p and N_Y is very small so they may be considered to take the same value. In point C the *hardening* of the material starts. In region DE a decrease of the axial force with deformation increase (softening) takes place. This softening is only apparent, since it is a consequence of a reduction of the cross-section (necking) which takes place prior to rupture: in fact, the force per area unit in the necking cross-section increases until rupture (cf. Sect. V.3). If the loading process is stopped at any stage after point B and the axial force is reduced until zero, the N-Δl relation follows a linear path, with the same angle as the initial straight line OA, and a residual deformation remains, as represented by the line $B'B''$. In a subsequent reloading the N-Δl diagram follows the path $B''B'CDE$. If the unloading occurs when the load is already in the hardening region (point C'), the material behaviour is the same: the unloading is linear (line $C'C''$) and the reloading follows the path $C''C'DE$. This means that, in the reloading, the first plastic deformations appear for a higher value of the axial force than in the first loading. This is why the curve $CC'D$ is called the hardening region.

If, instead of a tensile force N, a compressive one is applied, the obtained N-Δl diagram is approximately the same until point C'. Since no necking occurs in compression, the hardening continues indefinitely and no rupture takes place, even for very large deformations. The proportionality and yielding forces take the same value as in the tensile test. This behaviour is characteristic of ductile materials.

In the case of brittle materials as cast iron, concrete, glass, rock, ceramic materials, etc., the obtained force-elongation diagram takes a form of the type represented in Fig. 48. The main differences between the diagrams for brittle and ductile materials are: the linear elastic zone is less defined, i.e., the tangent to the curve decreases steadily until rupture, which occurs with little plastic deformations, and the behaviour under tensile and compressive forces is different. Generally, these materials display more stiffness and strength under compressive loading.

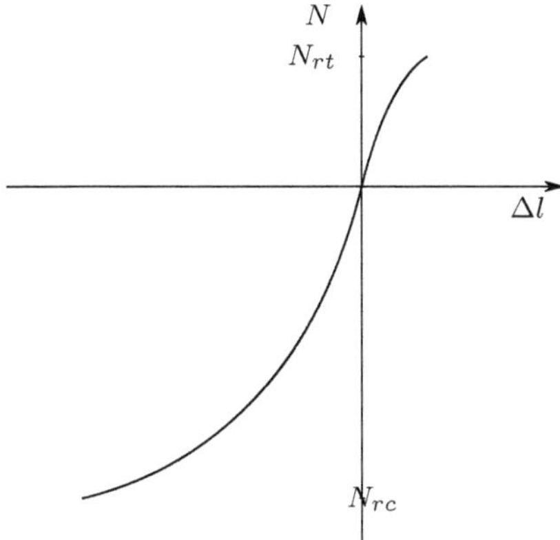

Fig. 48. Force-elongation diagram in a brittle material

In many materials, especially metallic materials, ductility changes with temperature, with less ductility for low temperatures.

V.3 Stress and Strain

In the previous section the relation between axial force and elongation was described. This relation obviously depends on the dimensions of the test specimen. To be more specific, the larger the cross-section area and the smaller the specimen's length, the smaller the elongation will be, for the same axial force N. It is, however, more convenient to express the material properties independently of the specimen's dimensions. This objective may be achieved by means of the *stress* and *strain* definitions. Thus, we may define stress σ as the force per cross-section unit Ω, $\sigma = \frac{N}{\Omega}$[2] and strain ε as the elongation per unit length l, $\varepsilon = \frac{\Delta l}{l}$. It is evident that the stress-strain relation has the same shape as the N-Δl diagram if Ω and l are approximately constant, which happens while the specimen's deformation is small. This is true in the diagrams presented above, with exception of the necking zone (curve DE) in Fig. 47. In the necking cross-section, although the axial force decreases with the deformation, the stress increases.

The coefficient of proportionality E between stress and strain in the linear elastic region (line OA, Fig. 47) is a *rheological parameter* of the material and is called the *longitudinal modulus of elasticity* or *Young's modulus*. In all kinds of steel this parameter takes the value $E = 206 \times 10^9 N/m^2$. In the curved regions of the σ-ε diagram the stress is no longer proportional to the strain. In this case a *tangent elasticity modulus* may be defined: $E_t = \frac{\mathrm{d}\sigma}{\mathrm{d}\varepsilon}$. A longitudinal elongation is usually accompanied by a reduction of the transversal dimensions and vice versa. The relation ν between the the transversal and longitudinal strains, ε_t and ε, respectively, multiplied by -1 ($\varepsilon_t = -\nu\varepsilon$), defines another rheological parameter and is called the *Poisson's coefficient* of the material. In steel, as in most metals, this parameter takes a value of 0.3 in elastic deformations and 0.5 in plastic deformations.

If we now consider an inclined section at an angle θ with the cross-section, we can define a stress $T = \frac{N}{\Omega_{inc}}$ (Ω_{inc} is the area of the inclined section). This stress must have the direction of the axial force N, in order to be able to balance it, as represented in Fig. 49. Thus, this stress has a normal component σ_θ (to the inclined section) and a tangential or shearing component τ_θ. Denoting the cross-section area by Ω, we get

$$\Omega_{inc} = \frac{\Omega}{\cos\theta} \Rightarrow T = \frac{N}{\Omega_{inc}} = \frac{N}{\Omega}\cos\theta.$$

The normal and shearing components are then, respectively,

$$\sigma_\theta = T\cos\theta = \frac{N}{\Omega}\cos^2\theta \quad \text{and} \quad \tau_\theta = T\sin\theta = \frac{N}{\Omega}\sin\theta\cos\theta = \frac{1}{2}\frac{N}{\Omega}\sin 2\theta\,.$$

The maximum value of σ_θ clearly occurs for $\theta = 0$ (cross-section). The shearing stress attains its maximum value in an inclined section at a 45° angle with the cross-section, as may be easily verified[3]

---

[2]In this expression a uniform distribution of the stress in the cross-section is assumed. This hypothesis will be proved in Sect. VI.1).

[3]These conclusions may also be drawn by means of the Mohr circle or from (35) and (36) ($\sigma_x = \frac{N}{\Omega}$, $\sigma_y = \tau_{xy} = 0$, $\theta = 45°$).

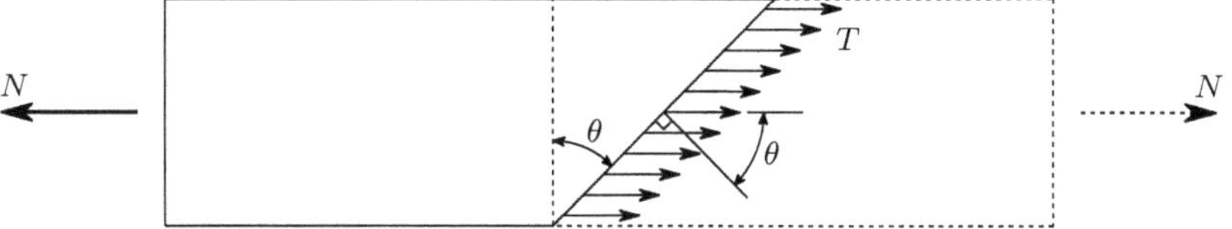

Fig. 49. Stresses in an inclined section

$$\frac{\mathrm{d}\tau_\theta}{\mathrm{d}\theta} = \frac{N}{\Omega} \cos 2\theta = 0 \Rightarrow \theta = \frac{\pi}{4} \ .$$

If, in the one-dimensional experimental test of a ductile material, a specimen with a flat and polished lateral surface is used, careful observation shows lines at a 45° angle with the longitudinal direction, when yielding takes place. These lines, called Lüder-Hartman's lines, have the directions corresponding to the maximum shearing stress and indicate that the plastic deformation is mainly a shearing deformation. This explains the same material behaviour observed in tensile and compressive experiments, especially for the yielding stress $\sigma_Y = \frac{N_Y}{\Omega}$. In fact, there is no physical difference between the shearing deformation in compressive and tensile tests. In brittle materials deformation and rupture are mainly influenced by cohesion, contact and friction forces between the material particles. These forces are obviously different under tensile and compressive loadings. Concrete is one example. In this material the tensile strength is mainly influenced by the cohesion properties of the cement paste, while in compression the properties of the aggregates play an important role, because of the contact and friction forces between the rock particles.

V.4 Work of Deformation. Resilience and Tenacity

When a body deforms under the action of external forces, their points of application suffer displacements and the forces do work. Physics defines the work of a constant force in a straight displacement as the scalar product of the vectors defining the force and the displacement of its point of application.

In the example of the prismatic steel bar under a tensile axial force (Fig. 47-a) the work W_0 done by force N, for a given elongation Δl_i, may be given by the expression

$$W_0 = \int_0^{\Delta l_i} N\left(\Delta l\right) \mathrm{d}\left(\Delta l\right) \ . \tag{112}$$

The integral is necessary because the force N is not constant, but varies during the deformation as a function of Δl, as shown in Fig. 47. In fact, the definition of work given above is valid only for an infinitesimal displacement $\mathrm{d}\left(\Delta l\right)$. By introducing the definitions of stress and strain into (112), the work W done per volume unit may be obtained

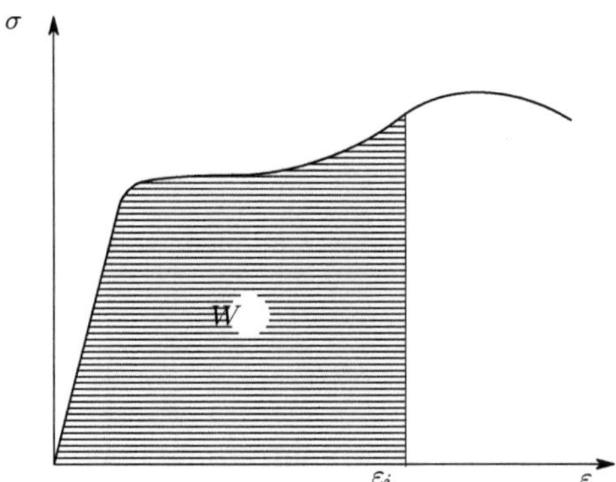

Fig. 50. Work done per volume unit in the deformation ε_i

$$W_0 = \int_0^{\varepsilon_i} \underbrace{\sigma\Omega}_{N} \overbrace{l\,\mathrm{d}\varepsilon}^{\mathrm{d}(\Delta l)} = V \int_0^{\varepsilon_i} \sigma\,\mathrm{d}\varepsilon \Rightarrow W = \frac{W_0}{V} = \int_0^{\varepsilon_i} \sigma\left(\varepsilon\right)\mathrm{d}\varepsilon \ ,$$

where $V = \Omega l$ represents the volume of the bar. This quantity takes the same value as the area under the stress-strain diagram, as represented in Fig. 50.

From Physics we know that for the production of a given amount of work an equal amount of energy U must be spent. If the strain ε_i is in the elastic region of the stress-strain diagram, this energy is totally stored by the deformed material as *elastic potential energy*. This energy is recovered during unloading. However, if the strain is larger than the value corresponding to the elasticity limit, the energy is partly dissipated (transformed in heat) during the plastic deformation. In this case, the elastic potential energy is only a fraction of the work done in the deformation and is given by the expression

$$U_e = \int_{\varepsilon_r}^{\varepsilon_i} \sigma'\left(\varepsilon\right)\mathrm{d}\varepsilon \ ,$$

where ε_r is the residual strain and $\sigma'\left(\varepsilon\right)$ is the stress corresponding to the strain ε in the unloading. As stated above, in the unloading of a steel bar the stress-strain relation is linear, even when the stress is larger than the proportionality limit. The dissipated and potential elastic parts of the deformation energy per volume unit (energy density) are represented in Fig. 51.

The amount of energy per volume unit needed to start plastic deformation, is called *resilience*. The amount of energy per volume unit needed to cause rupture is called *tenacity*. These quantities play important roles in the shock-absorbing capacity of a structure. Ductile materials have high tenacity, as opposed to brittle materials, which display low tenacity. In a ductile material the tenacity is much larger than the resilience, while in brittle materials these quantities are similar since the plastic deformations are small. Ductile materials usually have a much higher tenacity than brittle materials.

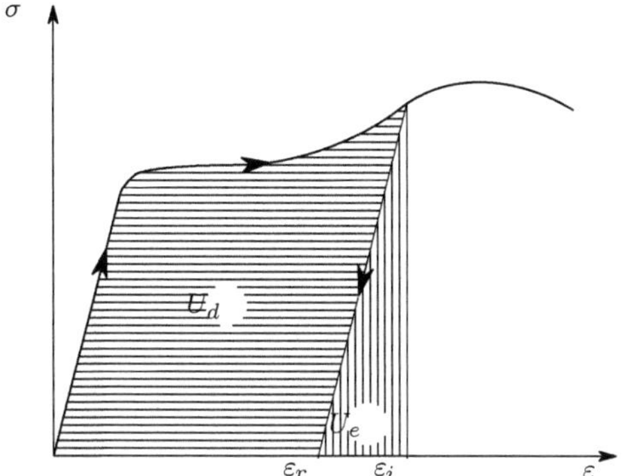

Fig. 51. Dissipated energy (U_d) and elastic potential energy (U_e) in the deformation of a steel bar

V.5 High-Strength Steel

As described in Sect. V.2, if a mild steel is deformed until the strain reaches the hardening zone and the loading is subsequently removed, this steel displays a higher elasticity limit in a later reloading, i.e., a larger linear elastic region in the stress-strain diagram. Since the elasticity limit is usually considered as the limit load under service conditions, the loading capacity of a steel may be increased in this way. This method of increasing the elasticity limit of a steel by means of pre-deformation, with the objective of increasing the admissible stress, is called *strain hardening*. Obviously this process causes a loss of tenacity, since part of the energy dissipation capacity of the material is consumed by the pre-deformation in the hardening process. Conversely, the resilience is increased since the elasticity limit is higher and the elasticity modulus remains unchanged, as depicted in Fig. 52.

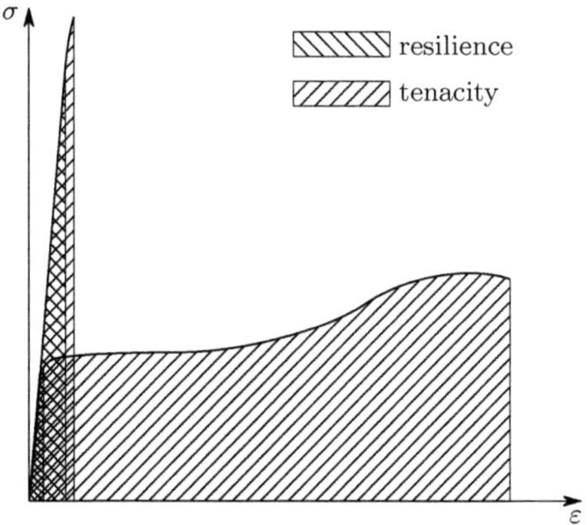

Fig. 52. Resilience and tenacity of a mild steel and of a high strength steel

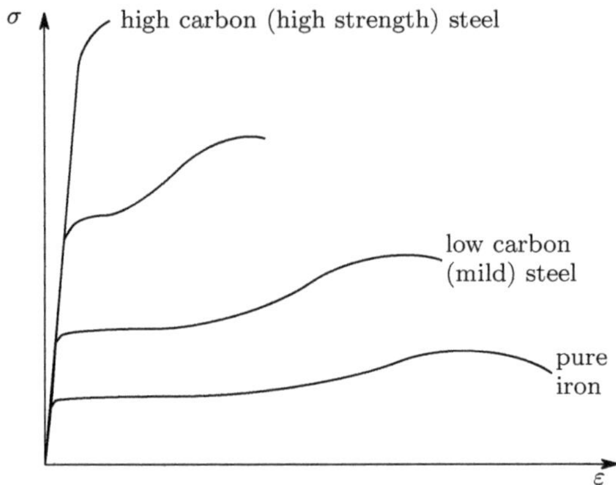

Fig. 53. Variation of the stress-strain diagram with the carbon percentage

The hardened steel is therefore more brittle. Furthermore, the yielding stress for a compressive axial force decreases in a steel bar that is hardened by means of a tensile axial force. This is why the strain hardened bars used in reinforced concrete are pre-deformed by torsion, which increases the tensile and compressive limits of elasticity to the same extent.

The elasticity limit of steel can also be increased by increasing the quantity of carbon added during the metallurgical process of steel production. This process, called *natural hardening*, has the advantage of not disturbing the isotropy of the material. Just as in the strain hardening process, the capacity of plastic deformation (ductility) decreases as the elastic limit increases, as indicated in Fig. 53.

High strength steels do not have a yielding zone where only plastic deformations take place (cf. Figs. 52 and 53). As a consequence, the onset of the plastic deformation is not clearly shown by the stress-strain diagram. For this reason, the elastic limit is defined with the help of a convention, which states that the elastic limit is the stress which causes an unrecoverable strain with the value $0.2\% = 0.002$ ($\sigma_{0.2}$, Fig. 54).

V.6 Fatigue Failure

In structural elements subjected to rapidly changing internal forces, such as bridge elements under vibration loads caused by traffic or wind loads, machine parts performing cyclic motions, aircraft structural elements, etc., *fatigue failure* may occur. This kind of failure usually takes place for substantially lower stresses than in a monotonically increasing loading, as in the experimental tests described in Sect. V.2.

The behaviour of structural materials under the action of loads varying with great frequency is investigated by means of fatigue tests. In these tests a

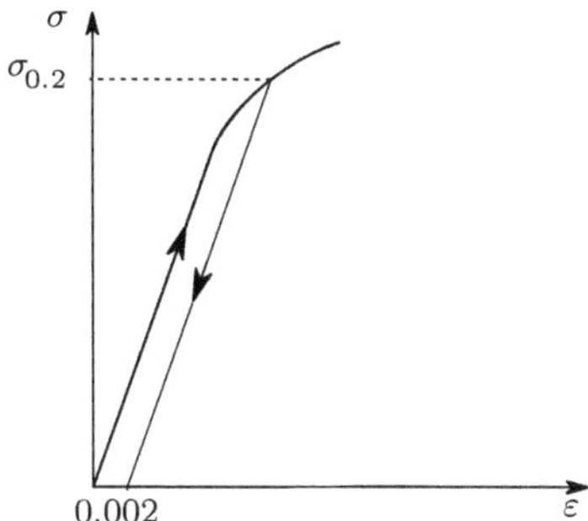

Fig. 54. Conventional elastic limit

specimen undergoes a generally one-dimensional loading which causes stresses varying cyclically between two given values, usually a compressive and a tensile stress. The loading cycles are repeated a large number of times until rupture takes place. The higher the given stress values are, the smaller is the number of cycles needed to cause rupture. In the simplest and commonest of these experimental tests equal compressive and tensile stresses ($\sigma_{\min} = -\sigma_{\max}$) are cyclically applied.

If, for example, several specimens of steel are tested under the same stress level, the results are generally found to be widely dispersed, i.e., the number of cycles necessary to cause rupture varies substantially from one test to another. However, if the number of tests is sufficiently high and the experiments are performed for different stress levels, results are obtained which may be approximated by a curve with the shape depicted in Fig. 55 [3]. This curve tends asymptotically for a stress value σ_f, which means that for a sufficiently low stress no fatigue failure occurs, irrespective of the number of loading cycles. This stress value is called *fatigue limit stress.* In iron-carbon steel this stress is approximately half the rupture stress, which is lower than the elastic limit stress. In other more ductile materials like lead, copper, zinc or pure iron the fatigue limit stress is higher then the elastic limit [3].

The value of the fatigue failure stress depends strongly on the specimen's surface, with a higher failure stress obtained when the surface is polished. This is a consequence of the fact that rupture is initiated by a crack. The crack starts at the surface and propagates to the interior of the specimen as the cyclic loading goes on, until the uncracked part of the specimen's cross-section becomes too small to carry the applied loading and failure takes place. The crack starts at a lower stress in the unpolished surface, since its imperfections cause higher stress concentration, as will be seen later (Sect. VI.9). This crack initiation mechanism explains the above-mentioned dispersion of the number of cycles required to cause fatigue failure for the same stress level. It is also due to the sensitivity to the imperfections that fatigue failure takes place for fewer

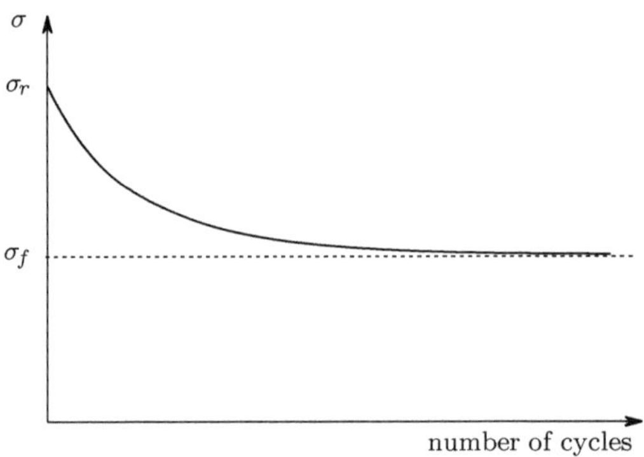

Fig. 55. Relation between the maximum stress σ and the number of loading cycles needed to cause rupture in fatigue tests with $\sigma_{\min} = -\sigma_{\max}$

cycles (or lower stresses) in the case of corrosive environments, since corrosion initially affects the surface of the material and causes larger imperfections.

From these considerations we conclude that fatigue failure has the same character as brittle failure, since it takes place without being preceded by large plastic deformations. It is a dangerous kind of failure, since there are no visible signs of the fatigue crack before rupture. Sensitivity to imperfections is also a characteristic of brittle failure. As will be seen later (Sect. VI.5), ductile structures are safer than structures made of brittle materials, since the capacity for plastic deformation allows a redistribution of internal forces, which automatically optimizes the distribution of internal forces until failure occurs. However, if there is a risk of fatigue failure, the advantages of structures made of ductile materials are lost, since fatigue-induced rupture is of a brittle nature, even when it occurs in ductile materials.

V.7 Saint-Venant's Principle

Saint-Venant's principle states that *in a body under the action of a system of forces which are applied in a limited region of its boundary, the stresses and strains induced by those forces in another region of the body, located at a large distance from the region where the forces are applied, do not depend on the particular way the forces are applied, but only on their resultant.* This "large distance" may be considered, in most cases, as the largest dimension of the region where the forces are applied.

This principle does not have a formal, general and exact demonstration as yet, but it has been verified in so many cases, both experimentally and numerically, that it is accepted as valid by the generality of authors on this subject. It is a very useful principle, since complex force systems may be reduced to their resultants, which substantially simplifies and reduces the computation effort in practical problems. Besides, it is a very helpful tool in

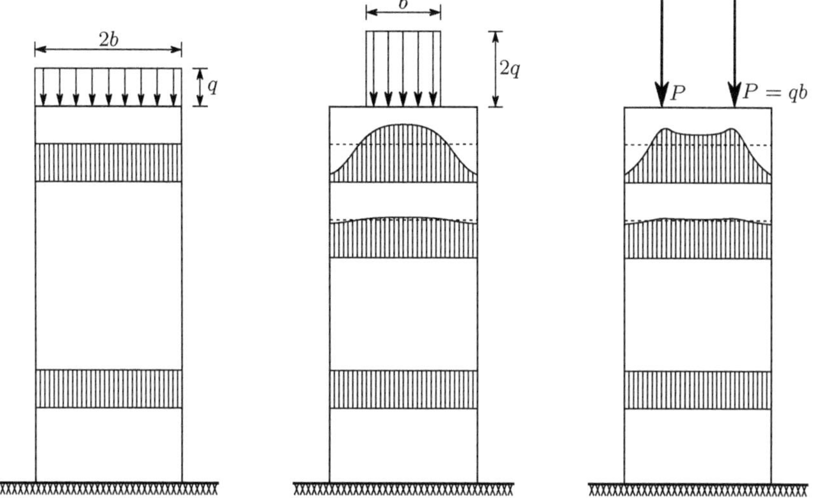

Fig. 56. Stress distribution in different cross-sections of a prismatic bar, caused by three force systems with the same resultant

the theoretical development of solutions for problems in Theory of Elasticity and Strength of Materials, as will be seen later.

As an example, let us consider the prismatic bar represented in Fig. 56 under the action of three systems of forces with equal resultants: the stresses at a grater distance than the transversal dimension $2b$ from the upper end of the bar may be accepted as equal in the three cases.

This principle is also valid in the cases of non-isotropic materials, non-linear material behaviour, plastic and viscous deformations and material heterogeneity. Furthermore, the validity of this principle is not limited to small deformations.

V.8 Principle of Superposition

In structures where the applied loading causes deformations and rotations which are sufficiently small to be considered as infinitesimal and where the rheological behavior of the material is linear (i.e., the proportionality limit stress is not exceeded), the relation between the intensity of a force and the effects it causes (stresses, strains, displacements) is linear, i.e., the effects are proportional to the intensity of the force which causes them.

The increase of displacement corresponding to the increase of the force which causes it is therefore independent of the intensity of the force before the increment. Furthermore, as the geometry of the structure, after the application of loading, is only infinitesimally different from the undeformed configuration, the initial geometry of the structure may always be used, regardless of the existence or not of other previously applied loads (geometrical linearity). Under these conditions (material and geometrical linearity) the *Principle of Superposition* is valid: *the effect of the application of a force to a structure is independent of the existence or not of other forces applied to the structure.* As a consequence, the effects of applying different loading systems to the structure may be computed separately and added.

This principle has a simple analytical demonstration. To this end, it suffices to take into consideration that, in the case of infinitesimal deformations and rotations, all the conditions relating applied forces, stresses, strains and displacements are linear. These conditions are:

– the differential equations of equilibrium (5),
– boundary balance equations (8),
– the relations between strains and displacements (50),
– the local (53) and integral conditions of compatibility (the integration of strains is a linear operation),
– the constitutive law (74), (75), (79), (81), (83) and (85).

As a consequence, the sum of two sets of forces, stresses, strains and displacements also obeys these conditions.

A temperature variation is also a kind of loading whose effect is generally defined by a linear law: the strain induced by a temperature variation is generally proportional to the value of that variation, with the thermal expansion coefficient playing the role of the proportionality constant. The effect of the temperature variation may be quantified by adding another element to the expressions, allowing the computation of the longitudinal strains for given stresses (e.g., (74)). Taking, for example, the longitudinal strain in direction x, we have $\varepsilon_x = \frac{1}{E}\left[\sigma_x - \nu\left(\sigma_y + \sigma_z\right)\right] + \alpha\Delta T$, where α is the coefficient of thermal expansion.

From a physical (and practical) point of view we may make the simplified statement that the effect of the application of a force to a supported body depends only on two components: the constitutive law of the material and the geometry of the body.

If the constitutive law is linear and if we can admit that its rheological behaviour does not depend on temperature, the first component does not change with the application of forces or with a temperature variation.

If, in addition, the deformations are small enough for it to be acceptable that the geometry of the body does not change, the second component does not change either. Thus, the effect of the force is independent of the previous application of forces and of temperature variations, which leads to the principle of superposition as stated above.[4]

---

[4]It must be noted that possible interactions between the deformations and the internal forces caused by the external loads are not taken into account in these considerations. This interaction may occur in presence of infinitesimal deformations and cause structural instability. As it will be seen later (Chap. XI), in the analysis of this phenomenon the principle of superposition is not valid.

This principle has many useful applications, both from a practical point of view, since it allows the separate consideration of single loading cases and any linear combination of their effects, and in theoretical analysis, as, for example, in the demonstration of energy theorems for linear elastic structures. It must, however, be noted that *it is only valid for structures with linear elastic force-displacement behaviour.*

V.9 Structural Reliability and Safety

V.9.a Introduction

A structure must resist all the loadings that will act on it, in its expected life and conditions of use. In the context of structural reliability "resist" means that the structure must be able to carry out safely all the functions for which it is designed.

We consider that the structure ceases to be able to carry out those functions when a *limit state* occurs. Two kinds of limit states are usually considered: *ultimate limit states* and *serviceability limit states.*

The first are associated with the rupture, collapse or failure of the entire structure or parts of it, such as failure of structural elements caused by material rupture, structural instability of compressed members, fatigue failure, displacements leading to loss of supports, etc.

The serviceability limit states include everything that may cause structural malfunction, although not inducing collapse. Examples of serviceability limit states in Civil Engineering structures include: excessive deformation of a structural frame which may cause cracking in non-structural walls, excessive vibration of a pedestrian bridge which may cause discomfort to the users, too large cracks in reinforced concrete members which may lead to corrosion of the reinforcing steel bars, etc.

V.9.b Uncertainties Affecting the Verification of Structural Reliability

When verifying the reliability of a structure it is not possible to use a totally deterministic approach, since the quantification of the problem data – the different kinds of loading (actions) and the rheological properties of the materials – is always affected by some uncertainty, which makes it impossible to define exact values. The main sources of uncertainty are:

– uncertainty in the *value of the actions*: all actions are characterized by a smaller or larger dispersion in relation to their mean value; besides, it is often impossible to define limiting values. Examples of loadings are those caused by wind, snow, temperature variation, earthquakes, etc. Furthermore, it is generally not economically defensible to use the limiting values of the actions, when they exist, since the probability of their occurrence is generally very small;

– *statistical dispersion of the rheological properties of structural materials*, especially rupture stress, elasticity modulus, resilience, tenacity, etc.;
– uncertainty introduced by the *dimensional tolerance* of pre-moulded structural elements;
– *execution imperfections*, especially in Civil Engineering structures as in concrete elements for example, introduce uncertainty into the geometrical dimensions, on the position of the reinforcing bars, the verticalness of columns, etc.;
– uncertainty introduced by the *methods of analysis and computation*, since they are always based on idealized models, resulting from *simplifying hypotheses*, like the consideration of a linear stress-strain relation and of the undeformed geometry of the structure, non-consideration of time-dependent effects, such as viscous deformation, etc.

As a consequence of these random factors, the verification of structural reliability necessarily has a probabilistic basis, since a zero probability of failure can never be guaranteed. The criteria of dimensioning and safety evaluation are based on the definition of a sufficiently low probability for the structure to reach a limit state. For ordinary Civil Engineering constructions the following values are considered acceptable [3]:

– serviceability limit states $< 5 \times 10^{-2}$;
– ultimate limit states $< 10^{-5}$.

In the case of structures with special safety requirements, like large dams or nuclear power plants, the maximum values of these parameters are much lower.

The above considerations lead to the conclusion, that a structure is safe if the probability of it reaching a limit state is sufficiently low.

V.9.c Probabilistic Approach

The probabilistic approach consists of the direct computation of the probability that the structure reaches a limit state. Thus, from a theoretical point of view, we can accept that either the actions and the resistance properties of the structure (strength) may be represented by parameters A and R which are described by probabilistic density curves $f_A(A)$ and $f_R(R)$, as represented in Fig. 57.

The probability of the simultaneous occurrence of a value of the action A within the interval dA and of a value of the strength R within the interval dR (dA and dR are infinitesimal quantities) is given by

$$d\left(dP_f\right) = f_A\, dA\, f_R\, dR\ .$$

The probability of strength R taking a lower value than action A, or, more precisely, of having the action in the interval dA and inferior values of the

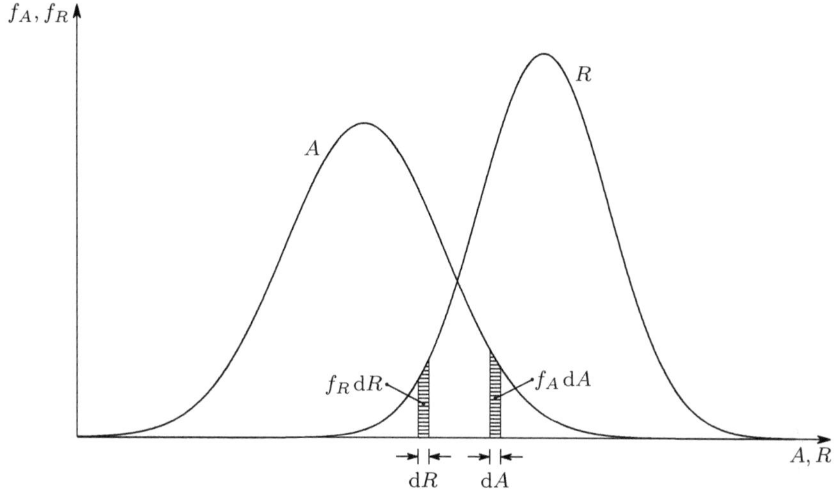

Fig. 57. Probabilistic density curves for action A and for strength R

strength R, may be obtained by integrating the previous expression for all values of $R < A$, which yields

$$dP_f = f_A \, dA \int_{R=0}^{A} f_R \, dR \ .$$

By integrating this expression for all possible values of the action, we get

$$P_f = \int_{A=0}^{\infty} dP_f = \int_{A=0}^{\infty} f_A \int_{R=0}^{A} f_R \, dR \, dA \ .$$

This expression represents the probability of the action exceeding the strength, that is, the probability of the structure reaching a limit state.

Using this methodology to verify the reliability of a structure is, however, not easy in practice, since it is generally very difficult and laborious to define and combine the multiple laws of probabilistic distribution for actions and strength parameters for a particular structure. In the practical verification of the reliability of ordinary structures, therefore, a *semi-probabilistic approach* is used instead, as described in the next sub-section.

V.9.d Semi-Probabilistic Approach

The semi-probabilistic approach is based on the definition, with a probabilistic basis, of nominal values for the actions and strength parameters, so that a sufficiently low probability of failure is guaranteed, without the need for the explicit computation of this probability.

The first step consists of defining *characteristic values* for the actions and strength parameters.

For actions these values are defined as values with a very low probability of being exceeded (upper quantile of the probabilistic density curve), except in the case of permanent actions with an advantageous effect on the safety of the structure. In the latter case the lower quantiles are used (values which are exceeded with a very high probability). However, in the case of permanent actions with low dispersion values, i.e., actions with close upper and lower quantiles, as the self-weight of structural materials, the mean value may be used. This considerably simplifies reliability verification, since it is not necessary to distinguish between permanent actions with beneficial and adverse effects on the structural safety.

For the strength parameters only the inferior quantiles of the probabilistic density curve are used as a rule (values which are exceeded with a very high probability). However, cases may be imagined where the failure of a structural element might be beneficial for the global failure safety. For these elements the use of the upper quantiles would be the logical choice.

Common values for the probabilities corresponding to the lower and upper quantiles are 5% and 95%, respectively. Generally, the values corresponding to these quantiles are defined in the official standards relating to the different types of structures and structural materials.

The second step consists of defining *nominal values* which are obtained from the characteristic values by multiplying them by *partial factors*.

In the case of actions, these factors take into account the probability of the characteristic values being exceeded, the reduced probability of all the actions present in a given loading case simultaneously reaching their characteristic values, the probability that the distribution of external forces resulting from a particular action (wind, for example) may be different from the assumed distribution, etc.

In the case of the strength parameters, the partial factors are intended to cover the reduction of the material's strength due to accidental material defects, the reduction of the strength parameters with the time (aging), small time-dependent deformations, the simplifying hypotheses used in the definition of the material's constitutive law, etc.

V.9.e Safety Stresses

Traditionally, the structural safety, used to be verified on the basis of *safety stresses*, especially in the fields of Civil and Mechanical Engineering. This method has been gradually abandoned and replaced by the semi-probabilistic approach. However, a short description of it is entirely justified, both for historical reasons and because it is still used.

The safety stress method has the same probabilistic basis as the semi-probabilistic approach, since the safety stresses are defined on the basis of the same characteristic values for the material strength parameters. The safety verification is performed by computing the stresses induced by the same characteristic values of the actions, which must not exceed the characteristic value of the strength (the yield stress in ductile metals), multiplied by a *safety coefficient*. This parameter plays the same role as the partial factors for the actions and strength parameters, simultaneously, in the semi-probabilistic approach.

Obviously, this method leads to the same degree of safety as the semi-probabilistic approach only if the structure behaves linearly until it reaches the limit states: in this case, multiplying the actions by a factor leads to the same result as dividing the allowable stress by the same factor. However, as seen earlier, a linear stress-strain relation is generally only acceptable in the initial loading phase and not until the limit states. For this reason, the results yielded by the semi-probabilistic approach are generally better, since the increase in safety due to multiplying the actions by a factor, whose objective is to guarantee that the structure resists a larger loading than that expected, is not "distorted" by the non-linear character of the relation between the external forces and the stresses in a close to the limit state loading situation.

Furthermore, treating the different actions separately allows different factors to be considered for each one, in accordance with the degree of uncertainty associated with it. As an example, let us consider two actions: the self-weight of the structure of a building and the wind acting on it. The uncertainty associated with the self-weight is very low, since, once the characteristic value of the density of the material has been defined, the computation of the corresponding internal forces is not affected by significant uncertainties. In the case of the wind action, on the other hand, the probabilistic analysis leading to the characteristic value takes only the statistical dispersion of the wind velocity into account, with a large degree of uncertainty remaining in relation to the distribution of pressures caused by a wind with that velocity on the surface of the building. This is why it is advisable to use a larger partial factor for the wind than for the self-weight.

V.10 Slender Members

V.10.a Introduction

As mentioned in Sect. V.1, the relations between external forces, stresses, strains and displacements are generally complex. The degree of complexity depends on two components: the rheological behaviour of the structural material and the geometry of the structure. For this reason, analytical solutions for these relations are only possible if both components have particularly simple forms, such as isotropy and linearity of the constitutive law, and a geometry (and loading) with a simple description in a given reference system (rectangular, spherical, cylindrical or polar coordinates). If the structure, or structural component, does not obey these conditions, the solution must be obtained numerically using, for example, the finite element method.

Slender members are an exception, since in these structural elements it is possible to find relatively simple analytical relations between the internal forces which act symmetrically in relation to the cross-sections (constant axial force and bending moment) and the corresponding stresses, especially if the stress-strain relation is linear. For the shear force and torsional moment this relation is not so simple, except for particular cross-section shapes: thin-walled sections under shear force and closed thin-walled and circular sections under torsion.

In Chaps. VI to VIII and X, these cases, where the solution may be considered as exact, are analysed and approximate solutions for other cases are indicated: cross-sections with a symmetry axis under shear force and rectangular and open thin-walled cross-sections under torsion.

Slender members are very often used as structural elements in the fields of Civil Engineering (structures in buildings and bridges contain very often slender members), Mechanical Engineering (many machine parts may be analysed as slender members) and Aeronautical Engineering (the wings of gliders and low speed airplanes, for example, may be considered as slender members). Structures made of slender members are called *framed structures*.

V.10.b Definition of Slender Member

A slender member is a bar-shaped body, i.e., a three-dimensional body, in which one dimension (the length of the bar) is considerably larger than any of the other two (at least five times). More precisely, a slender member may be understood as a solid body generated by a plane geometrical figure, when it moves along a straight line (or a curved line with a large curvature radius, in comparison to the dimensions of the figure), remaining perpendicular to that line. The shape and dimensions of the plane figure – which represents the cross-section of the slender member – may change during that motion, but only gradually. The discontinuities corresponding to a sudden change of the cross-section, or to a corner of the longitudinal line may be regarded as a link between two slender members. The theory of slender members is not valid in the region around these singularities.

Only prismatic bars are considered in the development of the theory of slender members. The generalization of the theory to curved bars or to members with a variable cross-section is only possible for bars with a large curvature radius (as compared with the dimensions of the cross-section), or for a gradually varying cross-section (cf. Sects. VI.7, VII.8 and VIII.5).

V.10.c Conservation of Plane Sections

The cross-sections of prismatic bars under the action of a constant axial force and a constant bending moment remain plane and perpendicular to the axis of the bar during the deformation. In order to demonstrate this statement, let us consider a piece of the bar, whose ends are sufficiently far from the ends of the bar and from the sections where the external forces are applied for Saint-Venant's principle to be valid (Fig. 58).

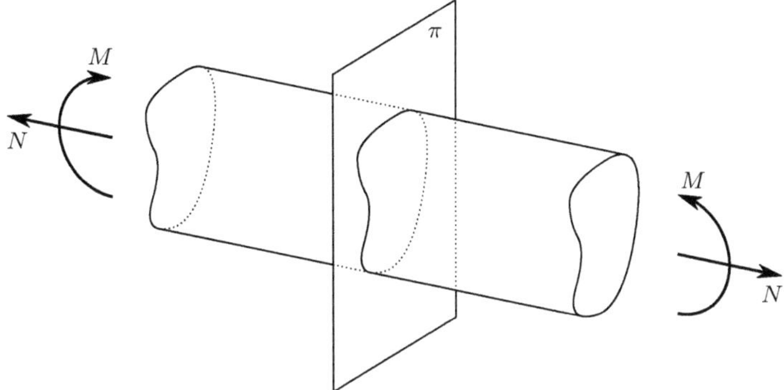

Fig. 58. Symmetrical internal force resultants in a piece of a prismatic bar

Plane π (Fig. 58) is a symmetry plane, both in relation to the geometry of the piece, and in relation to the applied forces (N and M). If, in addition, the material is isotropic, or has, at least, a rheological plane of symmetry parallel to plane π, then the problem is completely symmetric in relation to that plane.

Using the symmetry principle, we conclude that the deformation of the piece must also be symmetric in relation to plane π. Thus, the points of the piece which are on the plane π, will remain there after the deformation, which means, that this cross-section remains plane. The symmetry of the deformation also leads to the conclusion that, in an infinitesimal neighbourhood of the plane, the bar axis (and any longitudinal axis parallel to it) remains perpendicular to plane π. Furthermore, by choosing the piece of prismatic bar properly, any cross-section may be considered as the middle section of a piece. Thus, the above conclusions are valid for any cross-section which is sufficiently far from the above-mentioned singularities. Therefore, the following conclusion may be drawn: *in a prismatic bar under constant axial force and constant bending moment, the cross-sections remain plane and perpendicular to the axis of the bar during the deformation.*

This statement was formulated as an hypothesis by J. Bernoulli in 1705, [3], for the case of bending and is still known by his name in the literature on Strength of Materials. With the demonstration above, it may be considered as a law, which *is valid independently of any considerations about the material properties, with the exception of the symmetry considerations. It is also independent of the size of the deformation.* However, it not valid for non-symmetrical internal force resultants, such as the shear force, torsional moment and varying axial force and bending moment.

VI

Axially Loaded Members

VI.1 Introduction

A slender member which has a straight axis in the undeformed configuration, is said to be under purely axial loading if that axis remains a straight line after deformation, which may be caused by a constant axial force or other symmetrical actions, such as a uniform temperature variation. According to this definition and to the law of conservation of plane sections, in a prismatic bar under purely axial loading, any two cross-sections remain parallel after the deformation, i.e., only the distance between them varies.

Considering, in a prismatic bar under purely axial loading, two cross-sections at a distance l from each other in the undeformed bar, this distance will change to a value l' after the deformation and the strain defined by the variation of that distance is given by the expression

$$\varepsilon = \frac{l' - l}{l} \, . \tag{113}$$

This strain is constant in the cross-section. Therefore, if the bar is homogeneous, i.e., if it is made of a material with the same rheological properties in the whole member, the stress σ will also be constant. The position of the resultant of the system of forces defined by the stresses acting in the cross-section may be obtained by computing the moment of the stresses in relation to any axis of the cross-section's plane, which must be equal to the moment of the resultant in relation to the same axis (Fig. 59). The moment of the force acting in the infinitesimal area $\mathrm{d}\Omega$, $\mathrm{d}N = \sigma \, \mathrm{d}\Omega$, is $\mathrm{d}M = \mathrm{d}N x = \sigma \, \mathrm{d}\Omega x$. Integrating this expression to the whole area Ω of the cross-section gives the moment of the stresses. This moment must be equal to the moment of the resultant axial force $N = \sigma \, \Omega$, which takes the value $N d = \sigma \, \Omega d$. The distance d of the resultant to the reference axis r (Fig. 59) is then

$$\int_\Omega \sigma x \, \mathrm{d}\Omega = \sigma \Omega d \;\Rightarrow\; d = \frac{\int_\Omega x \, \mathrm{d}\Omega}{\Omega} \, . \tag{114}$$

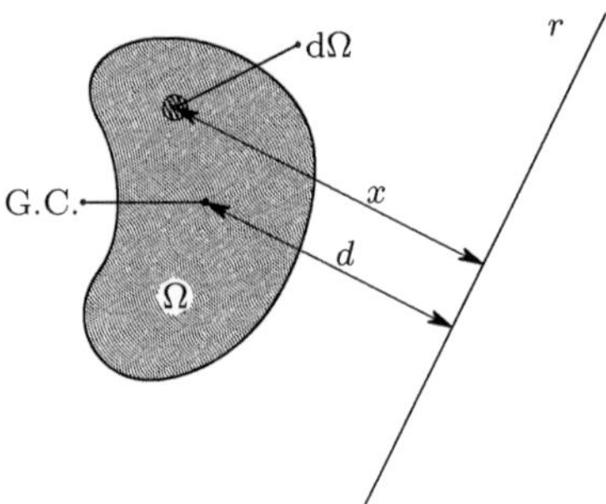

Fig. 59. Determination of the position of the stress resultant in the cross section of a slender member under purely axial loading

Equation (114) is also the expression used to compute the position of the center of gravity or centroid of a plane area Ω. Thus, we may conclude that *in a homogeneous prismatic bar under purely axial loading the line of action of the resultant of the applied forces contains the centroid of the cross-section.*

VI.2 Dimensioning of Members Under Axial Loading

If the axial force is tensile, the cross-section area Ω must be given dimensions which lead to a nominal value of the acting stress σ_{Ed} that is smaller than the nominal value of the material's resisting stress (allowable stress) σ_{all}. This condition may be expressed by the inequality

$$\sigma_{Ed} = \frac{N_{Ed}}{\Omega} \leq \sigma_{\text{all}} \ \Rightarrow \ \Omega \geq \frac{N_{Ed}}{\sigma_{\text{all}}} \ , \tag{115}$$

where N_{Ed} represents the nominal value of the axial force. In the following exposition the indices *"Ed"* of the acting forces and stresses are omitted, i.e., the values of internal or external forces and stresses written without indices are nominal values.

In compressed members (115) is a necessary but not a sufficient condition, since the phenomenon of *buckling* may occur. This kind of problems is analysed in Chap. XI. In this chapter we consider that the stability of the compressed member is guaranteed.

VI.3 Axial Deformations

As seen above, in a bar under pure axial loading the stress state may be considered as uniform, irrespective of how the forces are applied, provided that the material points under consideration are not close to the region of the member where the forces are applied (Saint-Venant's principle). In axially loaded slender members these regions are generally a small part of the member, so that a uniform distribution of the stresses may be accepted when the elongation of the bar is computed.

In the case of a material with linear elastic rheological behaviour the elongation Δl is given by the expression (cf. (113))

$$\Delta l = l' - l = \varepsilon l = \frac{\sigma}{E} l = \frac{Nl}{E\Omega} \, , \qquad (116)$$

where l and l' represent the length of the bar, before and after the deformation, respectively. The quantity $E\Omega$ represents the *axial stiffness*, since the larger this value, the smaller the member's deformation caused by the axial force.

If, in addition to the axial force a uniform temperature variation ΔT occurs, the total elongation may be computed by the expression

$$\Delta l = \left(\frac{N}{E\Omega} + \alpha \Delta T \right) l \, , \qquad (117)$$

where α represents the *coefficient of thermal expansion* of the material.

VI.4 Statically Indeterminate Structures

VI.4.a Introduction

Statically determinate structures are freely deformable, in the sense that their supports and internal connections do not restrict the deformations. This means that a small change in the geometry or size of the structural elements does not change the distribution of internal forces. For example, if the length of the left bar of the plane truss represented in Fig. 60-a increases Δl (e.g., due to a temperature increase), the truss adapts its geometry without the need of internal forces.

In the statically indeterminate structure represented in Fig. 60-b, however, there can be no change in the length of one of the bars without altering the

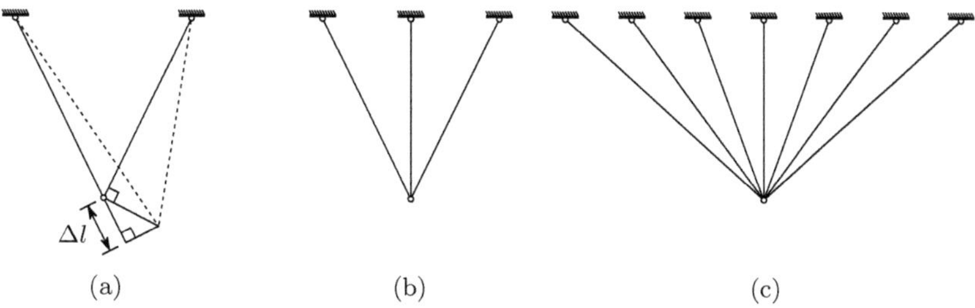

Fig. 60. Examples of structures with the same degree of kinematic indeterminacy and different degrees of statical indeterminacy

lengths of the other two: if, for example, the middle bar suffers a temperature increase, the three bars can only remain connected if the middle bar is compressed and the lateral bars are stretched, which requires tensile axial forces.

A statically determinate structure only has internal forces if external forces are applied, which means that they are insensitive to temperature change, material retraction, or any other actions that alter the dimensions of structural elements. Statically indeterminate structures, on the other hand, may have internal forces in the absence of external forces, as seen in the example above. Nevertheless, in these structures the release of a connection (for example, by the yielding or rupture of one bar) does not necessarily imply structural collapse, as statically indeterminate structures have a number of connections, which is greater than the minimum necessary to guarantee the static equilibrium.

VI.4.b Computation of Internal Forces

Since in statically indeterminate structures the internal forces are not independent of the deformation of structural elements, *conditions of compatibility of the deformations* must be taken in account in order to compute the internal forces. There are two main general methods to establish these conditions:

– *Direct*, by first releasing a number of connections equal to the degree of statical indeterminacy, and then computing the displacements which appear in the released connections (these displacements are zero in the real structure) and the forces necessary to eliminate these displacements. Taking as an example the truss represented in Fig. 60-b, the vertical restriction of the middle support, for example, may be released. The vertical displacement of the upper end of the middle bar (caused by external forces, temperature, etc.) is then computed. The force needed to cause a displacement in the opposite direction, i.e., to bring the upper end of the middle bar back to the initial position, corresponds to the vertical reaction force of the middle support. This force is the *hyperstatic unknown* or *redundant force*. Once this force is computed, the remaining reaction and internal forces may be obtained by means of static equilibrium considerations. This method is known as *the force method*, since the unknowns are the forces acting on the connections which were released in the first computation step.
– *Indirect*, taking as unknowns the displacements necessary to define fully the deformed configuration of the structure. These displacements – the *kinematic unknowns* – are computed by establishing the relations between them and the resultants of the internal forces in a deformed configuration. The conditions of equilibrium between these resultants and the external forces gives a system of equations, whose solution yields the unknown displacements. Once these displacements are known, all the internal forces may be computed. This method is known as *the displacement method*, as the

unknowns are displacements. The displacement method is more easily generalized to structures with non-linear behaviour than the force method.

The number of equations to be solved corresponds to the *degree of static indeterminacy* for the force method and to the *degree of kinematic indeterminacy* in the case of the displacement method. These two quantities are not related. For example, all the three trusses represented in Fig. 60 have a degree of kinematic indeterminacy of two (the two components of the displacement vector in the connection node of the bars) and degrees of static indeterminacy of zero, one and five, respectively. The detailed study of these two methods and their systematization belong to the scope of the Theory of Structures. Here they are only applied to simple structures in a general form.

VI.4.c Elasto-Plastic Analysis

The rheological behaviour of materials which display a well-defined yielding zone, such as mild steel, may be approximated by the idealized constitutive law represented in Fig. 61, provided that the the strain does not reach the hardening zone (cf. Fig. 47). As seen in Sect. IV.3, such a rheological behaviour is called elastic, perfectly plastic.

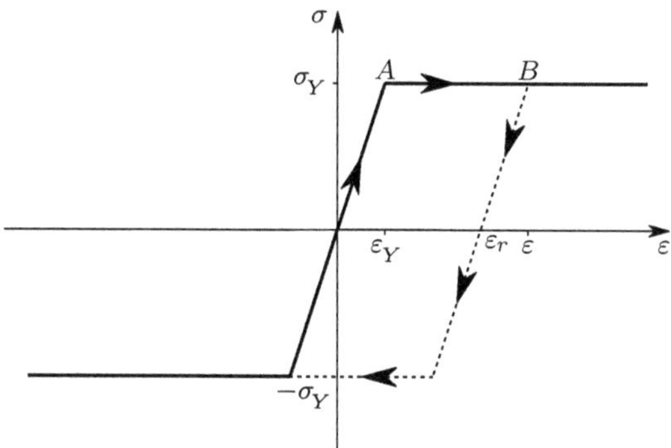

Fig. 61. Elastic, perfectly plastic rheological behaviour

The strain fraction corresponding to the yielding zone is generally much larger than the maximum elastic strain ε_Y. In mild steel, for example, yielding starts with a strain of approximately 0.1% and hardening starts with $\varepsilon \approx 1.5\%$. The yielding zone in this case is 14 times the maximum elastic strain. Thus, in a structure made of a ductile material with a yielding zone, we usually admit that in the structural elements where the yielding stress is first attained the stress keeps this value until collapse, unless there is a decrease in deformation, as defined by the constitutive law represented in Fig. 61.

Despite these simplifying assumptions, the computation of structures in the elasto-plastic range is substantially more complex than in the linear case.

The superposition principle is not valid any longer and the order of application of the external forces must be taken into account, as well as the different material behaviour when the strain changes from the elastic to the plastic range or vice versa.

To illustrate these considerations the elasto-plastic behaviour of the hyperstatic truss represented in Fig. 62 is described in detail. In order to keep the description as simple as possible, we assume that the three bars have equal cross-section areas Ω and that the material behaviour follows the simplest elasto-plastic constitutive law, as represented in Fig. 61. The displacement method is used for the analysis.

The conditions of static equilibrium are valid in any range (elastic, elasto-plastic or plastic). As a consequence of the symmetry of geometry and loading, only the vertical condition of equilibrium needs to be considered, yielding the relation

$$P = N_1 + 2N_2 \cos 60° \Rightarrow P = N_1 + N_2 \ . \tag{118}$$

The deformation's conditions of compatibility, when expressed in terms of displacements and strains are also valid in any range. In the present case the degree of kinematic indeterminacy is one, since the deformed configuration of the structure is completely defined by the vertical displacement of the point of application of load P. Denoting this displacement by δ (Fig. 62) and considering that it is sufficiently small to be considered as infinitesimal, the relation between δ and the strains in the bars is given by (119) (the dashed lines represent the deformed configuration with largely exaggerated displacements)

$$\delta = \Delta l_1 = \frac{1}{\Delta l_2}\cos 60° \Rightarrow \varepsilon_1 l = \frac{\varepsilon_2 \frac{l}{\cos 60°}}{\cos 60°} \Rightarrow \varepsilon_1 = 4\varepsilon_2 \ , \tag{119}$$

since, in the case of infinitesimal rotations, the arcs of circumference used to draw the undeformed length of bars 2 on the dashed lines may be substituted by the normals to those lines (Fig. 62). ε_1 and ε_2 represent the strains in bars 1 and 2, respectively.

If the deformations are caused only by the axial force, i.e., in absence of plastic deformations, temperature variation, residual strains, etc., the condition of compatibility may be expressed in terms of the axial forces, N_1 and

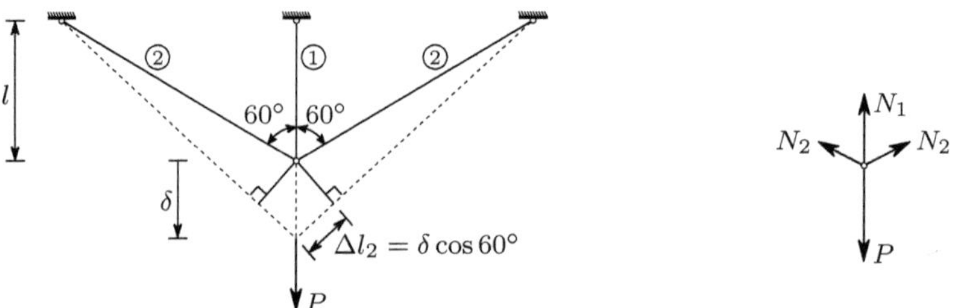

Fig. 62. Conditions of equilibrium and compatibility

N_2, by the equation

$$\delta = \frac{N_1 l}{E\Omega} = 4\frac{N_2 l}{E\Omega} \Rightarrow N_1 = 4N_2 \,, \tag{120}$$

Next the internal forces and the displacement δ in the different loading stages are analysed.

- *Elastic phase.* In this stage the axial forces may be computed by solving the system of two equations described by (118) and (120), yielding

$$\begin{cases} N_1 = \dfrac{4}{5}P \\[2ex] N_2 = \dfrac{1}{5}P \end{cases} \Rightarrow \delta = \frac{4}{5}\frac{l}{E\Omega}P \Rightarrow K_1 = \frac{\partial P}{\partial \delta} = \frac{5}{4}\frac{E\Omega}{l} \,, \tag{121}$$

where K_1 represents the structural stiffness corresponding to the displacement δ.

Since the strain ε_1 is always superior to the strain in bar 2 (119), bar 1 reaches the yielding strain at first. The values of the loading P and displacement δ corresponding to the yielding of bar 1 may be obtained from (121) and (120), taking the values

$$\sigma = \sigma_Y \Rightarrow N_1 = N_{1Y} = \Omega\sigma_Y$$
$$\Rightarrow P = P_1 = \frac{5}{4}\Omega\sigma_Y \Rightarrow \delta = \delta_1 = \frac{\sigma_Y l}{E} \,. \tag{122}$$

- *Elasto-plastic phase.* When the load P exceeds the value P_1, the axial force in bar 1 remains constant, with the value N_{1Y}, since the strain is in the yielding zone. As this internal force is known, the structure becomes statically determinate. Thus, the axial force N_2 may be obtained directly from the equilibrium condition (118), yielding

$$P = N_{1Y} + N_2 \Rightarrow N_2 = P - \Omega\sigma_Y \quad \text{with} \quad P > P_1 \,. \tag{123}$$

Taking in consideration that the lateral bars are still in the elastic range, the first of (120) may be used to compute the displacement and stiffness in the elasto-plastic phase, yielding

$$\delta = \frac{4N_2 l}{E\Omega} = (P - \Omega\sigma_Y)\frac{4l}{E\Omega} \Rightarrow K_2 = \frac{\partial P}{\partial \delta} = \frac{1}{4}\frac{E\Omega}{l} = \frac{1}{5}K_1 \,. \tag{124}$$

The structure collapses (yielding of the three bars) when the lateral bars reach the yielding strain. From (123), (122) and (124) we get

$$N_2 = N_{2Y} = \Omega\sigma_Y \Rightarrow \begin{cases} P = P_Y = 2\Omega\sigma_Y = \dfrac{8}{5}\dfrac{5}{4}\Omega\sigma_Y = \dfrac{8}{5}P_1 \\[2ex] \delta = \delta_Y = 4\dfrac{\sigma_Y l}{E} = 4\delta_1 \,. \end{cases} \tag{125}$$

At the moment of collapse the strain in bar 1 is therefore four times the yielding strain of the material, $\varepsilon_Y = \frac{\sigma_Y}{E}$, and the corresponding load P_Y is $\frac{8}{5}$ times the maximum load in in elastic range. Assuming the structure is made of mild steel, bar 1 is still at the beginning of the yielding zone when the structure collapses.

If the structure is unloaded in the elasto-plastic phase, the middle bar, which has already suffered some plastic deformation, has an elastic behaviour in the unloading, but has a residual strain $\varepsilon_{1r} = \varepsilon_{1\mathrm{max}} - \frac{\sigma_Y}{E}$ ($\varepsilon_{1\mathrm{max}}$ is the maximum value of the strain in the loading phase), as represented in Fig. 61 (dashed line). Let us now analyse the behaviour of the structure, as a function of this residual strain. The strain in bar 1 is now given by

$$\varepsilon_1 = \frac{N_1}{E\Omega} + \varepsilon_{1r} \ .$$

From this relation and the compatibility condition (119), we get

$$\varepsilon_1 = 4\varepsilon_2 \ \Rightarrow \ \frac{N_1}{E\Omega} + \varepsilon_{1r} = 4\frac{N_2}{E\Omega} \ \Rightarrow \ N_1 = 4N_2 - E\Omega\varepsilon_{1r} \ .$$

This expression and the equilibrium condition (118) define a system of equations which allows the computation of the axial forces N_1 and N_2, yielding

$$\begin{cases} N_1 = \dfrac{4}{5}P - \dfrac{1}{5}E\Omega\varepsilon_{1r} \\[2mm] N_2 = \dfrac{1}{5}P + \dfrac{1}{5}E\Omega\varepsilon_{1r} \end{cases} \Rightarrow \ \begin{vmatrix} \delta = 4\varepsilon_2 l = 4\dfrac{N_2}{E\Omega}l = \dfrac{4}{5}\dfrac{l}{E\Omega}P + \dfrac{4}{5}l\varepsilon_{1r} \\[2mm] \Rightarrow \ K_3 = \dfrac{\partial P}{\partial \delta} = \dfrac{5}{4}\dfrac{E\Omega}{l} = K_1 \ . \end{vmatrix} \tag{126}$$

From this expression we verify that, in the unloading, the stiffness of the structure is equal to the stiffness in the initial elastic phase. This is a consequence of having all bars in the elastic phase.

Equation (126) remain valid while the middle bar has the residual strain ε_{1r} and the lateral bars are in the elastic phase. This situation changes only if P exceeds the value which caused the the residual strain ε_{1r} in the loading phase ($P = P_{\mathrm{max}}$, Fig. 63), or if bar 1 attains the compressive yielding stress ($\sigma = -\sigma_Y$, Fig. 61). The value of P corresponding to the last situation may be computed by means of the axial force in the middle bar, expressed as a function of the maximum value of P in the loading phase P_{max} ($P_1 < P_{\mathrm{max}} < P_Y$). From (124) we get

$$\varepsilon_{1\mathrm{max}} = \frac{\delta_{\mathrm{max}}}{l} = 4\frac{P_{\mathrm{max}} - \Omega\sigma_Y}{E\Omega} \ \Rightarrow \ \varepsilon_{1r} = \varepsilon_{1\mathrm{max}} - \frac{\sigma_Y}{E} = 4\frac{P_{\mathrm{max}}}{E\Omega} - 5\frac{\sigma_Y}{E} \ . \tag{127}$$

Substituting this value of ε_{1r} in the first of (126), we get

$$N_1 = \Omega\sigma_Y - \frac{4}{5}(P_{\mathrm{max}} - P) \ . \tag{128}$$

Bar 1 attains the compressive yielding stress, when the axial force N_1 reaches the value $N_1 = -\Omega\sigma_Y$. From this condition and from (128) we get, considering the value of P_1 given by (122),

$$N_1 = -\Omega\sigma_Y \Rightarrow P = P_4 = P_{\max} - \frac{5}{2}\Omega\sigma_Y = P_{\max} - 2P_1 \ . \tag{129}$$

If the maximum value of P in the loading phase is P_1, the middle bar yields in compression for $P_4 = -P_1$, as might be expected, since no residual deformations were caused in the loading phase and the material has the same behaviour for tensile and compressive stresses. However, if a tensile plastic deformation takes place in bar 1, which happens in the above-described elasto-plastic phase ($P_1 < P_{\max} < P_Y$), there is a residual elongation of bar 1. As a consequence, the structural behaviour for the positive and negative values of P becomes different ($P_4 \neq -P_1$).

The behaviour of the structure in the different loading stages analysed above may be summarized by the force-displacement (P-δ) diagram presented in Fig. 63. In this diagram the line $OABC$ represents the load-displacement relation in a first loading. Line $\overline{OA}$ represents the elastic phase, line $\overline{AB}$ the elasto-plastic phase (the middle bar is in the yielding zone and the lateral bars are still in linear-elastic regime) and line $\overline{BC}$ the plastic stage (structural yielding or collapse). Line $OGHI$ represents the structural behaviour in a first loading with negative values of the force P.

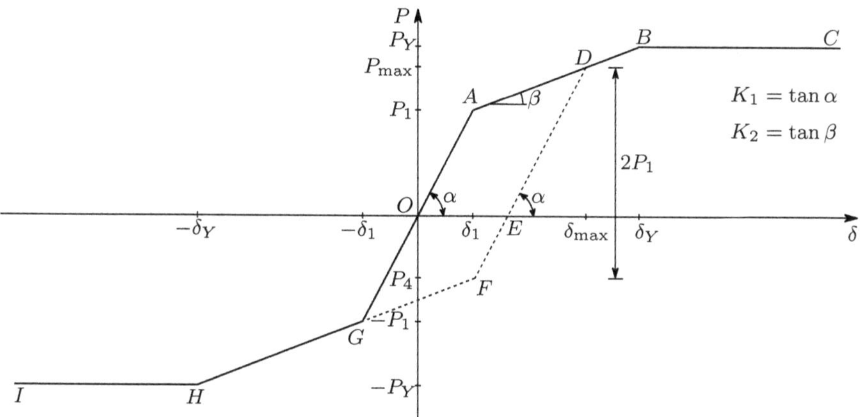

Fig. 63. Load-displacement diagram of the truss represented in Fig. 62

The structural behavior in the unloading from a load $P_{\max}$ in the elasto-plastic phase is described by the line $\overline{DE}$ which is parallel to line $\overline{OA}$, as seen above (same structural stiffness: $K_3 = K_1$, (126)). If, after unloading from $P_{\max}$ (point E), a negative load is applied (reversed force P), the structure behaves elastically until point F, which represents the yielding of the middle bar in compression (129). Yielding of the lateral bars under compression takes place for a displacement $\delta = -\delta_Y$ (point H), since the load $P_{\max}$ did not cause yielding of these bars, which means that its behaviour is the same as in a first compressive loading (line $OGHI$).

In the reloading from any point of line FD the structure follows the line $EDBC$, since at point D the structure re-enters in the elasto-plastic phase, with a strain in the middle bar of $\varepsilon_1 = \varepsilon_{1r} + \frac{\sigma_Y}{E}$ ((127) and Fig. 61, point B).

In comparison with the first loading, the behaviour of the structure in the reloading, which is represented by lines $EDBC$ and $EFGHI$ for positive and negative values of P, respectively, shows an increase of the elastic loading capacity (hardening: $P_{\max} > P_1$) for positive values of P (line ED) and softening (reduction of the elastic loading capacity: $|P_4| < |P_1|$) for negative values (line EF).

The hardening for positive loads is accompanied by a reduction of the deformation capacity. Conversely, for negative P-values, softening and increase of the deformation capacity took place (horizontal projection of $EFGHI$).

There is an analogy between the behaviour of this structure and the strain hardening process of steel described in Sect. V.5. In fact, in both cases the pre-deformation introduced by tensile forces increases the tensile elasticity limit and the deformation capacity in compression and vice versa for a compressive pre-deformation. The structure represented in Fig. 62 may therefore be seen as a *physical model* for the strain hardening process of ductile materials. Figure 64 gives two other models. The second one includes a yielding zone. In this figure the numbered corners in the diagrams represent a starting yielding of the corresponding bars. In unloading followed by a reloading, model a follows the dashed line which is parallel to line $\overline{01}$. Model b also has an elastic behaviour in the unloading, which, however, is more complex than in the case of model a. The analysis of the unloading in model b is left as an exercise for the reader.

VI.5 An Introduction to the Prestressing Technique

Let us consider now that the truss represented in Fig. 62 is made of a brittle material with linear elastic behaviour until rupture, which occurs when the stress attains the value σ_r. In order to make the comparisons easier, let us consider $\sigma_r = \sigma_Y$. Expressions 121 are still valid for representing the load-displacement relation, since, in the elastic phase, the only material parameter needed is the elasticity modulus E. However, when P reaches the value which causes the rupture of the middle bar, $P = P_r = \frac{5}{4}\Omega\sigma_r = \frac{5}{4}N_r$ ((122); N_r is the rupture load of a bar), the structure collapses totally. In fact, the loading capacity of the lateral bars alone is not sufficient to sustain the load P_r, as can be easily verified by the condition of equilibrium

$$P = 2N_2 \cos 60° \;\Rightarrow\; N_2 = P_r = \frac{5}{4}N_r > N_r \,.$$

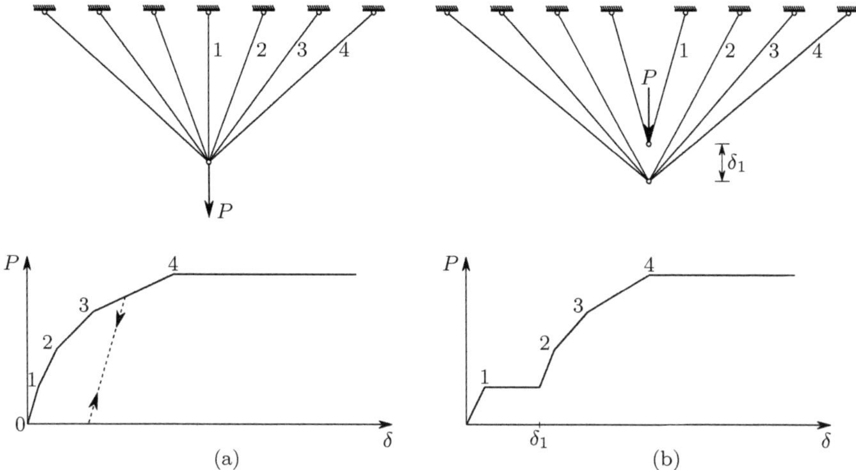

Fig. 64. Physical models for the behaviour of a ductile material with hardening (**a**) without yielding zone; (**b**) with yielding zone

This example shows that the strength reserves of a ductile structure, in relation to the load causing the first plastic deformations ($P_Y > P_1$ in the ductile truss, Fig. 63), may not exist if the structure is made of a brittle material.

However, the loading capacity of the brittle structure may also be increased, by introducing *prestressing* residual forces. In order to introduce this technique, let us assume that the undeformed middle bar does not have exactly the length l, but is slightly longer, with a length $l + l_r$. Under these conditions the three bars can only be connected, if the middle bar is compressed, which introduces tensile axial forces in the lateral bars. This means that the structure will have residual internal forces, i.e., the internal forces will not be zero, when the external forces vanish. In the analysis of the structure l_r may be treated as a residual elongation. Thus, the structural behaviour may be defined by (126), if ε_{1r} is substituted by $\frac{l_r}{l}$, yielding

$$
\begin{cases}
N_1 = \dfrac{4}{5}P - \dfrac{1}{5}\dfrac{E\Omega l_r}{l} \\[2mm]
N_2 = \dfrac{1}{5}P + \dfrac{1}{5}\dfrac{E\Omega l_r}{l}
\end{cases}
\Rightarrow \delta = \frac{4}{5}\frac{l}{E\Omega}P + \frac{4}{5}l_r \Rightarrow K = \frac{\partial P}{\partial \delta} = \frac{5}{4}\frac{E\Omega}{l} \, . \tag{130}
$$

The maximum loading capacity of the structure is achieved if the three bars reach the rupture stress simultaneously, as may be easily concluded from the equilibrium condition (118). This means that, at the collapse, we have $N_1 = N_2 = \Omega\sigma_r$. Introducing this condition in the first of (130), we get

$$\begin{cases} \dfrac{4}{5}P - \dfrac{1}{5}\dfrac{E\Omega l_r}{l} = \Omega\sigma_r \\[2mm] \dfrac{1}{5}P + \dfrac{1}{5}\dfrac{E\Omega l_r}{l} = \Omega\sigma_r \end{cases} \Rightarrow \begin{cases} P = 2\Omega\sigma_r \\[2mm] l_r = \dfrac{3l\sigma_r}{E} \end{cases}. \qquad (131)$$

From this result we conclude that the optimal prestressing is obtained by a residual elongation $l_r = \frac{3l\sigma_r}{E}$, which increases the loading capacity of the structure from $\frac{5}{4}\Omega\sigma_r$ to $2\Omega\sigma_r$. The displacement δ just before the collapse, taking as reference configuration the structure without prestressing, i.e., when the vertical distance between the connection node and the middle support is exactly l, may be obtained from the elongation of the lateral bars, since they have a zero axial force for $\delta = 0$ (reference configuration). Thus, we obtain from (119)

$$\varepsilon_2 = \frac{\sigma_r}{E} \Rightarrow \delta = 4\varepsilon_2 l = 4\frac{\sigma_r l}{E}.$$

We conclude that, if the brittle and ductile structures have the same elasticity modulus E and ultimate stress $\sigma_r = \sigma_Y$, the deformation of the brittle structure at the rupture takes the same value as the deformation of the ductile structure when yielding starts (δ_Y, (125)).

The initial internal forces caused by the prestressing may be obtained from (130), by taking $P = 0$ and $l_r = 3l\frac{\sigma_r}{E}$, yielding

$$N_1 = -N_2 = \frac{3}{5}\Omega\sigma_r = \frac{3}{5}N_r.$$

In this case the initial axial forces are lower than the loading capacity of the bars. However, it very often happens that *minimum loads* are required in prestressed structures, in order to prevent the initial loads causing failure of structural elements (cf. Example VI.10).

These two examples (ductile and brittle structure of Fig. 62) illustrate the fundamental differences in the behaviour of structures made of ductile and brittle materials:

- In ductile structures, a *redistribution of the internal forces* takes place automatically, which alters the relations between the internal forces in the different elements and allows the most strained elements to keep their loading capacity until the structure collapses. In the example, the middle bar keeps its loading capacity until the lateral bars reach the yielding axial force.[1]
- In brittle structures, the full loading capacity of the structural elements can only be used by means of prestressing residual forces. The prestressing technique must be implemented in a carefully controlled way, since small changes in the prestressing parameters (in the example parameter l_r), caused by temperature differences, creep, shrinkage, etc., may cause

---

[1]If the ductile structure is prestressed with $l_r = \frac{3l\sigma_r}{E}$ (131) the three bars reach the yielding strain simultaneously.

substantial changes in the loading capacity of the structure. Furthermore, the increase of loading capacity only takes place for the loading case under consideration: in the example, the prestressing considered causes a decrease in the compressive loading capacity of the structure. The above-mentioned minimum loads may also be a drawback of prestressed structures.

From the above considerations we may conclude that ductile structures are safer, since their capacity for internal force redistribution makes them less sensitive to imperfections and construction errors, and also because their failure does not happen unannounced, since it is usually preceded by large plastic deformations.

VI.6 Composite Members

VI.6.a Introduction

In prismatic bars made of two or more materials, with a constant cross-section, the law of conservation of plane sections remains valid, since the symmetry conditions in relation to the cross-section's plane still hold (Fig. 58). As a consequence, in the case of purely axial loading, the strain is constant in the cross-section. However, as the material is not the same throughout the cross-section, the stresses will not be constant and will depend on the rheological properties of the materials. We therefore have a statically indeterminate problem, which means that we may have internal forces without external loads, caused, for example, by a temperature variation. The degree of static indeterminacy is equal to the number of materials minus one. The forces in the connections between the different materials may be considered as the hyperstatic unknowns. The degree of kinematic indeterminacy is one, regardless of the number of materials. The strain in the cross-section may be taken as the kinematic unknown, since, once it is known, the stresses in all materials may be computed directly.

In the present analysis the displacement method will be used. For the sake of simplicity, members made of two linear elastic materials are considered. The generalization of the analysis to any number of materials is straightforward (cf. Example VI.14)

VI.6.b Position of the Stress Resultant

In a prismatic bar under purely axial loading the strain ε is constant in the cross-section, as shown above. The stresses in materials a and b are then $\sigma_a = E_a \varepsilon$ and $\sigma_b = E_b \varepsilon$,[2] where E_a and E_b represent the Young's moduli of materials a and b, respectively. By using the same line of reasoning as

---

[2] If the Poisson's coefficients in the two materials are different, the axial loading will generally cause stresses in longitudinal facets, since the transversal deforma-

for homogeneous members (Fig. 59) and denoting by Ω_a and Ω_b the areas occupied by the two materials in the cross-section, we get

$$\overbrace{\int_{\Omega_a} \sigma_a x\,\mathrm{d}\Omega_a + \int_{\Omega_b} \sigma_b x\,\mathrm{d}\Omega_b}^{\text{moment of the stresses}} = \overbrace{(\sigma_a\Omega_a + \sigma_b\Omega_b)d}^{\text{moment of the axial force}}$$

$$\Rightarrow d = \frac{E_a \int_{\Omega_a} x\,\mathrm{d}\Omega_a + E_b \int_{\Omega_b} x\,\mathrm{d}\Omega_b}{E_a\Omega_a + E_b\Omega_b} \,. \tag{132}$$

Thus, the axial force will not cause bending, i.e., the axis of the bar will remain a straight line, if its line of action contains the point defined by two distances d, computed by means of (132), considering two non-parallel reference axes of the cross-section's plane. This expression also defines the position of the section's centroid, *if the area occupied by each material is weighted with its Young's modulus*. A fundamental difference between this expression and (114) is that the latter does not depend on the material behaviour, as opposed to this one, which *is only valid for materials with linear elastic behaviour*.

VI.6.c Stresses and Strains Caused by the Axial Force

As explained in Sect. VI.6.a, the determination of stresses in composite members is a statically indeterminate problem, since a condition of deformation compatibility must be taken into account. In the case of purely axial loading that condition is represented by the relation $\varepsilon_a = \varepsilon_b \Rightarrow \frac{\sigma_a}{E_a} = \frac{\sigma_b}{E_b}$, as seen in the previous section. The last of these relations and the equilibrium condition $N = \sigma_a\Omega_a + \sigma_b\Omega_b$ define a system of two equations, whose solution yields the stresses in the two materials

$$\left\{ \begin{array}{l} \sigma_a = \dfrac{N}{E_a\Omega_a + E_b\Omega_b} E_a \\[2mm] \sigma_b = \dfrac{N}{E_a\Omega_a + E_b\Omega_b} E_b \end{array} \right. \Rightarrow \left\{ \begin{array}{l} \sigma_a = \dfrac{N}{\Omega_a + m_a\Omega_b} \\[2mm] \sigma_b = \dfrac{N}{m_b\Omega_a + \Omega_b} \end{array} \right. \tag{133}$$

In the last expressions $m_a = \frac{E_b}{E_a}$ and $m_b = \frac{E_a}{E_b}$ are called *homogenizing coefficients*, which allow the computation of the stresses in almost the same way, as in the case of homogeneous sections. For example, the stress in material a may be obtained by considering the cross-section area obtained by multiplying the

---

tions of the two materials will be different, which means that additional conditions of compatibility would be necessary. The same happens in the case of a temperature variation, if the coefficients of thermal expansion of the two materials are different. However, it can be shown that these stresses are sufficiently small to be ignored, which allows the use of the constitutive law in this one-dimensional form. Section VII.6 gives an analysis of the error introduced by this approximation in the case of axial force and bending.

area of material b by the homogenizing coefficient of material b in material a, m_a.

The elongation of a bar of length l may be computed from the stress in any material, since the strain is constant, yielding

$$\varepsilon = \frac{\sigma_a}{E_a} = \frac{\sigma_b}{E_b} \Rightarrow \Delta l = \varepsilon l = \frac{\sigma_a}{E_a} l = \frac{\sigma_b}{E_b} l = \frac{Nl}{E_a\Omega_a + E_b\Omega_b} \,. \tag{134}$$

VI.6.d Effects of Temperature Variations

As shown in Sect. VI.6.a, a temperature variation introduces stresses in a prismatic bar made of two materials with different thermal expansion coefficients. Generally, even a uniform temperature variation causes bending, i.e., the axis of the bar does not remain a straight line, as in the example given in Fig. 65 (α_a and α_b are the coefficients of thermal expansion of materials a and b).

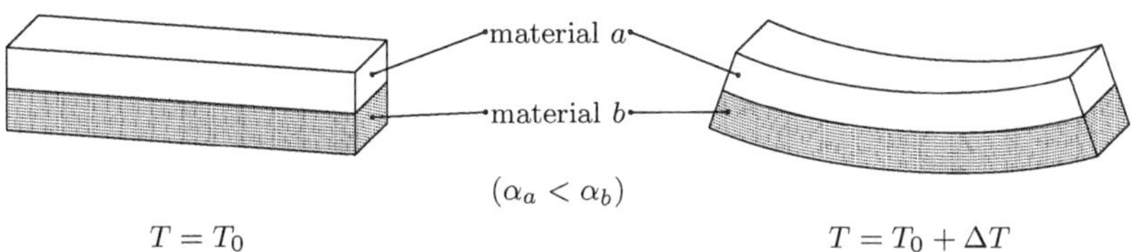

Fig. 65. Effect of a uniform temperature variation on a composite bar

The study of the stresses generated by that bending requires the bending theory which is explained in Chap. VII. However, if the cross section has two symmetry axes, i.e., if the prismatic bar has two longitudinal symmetry planes, its axis must remain a straight line, which means that the deformation is purely axial.[3]

In order to analyse the internal forces in the connection between the two materials, let us analyse the whole bar and not just a cross-section. The symmetry principle (Fig. 58) only permits the conclusion that the middle section of the bar remains plane in the deformation. Thus, the strain is constant in this section, which allows the computation of the stresses in the two materials, using the same equilibrium and compatibility conditions as in the case of the axial force, yielding

$$\begin{cases} \varepsilon = \dfrac{\sigma_a}{E_a} + \alpha_a\Delta T = \dfrac{\sigma_b}{E_b} + \alpha_b\Delta T \\[2ex] N = \sigma_a\Omega_a + \sigma_b\Omega_b = 0 \end{cases} \Rightarrow \begin{cases} \sigma_a = \dfrac{E_a E_b \Omega_a \Omega_b}{E_a\Omega_a + E_b\Omega_b}\Delta T\dfrac{\alpha_b - \alpha_a}{\Omega_a} \\[2ex] \sigma_b = \dfrac{E_a E_b \Omega_a \Omega_b}{E_a\Omega_a + E_b\Omega_b}\Delta T\dfrac{\alpha_a - \alpha_b}{\Omega_b} \,. \end{cases} \tag{135}$$

---

[3]The double symmetry of the cross-section is a sufficient condition for purely axial deformation. This condition may, however, not be necessary (see examples VI.20 and VI.21).

If an axial force is also acting, the total stresses may be computed by adding the stresses given by (133) and (135), since the superposition principle may be applied in the case of temperature variation (cf. Sect. V.8).

The cross-sections located near by the ends of the bar can not remain plane. In fact, if the end cross-sections remain plain, the stresses given by (135) act on them. This is not possible, since no external forces are applied. In order to clarify this issue, let us consider a rigid block connected to each end of the bar, as represented in Fig. 66-a. In this case, all cross-sections remain plain and (135) is valid for the whole bar.[4]

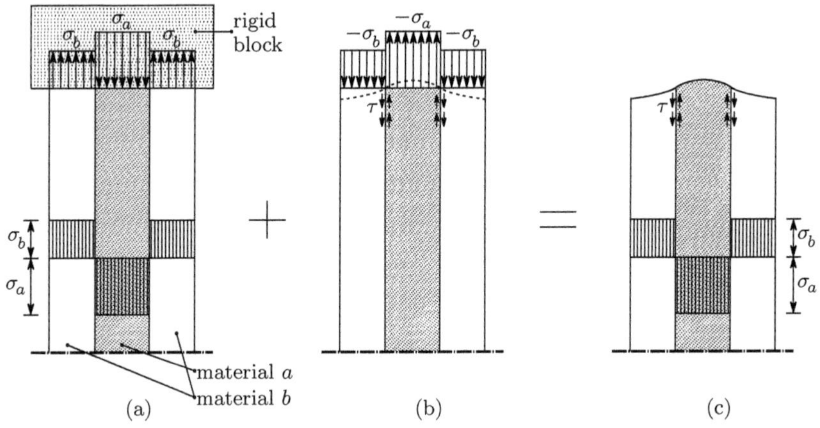

Fig. 66. Transmission of internal forces between the two materials in a composite member under a uniform temperature variation ($\Delta T > 0$ and $\alpha_a > \alpha_b$ or $\Delta T < 0$ and $\alpha_a < \alpha_b$)

The real stress distribution in the bar may be obtained by the superposition principle. To this end, let us superpose upon the situation corresponding to Fig. 66-a the loading situation illustrated in Fig. 66-b, where no temperature variation takes place and only end distributed loads act, corresponding to the stress distribution given by (135) with reversed direction. These stresses have a vanishing resultant ($N = 0$, cf. (135)). Thus, the Saint-Venant's principle leads to the conclusion that its effect is restricted to the region of the bar around its ends. The stress distribution in this region depends on the shape of the cross-section and cannot be obtained by means of the theory of slender members. However, a qualitative analysis of Fig. 66-b leads to the conclusion that shearing stresses τ must appear in the interface between the two materials. By superposing the stresses corresponding to the loading situa-

---

[4]This conclusion is easily arrived at by symmetry considerations: considering the part of the bar defined by the end and the central cross sections, we conclude that the middle section of this piece (the quarter length cross-section) remains plane. Further sub-division leads to the conclusion that every section must remain plane.

tions represented in Figs. 66-a and 66-b, we may conclude that these shearing stresses transmit the internal forces from one material to the other (Fig. 66-c). This analysis allows the conclusion that, irrespective of the member's length, these shearing stresses appear only near by the ends of the bar. Furthermore, they do not depend on the length of the bar.

VI.7 Non-Prismatic Members

VI.7.a Introduction

In the development of the expressions presented in the previous sections for the stresses in members under axial loading, only prismatic bars were considered, since the symmetry conditions leading to the law of conservation of plane sections are only valid for this special kind of slender members. However, these expressions are generally used for non-prismatic members, i.e. bars with a curved axis or with a non-constant cross-section. This leads to errors, since those expressions are only exact for prismatic bars. The magnitude of these errors depends on the relation between the curvature radius and the cross-section's dimension in the direction of that radius, in the case of curved members, and on the rate of variation of the cross-section's dimensions, in the case of a non-constant cross-section. In order to get an idea of the importance of these errors, we can compare approximate solutions obtained by means of the theory of prismatic bars with the exact results furnished by the Theory of Elasticity in particular cases.

VI.7.b Slender Members with Curved Axis

As an example of a slender member with a curved axis, we may take a ring defined by a slice of length b of a tube with constant wall thickness under an internal pressure p. The tube's dimensions are defined by the thickness e and the mean radius r_m, as shown in Fig. 67.

The mean stress acting on the tube's wall σ_{med} may be obtained by equilibrium considerations on a half ring (Fig. 67)

$$
p\,2\left(r_m - \frac{e}{2}\right)b = 2\sigma_{\mathrm{med}}eb \quad \Rightarrow \quad \left| \begin{array}{l} \sigma_{\mathrm{med}} = \dfrac{p\left(r_m - \frac{e}{2}\right)}{e} = p\left(\dfrac{1}{\alpha} - \dfrac{1}{2}\right) \\[2mm] \qquad\quad \text{with} \quad \alpha = \dfrac{e}{r_m}\,. \end{array}\right. \tag{136}
$$

This stress coincides with the solution obtained by considering the ring as a slender member, since, according to the theory of prismatic members, the stress is constant in the cross-section.

This problem is solved by Theory of Elasticity (Lamé's problem [4]) using polar coordinates. The solution is described by radial and circumferential stresses. The latter coincide with the stresses in the cross-section of the ring.

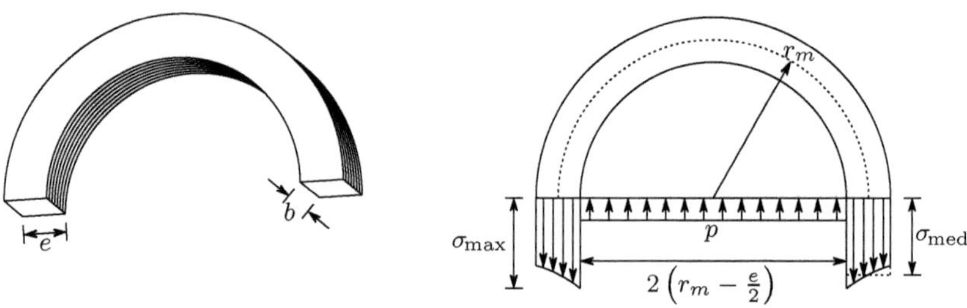

Fig. 67. Forces acting on a tube under internal pressure

The maximum circumferential stress occurs at the inner surface of the tube (σ_{max}, Fig. 67), taking the value

$$\sigma_{max} = \frac{p}{\alpha}\left(1 + \frac{\alpha^2}{4}\right) .$$

The relation between the maximum and mean stress (136) is given by

$$\frac{\sigma_{max}}{\sigma_{med}} = \frac{1 + \frac{\alpha^2}{4}}{1 - \frac{\alpha}{2}} .$$

The following table gives the error of the approximate solution furnished by the theory of prismatic members, as a function of parameter α.[5]

$\alpha = e/r_m$	0.01	0.02	0.05	0.1	0.2
$\sigma_{max}/\sigma_{med}$	1.0051	1.0102	1.0263	1.0553	1.1222
Error	0.51%	1.02%	2.63%	5.53%	12.22%

If the value of 5% is accepted as the maximum admissible error, we conclude, generalizing to other cases of curved bars under axial force, that the theory of prismatic bars does not introduce a significant error, while the dimension of the cross-section in the curvature plane is less than 0.1 times the dimension of the mean radius of the member.

The cross-sections of the ring still remain plane, although the bar is not prismatic, since the plane containing any cross-section of the ring is a symmetry plane. The inner pressure causes a diameter increase in the tube. If

---

[5] Equation (136), although similar, is not the same as the well known formula for the computation of the mean stress in a thin-walled tube under internal pressure $\sigma_{med} = pr_m/e$. This expression is obtained in the same way as (136), with the difference that the mean radius r_m is used in place of the inner radius $r_m - e/2$. If this expression is used instead of (136), the error on the computation of the maximum stress will be significantly lower. For example, for $\alpha = 0.2$ an error of about 1% is obtained. This expression was not used here, however, since the purpose of the analysis is to investigate the error introduced by the use of the theory of prismatic bars in curved members and not to compute the stresses in tubes.

the deformation caused by the radial stresses is ignored, the elongation at the inner and outer sides of the wall is equal. However, as the undeformed distance between two cross-sections is smaller at the inner side, the strain and, consequently, the stress is larger there.

VI.7.c Slender Members with Variable Cross-Section

As a simple example of a member with variable cross-section, with an exact solution furnished by the Theory of Elasticity, the wedge shaped element represented in Fig. 68 is considered.

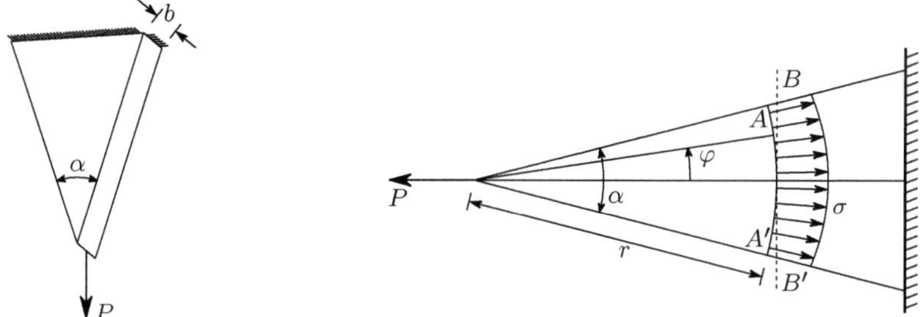

Fig. 68. Stresses in a slender member with variable cross-section

The solution of Theory of Elasticity is obtained for this problem by means of polar coordinates [4] and indicates that in a cylindrical section (AA', Fig. 68) the tangential (shearing) stress vanishes and the radial stress σ is given by the expression below (r and φ are the polar coordinates)

$$\sigma = \frac{2}{\alpha + \sin \alpha} \frac{P \cos \varphi}{br} .$$

For a given value of r, this normal stress reaches its maximum for $\varphi = 0$, taking the value

$$\varphi = 0 \Rightarrow \sigma = \sigma_{\max} = \frac{2}{\alpha + \sin \alpha} \frac{P}{br} .$$

The theory of prismatic bars yields for the stress in the cross-section BB' the value

$$\sigma_{\text{med}} = \frac{P}{\Omega} = \frac{P}{2br \tan \frac{\alpha}{2}} .$$

The error affecting the last solution may be expressed by the relation between $\sigma_{\max}$ and σ_{med}

$$\frac{\sigma_{\max}}{\sigma_{\text{med}}} = \frac{4 \tan \frac{\alpha}{2}}{\alpha + \sin \alpha} .$$

This relation depends only on angle α and takes the values

α	$10°$	$20°$	$30°$	$45°$	$60°$
$\sigma_{\max}/\sigma_{\mathrm{med}}$	1.0051	1.0206	1.0471	1.1101	1.2071
Error	0.51%	2.06%	4.71%	11.01%	20.71%

We conclude that for small values of angle α, which expresses in this case the rate of variation of the cross-section's dimensions, the error is very small.

VI.8 Non-Constant Axial Force – Self-Weight

The symmetry considerations used to demonstrate the law of conservation of plane sections are not satisfied in the case of a non-constant axial axial force, caused, for example, by the self-weight in non-horizontal bars, which means that there is no guarantee that the cross-sections remain plane. Furthermore, experimental observation shows that in the deformation caused by self-weight the sections do not remain plane (Fig. 69).

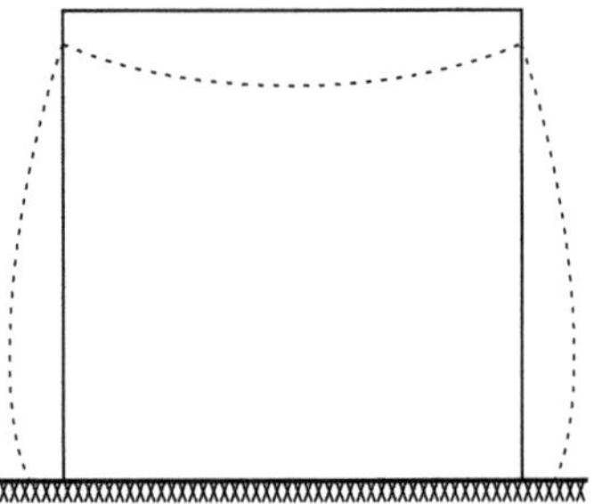

Fig. 69. Deformation of a prism of jelly under its self-weight

However, the solution of the Theory of Elasticity for a vertical homogeneous prism under its self-weight, shows that the stress is constant in the cross-section, although it does not remain plane (see example VI.12).[6] Although the generalization of this solution to other cases is not straightforward, it shows that a uniform distribution of the stress in the cross-section is possible, even without the conservation of plane sections. Besides, the self-weight of axially loaded structural members usually causes only a very small fraction of the total axial force, so that only a very small loss of symmetry occurs in the forces acting in a piece of the prismatic bar. For these reasons, the elongation Δl of a bar with length l, caused by a variable axial force $N(z)$, may be computed by the expression

---

[6]The law of conservation of plane sections is a sufficient condition to have a constant stress in the cross-section of a bar under purely axial force. This condition may, however, not be necessary, as the solution referred to shows.

$$\mathrm{d}\Delta l = \frac{N}{E\Omega}\,\mathrm{d}z \;\Rightarrow\; \Delta l = \frac{1}{E}\int_0^l \frac{N}{\Omega}\,\mathrm{d}z \;, \tag{137}$$

where z is a coordinate with the direction of the bar's axis. This expression is also valid in the case of a non-constant cross-section, provided that the rate of variation is small, as seen in Sect. VI.7.

VI.9 Stress Concentrations

In the neighbourhood of discontinuities in the bar, such as holes, notches, sudden changes of the cross-section, etc., the theory of prismatic bars is no longer valid, the cross-sections do not remain plane and the stress distribution is not uniform, which means that the maximum stress is larger than the mean value furnished by the theory of prismatic members. In the example depicted in Fig. 70 of a bar of constant thickness with two small semi-circular cuts ($r \ll b$), the maximum stress in the cross-section which contains the centers of the cuts is approximately twice the mean stress. We have in this case a *stress concentration factor* of 2.

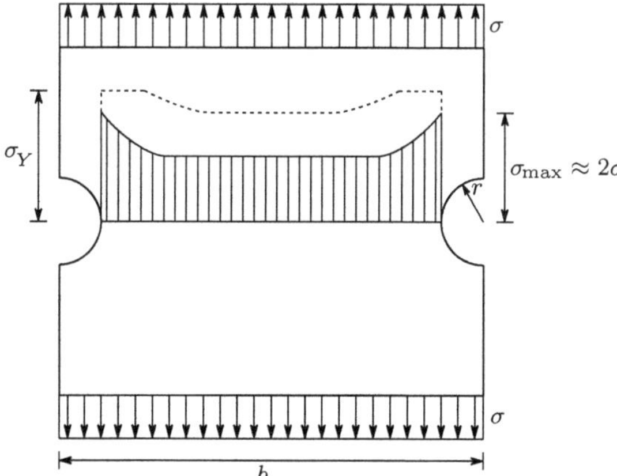

Fig. 70. Example of stress concentration caused by two semi-circular cuts ($r \ll b$).
_____ redistribution of stresses in the case of a ductile material by plastic deformation

The consideration of the stress concentrations in the safety evaluation is especially important in the case of brittle materials, since in this case there is no possibility of stress redistribution. The rupture process is similar to the one described in Sect. V.6: a crack starts at the points of stress concentration and propagates immediately to the rest of the cross-section, since the stress concentration at the tip of the crack is even larger, as will be seen later.

In the case of a ductile material, stress redistribution takes place, since the zones where the yielding strain is attained at first retain the loading capacity

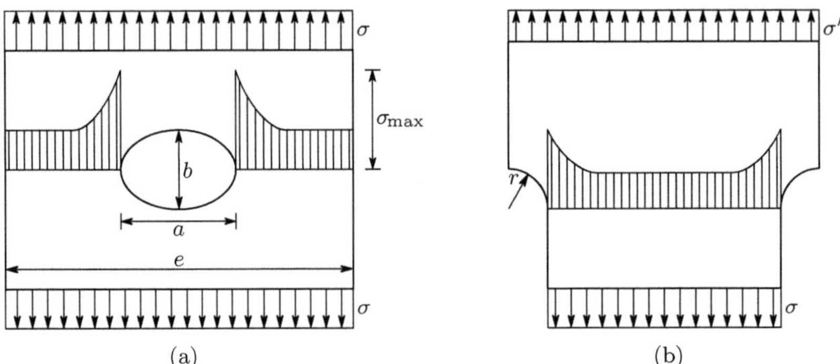

Fig. 71. Examples of stress concentration

described by the yielding stress σ_Y, until the whole cross-section yields, as indicated by the dashed line in Fig. 70. This is why the stress concentration is often not considered in the case of ductile materials. However, if the member is under cyclically varying loads, the stress concentration must be taken into account, due to the risk of fatigue failure, as seen in Sect. V.6.

In Fig. 71 two other examples of stress concentration are presented. In the first one (Fig. 71-a) a bar with a rectangular cross-section and a small thickness has an elliptical hole, with one of the ellipse's principal axes coinciding with the axis of the bar. The Theory of Elasticity furnishes a solution for the stress distribution in this case. The maximum stress takes the value [4]

$$\sigma_{\max} = \left(1 + 2\frac{a}{b}\right)\sigma \, ,$$

provided that the transversal dimension of the hole is very small, as compared with the width of the bar ($a \ll e$). From this expression we conclude that for $a = b$ (circular hole) the stress concentration factor is 3 ($\sigma_{\max} = 3\sigma$). The stress concentration increases if the larger axis has the transversal direction ($a > b$) and decreases in the opposite case. In particular, for $a = 0$ (a crack parallel to the axis of the bar) there will be no stress concentration. In the opposite case $b = 0$ (perpendicular crack in relation to axial force) we have an infinite stress concentration factor: $\sigma_{\max} = \infty$!!! Obviously this value is a purely theoretical one, since it is computed by considering that the material is perfectly continuous and has a linear elastic behaviour. However, the continuity hypothesis ceases to be acceptable when b takes a value similar to the dimension of the metal crystals or other material particles. Besides, the material does not retain linear elastic behaviour for large stress. It can be proved that a very small plastic deformation substantially reduces the stress concentration. Nevertheless, this example illustrates the danger of a transversal crack to the safety of a structure, in the case of brittle material, or cyclic loads (fatigue failure).

In Fig. 71-b an example of a suddenly changing cross-section is presented, where the transition is made by means of circular fillets. The stress concentration increases as the difference between the two cross-sections increases and as the radius of the fillet decreases.

VI.10 Examples and Exercises

VI.1. The truss described in Fig. VI.1-a is made of a material with linear elastic behaviour with a modulus of elasticity E.
  (a) Considering a nominal strength σ_{all} and knowing that buckling is prevented, determine the minimum cross-section area of bars $\overline{AB}$ and $\overline{BC}$.
  (b) Considering these cross-section areas compute the displacement of point B.

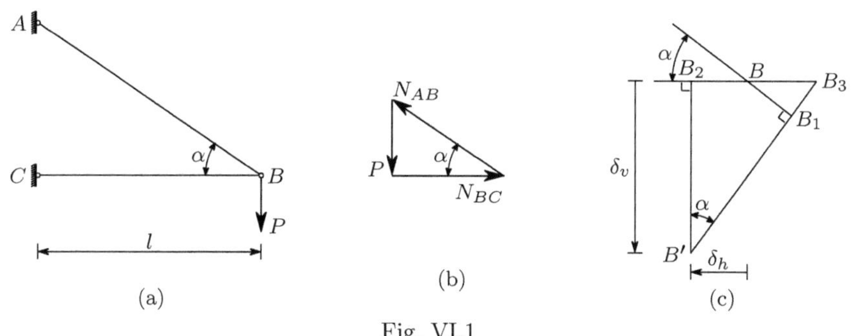

(a)   (b)   (c)

Fig. VI.1

Resolution

(a) The axial forces in the bars may be computed by means of the vertical and horizontal equilibrium conditions of the forces acting in node B (Fig. VI.1-b), yielding

$$N_{AB} = \frac{P}{\sin\alpha} \qquad N_{BC} = \frac{P}{\tan\alpha} \quad \text{(compression)}.$$

Equation (115) gives the minimum areas of the cross-sections

$$\Omega_{AB} = \frac{N_{AB}}{\sigma_{\text{all}}} = \frac{P}{\sigma_{\text{all}}\sin\alpha} \quad \text{and} \quad \Omega_{BC} = \frac{N_{BC}}{\sigma_{\text{all}}} = \frac{P}{\sigma_{\text{all}}\tan\alpha}.$$

(b) The displacement of point B is caused by the changes in the length of the two bars. The corresponding values may be computed by means of (116), yielding

$$\Delta l_{AB} = \frac{N_{AB} l_{AB}}{E\Omega_{AB}} = \frac{\frac{P}{\sin\alpha}\frac{l}{\cos\alpha}}{E\frac{P}{\sigma_{all}\sin\alpha}} = \frac{l\sigma_{all}}{E\cos\alpha}$$

and

$$\Delta l_{BC} = \frac{N_{BC} l_{BC}}{E\Omega_{BC}} = \frac{\frac{P}{\tan\alpha}l}{E\frac{P}{\sigma_{all}\tan\alpha}} = \frac{l\sigma_{all}}{E} \; .$$

As a consequence of the bars' variations in length, the position of point B changes to the intersection of the two circumference arcs, whose centers are points A and C and whose radii are the lengths of the deformed bars. Since the deformations are very small, as compared with the truss' dimensions, the rotation of the bars will also be very small, which allows the substitution of the circumference arcs by straight lines, as shown in the graphic construction presented in Fig. VI.1-c. In this Figure the elongation of bar $\overline{AB}$ is represented by the line segment $\overline{BB_1}$, the length reduction of bar $\overline{BC}$ is $\overline{BB_2}$ and the circumference axes are approximated by the line segments $\overline{B_1 B'}$ and $\overline{B_2 B'}$. The intersection of these segments (point B') defines the position of point B after the deformation.

The horizontal component of the displacement is then

$$\delta_h = \overline{BB_2} = \Delta l_{BC} = \frac{l\sigma_{all}}{E} \; .$$

The vertical component may be computed by means of the auxiliary distance $\overline{BB_3} = \frac{\overline{BB_1}}{\cos\alpha} = \frac{\Delta l_{AB}}{\cos\alpha}$, yielding

$$\delta_v = \overline{B_2 B'} = \frac{\overline{B_2 B_3}}{\tan\alpha} = \frac{\Delta l_{BC} + \frac{\Delta l_{AB}}{\cos\alpha}}{\tan\alpha} = \frac{l\sigma_{all}}{E}\left(\frac{1}{\tan\alpha} + \frac{1}{\sin\alpha\cos\alpha}\right) \; .$$

VI.2. The truss represented in Fig. VI.2-a is made of a material with an elasticity modulus E and a thermal expansion coefficient α. Determine the expressions required to compute the displacement of point B, caused by the force P and by a temperature increase ΔT.

Resolution

The axial forces in the bars may be computed by means of the equilibrium conditions of the forces acting in node B, which are represented by the system of equations (Fig. VI.2-b)

$$\begin{cases} N_1 \cos\alpha_1 = N_2 \cos\alpha_2 \\ N_1 \sin\alpha_1 + N_2 \sin\alpha_2 = P \; . \end{cases}$$

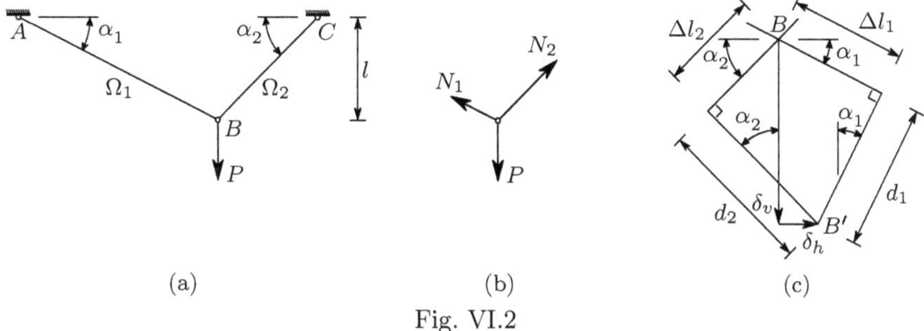

(a)   (b)   (c)

Fig. VI.2

The deformation of the bars under the axial forces and the temperature variation may be computed by means of (117), yielding

$$\Delta l_1 = \left(\frac{N_1}{E\Omega_1} + \alpha\Delta T\right)\frac{l}{\sin\alpha_1} \qquad \text{and} \qquad \Delta l_2 = \left(\frac{N_2}{E\Omega_2} + \alpha\Delta T\right)\frac{l}{\sin\alpha_2} .$$

The horizontal and vertical components of the displacement of point B (δ_h and δ_v) may be computed by means of the graphical construction presented in Fig. VI.2-c. From this Figure the following relations may be established.

$$\delta_v = d_1\cos\alpha_1 + \Delta l_1\sin\alpha_1 = d_2\cos\alpha_2 + \Delta l_2\sin\alpha_2$$
$$\delta_h = \Delta l_1\cos\alpha_1 - d_1\sin\alpha_1 = d_2\sin\alpha_2 - \Delta l_2\cos\alpha_2 .$$

By solving the system formed by the second equalities of each one of these expressions the distances d_1 and d_2 may be obtained. After this, δ_v and δ_h are immediately obtained. As an alternative, the projections of δ_v and δ_h on the directions of Δl_1 and Δl_2 yield the following system of equations

$$\begin{cases} \delta_v\sin\alpha_1 + \delta_h\cos\alpha_1 = \Delta l_1 \\ \delta_v\sin\alpha_2 - \delta_h\cos\alpha_2 = \Delta l_2 , \end{cases}$$

whose solution directly gives δ_v and δ_h.

VI.3. Consider a bar of cross-section area Ω, with built-in supports in both ends, made of a material with an elasticity modulus E and a thermal expansion coefficient α. Using the force and the displacement methods, compute the axial force introduced into the bar by a uniform temperature reduction ΔT, in relation to the construction temperature.

Resolution

Force method

Since it is known that all internal forces, except the axial force, vanish, the degree of static indeterminacy is one, although the bar has twelve external connections (six at each end) and only six are necessary to guarantee the static equilibrium.

If we suppose that one of the supports of the bar is removed, it becomes statically determined and, as a consequence, the temperature variation does not cause internal forces, but only a length reduction with the value

$$\Delta l = \alpha \Delta T l \, ,$$

where l represents the bar's length. The axial force in the bar with the two supports is the force needed to prevent the length reduction. In other words, it is the force which causes an elongation with the same value as the length reduction caused by the temperature variation. Using (116) and this condition, we get

$$\alpha \Delta T l = \frac{N l}{E \Omega} \;\Rightarrow\; N = E \Omega \alpha \Delta T \, .$$

Displacement Method

The degree of kinematic indeterminacy of this structure is zero, since the displacement of any cross-section is zero. We easily reach this conclusion by symmetry considerations: the middle cross-section has zero displacement; taking half the bar, we conclude that its middle section (the quarter length section) does not move, and so on. As a consequence, the total strain caused by the axial force and by the temperature variation is also zero. From (117) we get, since the temperature variation is negative ($\Delta T < 0$)

$$\varepsilon = \frac{\Delta l}{l} = \frac{N}{E \Omega} - \alpha \Delta T = 0 \;\Rightarrow\; N = E \Omega \alpha \Delta T \, .$$

VI.4. Consider the bar represented in Fig. VI.4, which is made of a material with elasticity modulus E and thermal expansion coefficient α and has three zones with different cross-section areas. Using the force and the displacement methods, compute the axial force introduced into the bar by a uniform temperature reduction ΔT, in relation to the construction temperature.

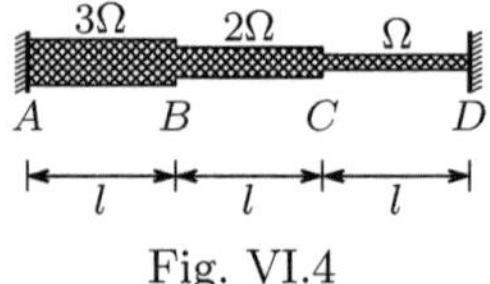

Fig. VI.4

Resolution

Force Method

Using the same reasoning as in example VI.3, we conclude immediately that the degree of static indeterminacy is one. The axial force may be computed in the same way. The length reduction caused by the temperature variation takes the value

$$\Delta l = 3\alpha\Delta Tl \, .$$

The elongation caused by the axial force is the sum of the elongations in the three zones. Since this elongation must compensate for the length reduction Δl, we have

$$\begin{cases} \Delta l_{AB} = \frac{Nl}{3E\Omega} \\ \Delta l_{BC} = \frac{Nl}{2E\Omega} \\ \Delta l_{CD} = \frac{Nl}{E\Omega} \end{cases} \Rightarrow 3\alpha\Delta Tl = \frac{Nl}{E\Omega}\left(\frac{1}{3} + \frac{1}{2} + 1\right) \Rightarrow N = \frac{18}{11}E\Omega\alpha\Delta T \, .$$

Displacement Method

Resolving this problem by means of the displacement method is a little more lengthy than by using the force method, since the degree of kinematic indeterminacy is two. In fact, if the displacements of points B and C are known, the strains and stresses in each zone may be immediately computed. Denoting by δ_1 and δ_2 the displacements of sections B and C, respectively, considered as positive from left to right, we get from (117), taking into consideration that the axial force N is the same in the whole bar

$$\Delta l_{AB} = \delta_1 = \frac{Nl}{3E\Omega} - \alpha\Delta Tl \qquad \Rightarrow N = \frac{E\Omega}{l}\left(3\delta_1 + 3\alpha\Delta Tl\right)$$

$$\Delta l_{BC} = \delta_2 - \delta_1 = \frac{Nl}{2E\Omega} - \alpha\Delta Tl \Rightarrow N = \frac{E\Omega}{l}\left[2\left(\delta_2 - \delta_1\right) + 2\alpha\Delta Tl\right]$$

$$\Delta l_{CD} = -\delta_2 = \frac{Nl}{E\Omega} - \alpha\Delta Tl \qquad \Rightarrow N = \frac{E\Omega}{l}\left(-\delta_2 + \alpha\Delta Tl\right) \, .$$

By eliminating N from these equations we get a system of two equations, which allows the computation of δ_1 and δ_2, yielding

$$\begin{cases} 3\delta_1 + 3\alpha\Delta Tl = 2\left(\delta_2 - \delta_1\right) + 2\alpha\Delta Tl \\ 3\delta_1 + 3\alpha\Delta Tl = -\delta_2 + \alpha\Delta Tl \end{cases} \Rightarrow \begin{cases} \delta_1 = -\frac{5}{11}\alpha\Delta Tl \\ \delta_2 = -\frac{7}{11}\alpha\Delta Tl \, . \end{cases}$$

Any of the three relations between the axial force N and the displacements δ_1 and δ_2 allows the computation of N. Using the first, for example, we get

$$N = \frac{3E\Omega}{l}\left(\delta_1 + \alpha\Delta Tl\right) = \frac{3E\Omega}{l}\left(-\frac{5}{11} + 1\right)\alpha\Delta Tl = \frac{18}{11}E\Omega\alpha\Delta T \, .$$

VI.5. The bar represented in Fig. VI.5 is made of a material with an elasticity modulus E and a thermal expansion coefficient α. The bar has a rectangular cross-section with a constant thickness a. Determine the axial force caused by a uniform temperature reduction ΔT.

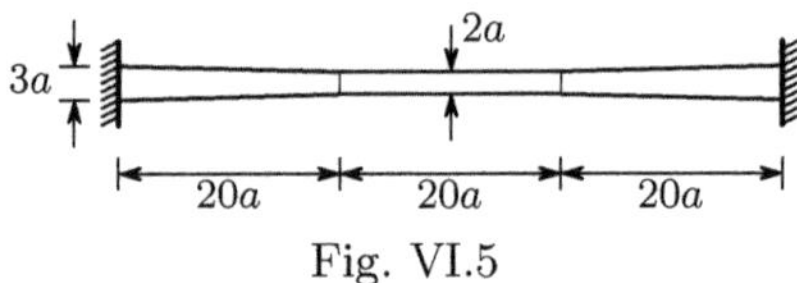

Fig. VI.5

Resolution

The resolution of this problem by the displacement method requires the prior computation of the relation between the axial force and the elongation in a bar with variable cross-section. The problem is, however, easily solved by the force method. The statically determinate base structure may be obtained by removing the right support, for example. In this situation the temperature variation causes the length reduction

$$\Delta l_T = \alpha l \Delta T = \alpha 60 a \Delta T \ .$$

The axial force N must cause an elongation which compensates for this length reduction. Considering a coordinate z with its origin on the left support, with increasing values from left to right, we may express the cross-section area as a function of z. For values of z between 0 and $20a$ we get

$$\Omega(z) = 3a^2 - \frac{a}{20}z \ .$$

As the middle section of the bar is in a symmetry plane, the total elongation caused by the axial force N is given by

$$\Delta l_N = \int_0^l \frac{N}{E\Omega}\,\mathrm{d}z = 2\int_0^{20a} \frac{N}{E\left(3a^2 - \frac{a}{20}z\right)}\,\mathrm{d}z + \int_0^{20a} \frac{N}{2Ea^2}\,\mathrm{d}z$$

$$= 2\frac{N}{E}\left[-\frac{20}{a}\ln\left(3a^2 - \frac{a}{20}z\right)\right]_0^{20a} + \frac{10N}{Ea} = \left[40\ln\left(\frac{3}{2}\right) + 10\right]\frac{N}{Ea} \ .$$

The equality $\Delta l_T = \Delta l_N$ describes the condition of deformation compatibility. The solution of this equation yields the value of the axial force

$$\Delta l_T = \Delta l_N \Rightarrow N = \frac{60\alpha E\Delta T a^2}{40\ln\left(\frac{3}{2}\right) + 10} \approx 2.28845\,\alpha E\Delta T a^2 \ .$$

VI.6. The bar $\overline{AB}$ of the structure represented in Fig. VI.6-a has a sufficiently high stiffness to be considered as rigid. The three suspension cables

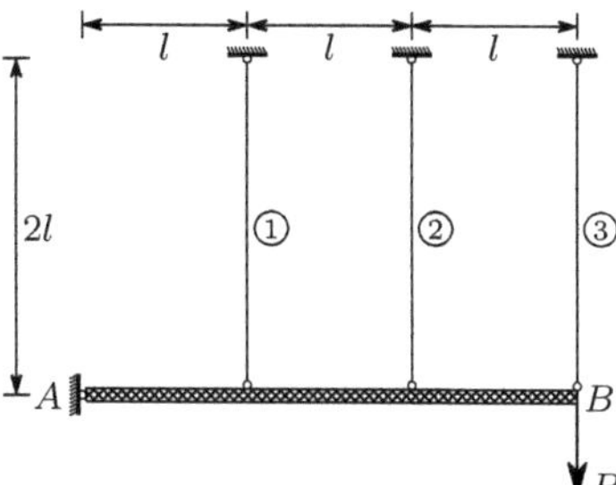

Fig. VI.6-a

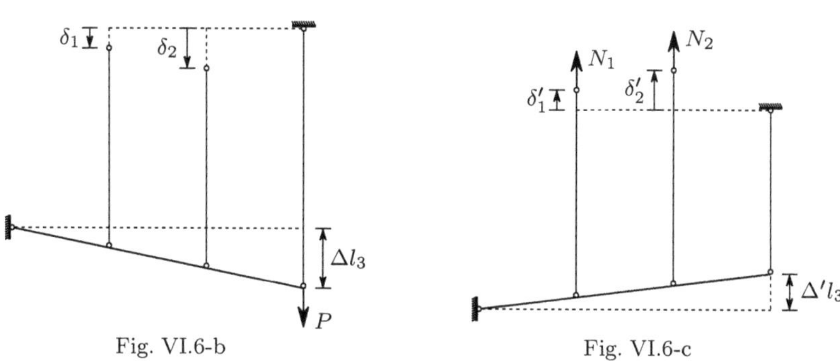

Fig. VI.6-b   Fig. VI.6-c

have the same cross-section area Ω and are made of a material with an elasticity modulus E.

Compute the axial forces in the cables using:
(a) the force method;
(b) the displacement method.

Resolution

(a) The structure has a degree of static indeterminacy of two. Thus, the statically determinate base structure is obtained by releasing two connections. In this case, the vertical connections on the top of bars 1 and 2 are released (Fig. VI.6-b). Under these conditions the axial force in bar 3 takes the value P. The corresponding elongation and the displacements of the top ends of bars 1 and 2 are

$$N_3 = P \Rightarrow \Delta l_3 = \frac{P2l}{E\Omega} \Rightarrow \begin{cases} \delta_1 = \dfrac{1}{3}\Delta l_3 = \dfrac{2}{3}\dfrac{Pl}{E\Omega} \\[2mm] \delta_2 = \dfrac{2}{3}\Delta l_3 = \dfrac{4}{3}\dfrac{Pl}{E\Omega} . \end{cases}$$

The displacements caused by axial forces in bars 1 and 2, N_1 and N_2, in the released connections of the statically determinate base structure, δ_1' and δ_2' (Fig. VI.6-c), must compensate for the displacements δ_1 and δ_2 caused by force P. The axial compressive force in bar 3 and the corresponding length reduction $\Delta' l_3$ (Fig. VI.6-c) take the values

$$N_3' = \frac{1}{3}N_1 + \frac{2}{3}N_2 \qquad\qquad \Delta' l_3 = \frac{N_3' 2l}{E\Omega} = \frac{2}{3}\frac{l}{E\Omega}(N_1 + 2N_2) \ .$$

The displacements δ_1' and δ_2' are caused by the rotation of the rigid bar $\overline{AB}$ and by the elongations in bars 1 and 2, respectively

$$\delta_1' = \frac{1}{3}\Delta' l_3 + \frac{N_1 2l}{E\Omega} = \frac{l}{E\Omega}\left(\frac{20}{9}N_1 + \frac{4}{9}N_2\right)$$

$$\delta_2' = \frac{2}{3}\Delta' l_3 + \frac{N_2 2l}{E\Omega} = \frac{l}{E\Omega}\left(\frac{4}{9}N_1 + \frac{26}{9}N_2\right) \ .$$

The conditions of compatibility of the deformations in the released connections define a system of two equations, whose solution gives the values of N_1 and N_2

$$\begin{cases} \delta_1 = \delta_1' \\[2mm] \delta_2 = \delta_2' \end{cases} \Rightarrow \begin{cases} \dfrac{2}{3}\dfrac{Pl}{E\Omega} = \dfrac{l}{E\Omega}\left(\dfrac{20}{9}N_1 + \dfrac{4}{9}N_2\right) \\[4mm] \dfrac{4}{3}\dfrac{Pl}{E\Omega} = \dfrac{l}{E\Omega}\left(\dfrac{4}{9}N_1 + \dfrac{26}{9}N_2\right) \end{cases} \Rightarrow \begin{cases} N_1 = \dfrac{3}{14}P \\[3mm] N_2 = \dfrac{6}{14}P \ . \end{cases}$$

The axial force in bar 3 may now be obtained by equilibrium considerations. The condition of equilibrium of moments in relation to point A gives

$$N_1 l + N_2 2l + N_3 3l = P3l \Rightarrow N_3 = P - \frac{1}{3}N_1 - \frac{2}{3}N_2 = \frac{9}{14}P \ .$$

(b) The solution of the problem by means of the displacement method is substantially easier, since the degree of kinematic indeterminacy is one. In fact, once the rotation of the rigid bar $\overline{AB}$ is known, the elongation of the bars is defined, which allows the immediate computation of the axial forces. Taking as kinematic parameter the displacement $\delta = \Delta l_3$ of point B (Fig. VI.6-d), we get the following values for the axial forces in the three bars

$$\Delta l_1 = \frac{N_1 2l}{E\Omega} = \frac{1}{3}\delta \Rightarrow N_1 = \frac{E\Omega}{6l}\delta \ ;$$

$$\Delta l_2 = \frac{N_2 2l}{E\Omega} = \frac{2}{3}\delta \Rightarrow N_2 = \frac{E\Omega}{3l}\delta \ ;$$

$$\Delta l_3 = \frac{N_3 2l}{E\Omega} = \delta \Rightarrow N_3 = \frac{E\Omega}{2l}\delta \ .$$

The condition of equilibrium of moments in relation to point A furnishes a relation between P and δ, yielding

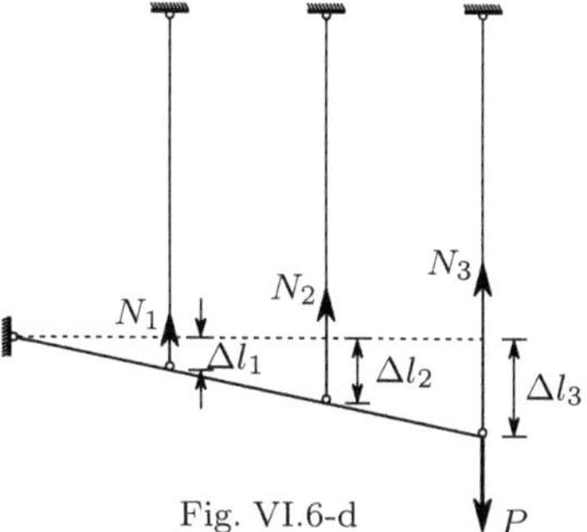

Fig. VI.6-d

$$P3l = N_1 l + N_2 2l + N_3 3l = \left(\frac{1}{6} + \frac{2}{3} + \frac{3}{2}\right) E\Omega\delta \;\Rightarrow\; \delta = \frac{18}{14}\frac{Pl}{E\Omega}\,.$$

Substitution of this value in the relations above, yields the same values of N_1, N_2 and N_3, as obtained by the force method.

VI.7. In the structure represented in Fig. VI.7-a the bar $\overline{AB}$ may be considered as rigid. The suspension bars are made of an elastic, perfectly plastic material with a modulus of elasticity E and a yielding stress σ_Y.

   (a) Determine the sequence of yielding of the bars, when the load P varies from zero to the value which causes the collapse of the structure.

   (b) Compute the value of P which causes the yielding of the structure and the displacement of point C just before the collapse.

   (c) Compute the values of the loads and displacements corresponding to the yielding of the remaining bars.

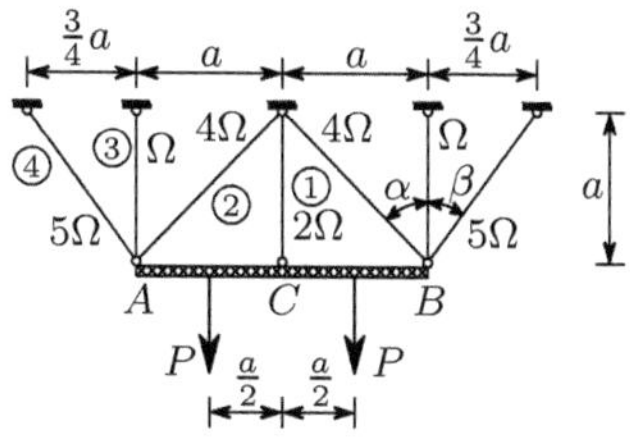

Fig. VI.7-a

Resolution

(a) The degree of kinematic indeterminacy is one, since the symmetry of the structure leads to the conclusion that both the horizontal displacement and the rotation of bar $\overline{AB}$ are zero. Choosing as kinematic parameter the vertical displacement δ of the rigid bar, we may express the strains in the bars as functions of δ, obtaining (Fig. VI.7-b)

$$\begin{cases} \Delta l_1 = \Delta l_3 = \delta \\ l_1 = l_3 = a \end{cases} ; \Delta l_1 = \varepsilon_1 a = \delta \Rightarrow \varepsilon_1 = \varepsilon_3 = \frac{\delta}{a}$$

$$\begin{cases} \Delta l_2 = \delta \cos \alpha \\ l_2 = \frac{a}{\cos \alpha} \end{cases} ; \Delta l_2 = \varepsilon_2 l_2 = \varepsilon_2 \frac{a}{\cos \alpha} = \delta \cos \alpha \Rightarrow \varepsilon_2 = \frac{1}{a}\cos^2 \alpha \, \delta = 0.5\frac{\delta}{a}$$

$$\begin{cases} \Delta l_4 = \delta \cos \beta \\ l_4 = \frac{a}{\cos \beta} \end{cases} ; \Delta l_4 = \varepsilon_4 l_4 = \varepsilon_4 \frac{a}{\cos \beta} = \delta \cos \beta \Rightarrow \varepsilon_4 = \frac{1}{a}\cos^2 \beta \, \delta = 0.64\frac{\delta}{a} .$$

Since we have $\varepsilon_1 = \varepsilon_3 > \varepsilon_4 > \varepsilon_2$, we conclude that bars 1 and 3 yield at first and are followed by bars 4. Bars 2 yield last, leading to the collapse of the structure.

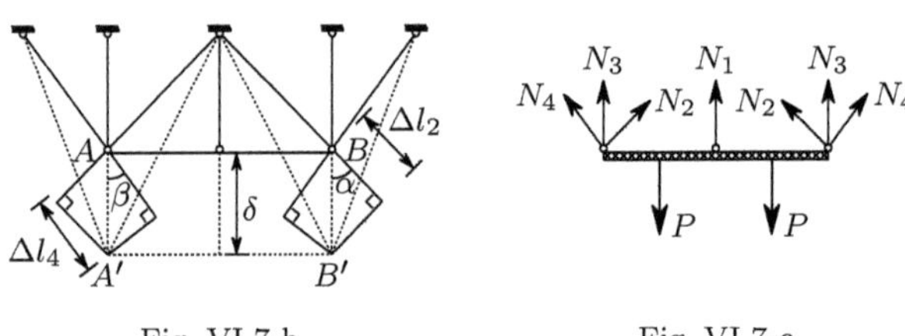

Fig. VI.7-b   Fig. VI.7-c

(b) The structural collapse occurs when all suspension bars yield. At this moment the stresses in all bars take the value of the yielding stress σ_Y. The corresponding value of P may be computed by means of the condition of vertical equilibrium of the forces acting on the rigid bar $\overline{AB}$ (Fig. VI.7-c), yielding

$$P = \frac{N_1}{2} + N_2 \cos \alpha + N_3 + N_4 \cos \beta$$
$$= \Omega \sigma_Y + 4\Omega \sigma_Y \cos \alpha + \Omega \sigma_Y + 5\Omega \sigma_Y \cos \beta = 8.8284 \Omega \sigma_Y .$$

As bars 2 yield last, the displacement of point C just before the collapse is the value of δ corresponding to $\varepsilon_2 = \varepsilon_Y$. Thus, we have

$$\varepsilon_2 = \frac{\sigma_Y}{E} = \frac{1}{2a}\delta_Y \Rightarrow \delta_Y = 2\frac{a\sigma_Y}{E} .$$

(c) When bars 1 and 3 start yielding, the kinematic parameter δ takes the value corresponding to the yielding strain in these bars, which is $\delta = a\varepsilon_Y = \frac{a\sigma_Y}{E}$. The axial forces in the remaining bars may be obtained from the relations between the strains and the displacement δ. The corresponding value of P is then obtained by means of the vertical equilibrium condition. This computation sequence is summarized by the following expressions

$$\delta = a\frac{\sigma_Y}{E} \Rightarrow \begin{cases} N_1 = 2\Omega\sigma_Y \\ N_2 = 4\Omega E\varepsilon_2 = 4\Omega E\frac{1}{a}\cos^2\alpha\,\delta = 2\Omega\sigma_Y \\ N_3 = \Omega\sigma_Y \\ N_4 = 5\Omega E\varepsilon_4 = 5\Omega E\frac{1}{a}\cos^2\beta\,\delta = 3.2\Omega\sigma_Y \end{cases} \Rightarrow P = 5.9742\Omega\sigma_Y \ .$$

Yielding of bars 4 takes place for $\delta = \frac{a\varepsilon_Y}{\cos^2\beta} = 1.5625\frac{a\sigma_Y}{E}$. At this stage bars 1 and 3 are already over the yielding strain. Only bars 2 are still in the elastic range. The axial force in these bars takes the value

$$N_2 = 4\Omega E\frac{1}{a}\cos^2\alpha\,1.5625\frac{a\sigma_Y}{E} = 3.125\Omega\sigma_Y \ .$$

The vertical equilibrium condition is then used to compute the corresponding value of P, yielding $P = 8.2097\Omega\sigma_Y$.

We conclude therefore that the load-displacement (P-δ) diagram of this structure has 4 straight line segments. The coordinates of the three corners, which correspond to the yielding of the different bars, are the (P,δ) pairs $(5.9742\Omega\sigma_Y, \frac{a\sigma_Y}{E})$, $(8.2097\Omega\sigma_Y, 1.5625\frac{a\sigma_Y}{E})$ and $(8.8284\Omega\sigma_Y, 2\frac{a\sigma_Y}{E})$.

VI.8. In the structure represented in Fig. VI.8-a bars AB and BC may be considered as rigid. The wires BD, EF and AC are made of a ductile material with elastic, perfectly plastic behaviour, with an elasticity modulus E and a yielding stress σ_Y. Determine the yielding sequence of the three wires, when the value of force P is gradually increased from zero until the value which causes the collapse of the structure.

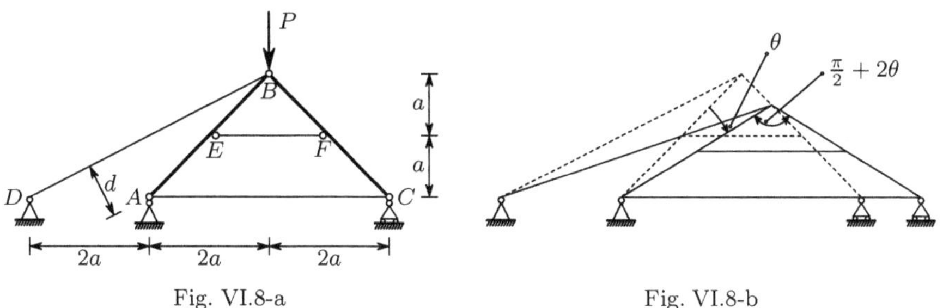

Fig. VI.8-a   Fig. VI.8-b

Resolution

The degree of kinematic indeterminacy of the structure is one, since once the horizontal displacement of the support C is known, for example, the deformed configuration of the structure is completely defined.

Taking the rotation θ (Fig. VI.8-b) as the kinematic parameter, we get the elongations given below for the wires BD, EF and AC

$$\Delta l_{BD} = \theta \times d = \theta \times 2a \sin\left(\arctan\tfrac{2a}{4a}\right) = 2\sin\left(\arctan 0.5\right) a\theta$$

$$\Delta l_{EF} = 2\theta \times a = 2a\theta$$

$$\Delta l_{AC} = 2\theta \times 2a = 4a\theta \ .$$

The undeformed lengths of these bars are

$$l_{BD} = 2\sqrt{5}\,a \qquad l_{EF} = 2a \qquad l_{AC} = 4a \ .$$

The corresponding strains are then

$$\varepsilon_{BD} = \frac{2\sin\left(\arctan 0.5\right)a\theta}{2\sqrt{5}\,a} = 0.2\,\theta \qquad \varepsilon_{EF} = \frac{2a\theta}{2a} = \theta \qquad \varepsilon_{AC} = \frac{4a\theta}{4a} = \theta \ .$$

We conclude that the wires EF and AC yield at the same time for a value $\theta = \varepsilon_Y = \frac{\sigma_Y}{E}$. Yielding of bar DB takes place only for $\theta = 5\varepsilon_Y$.

VI.9. In the structure represented in Fig. VI.9-a the horizontal bar may be considered as rigid. The inclined bars have different cross-section areas and are made of a material with elastic, perfectly plastic behaviour. Determine the yielding sequence of the inclined bars when the value of force P is gradually increased from zero until the value which causes the collapse of the structure.

Resolution

As in example VI.8, we have a structure with a degree of kinetic indeterminacy one, so that the elongation in each inclined bar may be directly related to the kinematic parameter. The cross-section areas do not play any role in that relation nor, as a consequence, in the bars' yielding sequence.

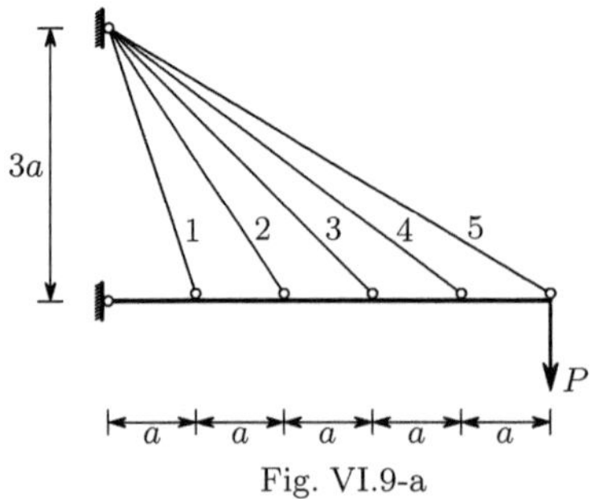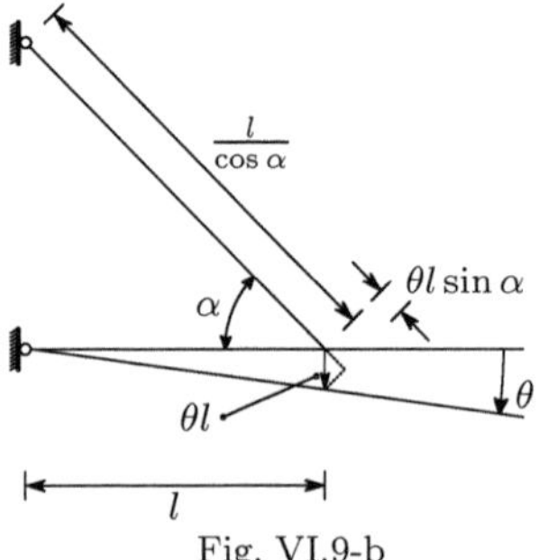

Fig. VI.9-a   Fig. VI.9-b

Taking the rotation θ (Fig. VI.9-b) as kinematic parameter, the strain in a generic bar whose position is defined by the distance l from the support takes the value

$$\varepsilon = \frac{\theta l \sin \alpha}{\frac{l}{\cos \alpha}} = \theta \sin \alpha \cos \alpha = \frac{\sin 2\alpha}{2}\,\theta\ .$$

Particularizing this expression for each of the five inclined bars, we get

$$\text{bar 1:}\quad l = \ a \Rightarrow \alpha = \arctan\frac{3a}{a} \Rightarrow \varepsilon = 0.3\,\theta$$

$$\text{bar 2:}\quad l = 2a \Rightarrow \alpha = \arctan\frac{3a}{2a} \Rightarrow \varepsilon \approx 0.46154\,\theta$$

$$\text{bar 3:}\quad l = 3a \Rightarrow \alpha = \arctan\frac{3a}{3a} \Rightarrow \varepsilon = 0.5\,\theta$$

$$\text{bar 4:}\quad l = 4a \Rightarrow \alpha = \arctan\frac{3a}{4a} \Rightarrow \varepsilon = 0.48\,\theta$$

$$\text{bar 5:}\quad l = 5a \Rightarrow \alpha = \arctan\frac{3a}{5a} \Rightarrow \varepsilon \approx 0.44118\,\theta\ .$$

The yielding sequence of the inclined bars is defined by the descending order of these strains, i.e., $3 \to 4 \to 2 \to 5 \to 1$.

VI.10. The bar $\overline{AB}$ of the structure represented in Fig. VI.10 is sufficiently stiff to be considered as rigid. The vertical bars are made of a brittle material with linear elastic behaviour until rupture defined by a Young's modulus E and a rupture stress σ_r.

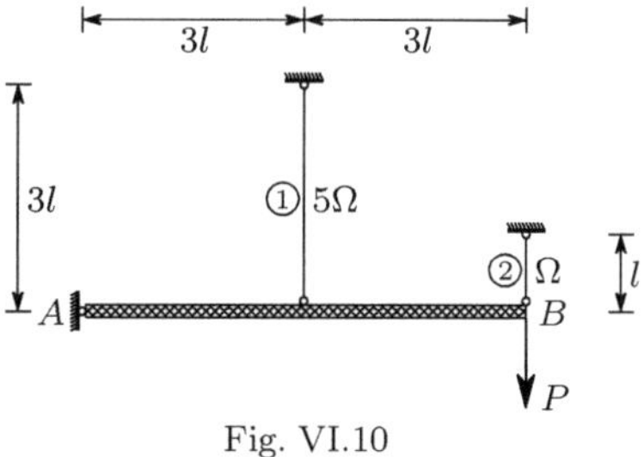

Fig. VI.10

(a) Determine the value of P which causes the collapse of the structure.
(b) Determine the increase in loading capacity that can be obtained by prestressing the structure, so that the two vertical bars reach the rupture stress simultaneously.
(c) Assuming that the prestressing is achieved by fabricating bar 1 with a length which is slightly different from the design length $3l$ and denoting by $3l + l_r$ the undeformed length of this bar, determine the value of l_r which maximizes the loading capacity of the structure.
(d) Ascertain whether the structure resists the initial internal forces ($P = 0$) caused by the prestressing. If it does not resist, compute the minimum value of force P.

Resolution

The equilibrium condition of moments in relation to point A yields the relation

$$6lP = 3lN_1 + 6lN_2 \Rightarrow P = \frac{N_1}{2} + N_2 .$$

Since the degree of kinematic indeterminacy is one, an infinitesimal rotation of the rigid bar around point A yields the condition of deformation compatibility of the vertical bars

$$\Delta l_1 = \frac{\Delta l_2}{2} \Rightarrow \varepsilon_2 = 6\varepsilon_1 .$$

(a) The axial forces in the vertical bars may be computed from the previous two conditions. The relations between the axial forces and the strains in the two vertical bars are

$$\varepsilon_1 = \frac{N_1}{E5\Omega} \quad \text{and} \quad \varepsilon_2 = \frac{N_2}{E\Omega} .$$

Substituting these values in the compatibility relation, we get

$$N_1 = \frac{5}{6}N_2 .$$

This equation and the equilibrium condition define a system of two equations, whose solution yields the axial forces and the stresses

$$\begin{cases} N_1 = \dfrac{10}{17}P \\[2mm] N_2 = \dfrac{12}{17}P . \end{cases} \Rightarrow \begin{cases} \sigma_1 = \dfrac{N_1}{5\Omega} = \dfrac{2}{17}\dfrac{P}{\Omega} \\[2mm] \sigma_2 = \dfrac{N_2}{\Omega} = \dfrac{12}{17}\dfrac{P}{\Omega} . \end{cases}$$

The rupture of bar 2 occurs at first and takes place when force P reaches the value

$$\sigma_2 = \sigma_r \Rightarrow P = \frac{17}{12}\Omega\sigma_r .$$

The rupture of this bar does not cause the collapse of the structure, since bar 1 alone is able to support this value of P, as may be easily verified from the equilibrium condition with $N_2 = 0$

$$N_1 = 2P = \frac{34}{12}\Omega\sigma_r \Rightarrow \sigma_1 = \frac{N_1}{5\Omega} = \frac{17}{30}\sigma_r < \sigma_r .$$

(The dynamic effects associated with the shock caused by the rupture of a bar on the other structural elements is not considered here. The study of this kind of problem is introduced in Chap. XII). The value of P which causes the rupture of bar 1 and, as a consequence, the structural failure is then

$$\sigma_1 = \sigma_r \Rightarrow \frac{2P}{5\Omega} = \sigma_r \Rightarrow P = \frac{5}{2}\Omega\sigma_r = 2.5\Omega\sigma_r .$$

(b) The loading corresponding to the simultaneous rupture of the two bars in the prestressed structure may be obtained directly from the equilibrium condition and takes the value

$$\begin{cases} N_1 = 5\Omega\sigma_r \\ N_2 = \Omega\sigma_r \end{cases} \Rightarrow P = \frac{1}{2}5\Omega\sigma_r + \Omega\sigma_r = 3.5\Omega\sigma_r \ .$$

We find that, prestressing the structure, its loading capacity may be increased by 40%.

(c) The value of l_r which introduces the optimum residual forces may be obtained by considering it as a residual deformation in the constitutive law of the material of bar 1. Under these conditions, the stress-strain relation in this bar takes the form

$$\varepsilon_1 = \frac{l_r}{3l} + \frac{\sigma_1}{E} \ .$$

Substituting this value in the condition of compatibility in terms of strains and considering $\sigma_1 = \sigma_2 = \sigma_r$, we get

$$\varepsilon_2 = 6\varepsilon_1 \ \Rightarrow \ \frac{\sigma_r}{E} = 6\left(\frac{l_r}{3l} + \frac{\sigma_r}{E}\right) \ \Rightarrow \ l_r = -\frac{5}{2}\frac{l\sigma_r}{E} \ .$$

We conclude that, in order to have a simultaneous rupture of the two bars, the undeformed length of bar 1 must be $l_1 = 3l - 2.5\frac{l\sigma_r}{E}$.

(d) The axial forces in the vertical bars under the actions of the prestressing internal forces and load P may be computed by introducing the value of l_r into the compatibility condition, which yields

$$\varepsilon_2 = 6\varepsilon_1 \ \Rightarrow \ \frac{N_2}{E\Omega} = 6\left(\frac{N_1}{E5\Omega} - \frac{5}{2}\frac{l\sigma_r}{E}\frac{1}{3l}\right) \ \Rightarrow \ \frac{6}{5}N_1 - N_2 = 5\Omega\sigma_r \ .$$

Solving the system formed by this equation and the equilibrium condition, we get the axial forces

$$N_1 = \frac{10}{17}P + \frac{50}{17}\Omega\sigma_r \qquad \text{and} \qquad N_2 = \frac{24}{34}P - \frac{50}{34}\Omega\sigma_r \ .$$

The residual stresses are the stresses corresponding to $P = 0$, taking the values

$$P = 0 \ \Rightarrow \ \begin{cases} N_1 = \dfrac{50}{17}\Omega\sigma_r \\ N_2 = -\dfrac{50}{34}\Omega\sigma_r \end{cases} \ \Rightarrow \ \begin{cases} \sigma_1 = \dfrac{N_1}{5\Omega} = \dfrac{10}{17}\sigma_r \\ \sigma_2 = \dfrac{N_2}{\Omega} = -\dfrac{50}{34}\sigma_r \approx -1.47\sigma_r \ . \end{cases}$$

We conclude that bar 2 does not resist the initial internal forces, since for $P = 0$ the stress exceeds the rupture stress. The minimum load necessary for the internal force in bar 2 not to exceed its strength, is then

$$\sigma_2 = -\sigma_r \ \Rightarrow \ N_2 = \frac{24}{34}P - \frac{50}{34}\Omega\sigma_r = -\Omega\sigma_r \ \Rightarrow \ P = \frac{2}{3}\Omega\sigma_r \approx 0.19 P_{\max} \ .$$

VI.11. In the symmetrical structure depicted in Fig. VI.11-a the bending stiffness of the horizontal bar is sufficiently high to consider the bar as
rigid. The vertical bars are cables made of a material with linear elastic behaviour until rupture, defined by a modulus of elasticity E and
a rupture stress σ_r. In order to optimize the loading capacity of the
structure, the middle cable is slightly longer than the design length l.
Compute the exact undeformed length of this cable, so that the three
cables reach rupture stress at the same time.

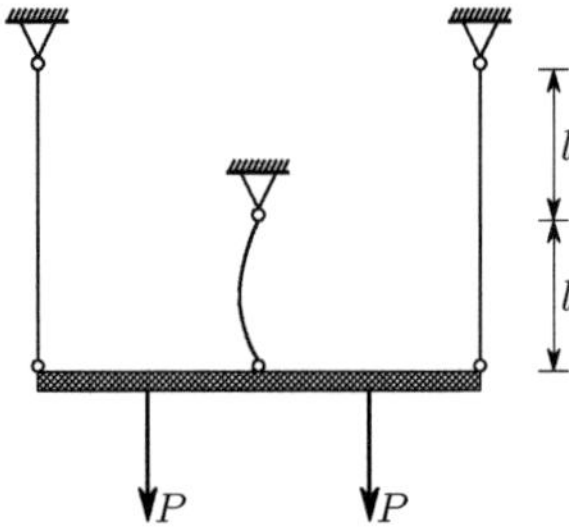

Fig. VI.11-a

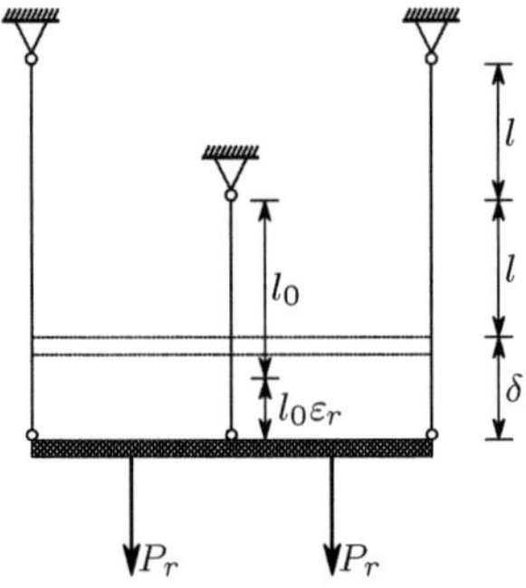

Fig. VI.11-b

Resolution

As the cables are made of the same material, they must have the same strain
$\varepsilon_r = \frac{\sigma_r}{E}$ at the moment of structural collapse. The lateral cables reach the
rupture strain, when the vertical displacement of the rigid bar attains the
value

$$\delta = 2l\varepsilon_r = 2l\frac{\sigma_r}{E} \ .$$

The situation of simultaneous rupture of the three cables is depicted in Fig. VI.11-b. Denoting the undeformed length of the middle cable by l_0, this value may be obtained by the following equation

$$\delta = 2l\varepsilon_r = l_0 + l_0\varepsilon_r - l \Rightarrow l_0 = \frac{1 + 2\frac{\sigma_r}{E}}{1 + \frac{\sigma_r}{E}} l \ .$$

VI.12. Consider a homogeneous vertical prismatic bar supported in its bottom cross-section, under its self-weight. The bar has height h and is made of an isotropic material with linear elastic behaviour. Disregarding the end effect introduced by the support, i.e., considering only the part of the bar which is sufficiently far from the support to accept the validity of Saint-Venant's principle, show that a uniform stress distribution in the cross-sections obeys every condition of equilibrium and compatibility.

Resolution

Considering a reference frame with the axes x and y in the horizontal plane containing the upper cross-section of the bar and axis z pointing from top to bottom, the assumed stress distribution corresponds to the following components of the stress tensor

$$\sigma_x = \sigma_y = \tau_{xy} = \tau_{xz} = \tau_{yz} = 0 \qquad \text{and} \qquad \sigma_z = -qz \ ,$$

where q represents the weight of the material per unit of volume.

We conclude at once that the differential equations of equilibrium (5) are satisfied, since the only body force is $Z = q$.

We also easily verify that the conditions of equilibrium at the boundary (8) are satisfied at the upper cross-section (in this section both the stresses and the boundary forces vanish) and at the lateral boundary (in this surface we have $\sigma_z \neq 0$ but also $n = 0$, so that the product $n\sigma_z$ vanishes). At the bottom cross-section, however, the conditions of equilibrium at the boundary ($\overline{Z} = \sigma_z$, (8) with $l = m = 0$ and $n = 1$) are only satisfied if the reaction force is uniformly distributed on the contact surface ($\overline{Z} = -qh$). Generally this does not happen, since in order to have this stress distribution, the support would have to deform in the same way as the bar at the bottom cross-section with the assumed stress distribution (it may be shown that the cross-sections take a spherical shape). However, in accordance with Saint-Venant's principle, we may assume a uniform stress distribution if we consider only points which are not close to this section.

With respect to the equations of strain compatibility, only (53) have to be considered if the bar is simply-connected, i.e., if the cross-section does not have holes. From Hooke's law for isotropic materials (74) and (75) we get the strain functions in the bar, which take the forms

$$\varepsilon_x = \varepsilon_z = -\frac{\nu}{E}\sigma_z = \frac{\nu q}{E}z \qquad \varepsilon_z = \frac{1}{E}\sigma_z = -\frac{q}{E}z \qquad \gamma_{xy} = \gamma_{xz} = \gamma_{yz} = 0 \ .$$

We conclude immediately that the equations of strain compatibility are satisfied, since these expressions are linear, which means that their second derivatives are zero (74) and (75) contain only second derivatives of the strain functions).

Since the bar is statically determinate, in terms of the way as it is supported, it is not necessary to verify whether the displacements are compatible with the support conditions (which would be an integral condition of compatibility). Thus, the assumed stress distribution obeys all conditions of equilibrium and compatibility. According to the Theorem of Uniqueness (cf. e.g. [2]) the linear problems of Continuum Solid Mechanics (materials obeying the Hooke's law and linear strain displacement relations) admit only one solution. This means that the assumed stress distribution is the actual solution of the problem.

VI.13. Obtain Expression 133 by means of the force method.

Resolution

Let us first consider that, in order to get a statically determined base structure, the connection between the two materials is released and that the axial force N is a tensile one and is applied to the part of the bar made of material a. Under these conditions, only the part of the bar which is made of material a is deformed. The corresponding elongation takes the value

$$\Delta l = \frac{Nl}{E_a \Omega_a} \, ,$$

where l represents the length of the bar. This deformation introduces a discontinuity in the cross-sections of the bar, which must be eliminated. To this end, let us consider a tensile axial force N' acting on the part of the bar made of material b and a compressive axial force with the same value N' acting on the part of the bar made of material a. Since this pair of forces has a zero resultant, it does not affect the total axial force considered in the first step. The pair of axial forces N', which represents the hyperstatic unknown, causes the deformations

$$\Delta l'_a = \frac{N'l}{E_a \Omega_a} \quad \text{(shortening)} \qquad \text{and} \qquad \Delta l'_b = \frac{N'l}{E_b \Omega_b} \quad \text{(elongation)} \, .$$

The discontinuity in the cross-sections of the bar is eliminated if $\Delta l'_a + \Delta l'_b = \Delta l$. This condition defines an equation, from which the value of the hyperstatic unknown may be obtained, yielding

$$\Delta l = \Delta l'_a + \Delta l'_b \Rightarrow \frac{Nl}{E_a \Omega_a} = \frac{N'l}{E_a \Omega_a} + \frac{N'l}{E_b \Omega_b} \Rightarrow N' = \frac{N E_b \Omega_b}{E_a \Omega_a + E_b \Omega_b} \, .$$

The stresses in each material are then

$$\sigma_a = \frac{N - N'}{\Omega_a} = \frac{N E_a}{E_a \Omega_a + E_b \Omega_b} \quad \text{and} \quad \sigma_b = \frac{N'}{\Omega_b} = \frac{N E_b}{E_a \Omega_a + E_b \Omega_b} \ .$$

The axial force N' is the resultant of the tangential stresses appearing in the connection between the two materials, when the axial force is applied in only one of the them. By means of the Saint-Venant's principle, it can be easily proved that these stresses appear only near to the cross-section where the external forces are applied and that they are independent of the length of the bar.

VI.14. Generalize (133) to a composite prismatic bar made of n materials.

Resolution

Since the degree of static indeterminacy is $n - 1$ and the degree of kinematic indeterminacy is 1, it is more convenient to use the displacement method.

As the strain is the same in all materials, the stresses in each of them may be expressed as functions of the stress in one of them. Thus, we have

$$\varepsilon = \frac{\sigma_i}{E_i} = \frac{\sigma_j}{E_j} \Rightarrow \sigma_i = \frac{\sigma_j}{E_j} E_i \ .$$

The condition of static equilibrium requires that the resultant of the stresses is equal to the axial force. Thus, we must have

$$N = \sum_{i=1}^{n} \sigma_i \Omega_i \ .$$

Expressing all the stresses σ_i as functions of the stress σ_j and substituting in the previous expression, we get the stress in material j

$$N = \frac{\sigma_j}{E_j} \sum_{i=1}^{n} E_i \Omega_i \Rightarrow \sigma_j = \frac{N E_j}{\sum_{i=1}^{n} E_i \Omega_i} \ .$$

VI.15. Figure VI.15-a represents the cross-section of a bar made of two materials, whose constitutive laws are defined by the stress-strain diagrams presented in Fig. VI.15-b.

  (a) Compute the maximum axial force that may be applied to the bar with the two materials in the elastic regime.

  (b) Compute the axial force which causes yielding of the bar.

  (c) In order to maximize the tensile loading capacity of the bar with both materials in the elastic regime, the bar is prestressed, by applying a tensile axial force to the interior part of the bar (material a) before the connection between the two materials is established. This force is removed after the connection's bonding.

      Compute the value of the prestressing axial force, which leads to the simultaneous yielding of the two materials.

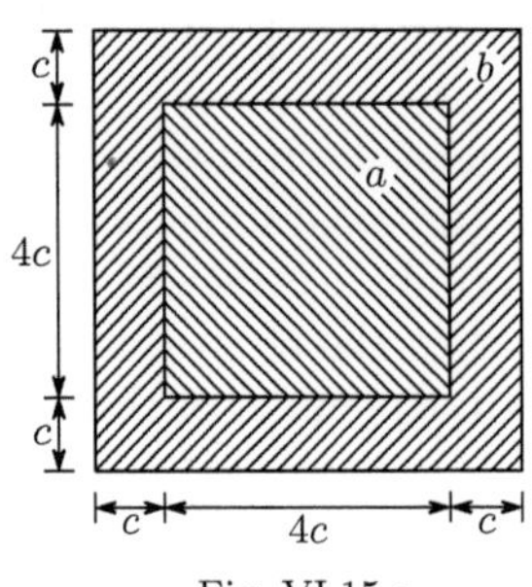

Fig. VI.15-a

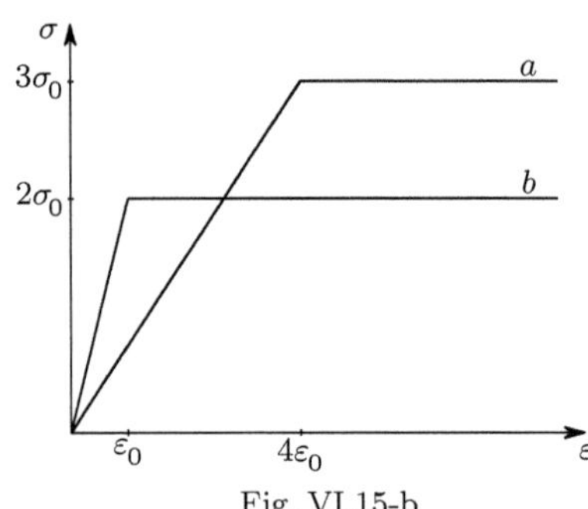

Fig. VI.15-b

Resolution

The areas occupied by each material in the cross-section and the elasticity moduli of the two materials take the values

$$\Omega_a = (4c)^2 = 16c^2; \quad \Omega_b = (6c)^2 - 16c^2 = 20c^2; \quad E_a = \frac{3\sigma_0}{4\varepsilon_0}; \quad E_b = \frac{2\sigma_0}{\varepsilon_0}.$$

(a) The maximum strain with both materials in the elastic regime is the yielding strain of material b, $\varepsilon = \varepsilon_0$, as is easily concluded from the stress-strain diagrams (Fig. VI.15-b). The stresses corresponding to this strain in the two materials are

$$\sigma_a = \varepsilon_0 E_a = \frac{3}{4}\sigma_0 \qquad \text{and} \qquad \sigma_b = 2\sigma_0 .$$

The axial force resulting from these stresses is then

$$N = \sigma_a \Omega_a + \sigma_b \Omega_b = \frac{3}{4}\sigma_0 \times 16c^2 + 2\sigma_0 \times 20c^2 = 52\sigma_0 c^2 .$$

(b) When the bar yields, the stresses in the two materials are the corresponding yielding stresses. Thus the yielding axial force of this bar takes the value

$$N = \sigma_a \Omega_a + \sigma_b \Omega_b = 3\sigma_0 \times 16c^2 + 2\sigma_0 \times 20c^2 = 88\sigma_0 c^2 .$$

(c) The prestressing force must take a value that leads to a strain in material b with the value $\varepsilon_b = \varepsilon_0$, when the strain in material a takes the value $\varepsilon_a = 4\varepsilon_0$. As at the moment of application of the prestressing force we have $\sigma_b = 0 \Rightarrow \varepsilon_b = 0$, the strain needed in material a at the same time is

$$\varepsilon_a = 3\varepsilon_0 \Rightarrow \sigma_a = E_a \varepsilon_a = \frac{3}{4}\frac{\sigma_0}{\varepsilon_0} \times 3\varepsilon_0 = \frac{9}{4}\sigma_0 .$$

The force which introduces the required prestressing is then

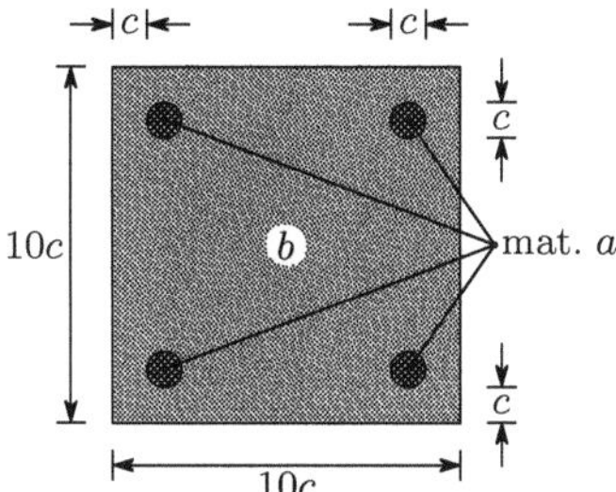

Fig. VI.16

$$N = \sigma_a \Omega_a = \frac{9}{4}\sigma_0 \times 16c^2 = 36\sigma_0 c^2 \ .$$

This prestressing force raises the maximum load in the elastic regime from $52\sigma_0 c^2$ to the yielding stress of the bar $(88\sigma_0 c^2)$.

VI.16. Figure VI.16 shows the cross-section of a prismatic bar made of two materials, a and b, with linear elastic behaviour until rupture. The elasticity moduli and the rupture stresses of the two materials take the values

$$\begin{cases} E_a = 30E \\ E_b = E \end{cases} \qquad \begin{cases} \sigma_{ra} = 150\sigma_r \\ \sigma_{rb} = \sigma_r \ . \end{cases}$$

(a) Disregarding dynamic effects compute the value of the tensile axial force which causes the rupture of each material.

(b) In order to maximize the tensile axial loading capacity, the bar is prestressed. The prestressing residual forces are introduced by applying a tensile axial force to each of the bars of material a before the connection between the two materials is established. These forces are removed after the two materials are bonded together.

Compute the force to be applied to each bar of material a, in order to obtain a simultaneous rupture of the two materials under a tensile axial force applied to the composite bar.

(c) Under the conditions specified in question b), compute the value of the tensile axial force which causes the rupture of the bar.

(d) Under the conditions specified in question b), compute the stresses in the two materials, when the axial force is zero.

Resolution

(a) The areas occupied by each material and the corresponding rupture strains take the values

$$\Omega_a = \pi c^2 \qquad \Omega_b = (100 - \pi)\,c^2 \qquad \varepsilon_{ra} = 5\frac{\sigma_r}{E} \qquad \varepsilon_{rb} = \frac{\sigma_r}{E} \ .$$

Since the strain is constant in the cross-section, material b reaches the rupture strain at first. This rupture takes place when the stresses take the values

$$\varepsilon = \frac{\sigma_r}{E} \Rightarrow \begin{cases} \sigma_a = E_a\varepsilon = 30E\frac{\sigma_r}{E} = 30\sigma_r \\ \sigma_b = E_b\varepsilon = E\frac{\sigma_r}{E} = \sigma_r \ . \end{cases}$$

The axial force which causes rupture of material b is then

$$N = \sigma_a\Omega_a + \sigma_b\Omega_b = 30\sigma_r\pi c^2 + \sigma_r\left(100 - \pi\right)c^2$$
$$= \left(29\pi + 100\right)c^2\sigma_r \approx 191.106\,c^2\sigma_r \ .$$

Since dynamic effects may be ignored, this axial force does not cause the rupture of the bar. In fact, when the value of the axial force exceeds this value, material b ceases to contribute to the resistance of the bar, but material a only attains its rupture stress, when the axial force reaches the value

$$N = \Omega_a\sigma_{ra} = \pi c^2 150\sigma_r \approx 471.239\,c^2\sigma_r \ .$$

(b) In order to achieve the simultaneous rupture of the two materials, material a must already have a strain $\varepsilon_a = 4\frac{\sigma_r}{E}$, when the strain in material b is zero. In this way, when the strain in the cross-section increases $\varepsilon_{rb} = \frac{\sigma_r}{E}$, we will have $\varepsilon_b = \frac{\sigma_r}{E}$ and $\varepsilon_a = 5\frac{\sigma_r}{E} = \varepsilon_{ra}$. The force needed to introduce the required prestressing, i.e. to introduce the strain $\varepsilon_a = 4\frac{\sigma_r}{E}$ into a bar of material a, then takes the value

$$N = \varepsilon_a E_a \frac{\Omega_a}{4} = 4\frac{\sigma_r}{E} 30E\frac{\pi c^2}{4} = 30\pi c^2\sigma_r \approx 94.248\,c^2\sigma_r \ .$$

(c) Since, under the conditions defined in the previous answer, the rupture of both materials occurs for the same strain, the tensile axial force to rupture the bar takes the value

$$N = \sigma_{ra}\Omega_a + \sigma_{rb}\Omega_b$$

$$= 150\sigma_r\pi c^2 + \sigma_r\left(100 - \pi\right)c^2 = \left(100 + 149\pi\right)c^2\sigma_r \approx 568.097c^2\sigma_r \ .$$

We conclude that the residual forces introduced by the prestressing procedure increase the tensile loading capacity of the bar from $471.239\,c^2\sigma_r$ to $568.097c^2\sigma_r$.

(d) Taking as reference the strain in material b, the constitutive laws of the two materials are defined by the expressions below (note that, when the strain in material b is zero, the stress in material a is $\sigma_a = 4\frac{\sigma_r}{E}30E = 120\sigma_r$)

$$\sigma_a = 120\sigma_r + 30E\varepsilon \qquad \text{and} \qquad \sigma_b = E\varepsilon \ .$$

The strain corresponding to a zero axial force may be computed from the condition

$$N = \sigma_a \Omega_a + \sigma_b \Omega_b = 0 \Rightarrow (120\sigma_r + 30E\varepsilon)\,\pi c^2 + E\varepsilon\,(100 - \pi)\,c^2 = 0$$

$$\Rightarrow \varepsilon = -\frac{120\pi}{100 + 29\pi}\frac{\sigma_r}{E}\ .$$

The stresses corresponding to this strain in each material are then

$$\sigma_a = E_a\varepsilon = 120\sigma_r + 30E\left(-\frac{120\pi}{100 + 29\pi}\frac{\sigma_r}{E}\right) \approx 60.820\,\sigma_r$$

$$\sigma_b = E_b\varepsilon = E\left(-\frac{120\pi}{100 + 29\pi}\frac{\sigma_r}{E}\right) \approx -1.973\,\sigma_r\ .$$

These stresses could also be computed by means of the superposition principle, taking the stresses acting in the situation defined in question b) ($\sigma_a = 120\sigma_r$ and $\sigma_b = 0$) and adding to them the stresses caused by the elimination of the axial force, i.e., by a compressive axial force $N = -\sigma_a\Omega_a = -120\pi c^2\sigma_r$. These stresses could be computed by means of (133).

Note that the absolute value of the initial strain obtained for material b exceeds its tensile strength. If the material has the same rupture stress for tensile and compressive forces, it will be necessary to have a minimum tensile axial force, in order to prevent this material from failing under the action of the prestressing forces. However, brittle materials usually have a much higher strength in compression than under tensile forces.

This example describes a prestressing technique widely used in prefabricated concrete structural elements called *pre-tensioning*. Another technique called *post-tensioning* is more used in on-site prestressing, once the structure is built. In this technique channels are left in the concrete element, where the prestressing steel wires are introduced later. The prestressing forces in these wires are usually introduced by means of hydraulic jacks which are supported by the concrete element to be prestressed itself. In this situation the reaction forces of the jacks are transmitted to the concrete element, so that the total axial force is zero. In this situation, the axial force to be applied to the prestressing wire is lower, corresponding to the stress computed in answer d). In the present example, the force to be applied in each bar of material a would be $N = 60.82\,\sigma_r\frac{\pi c^2}{4} = 47.768\,c^2\sigma_r$.

VI.17. Determine the displacement of the cross-sections of a vertical cable with length l and cross-section area Ω, supported in its upper end, under the action of its self-weight and of a downwards vertical force P applied in its bottom cross-section.

Resolution

Let us consider a vertical coordinate z originating in the bottom end of the cable and pointing upwards. Denoting by q the self-weight of the cable per unit length, the distribution of the axial force is defined by the expression

$$N = P + qz \ .$$

The displacement of the cross-section at the distance z from the bottom end may be computed by integrating the elongation of the infinitesimal pieces of length $\mathrm{d}z'$ ($z' \equiv z$), between that section and the upper end (137), yielding

$$\Delta l(z) = \frac{1}{E\Omega} \int_z^l (P + qz')\,\mathrm{d}z' = \frac{P}{E\Omega}(l - z) + \frac{q}{2E\Omega}(l^2 - z^2) \ .$$

The displacement of the bottom end of the cable is then

$$z = 0 \;\Rightarrow\; \Delta l = \frac{1}{E\Omega}\left(Pl - \frac{ql^2}{2}\right) \ .$$

VI.18. Determine the longitudinal shape of a bar with square cross-section, with a vertical axis, under the action of a force P and of its self-weight q per volume unit, as represented in Fig. VI.18-a, in order to have the same stress σ in the whole bar.

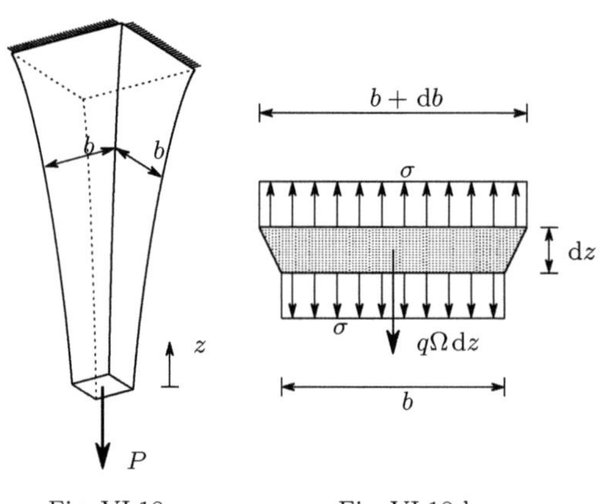

Fig. VI.18-a   Fig. VI.18-b

Resolution

The condition of equilibrium of the vertical forces acting in a piece of bar with an infinitesimal length $\mathrm{d}z$ (Fig. VI.18-b, $\Omega = b^2$) yields the equation

$$\sigma\Omega + q\Omega\,\mathrm{d}z = \sigma(\Omega + \mathrm{d}\Omega)$$
$$\Rightarrow \sigma\,\mathrm{d}\Omega = q\Omega\,\mathrm{d}z$$
$$\Rightarrow \sigma\frac{\mathrm{d}\Omega}{\Omega} = q\,\mathrm{d}z \ .$$

This differential equation is easily integrated, yielding

$$\sigma \ln \Omega = qz + C \Rightarrow \ln \Omega = \frac{qz + C}{\sigma} \Rightarrow \Omega = C' e^{\frac{qz}{\sigma}} \quad \text{with} \quad C' = e^{\frac{C}{\sigma}}.$$

As the bottom cross-section $(z = 0)$ only has to resist to the load P, this cross-section has the area

$$z = 0 \Rightarrow \Omega = \frac{P}{\sigma}.$$

The integration constant and the value of b as function of z are then

$$C' = \frac{P}{\sigma} \Rightarrow \Omega = \frac{P}{\sigma} e^{\frac{qz}{\sigma}} \Rightarrow b = \sqrt{\Omega} = \sqrt{\frac{P}{\sigma}} e^{\frac{qz}{2\sigma}}.$$

We find that the cross-section area grows exponentially with the length of the bar (Fig. VI.18-a). However, the growth rate is very low for the usual structural materials, since the number representing q is generally much smaller than the number representing σ, so that the numerical value of $e^{\frac{qz}{2\sigma}}$ is remains very close to 1, even for significant values of the length of the bar. This means that it is perfectly correct to apply the theory of prismatic members to this bar, with variable cross-section.

VI.19. Compute the axial forces in the column represented in Fig. VI.19.

VI.20. Figure VI.20 represents the cross-section of a prismatic bar made of two materials with linear elastic behaviour. The materials, a and b, have elasticity moduli $E_a = 2E$ and $E_b = 5E$ and coefficients of thermal expansion $\alpha_a = 3\alpha$ and $\alpha_b = \alpha$, respectively.

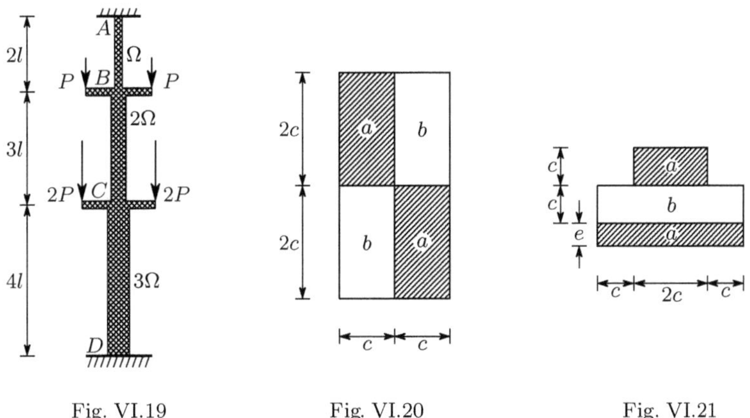

Fig. VI.19   Fig. VI.20   Fig. VI.21

Determine the stresses induced in this cross-section by a temperature rise ΔT. Justify the analytical methodology used.

VI.21. Figure VI.21 represents the cross-section of a prismatic bar made of two materials, a and b, which have different thermal expansion coefficients. Determine the value of the thickness e, so that a temperature variation does not introduce curvature in the bar.

VII

Bending Moment

VII.1 Introduction

As mentioned in Subsect. V.10.c, a prismatic bar under a constant bending moment is a symmetric problem in relation to any plane containing a cross-section and, as a consequence, the law of conservation of plane sections is valid. Therefore, although the axis of a bar under a bending moment does not remain a straight line – it acquires *curvature* – the cross-sections remain plane and perpendicular to the bar axis, provided that the bending moment is constant.

In the case of a varying bending moment the symmetry is lost not only because the moments acting in both ends of a piece of bar are different, but also due to the appearance of a shear force, which is equal to the derivative of the function describing the bending moment, in relation to a coordinate with the direction of the bar axis.

However, the stresses induced in a slender member by a constant bending moment, are not changed if a constant shear force is applied, and are a very close approximation to the actual stress distribution caused by the bending moment in the case of a non-constant shear force, as will be seen later (Sect. VII.7 and Chap. VIII). For these reasons, the stresses induced by bending are studied by considering a zero shear force, which means a constant bending moment.

It is usual to distinguish between the following types of bending:

- *Pure or circular bending.* This type of bending occurs when the only internal force in the bar is a constant bending moment, i.e., the axial (N) and shear (V) forces and the torsional moment (T) are zero. The designation of circular bending is suggested by the fact that the deformed axis of the initially prismatic bar is an arc of circumference, when the bending moment is constant (a constant moment implies a constant curvature).

- *Non-uniform bending*. This designation is normally used for a loading causing bending moment and shear force, that is, for a non-constant bending
  moment. The axial force and the torsional moment are zero.
- *Composed bending*. This designation is used in this book for a loading causing bending moment and axial force. The bending moment may be constant
  (circular composed bending: $M \neq 0$, $N \neq 0$, $V = 0$, $T = 0$) or variable
  (non-uniform composed bending: $M \neq 0$, $N \neq 0$, $V \neq 0$, $T = 0$).

Each of these three types of bending may be sub-divided into *plane* and
inclined bending. In the first case the plane containing the deformed bar is
parallel to the plane containing the couple of forces which defines the bending
moment. In the second case, the first plane is inclined in relation to the second
one.

VII.2 General Considerations

When a bar is under the action of symmetrical internal forces (constant axial
force and bending moment) the symmetry of the problem leads to the conclusion that the cross-sections remain plane and perpendicular to the bar's axis,
as seen before. The same symmetry conditions allow the conclusion that the
shearing stress in the cross-section vanishes.[1] We may also easily demonstrate
that, *if Poisson's coefficient is constant*, the normal and shearing stresses
acting in facets which are perpendicular to the cross-section's plane vanish
(cf. Sect. VII.6). Here we are considering the geometry of the member defined in relation to a rectangular Cartesian reference frame x, y, z, with axis
z parallel to the bar's axis. In accordance with these considerations, we will
have a one-dimensional stress state, i.e., a stress tensor with the components
$\sigma_x = \sigma_y = \tau_{xy} = \tau_{xz} = \tau_{yz} = 0$ and $\sigma_z \neq 0$.

The analysis of the normal stresses in the cross-section ($\sigma_z \neq 0$) may
be carried out directly from the law of conservation of plane sections and
equilibrium considerations. To this end, let us consider two cross-sections of
a prismatic bar at an infinitesimal distance l_0 from each other. Considering
a reference system with its origin in the centroid of the left cross-section,
the position of the points pertaining to the right section in the deformed
configuration may be defined by the equation of an inclined plane (Fig. 72)

$$z\left(x, y\right) = l_0 + a_1 x + b_1 y + c_1 .$$

In fact, it is easily demonstrated that, as l_0 is an infinitesimal distance, the
displacements in the plane (x, y) are infinitesimal quantities of higher order,
as compared with the displacements in direction z, so they may be neglected.

---

[1]If there were any shearing stresses, they would be represented by vectors with
opposite directions in the end sections of a small piece of bar, which is not compatible
with the complete symmetry of the problem in relation to the middle cross-section
of any piece of the bar.

The strain ε_z at the point of the cross-section defined by the coordinates x and y is then

$$\varepsilon_z\left(x,y\right) = \frac{l - l_0}{l_0} = \frac{z - l_0}{l_0} = \frac{a_1}{l_0}x + \frac{b_1}{l_0}y + \frac{c_1}{l_0}\;. \qquad (138)$$

The validity of this expression does not depend on the rheological behaviour of the material of which the bar is made, since it depends directly on the law of conservation of plane sections. The application of the law of conservation of plane sections reduces to three (a_1, b_1 and c_1) the number of parameters needed to define completely the relative motion of two cross-sections at an infinitesimal distance from each other. As this is exactly the number of equilibrium conditions which may be established for a system of parallel forces in a three-dimensional space (the stresses σ_z) the problem of computation of the stresses induced by the bending moment becomes statically determinate. If the bar is homogeneous and is made of a material with

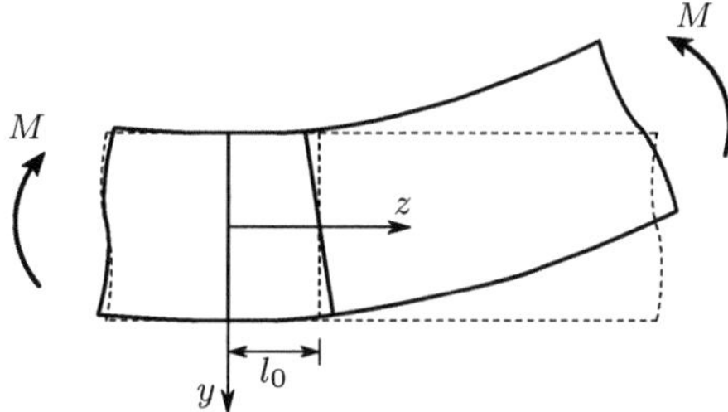

Fig. 72. Relative motion of two cross-sections in the bending deformation: ------ original configuration; —— deformed configuration

linear elastic behaviour, we have $\sigma_x = \sigma_y = 0$, as mentioned above. Therefore, the normal stress in the cross-section, σ_z, may be obtained by means of the one-dimensional Hooke's law, $\sigma = E\varepsilon$[2]

$$\sigma = \sigma_z = E\varepsilon_z = ax + by + c \quad \text{with} \quad \begin{cases} a = \dfrac{a_1}{l_0}E \\[2mm] b = \dfrac{b_1}{l_0}E \\[2mm] c = \dfrac{c_1}{l_0}E \;. \end{cases}$$

The constants a, b and c may be obtained by means of the aforementioned equilibrium conditions. Considering the components M_x and M_y of the bending moment as positive when they take the direction defined by a positive (tensile) stress acting in a point with positive x and y coordinates, we get

---

[2]In this expression and in the following account the index z is omitted ($\sigma = \sigma_z$), since σ_z is the only normal stress which needs to be considered in the theory of bending.

$$\begin{cases} N = \int_\Omega \sigma \, d\Omega = \int_\Omega (ax + by + c) \, d\Omega = c\Omega \\ M_x = \int_\Omega \sigma y \, d\Omega = \int_\Omega (axy + by^2 + cy) \, d\Omega = aI_{xy} + bI_x \\ M_y = \int_\Omega \sigma x \, d\Omega = \int_\Omega (ax^2 + bxy + cx) \, d\Omega = aI_y + bI_{xy} \, . \end{cases} \qquad (139)$$

The first area moments $\int_\Omega x \, d\Omega$ and $\int_\Omega y \, d\Omega$ vanish, since the axes x and y pass through the centroid of the cross-section. The quantities $I_x = \int_\Omega y^2 \, d\Omega$, $I_y = \int_\Omega x^2 \, d\Omega$ and $I_{xy} = \int_\Omega xy \, d\Omega$ are the moments and the product of inertia with respect to the central x and y axes. Solving this system of equations, we get

$$\begin{cases} a = \dfrac{M_y I_x - M_x I_{xy}}{I_x I_y - I_{xy}^2} \\[2mm] b = \dfrac{M_x I_y - M_y I_{xy}}{I_x I_y - I_{xy}^2} \\[2mm] c = \dfrac{N}{\Omega} \end{cases} \Rightarrow \sigma = \dfrac{M_y I_x - M_x I_{xy}}{I_x I_y - I_{xy}^2} x + \dfrac{M_x I_y - M_y I_{xy}}{I_x I_y - I_{xy}^2} y + \dfrac{N}{\Omega} \, .$$

$$(140)$$

This expression furnishes the stresses induced in the cross-section of a homogeneous prismatic member, made of a material with a linear elastic constitutive law, under the action of a bending moment and an axial force. It is important to note that the validity of this expression is not restricted to infinitesimal relative rotations of the cross-sections (i.e., to an infinitesimal curvature of the deformed member), since it was not necessary to use this approximation to deduce (140).

The analysis of the different types of bending referred to in Sect. VII.1 could be performed from (140), by particularizing it to the different cases, namely pure or composed and plane or inclined bending. However, in order to make the physical understanding easier, we expound the bending theory in the opposite sequence, i.e., we start with the most simple case (pure plane bending) and progressively generalize the conclusions to more complex cases. Equation (140) will however still be used for some particular problems. The last part of this chapter contains some problems where (140) is not valid: prismatic members made of two materials with linear elastic behaviour, and members made of materials with nonlinear behaviour in particularly simple cases.

In order to systematize the exposition of the theory of bending, some frequently used concepts are first defined:

– *action axis of the bending moment*: axis defined in the plane of the cross-section which is perpendicular to the vector representation of the bending moment (moments are usually represented by double-headed arrows); an equivalent definition is the intersection of the cross-section plane with the plane containing the couple of forces which defines the bending moment; this axis is simply called *action axis*, if it is not necessary to distinguish it from the action axis of the shear force;

- *action axis of the shear force*: line of action of the shear force acting on the cross-section;[3]
- *fibre*: prism with an infinitesimal cross-section area ($d\Omega$) with its axis parallel to the axis of the slender member;
- *neutral axis*: axis of rotation of a cross-section in relation to another, infinitesimally close, cross-section, in the deformation caused by the bending moment (and by the axial force in the case of composed bending); this name (neutral) comes from the fact, that the normal stress vanishes in the points of the cross-section belonging to this axis, since there is no elongation of the fibres passing through it; if the axial force is zero, or takes a sufficiently small value, the neutral axis divides the cross-section in compression and tension zones;
- *neutral surface*: surface defined by the points contained in the neutral axes of the cross-sections in the deformed member; it is also the surface defined by the fibres which do not suffer elongation or shortening (neutral fibres);
- *deflection curve*: line defining the shape of the bar's axis after the deformation; it may or may not be contained in a plane;
- *deflection plane*: plane containing the deflection curve; if this curve is not contained in a plane, the deflection plane varies along the deflection curve and is defined in an infinitesimal axis' length; this plane is perpendicular to the neutral axis.

VII.3 Pure Plane Bending

A prismatic bar is said to be under pure plane bending if the bending moment is constant and there is no axial force (pure bending) and the deflection plane is perpendicular to the vector representing the bending moment (plane bending). In this case, the cross-section rotates around an axis parallel to this vector, which means that the action axis of the bending moment is perpendicular to the neutral axis. This kind of bending takes place, for example, in a bar whose cross-section has a symmetry axis, if the action axis coincides with that axis, as represented in Fig. 73.

In the deformation of the bar, the neutral fibre AB did not change its length l. As the bending moment is constant, the curvature of the deflection curve is also constant, which means that this curve has the shape of a circumference arc. The angle φ, defining the relative rotation of the end cross-sections, may be related to the curvature radius ρ by the expression

$$\rho\varphi = l \;\Rightarrow\; \varphi = \frac{l}{\rho}\,. \tag{141}$$

---

[3]If the forces causing the non-uniform bending are all in the same plane and are perpendicular to the axis of the bar, this plane may be called *plane of actions*. In this case, the action axes of the bending moment and the shear force coincide with the intersection of the plane of actions and the cross-section plane.

Fibre CD, located at a distance y from the neutral axis, suffers a strain, which may be related to the curvature $\frac{1}{\rho}$ by the expression (Fig. 73)

$$\Delta l = \varphi\,(\rho + y) - \varphi\rho = \varphi y = \frac{l}{\rho}y \Rightarrow \varepsilon = \frac{\Delta l}{l} = \frac{1}{\rho}y\,. \tag{142}$$

This relation between the curvature $\frac{1}{\rho}$ and the strain at the point defined by coordinate y has been obtained directly by means of geometrical considerations based of the law of conservation of plane sections. It is therefore valid independently of the rheological properties of the material of the bar. Nor is its validity limited by the size of the deformations.

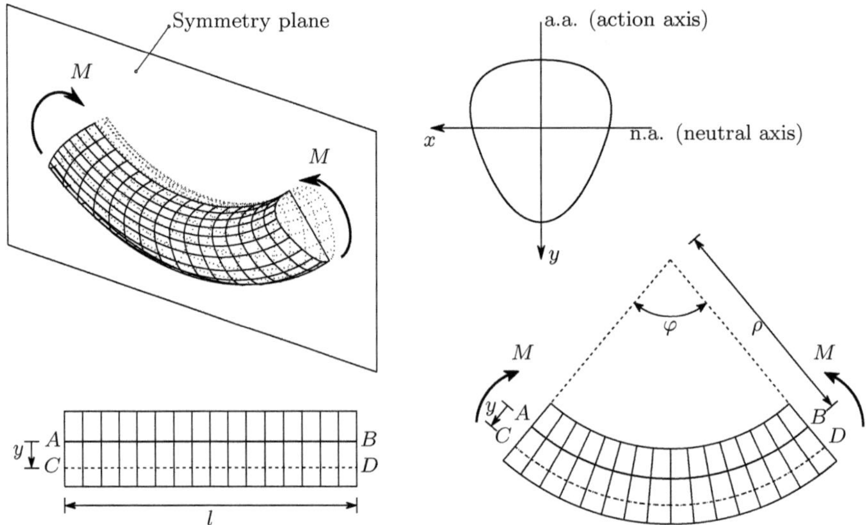

Fig. 73. Plane circular bending of a bar with a symmetric cross-section

If the bar is homogeneous and is made of a material with linear elastic behaviour, the stress may be found using the one-dimensional Hooke's law

$$\sigma = E\varepsilon = \frac{Ey}{\rho}\,. \tag{143}$$

The position of the neutral axis is obtained by the condition of equilibrium of the normal forces to the cross-section. As the axial force is zero, the resultant of the normal stresses must vanish. This condition yields

$$\int_\Omega \sigma\, d\Omega = 0 \Rightarrow \int_\Omega \frac{Ey}{\rho}\, d\Omega = 0 \Rightarrow \int_\Omega y\, d\Omega = 0\,. \tag{144}$$

This integral represents the first area moment of the cross-section with respect to the neutral axis (the distance y is defined in relation to this axis).

As this moment is zero, the neutral axis must pass through the centroid of the cross-section.

The resulting moment of the stresses acting in the cross-section must be equal to the bending moment M. From this condition, a relation between the curvature and the bending moment may be obtained

$$M = \int_\Omega \sigma y \, \mathrm{d}\Omega = \frac{E}{\rho} \int_\Omega y^2 \, \mathrm{d}\Omega \;\Rightarrow\; \frac{1}{\rho} = \frac{M}{EI} \quad \text{with} \quad I = \int_\Omega y^2 \, \mathrm{d}\Omega \;. \tag{145}$$

In this expression I represents the moment of inertia of the cross-section in relation to the neutral axis. The quantity $EI = \frac{\mathrm{d}M}{\mathrm{d}\left(\frac{1}{\rho}\right)}$ is called the *bending stiffness*, since it relates the amount of bending deformation (the curvature) to the internal force causing it (the bending moment).

By substituting (145) in (143), we get a relation between the stress and the bending moment

$$\sigma = \frac{My}{I} \;. \tag{146}$$

It is obvious from (143) or (146), that the stress attains its maximum value at the most distant point from the neutral axis. Denoting this distance by v ($v = |y|_{\max}$), we get, for the maximum absolute value of the stress in the cross-section,

$$|\sigma|_{\max} = \frac{M}{\left(\frac{I}{v}\right)} \;. \tag{147}$$

The quantity $\left(\frac{I}{v}\right)$, which directly relates the bending moment to the maximum stress is called the *section modulus*. It depends only on the geometry of the cross-section and allows the direct design of it from a given bending moment and a given nominal value for strength of the used material.[4] If the maximum allowable stress is the nominal strength of the material, σ_{all}, the section modulus must obey the design condition

$$\sigma_{\max} \leq \sigma_{\mathrm{all}} \;\Rightarrow\; \left(\frac{I}{v}\right) \geq \frac{M}{\sigma_{\mathrm{all}}} \;.$$

For two reasons the part of the cross-section which is close to the neutral axis plays a very small role in the resistance to the bending moment:

– the stress at a given point of the cross-section is proportional to the distance to the neutral axis (146); if the highest allowable stress is installed in the farthest fibres, in the points which are close to the neutral axis the stress will be proportionally lower, which means that the material is only used in a small part of its loading capacity;

---

[4]Usually the design problem is indeterminate, i.e., there is an infinite number of solutions which obey the design condition represented by the maximum acting stress. In the present case (as in the case of the axial force) the geometrical component of the cross-section's strength may be defined by only one parameter, which means that it may be directly computed from this design condition.

– the contribution of a given stress to the resistance to the bending moment depends on the distance to the neutral axis, since the moment of the stress increases with the distance to that axis; in fact, if the distance between the lines of action of the tensile and compressive stress resultants increases, the moment of the couple formed by these two forces also increases.[5]

For these reasons, from the point of view of economizing material, it is better if the cross-section has a shape with as little material as possible placed in the neighbourhood of the neutral axis. This may be achieved by increasing the height of the cross-section (taken as the dimension perpendicular to the neutral axis) or by giving it an appropriate form, such as an I-shape. In the first case the height increase may be limited by a maximum allowable size (for architectural reasons for example, in the case of Civil Engineering constructions), or by the possibility of structural instability caused by the compressive stresses (lateral buckling), if the width/height ratio is too small. In the second case, the minimum thickness of the vertical element of the I-shaped section (the web) may be imposed for stability reasons in the compressed zone as well, or by the shearing stresses caused by a shear force, as it will be seen later (Chap. VIII). Considering, for example, a rectangular cross-section and an I-beam, the section moduli, expressed in terms of the section's height h and the cross-section's area Ω are, respectively, (see example VII.5)

$$\left(\frac{I}{v}\right) = \frac{bh^2}{6} \approx 0.167\Omega h \qquad \text{and} \qquad \left(\frac{I}{v}\right) \approx 0.33\Omega h \ .$$

We conclude that, for the same area and the same height, the bending strength of the beam with the I-shaped cross-section is approximately twice the strength of the beam with rectangular cross-section.

VII.4 Pure Inclined Bending

In the general case of an action axis which is not a symmetry axis of the cross-section the angle between the neutral axis and the action axis is not known a priori. In order to analyse this more general case, let us consider the cross-section represented in Fig. 74, under the action of a bending moment M with a vertical action axis.

The law of conservation of plane sections leads to the conclusion that, in exactly the same way, as in the case of plane bending, the stress is proportional to the distance to the neutral axis, as defined by (143). The condition of equilibrium of the forces acting in the axial direction (z), used in the plane

---

[5]This conclusion becomes more obvious, if the stress does not vary with the distance to the neutral axis, as happens in the case of materials with elastic perfectly plastic behaviour, when the yielding strain is exceeded. This kind of problems is introduced in Sect. VII.10.c (see example VII.26).

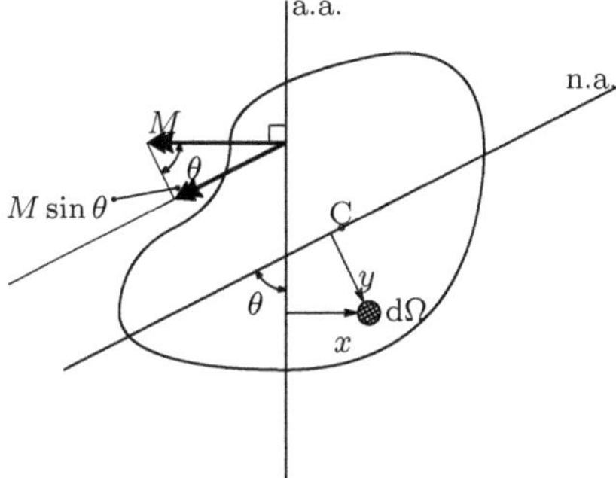

Fig. 74. Inclined bending of a bar without symmetry axes

bending case, is valid in exactly the same way for inclined bending, leading to the conclusion that the neutral axis passes through the centroid C of the cross-section (Fig. 74), as expressed by (144).

As in plane bending, we may establish a relation between the curvature of the deformed bar and the bending moment by means of the equilibrium condition of the moments around the neutral axis. This condition and (143) yield

$$M \sin \theta = \int_\Omega \sigma y \, d\Omega = \frac{EI_n}{\rho} \Rightarrow \frac{1}{\rho} = \frac{M}{EI_\theta} \quad \text{with} \quad I_\theta = \frac{I_n}{\sin \theta} \, , \qquad (148)$$

where I_n represents the moment of inertia of the cross-section in relation to the neutral axis. The curvature may be eliminated from this expression by substituting (148) in (143), yielding

$$\sigma = \frac{My}{I_\theta} \, . \qquad (149)$$

The orientation of the neutral axis i.e., the angle θ between the action and neutral axes, is still unknown. In order to define it, the condition of equilibrium of moments around the action axis may be used. Since the moment of the stresses in relation to this axis must vanish, using (143) we get

$$\int_\Omega \sigma x \, d\Omega = 0 \Rightarrow \frac{E}{\rho} \int_\Omega xy \, d\Omega = 0 \Rightarrow I_{xy} = 0 \, ,$$

where $I_{xy} = \int_\Omega xy \, d\Omega$ is the product of inertia with respect to the action and neutral axes. The condition $I_{xy} = 0$ means that those axes are conjugate in relation to the ellipse of inertia of the point defined by its intersection.

If the action axis is displaced, remaining parallel to the original position, the product of inertia does not change. In fact, denoting the translation of the axis by a, the product of inertia considering the new position of the action axis is given by

$$\int_{\Omega} (x + a)\, y \, \mathrm{d}\Omega = I_{xy} + a S_n = 0 \,,$$

where $S_n = \int_{\Omega} y \, \mathrm{d}\Omega$. This integral represents the first area moment of the cross-section in relation to the neutral axis. This quantity vanishes, since the neutral axis passes through the section's centroid. As the product of inertia with respect to the neutral axis and any axis parallel to the action axis is zero, we may conclude that the action and neutral axis are conjugate in relation to the centroidal ellipse of inertia.

The conjugate of a principal axis of inertia is the other principal axis.[6] Thus, we may conclude that *bending will be plane, if the action axis is parallel to a principal centroidal axis*. If the two principal moments of inertia have the same value, any axis passing through the centroid of the cross-section is a principal axis, which means that in bars with such cross-sections the bending is always plane. A square cross-section is an example of this kind of section.[7]

The angle θ could be obtained from the equation defined by $I_{xy} = 0$, by expressing I_{xy} as a function of θ. However, in most cases the principal moments of inertia and the corresponding principal directions are easily computed or may be obtained from tables with the geometrical characteristics of current cross-sections. For this reason, the inclined bending is usually analysed by decomposing the vector representing the bending moment (the moment vector) in the centroidal principal directions of inertia, as represented in Fig. 75. In this way, the inclined bending may be analysed as the superposition of two cases of plane bending. In fact, as the moment vector is perpendicular to the action axis, the bending will be plane, if this vector has the direction of a centroidal principal direction of inertia. Another possibility for the computation of the stresses in inclined bending is the use of non-principal reference axes (140).

Considering the separate action of the principal bending moments M_x and M_y, the stress acting in the point defined by the coordinate pair (x, y) is given by the expression

$$\sigma = \frac{M_x}{I_x} y - \frac{M_y}{I_y} x = M \left(\frac{\cos \alpha}{I_x} y - \frac{\sin \alpha}{I_y} x \right) \,. \tag{150}$$

The minus sign appearing in the stress caused by M_y results from the fact, that a bending moment M_y, positive when it has the same direction as

---

[6]Principal axes are two perpendicular axes, in respect of which the product of inertia is zero. Principal axes and principal directions of the inertial tensor are computed in the same way as the principal stresses and directions of the stress tensor, in the two-dimensional case. If the cross-section has a symmetry axis, it is one of the principal centroidal axes.

[7]These considerations are only valid if the material has a linear elastic behaviour. However, if the cross-section is symmetric in relation to the action axis, the bending will be plane, regardless of the linearity or nonlinearity of the material behaviour.

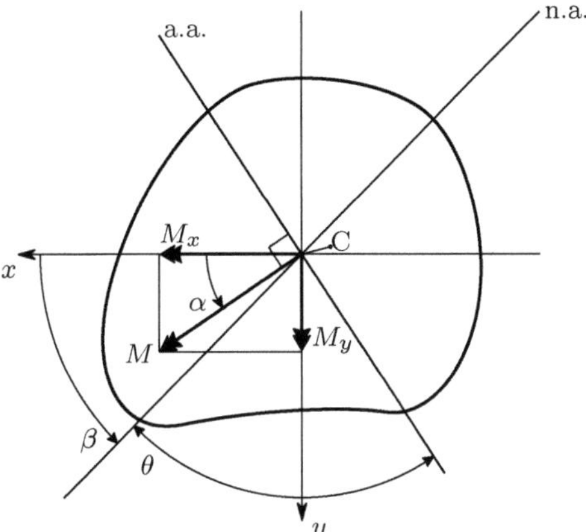

Fig. 75. Decomposition of inclined bending in two cases of plane bending

axis y, causes compression (negative stresses) in the points with a positive x-coordinate.[8] Since in the points belonging to the neutral axis the stress vanishes, the condition $\sigma = 0$ may be used to obtain its position, yielding

$$\sigma = 0 \Rightarrow y = x\frac{I_x}{I_y}\tan\alpha = x\tan\beta \quad \text{with} \quad \tan\beta = \frac{I_x}{I_y}\tan\alpha \ . \tag{151}$$

From this expression we see that, if $I_x > I_y$ then $\beta > \alpha$, i.e., in the inclined bending the neutral axis deviates from the perpendicular direction to the action axis, rotating in the direction of the principal axis with the smallest moment of inertia. In other words, the neutral axis is between the moment vector M and the principal axis with the smallest moment of inertia.

The maximum stress obviously occurs in the farthest point from the neutral axis. In many usual cross-sections, such as I-beams, C-channels, rectangular sections, etc., it is possible to identify those points without having to compute the orientation of the neutral axis. These cross-sections are particular cases of shapes with a rectangular convex contour and a symmetry axis. In this kind of sections, the farthest point from the neutral axis is one of the corners, which is also one of the farthest points in the two possible cases of plane bending ($M = M_x$ and $M = M_y$). In these cases, the maximum absolute value of the normal stress may be obtained directly from the two section moduli in plane bending, i.e., by adding the maximum stresses caused by M_x and M_y

$$|\sigma|_{\max} = \frac{|M_x|}{\left(\frac{I}{v}\right)_x} + \frac{|M_y|}{\left(\frac{I}{v}\right)_y} \ . \tag{152}$$

---

[8]This expression may also be obtained be particularizing (140) for pure bending ($N = 0$) and the principal axes of inertia ($I_{xy} = 0$). In this case, the minus sign does not appear, since, in the sign convention used in Sect. VII.2, the positive direction of M_y coincides with the negative direction of axis y.

In other types of cross-section, such as a curved contour or polygonal sections with a larger number of sides, the position of the farthest point is often not obvious, which means that the orientation of the neutral axis must be computed, in order to identify the farthest point from the neutral axis (151).

The total deformation may also be computed by superposing the curvatures around the principal axes x and y. Considering two sections located at an infinitesimal distance dl from each other, their relative rotations around axes x and y are (cf. (141) and (145))

$$\mathrm{d}\varphi_x = \frac{M_x\,\mathrm{d}l}{EI_x} \qquad \text{and} \qquad \mathrm{d}\varphi_y = \frac{M_y\,\mathrm{d}l}{EI_y}\,. \tag{153}$$

These rotations are infinitesimal (even for a large curvature), which mean that they may be treated as vectors with the directions of axes x and y, respectively. The resultant vector takes the direction of the neutral axis, as may be easily confirmed (cf. Fig. 75 and (151))

$$\frac{\mathrm{d}\varphi_y}{\mathrm{d}\varphi_x} = \frac{I_x}{I_y}\frac{M_y}{M_x} = \frac{I_x}{I_y}\tan\alpha = \tan\beta\,.$$

The curvature of the bar is then given by the expression

$$\mathrm{d}\varphi = \sqrt{\mathrm{d}\varphi_x^2 + \mathrm{d}\varphi_y^2} \ \Rightarrow\ \frac{1}{\rho} = \frac{\mathrm{d}\varphi}{\mathrm{d}l} = \frac{M}{E}\sqrt{\frac{\cos^2\alpha}{I_x^2} + \frac{\sin^2\alpha}{I_y^2}} = \frac{M}{EI_\theta}\,. \tag{154}$$

The last equality of this expression results from the two obtained expressions for the curvature, (148) and (154). The analytical proof of this equality is relatively lengthy, and so it is not presented here (see example VII.9).

As the quantities M_x, M_y, E, I_x and I_y, contained in (153) do not vary along the axis of the bar, the integration of $\mathrm{d}\varphi$ in a length l is straightforward, yielding the relative rotation of two sections located at a distance l from each other

$$\varphi = \int_l \mathrm{d}\varphi = \frac{Ml}{E}\sqrt{\frac{\cos^2\alpha}{I_x^2} + \frac{\sin^2\alpha}{I_y^2}} = \frac{Ml}{EI_\theta}\,.$$

It should be stressed that the validity of this expression is not restricted to infinitesimal rotations, i.e., it is valid for any value of the rotation φ.

VII.5 Composed Circular Bending

Let us consider a prismatic bar under the action of forces whose resultant is parallel to the axis of the bar. As seen in Sect. VI.1, if the line of action of this resultant passes through the centroid of the cross-section, the axis of the bar remains a straight line, which means that the bar is under pure axial

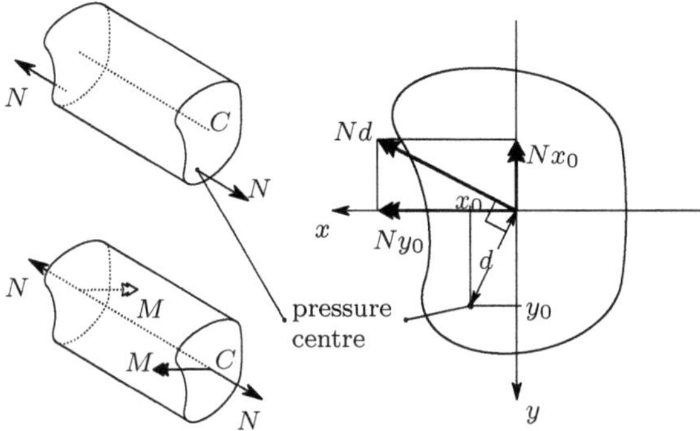

Fig. 76. Bending moments caused by an eccentric tensile axial force

loading. If that does not happen, additional bending takes place. The bending moment is given by the product of the axial force and the distance of its line of action to the centroid of the cross-section. The action axis is in this case the line joining the centroid to the point where the line of action of the axial force intersects the cross-section's plane. In the following description this point is called the *pressure centre*. We therefore have circular composed bending, since $M \neq 0$, $N \neq 0$, $V = 0$ and $T = 0$. This problem is usually analysed by superposition of the effects of the bending moment and the axial force. In the case of inclined bending the moment is decomposed in the principal axes of inertia of the cross-section, as shown in Fig. 76.

The stress acting in a point with coordinates (x, y) may be obtained by adding the stresses caused by the axial force N and by the bending moments $M_x = N y_0$ and $M_y = -N x_0$ (Fig. 76). Taking (150) into consideration, we get[9]

$$\sigma = \frac{N}{\Omega} + \frac{N y_0 y}{I_x} + \frac{N x_0 x}{I_y} = \frac{N}{\Omega}\left(1 + \frac{y_0 y}{i_x^2} + \frac{x_0 x}{i_y^2}\right) , \qquad (155)$$

where i_x and i_y represent the radii of gyration with respect to the principal axis x and y ($I_x = \Omega\, i_x^2$ and $I_y = \Omega\, i_y^2$). The position of the neutral axis may be obtained by the condition of zero stresses, yielding

$$\sigma = 0 \quad \Rightarrow \quad 1 + \frac{y_0 y}{i_x^2} + \frac{x_0 x}{i_y^2} = 0 \quad \Rightarrow \quad \begin{cases} x = 0 \Rightarrow y = y_1 = -\dfrac{i_x^2}{y_0} \\[2mm] y = 0 \Rightarrow x = x_1 = -\dfrac{i_y^2}{x_0} . \end{cases} \qquad (156)$$

The position of the neutral axis may be defined by the coordinates of the points where it intersects the principal axes of inertia $(0, y_1)$ and $(x_1, 0)$.

---

[9]This expression may also be determined by particularizing (140) to principal axes of inertia ($I_{xy} = 0$). The sign convention used for the bending moments does not play any role, since the eccentric axial force F is directly related to the stress.

VII.5.a The Core of a Cross-Section

In composed bending the neutral axis does not pass through the centroid, since there the stress corresponding to the axial force N is installed, as may easily be confirmed by making $x = y = 0$ in (155). From (156) we conclude that the distance of the neutral axis to the centroid increases when the coordinates of the pressure centre, x_0 and y_0 decrease, and vice versa. In the limit case, $x_0 = 0$ and $y_0 = 0$, which corresponds to $M = 0$, that distance is infinite. These considerations are illustrated in the example in Fig. 77.

When the pressure centre is sufficiently close to the centroid, the neutral axis does not intersect the cross-section, which means the the stresses are all tensile, or all compressive. This region around the centroid, where the pressure centre is located, when the neutral axis does not intersect the cross-section, is called the *core* of the cross-section. If the pressure centre is on the core's border, the neutral axis is tangent to the cross-section, as in example 3 of Fig. 77.

The determination of the cross-section core is important in the cases where a single linear elastic constitutive law for tensile and compressive stresses is not acceptable. This happens frequently with brittle materials, as for example, concrete, stone, soil, etc., and also in contact interfaces which are only resistant to compressive stresses. In these cases, (155) is only valid if the pressure centre is in the core of the cross-section, since otherwise tensile and compressive stresses will appear. This also applies to the material, whose constitutive law is shown in Fig. 78.

The core's limits may be directly computed from (156), by determining the position of the pressure centre which corresponds to a neutral axis that is tangent to the cross-section, and repeating this procedure for a sufficiently

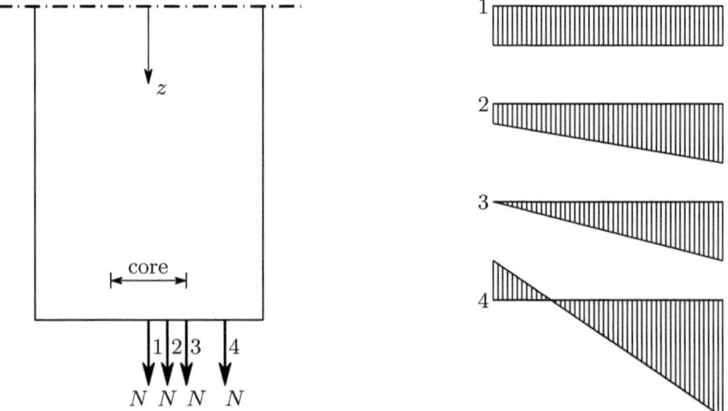

Fig. 77. Stress distribution in the cross-section of a prismatic bar, for different values of the eccentricity of the axial force N

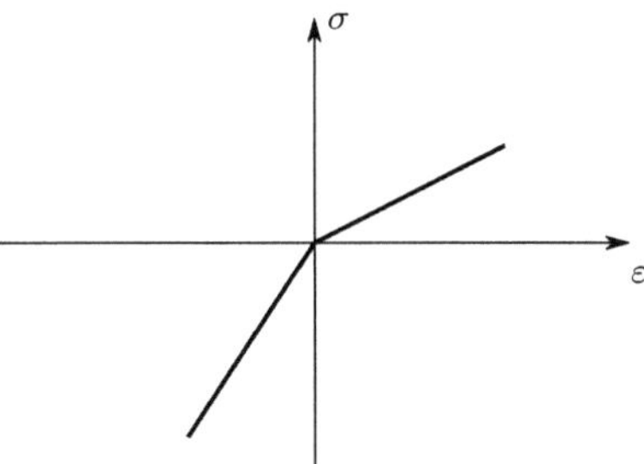

Fig. 78. Material with different linear elastic behaviour for tensile and compressive stresses

high number of pressure centres to define the core with acceptable precision. This technique is useful in the case of a cross-section with a curved boundary.

In the case of a polygonal border, it is more convenient to use another technique, which is based on the reciprocity of (155) in relation to the coordinates of the pressure centre (x_0, y_0) and of a generic point (x, y). In fact, we see immediately from (155) that applying an axial force N on the point with the coordinates (x_0, y_0), we get in the point with coordinates (x, y) the same stress that would be caused on point (x_0, y_0) by the same axial force applied on point (x, y), since the interchange of the roles of x with x_0 and of y with y_0 does not change the result.[10] Therefore, if the axial force is at first applied on (x_0, y_0) and the position of the corresponding neutral axis is computed and, subsequently, the load is applied on different points of this axis, neutral axes will be obtained, which will contain point (x_0, y_0). Thus, we may conclude that the displacement of the pressure centre along a straight line causes a rotation of the corresponding neutral axis around a point. This point corresponds to the pressure centre, whose neutral axis is that straight line.

The core of a polygonal cross-section may be determined by considering a pressure centre on a corner of the cross-section contour and computing the position of the corresponding neutral axis. Pressure centres located on a segment of this axis correspond to neutral axes passing through the corner of the cross-section contour, which do not intersect the cross-section. That line segment is therefore on the limit of the core. By repeating this procedure for all corners the complete core is obtained (Fig. 79-a). Another possibility consists of using the inverse procedure: given a neutral axis joining two corners of the section's contour, the corresponding pressure centre is a corner of the core.

It is obvious that, in the case of a cross-section whose boundary is not completely convex, the tangents to the boundary at the concavities cannot be used to define the core, since they intersect the cross-section. In this kind of

---

[10]This is a particular case of the Theorem of Maxwell, which is studied in Chap. XII.

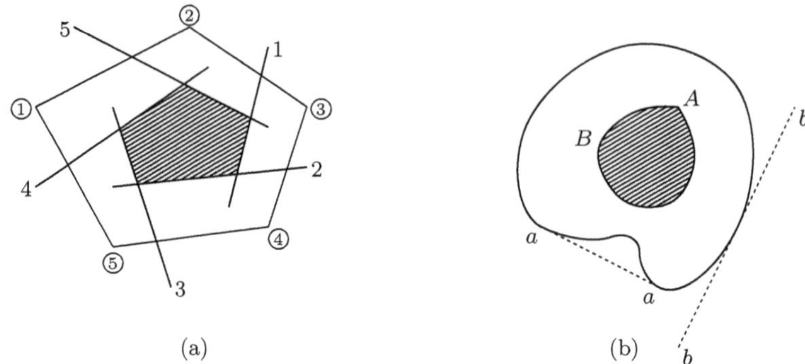

Fig. 79. Determination of the core of a cross-section: (a) polygonal cross-section boundary; (b) curved cross-section boundary

section the shape of the core is defined by the shape of the convex contour, as represented in Fig. 79-b. The corner A corresponds to the straight part of the convex contour (line $\overline{aa}$). For this reason, the core of an I-beam has the same shape (but not the same dimensions) as a rectangular cross-section (Fig. 80).

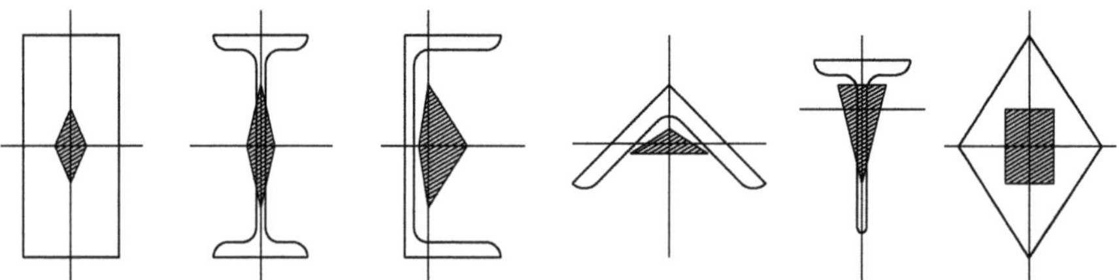

Fig. 80. Core shapes of some usual cross-sections

VII.6 Deformation in the Cross-Section Plane

In order to study the deformation of a cross-section in its plane, let us consider a piece of a prismatic bar whose end sections are sufficiently far from the points of application of the loading to accept the validity of the Saint-Venant principle. If the bar is under circular bending (with or without axial force), the strain distribution in the cross-section is defined by (142), i.e., it is linear, as represented in Fig. 81. Since the validity of what follows is not limited to materials with linear elastic behaviour, we may consider the more general case of a nonlinear stress distribution (Fig. 81).

If we now consider the cross-section divided in narrow strips which are perpendicular to the neutral axis and have an infinitesimal width, these strips may be considered as rectangles under composed circular *plane* bending. We

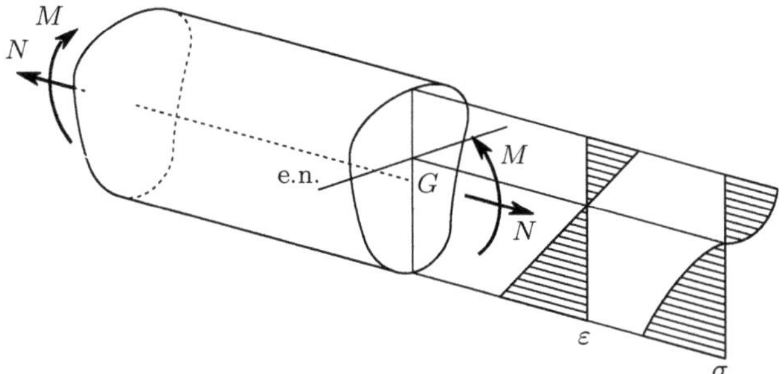

Fig. 81. Piece of a prismatic bar under composed circular bending

should now stress the fact that the division of the cross-section into strips does not change the bending analysis presented above. The only additional consideration is that the stresses σ_x, σ_y and τ_{xy}, which are admitted to vanish, without this being demonstrated (Sect. VII.1), are actually zero in the cross-section divided into strips, since no stresses are applied in its lateral faces, and in the facets parallel to the neutral surface the normal stresses are likewise zero, as there are no lateral forces applied in the piece of the bar under consideration.[11]

The transversal strain of a strip ε_t may be related to the longitudinal strain in the fibre's direction ε by means of the Poisson coefficient ν, yielding

$$\varepsilon_t = -\nu\varepsilon = -\frac{\nu}{\rho}y \ .$$

During the deformation the sides of the strips remain straight, but not parallel, since the transversal strain is proportional to the distance to the neutral axis, *provided that the Poisson's coefficient is constant*. The angle α between the sides of the strip may be related to the curvature of the bar $\frac{1}{\rho}$. The previous expression and the geometrical considerations depicted in Fig. 82 yield the expression

$$\frac{\alpha}{2} = \frac{1}{2}\frac{\mathrm{d}x\,\varepsilon_t}{y} \ \Rightarrow \ \alpha = -\frac{\nu}{\rho}\mathrm{d}x \ . \tag{157}$$

The displacement δ_y of the points of the strip in the direction perpendicular to the neutral axis may also be obtained from ε_t, and are given by the expression

$$\delta_y = \int_0^y \varepsilon_t\,\mathrm{d}y = -\frac{\nu}{\rho}\frac{y^2}{2} \ .$$

---

[11]There will be normal stresses in these facets, in the case of a large curvature introduced by bending. These stresses are the radial stresses which equilibrate the longitudinal stresses acting on the fibres when they acquire curvature. These stresses may, however, be neglected for small values of the curvature.

Since the sides of the strip remain straight and since the displacement δ_y only depends on the distance y (this means that δ_y is equal on both sides of the line dividing two contiguous strips), the strips may be assembled after the deformation, reconstructing the bar without any kind of discontinuities, which means that the deformations represented by the previous expressions are *compatible*. We may therefore conclude that the stresses σ_x, σ_y and τ_{xy} also vanish in the undivided bar.

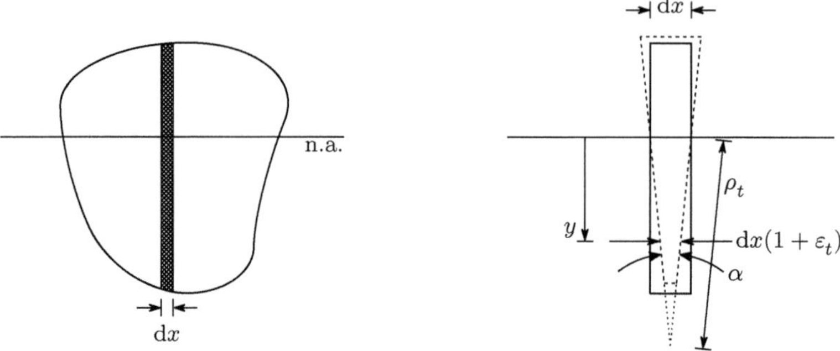

Fig. 82. Transversal deformation of a strip of infinitesimal width and perpendicular to the neutral axis

From Fig. 82 we conclude that, on assembling the deformed strips, the neutral axis is transformed into a circumference arc, whose curvature $\frac{1}{\rho_t}$ may be obtained from (157), yielding

$$\rho_t \alpha = \mathrm{d}x \;\Rightarrow\; \frac{1}{\rho_t} = \frac{\alpha}{\mathrm{d}x} = -\nu\frac{1}{\rho} \;.$$

The neutral surface therefore has an *anticlastic* shape, i.e., it has opposite curvatures in the cross-section plane and in the deflection plane (saddle shape), as represented in Fig. 83 for a bar with rectangular cross-section.

The considerations established in the present section are based only on the law of conservation of plane sections and on a constant value for Poisson's coefficient. Thus, we may consider as demonstrated the statement made in Sect. VII.2 without demonstration that *the normal and shearing stresses acting in facets which are perpendicular to the cross-section plane vanish if the Poisson coefficient is constant.* If this coefficient is not constant, as happens in prismatic bars made of two or more materials, or when plastic deformations take place (cf. Sect. V.3), compatibility conditions for the transversal deformations should be taken into account in the computation of stresses and deformations.

However, these compatibility conditions are usually not considered in the bending analysis of bars with a non-constant Poisson's coefficient. In order to

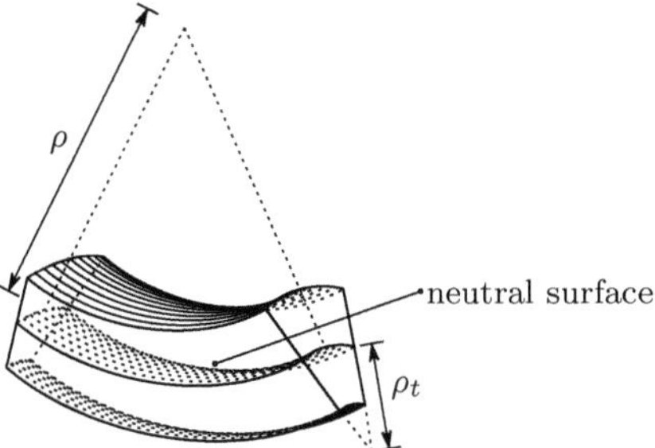

Fig. 83. Deformation caused by a positive bending moment in a bar with rectangular cross-section

get an idea about the importance of the error introduced by this approxima-tion, let us consider the bending of the composite bar represented in Fig. 84. This bar is made of alternate thin layers of two isotropic materials, a and b, which have the same modulus of elasticity E, but different Poisson's coeffi-cients, ν_a and ν_b. The layers all have the same thickness.

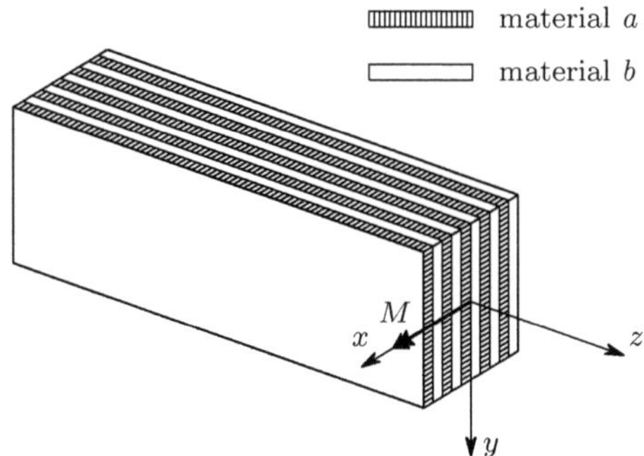

Fig. 84. Composite bar made of two isotropic materials

As the elasticity moduli of the two materials are equal, the solution ob-tained for the pure plane bending of homogeneous materials is valid, if the transversal compatibility conditions are not considered (143) and (145)

$$\frac{1}{\rho} = \frac{M}{EI} \Rightarrow {}^a\sigma_z = {}^b\sigma_z = \frac{Ey}{\rho} \,.$$

If the thickness of the layers is very small in relation to the section's dimensions, we may accept that the strain ε_y is the same in both materials. Besides, as the thickness is the same in all layers, the condition of equilibrium in direction y leads to the conclusion that the stresses σ_y are equal and have opposite signs in the two materials (${}^a\sigma_y = - {}^b\sigma_y$). For simplicity, only the extreme case of having the minimum value of the Poisson coefficient in one material ($\nu_a = 0$) and the maximum value in the other ($\nu_b = 0.5$), is analysed here. The stress-strain relations in the two materials are then (cf. (74))

$$\begin{cases} \varepsilon_z = \dfrac{y}{\rho} = \dfrac{{}^a\sigma_z}{E} \\[2ex] \varepsilon_y = \dfrac{{}^a\sigma_y}{E} \end{cases} \Rightarrow \begin{cases} {}^a\sigma_z = \dfrac{Ey}{\rho} \\[2ex] {}^a\sigma_y = E\varepsilon_y \end{cases} \tag{158}$$

for material a, and

$$\begin{cases} \varepsilon_z = \dfrac{y}{\rho} = \dfrac{1}{E}\left({}^b\sigma_z - \dfrac{1}{2}{}^b\sigma_y\right) \\[2ex] \varepsilon_y = \dfrac{1}{E}\left({}^b\sigma_y - \dfrac{1}{2}{}^b\sigma_z\right) \end{cases} \Rightarrow \begin{cases} {}^b\sigma_z = \dfrac{4}{3}\dfrac{Ey}{\rho} + \dfrac{2}{3}E\varepsilon_y \\[2ex] {}^b\sigma_y = \dfrac{2}{3}\dfrac{Ey}{\rho} + \dfrac{4}{3}E\varepsilon_y \end{cases}$$

for material b.

The strain ε_y may be obtained from these expressions and the equilibrium condition in direction y, yielding the stress ${}^b\sigma_z$ as function of the curvature

$$ {}^a\sigma_y = -{}^b\sigma_y \Rightarrow \varepsilon_y = -\dfrac{2}{7}\dfrac{y}{\rho} \Rightarrow {}^b\sigma_z = \dfrac{8}{7}\dfrac{Ey}{\rho}\,. \tag{159}$$

The moment-curvature relation may be obtained from the condition of equilibrium of moments and from the expressions relating the stresses in the two materials with the curvature (158) and (159), yielding

$$\begin{aligned} M &= \int_{\frac{\Omega}{2}} {}^a\sigma_z y \, d\Omega + \int_{\frac{\Omega}{2}} {}^b\sigma_z y \, d\Omega \\[2ex] &= \int_{\frac{\Omega}{2}} \dfrac{Ey^2}{\rho}\, d\Omega + \int_{\frac{\Omega}{2}} \dfrac{8}{7}\dfrac{Ey^2}{\rho}\, d\Omega \Rightarrow \dfrac{1}{\rho} = \dfrac{14}{15}\dfrac{M}{EI}\,. \end{aligned} \tag{160}$$

Substituting this value in (158) and (159), we get

$$ {}^a\sigma_z = \dfrac{14}{15}\dfrac{My}{I} \approx 0.933\dfrac{My}{I} \quad \text{and} \quad {}^b\sigma_z = \dfrac{16}{15}\dfrac{My}{I} \approx 1.067\dfrac{My}{I}\,. \tag{161}$$

From (160) and (161) we conclude that an error of about 6.7% is introduced into both the computation of the stresses and into the curvature, if the conditions of deformation compatibility in the cross-section plane are not taken into account. However, it should be remembered that this analysis was made for an extreme case ($\nu_a = 0$, $\nu_b = 0.5$). In current materials the error will be smaller. Taking the more usual values $\nu_a = 0.15$ and $\nu_b = 0.30$, for example, a similar analysis shows that the values of the errors decrease to 0.6% in the computation of the curvature and to 1.8% in the computation of the stresses.[12]

---

[12]In the development of (133) and (134) (stresses induced by the axial force in composite bars) the compatibility conditions of the deformations in the cross-section

VII.7 Influence of a Non-Constant Shear Force

As referred in Sect. VII.1, the stresses computed for the case of pure bending are not changed if a constant shear force is applied. However, in the case of a non-uniform shear force distribution, the stresses computed assuming pure bending are no more an exact description of the actual stress distribution.

In order to get an idea about the importance of the error affecting the computation of the normal stresses, when the expressions developed for circular bending are applied to a bar under a non-constant shear force, we compare the results obtained thereby, with the exact solution of a simple problem, given by the Theory of Elasticity. Let us therefore consider the prismatic bar with a rectangular cross-section, under a uniformly distributed loading p, as represented in Fig. 85. The cross-section has a height h and a width b, which is small compared with h, so that the stress distribution may be assumed as plane. The solution of this problem is given by the functions describing the elements of the stress tensor in plane yz [4]

$$\sigma_z = \frac{pz\,(l-z)}{2}\frac{y}{I} + \frac{p}{2I}\left(\frac{2}{3}y^3 - \frac{h^2}{10}y\right)$$

$$\sigma_y = -\frac{p}{2I}\left(\frac{1}{3}y^3 - \frac{h^2}{4}y + \frac{1}{12}h^3\right) \tag{162}$$

$$\tau_{yz} = p\left(\frac{l}{2} - z\right)\frac{1}{2I}\left(\frac{h^2}{4} - y^2\right)\ .$$

In the expression of σ_z the first term coincides with the solution given in (146), since $M = \frac{pz(l-z)}{2}$. Therefore, the second term, which depends only on y, represents the error introduced when the stress is computed by means of (146). It may be easily demonstrated, by making $\frac{\partial \sigma_z}{\partial z} = 0$ and $\frac{\partial \sigma_z}{\partial y} = 0$, that the maximum value of σ_z occurs in the extreme fibres of the cross-section ($z = \frac{l}{2}$, $y = \pm\frac{h}{2}$), if $l > \sqrt{0.4}h$. From the first of (162) we get

$$\begin{cases} y = \frac{h}{2} \\[2mm] z = \frac{l}{2} \end{cases} \Rightarrow \sigma_z = \sigma_{z-\max} = \frac{ph^3}{I}\left(\frac{\alpha^2}{16} + \frac{1}{60}\right) \quad \text{with} \quad \alpha = \frac{l}{h}\ . \tag{163}$$

The error introduced by (146) ($\sigma_{\max-\text{approx}} = \frac{M_{\max}h}{2I} = \frac{ph^3}{I}\frac{\alpha^2}{16}$), may then be expressed by the relation between $\frac{1}{60}$ and $\frac{\alpha^2}{16}$. This relation depends only on the value of α. In slender members we usually have $l \geq 10h$. For $\alpha = 10$ the second term is only 0.27% of the first one. We conclude therefore that the error introduced by (146) is negligible in this case.

---

plane were also not considered, so that an error with the same order of magnitude should be expected (see Footnote 29).

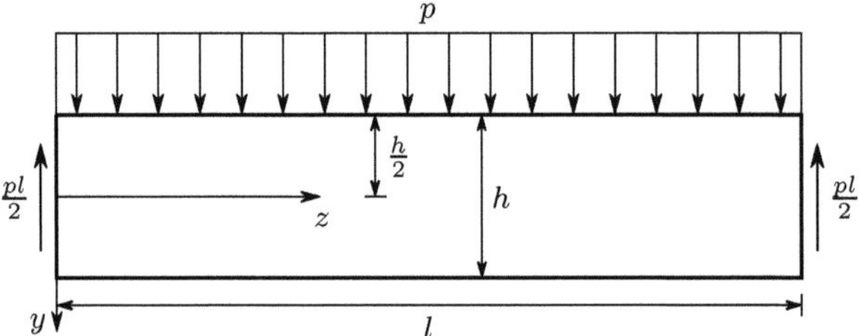

Fig. 85. Beam under a uniformly distributed load

VII.8 Non-Prismatic Members

VII.8.a Introduction

In the same way as in the study of the axial force, the error introduced, when the expressions developed for prismatic members are applied to bars with a non-constant cross-section or with a curved axis, is here investigated by comparing exact solutions given by the Theory of Elasticity with the approximate solutions obtained from the theory of prismatic members in very simple examples.

VII.8.b Slender Members with Variable Cross-Section

As an example of a bar with a non-constant cross-section, the wedge-shaped element with a rectangular cross-section (Fig. 68) is considered again. The force P is now perpendicular to the bar's axis (Fig. 86), so that it causes a bending moment Pr in the cross-section at the distance r of the point of application of the load.

The solution of the Theory of Elasticity is obtained by using polar coordinates (r and θ, Fig. 86) and it shows that in a cylindrical section at the

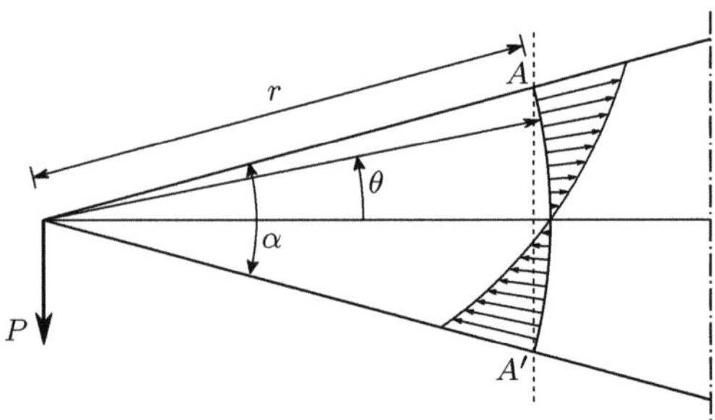

Fig. 86. Stresses caused by bending in a wedge shaped bar

distance r from the vertex the shearing stress vanishes and the radial stress[13] is given by the expression [4]

$$\sigma_r = \frac{2}{\alpha - \sin \alpha} \frac{P \sin \theta}{br} .$$ (164)

The maximum stress for a given value of r occurs for the maximum value of θ, $\theta = \frac{\alpha}{2}$. The theory of prismatic members gives, for the same points, the stress $\sigma_{\text{max−p}}$ (cross-section $AA' \rightarrow h = 2r \sin \frac{\alpha}{2}$, $M = Pr \cos \frac{\alpha}{2}$, (147))

$$\frac{I}{v} = \frac{bh^2}{6} = \frac{2}{3}br^2 \sin^2 \frac{\alpha}{2} \Rightarrow \sigma_{\text{max−p}} = \frac{M}{\frac{I}{v}} = \frac{P}{br} \frac{3}{2} \frac{\cos \frac{\alpha}{2}}{\sin^2 \frac{\alpha}{2}} .$$

The error affecting this approximate expression may be defined by the relation between the maximum stress obtained from (164), $\sigma_{r−\text{max}}$, and $\sigma_{\text{max−p}}$, which leads to

$$\frac{\sigma_{r−\text{max}}}{\sigma_{\text{max−p}}} = \frac{4}{3} \frac{\sin^3 \frac{\alpha}{2}}{(\alpha - \sin \alpha) \cos \frac{\alpha}{2}} .$$

We verify that the error depends only only α and takes the value

α	$10°$	$20°$	$30°$	$45°$	$60°$
$\sigma_{r−\text{max}}/\sigma_{\text{max−p}}$	1.0015	1.0062	1.0141	1.0331	1.0622
error	0.15%	0.62%	1.41%	3.31%	6.22%

We conclude that, for low values of angle α, the error is very small.

The Theory of Elasticity yields also a solution, if the force P is substituted by a moment M. In this case, the stress distribution is not purely radial, since the shearing stress $\tau_{r\theta}$ does not vanish. The solution is given by the expressions (M has a counterclockwise direction)

$$\sigma_r = \frac{2M}{\sin \alpha - \alpha \cos \alpha} \frac{\sin 2\theta}{br^2}$$
$$\sigma_\theta = 0$$
$$\tau_{r\theta} = -\frac{M}{\sin \alpha - \alpha \cos \alpha} \frac{\cos 2\theta - \cos \alpha}{br^2} .$$

We verify that the shearing stress attains a maximum for $\theta = 0$ and that, for values of θ_{max} under $45°$ ($\alpha \leq 90°$), the maximum normal stress occurs at the fibres farthest from the neutral axis and takes the value

$$\theta = \theta_{\text{max}} = \frac{\alpha}{2} \Rightarrow \sigma_r = \sigma_{r−\text{max}} = \frac{1}{br^2} \frac{2M \sin \alpha}{\sin \alpha - \alpha \cos \alpha} = \frac{M}{br^2} \frac{2}{1 - \frac{\alpha}{\tan \alpha}} .$$ (165)

---

[13]It may easily be verified, by evaluating the integral $\int_{-\frac{\alpha}{2}}^{\frac{\alpha}{2}} \sigma_r br \sin \theta \, d\theta$, that the vertical component of the radial stress balances the load P.

This is the maximum principal stress of the stress tensor in point A ($\tau_{r\theta} = 0$ for $\theta = \frac{\alpha}{2}$). The solution of the theory of prismatic members for the same point is

$$\sigma_{\text{max-p}} = \frac{M}{\left(\frac{I}{v}\right)} = \frac{M}{br^2} \frac{1}{\frac{2}{3}\sin^2\frac{\alpha}{2}} \,.$$

In the same way as in the previous case, the error introduced by the approximate solution $\sigma_{\text{max-p}}$ may be expressed by the relation

$$\frac{\sigma_{r\text{-max}}}{\sigma_{\text{max-p}}} = \frac{4}{3} \frac{\sin^2\frac{\alpha}{2}}{1 - \frac{\alpha}{\tan\alpha}} \,.$$

As in the previous case, the error depends only on α and takes the values

α	10°	20°	30°	45°	60°
$\sigma_{r\text{-max}}/\sigma_{\text{max-p}}$	0.9954	0.9818	0.9594	0.9099	0.8430
error	0.46%	1.82%	4.06%	9.01%	15.70%

The error is larger than in the case of the bending moment introduced by the load P but it is also small for low values of α. Besides, as in this case the approximate solution overestimates the actual value of the stress, the error is advantageous for the safety of the structure.

VII.8.c Slender Members with Curved Axis

The errors introduced by the theory of bending for prismatic members, when this is applied to slender members with a curved axis, are studied by comparing the approximate solution given by (147) with the exact solution furnished by the Theory of Elasticity in a curved bar with a circular axis and rectangular cross-section, under a constant bending moment, as represented in Fig. 87.

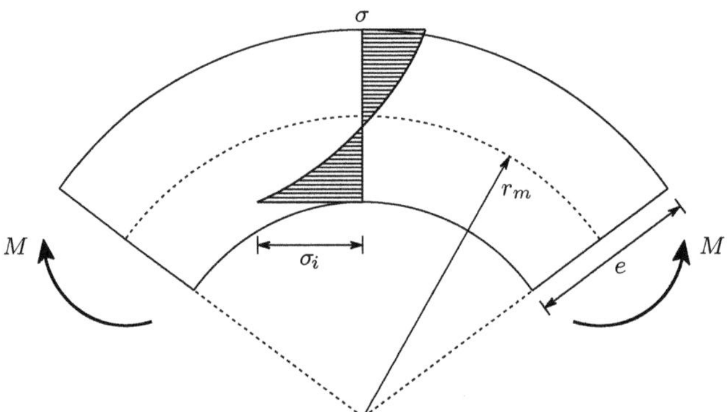

Fig. 87. Stresses introduced by bending in a curved bar

The maximum stress occurs in the extreme fibres of the concave side (σ_i, Fig. 87) and takes the value [4] (b is the cross-section width)

$$\sigma_{\max} = \sigma_i = \frac{6M}{be^2}\,\frac{4\alpha^3 - 4\alpha^2\beta\left(1 + \alpha + \frac{\alpha^2}{4}\right)}{3\beta^2\left(4 - 2\alpha^2 + \frac{\alpha^2}{4}\right) - 12\alpha^2} \quad \text{with} \quad \begin{cases} \alpha = \frac{e}{r_m} \\ \beta = \ln\frac{2+\alpha}{2-\alpha} \,. \end{cases}$$

As the approximate solution given by (147) is in this case $\sigma_{i-\text{approx}} = \frac{6M}{be^2}$, the error depends only on the relation α between the dimension e of the bar and the mean radius of curvature r_m and may be expressed by the relation

$$\gamma = \frac{\sigma_i}{\sigma_{i-\text{approx}}} = \frac{4\alpha^3 - 4\alpha^2\beta\left(1 + \alpha + \frac{\alpha^2}{4}\right)}{3\beta^2\left(4 - 2\alpha^2 + \frac{\alpha^2}{4}\right) - 12\alpha^2} \,.$$

The application of the theory of bending developed for prismatic bars to this curved bar thus leads to the errors

α	0.01	0.05	0.1	0.2	0.3	0.4
γ	1.00335	1.0170	1.0345	1.0717	1.112	1.155
error	0.335%	1.7%	3.45%	7.17%	11.2%	15.5%

Generalizing, we may conclude from these values that, for curved bars where the dimension of the cross-section in the plane containing the axis of the bar exceeds 0.1 times the mean radius of curvature of the bar, the expressions developed for prismatic bars may lead to considerable errors in the computation of bending stresses.

When the parameter α exceeds that value, a bending theory developed by Winkler [8] for curved bars may be used. This theory, although it is based on a simplifying hypothesis – it neglects the effect of the radial stresses (normal stresses acting in facets perpendicular to the curvature radius) – gives results which are very close to the solution furnished by the Theory of Elasticity. The theory of Winkler is also based on the law of conservation of plane sections, but, contrary to the theory of prismatic members, it considers the different initial length of the fibers (the fibres close to the concave face are considerably shorter than the outside fibres). As a consequence of this difference in the initial length of the fibers, the strain and stress distribution is not linear, even though the elongation is proportional to the distance to the neutral axis, but takes a form which is similar to the diagram represented in Fig. 87.

VII.9 Bending of Composite Members

The stresses and deformations induced by a bending moment in a prismatic bar made of more than one material (composite members) are analysed here for the simplest case of a bar made of two materials with linear elastic behaviour. The deformation compatibility conditions in the cross-section plane are not taken into consideration. As a consequence, the theory described here will lead to a small error in the computation of stresses and curvature if the Poisson coefficients of the two materials are different, as seen in Sect. VII.6. This error is not sufficiently high, however, to affect the practical application of the theory.

The law of conservation of plane sections is still valid, since the symmetry conditions used to demonstrate it (cf. Sect. V.10.c) are not affected by the fact that the bar is not homogeneous, provided that the distribution of the two materials in the cross-section is constant. The strain is therefore proportional to the distance to the neutral axis and the strain-curvature relation is still given by (142). Neglecting the stresses acting in facets perpendicular to the cross-section (in accordance with the considerations above), the stresses in the two materials may be related to the curvature of the bar by the expressions

$$\varepsilon = \frac{y}{\rho} \Rightarrow \begin{cases} \sigma_a = \frac{E_a y}{\rho} \\ \\ \sigma_b = \frac{E_b y}{\rho} \, , \end{cases} \tag{166}$$

where E_a and E_b are the elasticity moduli of the two materials and y is the distance of the point under consideration to the neutral axis. We consider here the more general case of inclined bending, as represented in Fig. 88.

As the axial force is zero, the resultant of the normal stresses must vanish. This condition is expressed by the relation

$$\int_\Omega \sigma \, d\Omega = 0 \Rightarrow E_a \int_{\Omega_a} y \, d\Omega_a + E_b \int_{\Omega_b} y \, d\Omega_b = 0 \, . \tag{167}$$

This expression represents the first moment of the areas occupied by each material in the cross-section, with the moment of each area weighted with the modulus of elasticity of the corresponding material, in relation to the neutral axis. Since this moment must vanish, we conclude that the neutral axis passes through the centroid of the cross-section, *computed by weighting the first area moment of each material with the corresponding modulus of elasticity.*

The moment of the stresses in relation to the neutral axis must be equal to the applied bending moment. This condition leads to the expression

$$M \sin \theta = \int_\Omega \sigma y \, d\Omega = \frac{1}{\rho} J_n \quad \text{with} \quad J_n = E_a \int_{\Omega_a} y^2 \, d\Omega_a + E_b \int_{\Omega_b} y^2 \, d\Omega_b \, . \tag{168}$$

J_n is the moment of inertia of the cross-section in relation to the neutral axis, computed by weighting the moment of inertia of the area occupied by each material with the corresponding elasticity modulus. The curvature and the stresses introduced into the bar by the bending moment may be obtained from (168) and (166), yielding

$$\frac{1}{\rho} = \frac{M}{J_\theta} \Rightarrow \begin{cases} \sigma_a = \frac{M E_a}{J_\theta} y \\ \\ \sigma_b = \frac{M E_b}{J_\theta} y \end{cases} \quad \text{with} \quad J_\theta = \frac{J_n}{\sin \theta} \, . \tag{169}$$

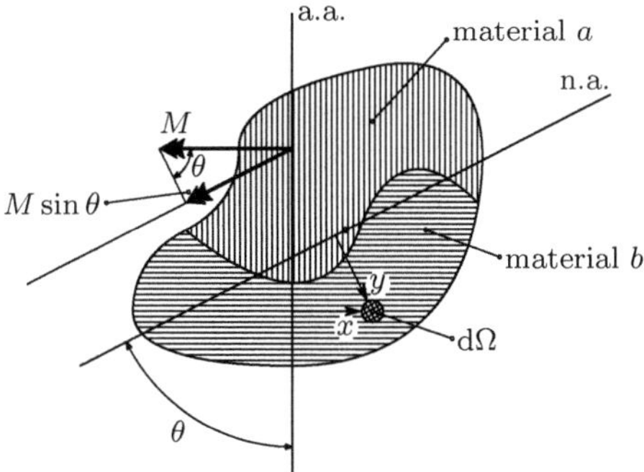

Fig. 88. Circular inclined bending of a prismatic bar made of two materials

The bending stiffness is obviously defined by J_θ. The moment of the stresses in relation to the action axis must vanish. This condition leads to the relation

$$\int_\Omega \sigma x \, d\Omega = 0 \Rightarrow E_a \int_{\Omega_a} xy \, d\Omega_a + E_b \int_{\Omega_b} xy \, d\Omega_b = 0 . \qquad (170)$$

This expression states that the weighted product of inertia with respect to the action and neutral axes vanishes. From this, we conclude, by establishing the same considerations as in the case of the inclined bending of homogeneous bars (Sect. VII.4), that the action and neutral axes are conjugate in relation to the central ellipse of inertia, if the centroid's position and the moments and product of inertia are computed weighting the areas of each material with the corresponding modulus of elasticity. Thus, plane bending will take place if the action axis is parallel to one of principal directions of inertia, computed with weighting. In the same way as in the homogeneous bars, the moment vector may be decomposed in the principal directions of inertia, allowing the inclined bending to be treated as the superposition of two cases of plane bending (Fig. 75).

In the practical applications we often deal with cross-sections with a symmetry axis and with a plane contact surface between the two materials that is either parallel or perpendicular to the symmetry axis. In these cases the concept of *homogenization* may be useful. This concept allows the composite cross-section to be treated as homogeneous. In order to introduce it, let us divide (167) by E_a and put this quantity in evidence in the expression of J_n (168). For simplicity, let us assume plane bending ($\theta = 90°$). We get

$$\int_{\Omega_a} y \, d\Omega_a + m_a \int_{\Omega_b} y \, d\Omega_b = 0 \qquad \text{with:}$$

$$m_a = \frac{E_b}{E_a} \quad \text{and}$$

$$M = \frac{1}{\rho} E_a I_{ha} \Rightarrow \frac{1}{\rho} = \frac{M}{E_a I_{ha}} \qquad I_{ha} = \int_{\Omega_a} y^2 \, d\Omega_a + m_a \int_{\Omega_b} y^2 \, d\Omega_b ,$$

$$(171)$$

where m_a is the homogenizing coefficient of material b in material a. Equations 171 show that, changing the cross-section's shape and dimensions, so that the area of material b is multiplied by a factor m_a, without altering the distances to the neutral axis, i.e., by multiplying the dimensions which are parallel to the neutral axis by m_a, the resulting cross-section may be analysed as homogeneous and made of material a when computing the centroid's position and the curvature caused by the bending moment. The stress in material a may also be computed using the same expression as in the case of a homogeneous bar. Only for the computation of the stress in material b must the homogenizing coefficient be used, as we conclude from (169) ($\theta = 90° \Rightarrow J_\theta = J_n$)

$$\sigma_a = \frac{ME_a}{J_n}y = \frac{ME_a}{E_a I_{ha}}y = \frac{My}{I_{ha}} \quad \text{and} \quad \sigma_b = \frac{ME_b}{J_n}y = \frac{ME_b}{E_a I_{ha}}y = m_a\frac{My}{I_{ha}} \;.$$

Obviously, the cross-section could also be homogenized in material b, as depicted in Fig. 89. In the case of inclined bending, the geometry of the homogenized section is different for the two principal directions, as exemplified in Fig. 90.

As in the case of homogeneous cross-sections, principal directions of inertia x and y may be defined whose orientation is computed from the weighted moments of inertia and the weighted product of inertia with respect to two orthogonal axes originating in the centroid of the weighted cross-section (see example VII.12). From (169) we easily conclude that the stresses in inclined bending may be computed from the components of the bending moment in the principal directions of inertia, M_x and M_y, by the expressions

$$\begin{cases} \sigma_a = \frac{M_x E_a}{J_x}y - \frac{M_y E_a}{J_y}x \\[2ex] \sigma_b = \frac{M_x E_b}{J_x}y - \frac{M_y E_b}{J_y}x \end{cases} \text{with} \quad \begin{cases} J_x = E_a \int_{\Omega_a} y^2\,\mathrm{d}\Omega_a + E_b \int_{\Omega_b} y^2\,\mathrm{d}\Omega_b \\[2ex] J_y = E_a \int_{\Omega_a} x^2\,\mathrm{d}\Omega_a + E_b \int_{\Omega_b} x^2\,\mathrm{d}\Omega_b \;. \end{cases}$$

$$(172)$$

The curvature of the bar may be computed by superposition of the curvatures around the principal axes, as in the case of the homogeneous members (154)

$$\frac{1}{\rho} = \sqrt{\left(\frac{1}{\rho_x}\right)^2 + \left(\frac{1}{\rho_y}\right)^2} = \sqrt{\frac{M_x^2}{J_x^2} + \frac{M_y^2}{J_y^2}} \;. \qquad (173)$$

The equation of the neutral axis may be obtained from any of the conditions $\sigma_a = 0$ or $\sigma_b = 0$, yielding $y = \frac{J_x}{J_y}\frac{M_y}{M_x}x$.

VII.9.a Linear Analysis of Symmetrical Reinforced Concrete Cross-Sections

When analysing reinforced concrete structural elements we generally neglect the tensile strength, since concrete has a much higher resistance to compressive

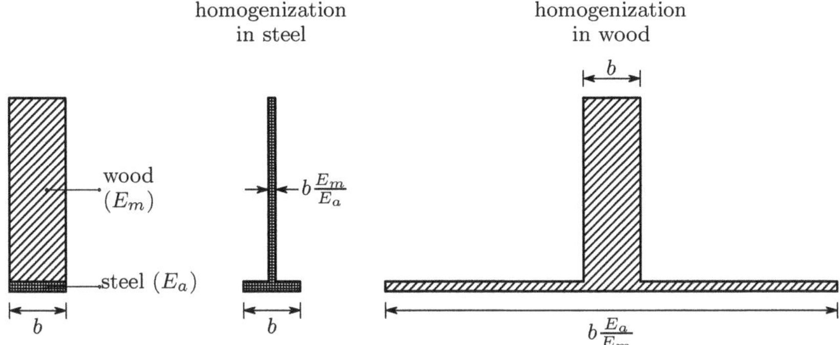

Fig. 89. Homogenization of a composite cross-section

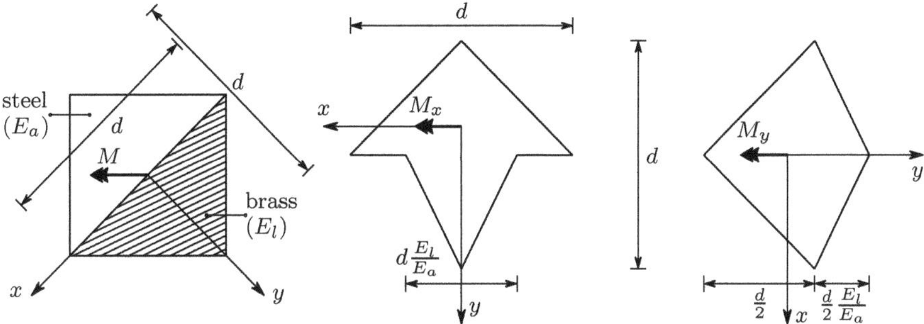

Fig. 90. Homogenization in inclined bending

than to tensile stresses, so that it is usually considered that these stresses are resisted only by the steel bars. In the linear analysis of concrete structures, we admit that the stress is proportional to the strain, as represented in Fig. 91.[14] Looking at this diagram we can immediately see that the above theory for composite bars is not directly applicable, since the stress-strain relation is not defined by the same linear law in tension and compression. However, as the internal tensile forces are supported by the steel bars, this difficulty may be circumvented by considering only the compressed concrete as active. This procedure raises the order of the equation to be solved to compute the position of the section's centroid. In the case of a cross-section that is symmetrical

---

[14]Practical design and safety evaluation of reinforced concrete slender members are currently done by computing the failure bending moment. Under these conditions, a linear stress-strain relation for the concrete is not admissible (see Subsect. VII.10.d). Therefore, the present analysis is only intended to be an example of the application of the theory of composite prismatic bars. However, in high strength concrete, the stress-strain relation remains very close to a straight line almost until rupture (Fig. 91), and so the better the concrete quality, the better the approximation of this analysis. The curvature of a bar under service loads may also be computed, by considering a linear law.

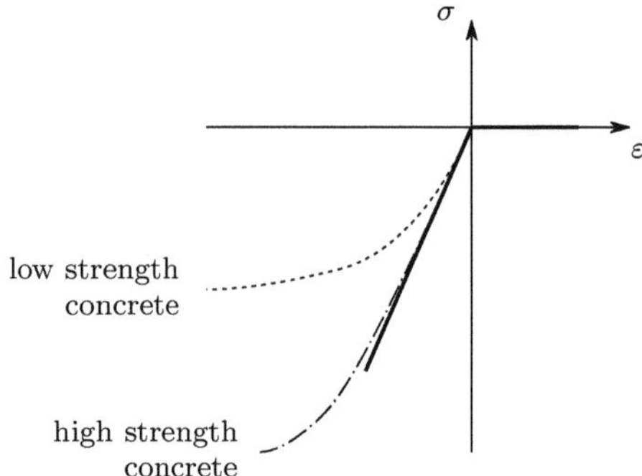

Fig. 91. Admitted stress-strain relation for concrete (*solid line*)

with respect to the action axis, the bending is always plane, regardless of
the material behaviour (see footnote 40). This means that the neutral axis
is perpendicular to the action axis and that the condition $N = 0$ suffices to
compute the position of the neutral axis.

In order to illustrate these considerations, the expressions needed to compute the stresses and curvature caused by plane bending in a rectangular
cross-section are developed. The simplest case of having only one layer of
steel bars is considered (Fig. 92).

Denoting the areas of steel and concrete by Ω_s and Ω_c respectively, the
position of the neutral axis may be obtained from 167, yielding

$$
E_s \Omega_s \left(1 - k\right) h = E_c \underbrace{bkh}_{\Omega_c} \frac{kh}{2} \;\Rightarrow\; k^2 = 2\left(1 - k\right)\lambda
$$

$$
\Rightarrow\quad k = -\lambda + \sqrt{\lambda^2 + 2\lambda} \quad\text{with}\quad \lambda = \frac{E_s}{E_c}\frac{\Omega_s}{bh}\,.
\tag{174}
$$

The weighted moment of inertia of the cross-section in relation to the neutral
axis is given by the expression (cf. (168))

$$
J_n = E_s \Omega_s \left(1 - k\right)^2 h^2 + E_c \frac{b\left(kh\right)^3}{3}\,.
$$

Using the first of (174), two other forms may be given to this expression

$$
J_n = E_s \Omega_s h^2 \left(1 - k\right)\left(1 - \frac{1}{3}k\right)
$$

$$
J_n = \frac{1}{2} E_c bh^3 k^2 \left(1 - \frac{1}{3}k\right)\,.
$$

The stress in the steel and the maximum stress in the concrete may then
be obtained by substituting these values of J_n in (169)

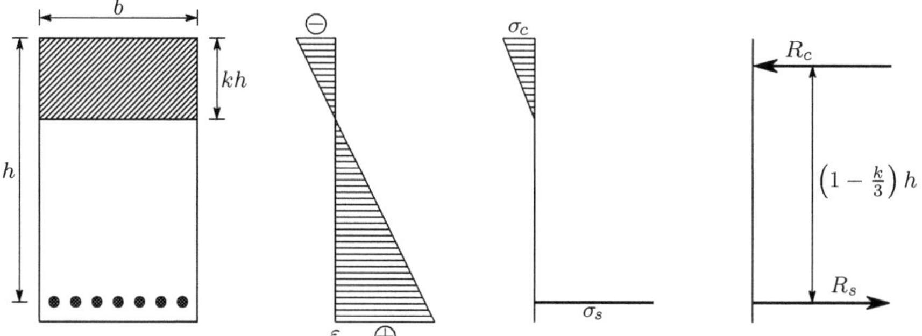

Fig. 92. Reinforced concrete cross-section under a positive bending moment

$$\begin{cases} \sigma_s = \dfrac{ME_s(1-k)h}{J_n} = \dfrac{M}{Z\Omega_s} \\[2ex] \sigma_{c-\max} = \dfrac{ME_c kh}{J_n} = \dfrac{2M}{Z kbh} \end{cases} \quad \text{with} \quad Z = \left(1 - \frac{1}{3}k\right)h \; .$$

Z is the arm of the couple of forces defined by the resultant of the compressive stresses in the concrete, R_c, and by the resultant of the tensile forces in the steel, R_s, (Fig. 92). The curvature of the bar may be computed from any of the above expressions given for J_n, since $\frac{1}{\rho} = \frac{M}{J_n}$.

The superposition principle may be applied to this problem, only if the active cross-section remains the same. Thus, the effects of two positive bending moments with parallel action axes may be superposed, but not the effects of a horizontal and a vertical bending moment. This is because of the lack of complete linearity of the constitutive law considered for the concrete. As a consequence, the analysis of inclined bending is more complicated and cannot be performed on the basis of the bending theory for composite beams described above.

VII.10 Nonlinear bending

VII.10.a Introduction

When the rheological behaviour of the material is not linear, there is no proportionality of stresses and strains and the system of equations obtained from the equilibrium conditions (139) is no longer linear, although the strain distribution remains linear owing to the law of conservation of plane sections (138). In the most general case, the computation of the stresses caused by a bending moment and an axial force in a prismatic bar made of a material, whose one-dimensional constitutive law is described by the function $\sigma = f(\varepsilon) = f(ax + by + c)$[15] requires the solution of the system of equations

---

[15] We still assume that the stresses in facets perpendicular to the cross-section are zero (see Section VII.6).

$$\begin{cases} N = \displaystyle\int_\Omega \sigma \, \mathrm{d}\Omega = \int_\Omega f\left(ax + by + c\right) \mathrm{d}\Omega \\[2ex] M_x = \displaystyle\int_\Omega \sigma y \, \mathrm{d}\Omega = \int_\Omega f\left(ax + by + c\right) y \, \mathrm{d}\Omega \\[2ex] M_y = \displaystyle\int_\Omega \sigma x \, \mathrm{d}\Omega = \int_\Omega f\left(ax + by + c\right) x \, \mathrm{d}\Omega \ . \end{cases} \qquad (175)$$

The degree of complexity of these equations depends on the shape of the cross-section and on the function $f\left(\varepsilon\right)$. The system of equations may admit more than one set of solutions (a, b, c) or be impossible. On the other hand, the computation of the set of internal forces, N, M_x and M_y which corresponds to a given deformation, defined by a set of parameters a, b, and c, is always a determinate problem, as (175) shows.[16]

From these considerations, we can see at once that nonlinear bending is a very wide field. The detailed general analysis of this kind of problems goes beyond the scope of this book, and so only some particularly simple cases are explained.

VII.10.b Nonlinear Elastic Bending

As an example of bending in a nonlinear elastic regime, let us consider a rectangular cross-section under pure plane bending (action axis coinciding with a symmetry axis of the cross-section) made of a material with the constitutive law which is schematically represented in Fig. 93.

Since the compressive and tensile stress-strain relations may be different, the neutral axis generally does not pass through the centroid of the cross-section. Its position may be determined from the condition $N = 0$, as in the linear case, yielding (y_1 and y_2 are defined in absolute value, Fig. 94)

$$\int_\Omega \sigma \, \mathrm{d}\Omega = 0 \Rightarrow \begin{cases} \displaystyle\int_{-y_1}^{y_2} \sigma b \, \mathrm{d}y = 0 \\[2ex] y = \rho\varepsilon \end{cases} \Rightarrow \int_{\varepsilon_1}^{0} \sigma\left(\varepsilon\right) \mathrm{d}\varepsilon = -\int_0^{\varepsilon_2} \sigma\left(\varepsilon\right) \mathrm{d}\varepsilon \ . \quad (176)$$

From (176) we conclude that the neutral axis must take a position which leads to equal compressive and tensile areas defined by the used region of the stress-strain diagram ($A_1 = A_2$, Fig. 94). As the shape of this diagram may not be the same in tension and compression, the position of the neutral axis is not independent of the magnitude of the bending moment, i.e., it may change as the moment increases.

---

[16]This conclusion is valid for most of the problems of Solid Mechanics. The computation of the internal forces corresponding to a given set of displacements is always a determinate problem, while the inverse problem is often not determinate. For this reason, the generalization of the displacement method to nonlinear problems is much easier than in the case of the force method.

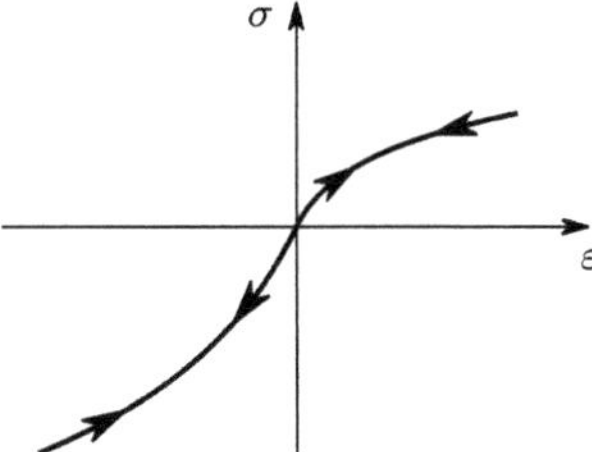

Fig. 93. Nonlinear elastic behaviour

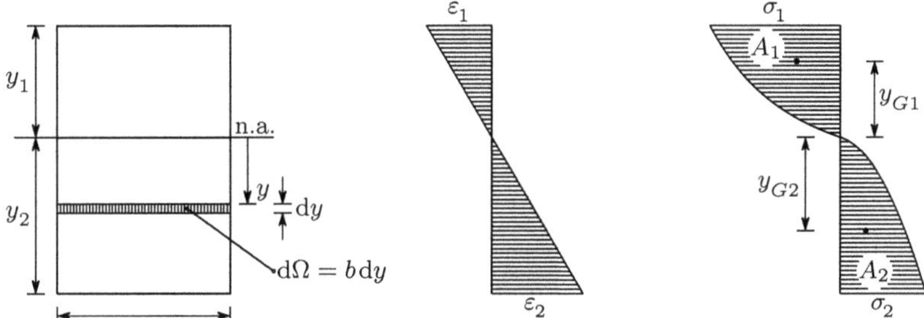

Fig. 94. Bending in nonlinear elastic regime

The bending moment corresponding to the curvature $\frac{1}{\rho} = \frac{\varepsilon}{y}$ (142) may be computed in two steps: first y_1 and y_2 are obtained from the curvature, which can be done by computing the difference between the maximum (tensile) and minimum (compressive) strains

$$\varepsilon_2 - \varepsilon_1 = \frac{y_2}{\rho} + \frac{y_1}{\rho} = \frac{h}{\rho}$$

and computing the values of ε_1 and ε_2 which lead to $A_1 = A_2$; the corresponding bending moment may then by computed by the expression

$$\begin{cases} y_1 = -\rho\varepsilon_1 \\[2mm] y_2 = \rho\varepsilon_2 \end{cases}$$

$$\Rightarrow\ M = \int_{\Omega} \sigma y \, d\Omega = b \int_{-y_1}^{y_2} \sigma y \, dy = A_1 b\, y_{G1} + A_2 b\, y_{G2} = A_1 b\, (y_{G1} + y_{G2}) \ .$$

In this expression y_{G1} and y_{G2} represent the distances from the neutral axis of the centroids of the areas defined by the stress-strain diagram in compression and in tension, respectively, as represented in Fig. 94.

VII.10.c Bending in Elasto-Plastic Regime

If the yielding strain of a ductile material is exceeded, the stress-strain relation becomes non-linear, which means that the linear bending theory is not valid anymore. Furthermore the material behaviour becomes different for loading and unloading. The analysis of this kind of problem is described in detail here for the example of the elasto-plastic plane bending of a bar with a rectangular cross-section, made of a material with elastic perfectly plastic behaviour (Fig. 61). In the last part of this Sub-section, the plastic analysis of cross-sections with one or two symmetry axes under plane bending is described.

Considering the rectangular cross-section represented in Fig. 95, we conclude that the neutral axis divides the cross-section in two equal parts, since the material behaviour is the same for compressive and tensile stresses.

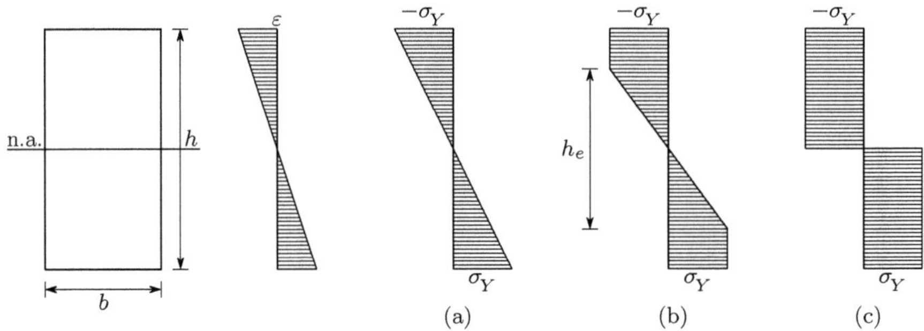

Fig. 95. Plane bending of a rectangular cross-section in the elasto-plastic regime

If the bending moment exceeds the value corresponding to the yielding strain in the fibres farthest from the neutral axis (Fig. 95-a), which is the highest possible bending moment in the elastic phase, the fibres undergoing more strain yield and the bar enters in the elasto-plastic regime (Fig. 95-b). In this phase, the relation between the bending moment and the curvature may be obtained from the strain in the fibres which are still under elastic deformation. With h_e being the height of the part of the section still under elastic deformations (Fig. 95-b), this quantity may be related to the curvature by the expression

$$y = \frac{h_e}{2} \Rightarrow \varepsilon = \varepsilon_Y = \frac{1}{\rho}\frac{h_e}{2} \Rightarrow h_e = 2\rho\varepsilon_Y = \frac{2\rho\sigma_Y}{E} . \qquad (177)$$

The moment of the stresses with respect to the neutral axis must be equal to the bending moment. From this condition, a relation between the bending moment M and h_e may be obtained

$$M = \frac{b}{4}\sigma_Y\left(h^2 - \frac{h_e^2}{3}\right) \Rightarrow M = M_e\left[\frac{3}{2} - \frac{1}{2}\left(\frac{h_e}{h}\right)^2\right] \quad \text{with} \quad M_e = \frac{bh^2}{6}\sigma_Y .$$
$$(178)$$

Substituting, in (178), h_e with the value given by (177), the relation between the curvature and the bending moment is obtained

$$\frac{1}{\rho} = \frac{2\sigma_Y}{hE} \frac{1}{\sqrt{3 - 2\frac{M}{M_e}}} \qquad \text{with} \quad M \geq M_e . \tag{179}$$

From this expression we conclude that the curvature $\frac{1}{\rho}$ goes to infinity, as the bending moment M goes to $\frac{3}{2}M_e$. The maximum bending moment supported by the bar is therefore 1.5 times the maximum bending moment in the elastic regime M_e, i.e., $M_p = 1.5M_e$. M_p represents the yielding bending moment. It corresponds to the limit case represented in Fig. 95-c, as may be easily verified by making $h_e = 0$ in (178). The relation between the yielding bending moment M_p and the maximum bending moment in the elastic regime M_e is called the *shape factor* of the cross-section $\varphi = \frac{M_p}{M_e}$. Thus, in a rectangular cross-section, we have $\varphi = 1.5$.

Let us now consider non-rectangular cross-sections with a symmetry axis. If the action axis is parallel to this axis we will have plane bending. The same happens if the action axis is perpendicular to the symmetry axis. In fact, as the material has the same behaviour in tension and compression, the neutral axis will be the symmetry axis, since, under these conditions, the moment of the stresses in relation to the action axis will vanish, as required by the equilibrium conditions (see Sect. VII.4). A general elasto-plastic analysis of any of these cases is, however, substantially more complex than the rectangular case, since the width of the cross-section is not constant.

Anyway, the most important issue in elasto-plastic analysis is the computation of the yielding bending moment M_p, also called the *plastic moment*. This problem is substantially simpler than the computation of the moment-curvature relation in the elasto-plastic phase, since only the limit case of having a constant tensile or compressive stress σ_Y (yielding stress) on each side of the neutral axis needs to be analysed. Considering the symmetrical cross-section represented in Fig. 96, the condition of equilibrium of the forces acting in the direction of the bar's axis leads to the relation

$$\sigma_Y \Omega_1 - \sigma_Y \Omega_2 = 0 \Rightarrow \Omega_1 = \Omega_2 = \frac{\Omega}{2} .$$

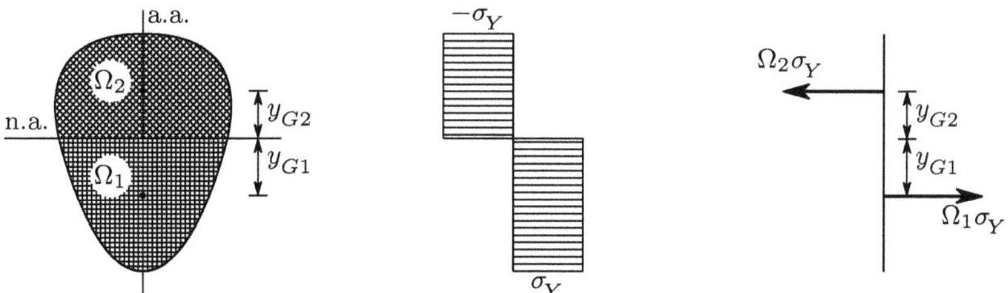

Fig. 96. Fully plastified symmetrical cross-section

Thus, the neutral axis divides the cross-section into two equal areas, which means that it will generally not pass through the centroid, unless it is a symmetry axis. The plastic moment M_p is equal to the moment of the stresses with respect to the neutral axis, taking the value below (y_{G1} and y_{G2} are the distances from the centroids of the tensioned and compressed zones of the cross-section to the neutral axis)

$$M_p = \int_\Omega \sigma y \, \mathrm{d}\Omega = \sigma_Y \int_\Omega |y| \, \mathrm{d}\Omega = \Omega_1 \, \sigma_Y \, y_{G1} + \Omega_2 \, \sigma_Y \, y_{G2} = \frac{\Omega}{2} \left(y_{G1} + y_{G2} \right) \sigma_Y \ .$$

$$(180)$$

The quantity $Z = \frac{M_p}{\sigma_Y} = \frac{\Omega}{2} \left(y_{G1} + y_{G2} \right)$ depends only on the geometry of the cross-section and is called the *plastic section modulus* or simply *plastic modulus*. The shape factor may be obtained from Z and the elastic section modulus $\left(\frac{I}{v} \right)$, $\varphi = \frac{M_p}{M_e} = \frac{Z}{\left(\frac{I}{v} \right)}$. In the following Table, the shape factors of some current symmetrical cross-sections are indicated.

Cross-section	shape factor – φ
rectangle	1.5
isosceles triangle	2.343
rhombus	2
circle	1.7
I-beam	≈ 1.15
channel	≈ 1.2

From these examples we conclude that, the less specialized the section for the resistance to the bending moment, the higher its shape factor. This is due to the fact that such sections have more material in the region around the centroid, whose contribution to the bending strength is not exhausted until the cross-section is totally plastified.

Plastification is a gradual process which, from a theoretical point of view, is only finished for an infinite curvature of the bar, as seen above. However, when the height of the elastic zone is small, we may consider the cross-section as totally plastified, since in that case the contribution of the elastic zone to the resistance to the bending moment is very small. The curvature may then be increased practically without any increase in the bending moment, i.e., *yielding of the entire cross-section* takes place. In this case, we say that a *plastic hinge* has been formed. In Fig. 97 the formation process of such a hinge is schematically represented.

If the bar is unloaded after the maximum bending moment in the elastic phase is exceeded, the internal stresses do not disappear totally, since the material behaves elastically in the unloading process (Fig. 61) and some residual deformation is left in the fibres where the yielding strain was exceeded.

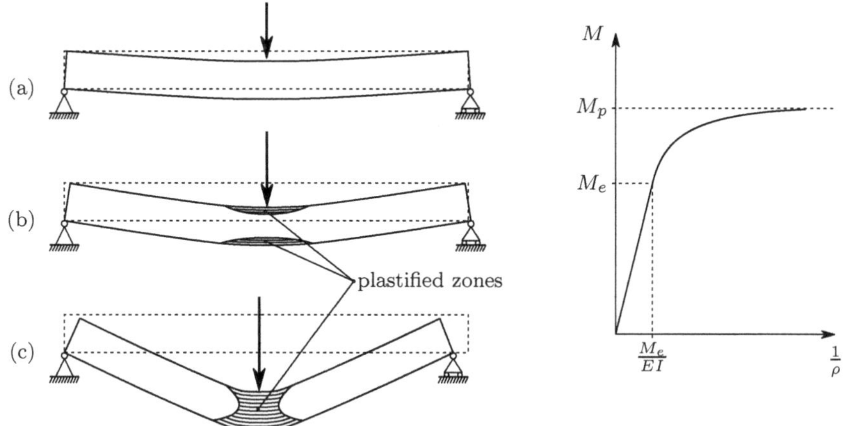

Fig. 97. Formation of a plastic hinge: (**a**) elastic phase; (**b**) elasto-plastic phase; (**c**) plastic phase

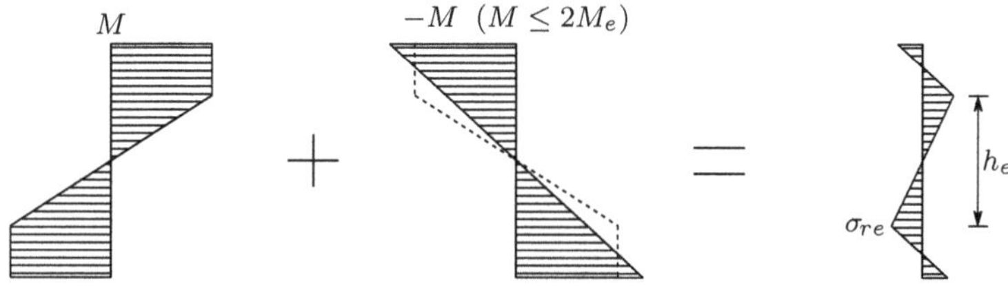

Fig. 98. Residual stresses after unloading in the elasto-plastic regime

The unloading may be understood as the application of a bending moment with the same magnitude and opposite direction. In terms of stresses, the unloading corresponds to the superposition of a linear elastic diagram on the elasto-plastic diagram resulting from the loading, as represented in Fig. 98.

The unloading bending moment will cause yielding of the fibres farthest from the neutral axis, only if the bending moment in the elastic phase exceeds twice the maximum bending moment in the elastic phase. In fact, after yielding under a tensile stress, a stress decrease of $2\sigma_Y$ is needed to cause yielding under compressive stress and vice versa in compression (Fig. 61). Obviously, this is only possible if the cross-section has a shape factor greater than 2, which does not happen in cross-sections used currently to absorb bending moments, as seen above.[17]

---

[17]This line of reasoning is only accurate if there is no displacement of the neutral axis in the elasto-plastic deformation, i.e., if the neutral axis is a symmetry axis. If it is not, the stress may not decrease in the unloading in part of the fibres located close to the neutral axis. However, in usual cross-sections the displacement of the neutral axis is small and the stresses in the region where it occurs are low, so that the error introduced by this procedure will be small, if any. Anyway, this error would be

The residual curvature of the bar may be computed from the residual stress in the fibres where yielding did not take place in the unloading. Thus, we have

$$\varepsilon = \frac{1}{\rho}y \;\Rightarrow\; \frac{1}{\rho} = \frac{\varepsilon}{y} = \frac{2\sigma_{re}}{Eh_e}\,. \tag{181}$$

As an alternative, if the curvature in the loading phase has been computed, the residual curvature may be computed by superposing the curvature recovered in the unloading on it. As the latter is elastic, we have

$$\left(\frac{1}{\rho}\right)_{\text{residual}} = \left(\frac{1}{\rho}\right)_{\text{loading}} - \frac{M}{EI}\,. \tag{182}$$

It should be noted that this operation is valid regardless of the size of the displacements and rotations caused by the deformation, since it is the addition of two angles (the curvature is the relative rotation, measured in radians, of two cross-sections at a unit distance from each other). In fact, the results given by (181) and (182) are exactly the same (see example VII.13).

VII.10.d Ultimate Bending Strength of Reinforced Concrete Members

As a final example of the application of the nonlinear bending theory, the ultimate bending strength of a reinforced concrete member, with a rectangular cross-section is computed. The cross-section of the bar is the same as considered in Sect. VII.9. In this example, the rheological behaviour of steel and concrete recommended in the Portuguese concrete norms is used: the steel is considered as elastic perfectly plastic and to have a limit strain of 0.01; the one-dimensional constitutive law for the concrete is described by a parabola for smaller strains ($0 < \varepsilon < 0.002$), followed by a yielding zone ($0.002 < \varepsilon < 0.0035$), as represented in Fig. 99.[18]

The limit bending moment is reached when the steel reaches a strain of 0.01, or when the maximum strain in the concrete reaches 0.0035. In accordance with the above defined constitutive laws, there are four analytical possibilities of ultimate bending strength, which are:[19]

---

immediately detected, since it would lead to larger stresses than the yielding stress σ_Y.

[18]In the recent European standards (Eurocode 2) different constitutive laws are recommended. Among other differences, the limit strains both for the steel and for the concrete vary with the type of material: for example, high strength concrete has a lower limit strain. For the steel, either an elastic perfectly plastic or a hardening elasto-plastic behaviour may be considered, with no limiting value for the strain in the first case. The constitutive law for concrete is described by a unique curve, which includes a softening zone.

[19]In these expressions and the ones that follow, we consider the compressive stresses and strains in the concrete as positive.

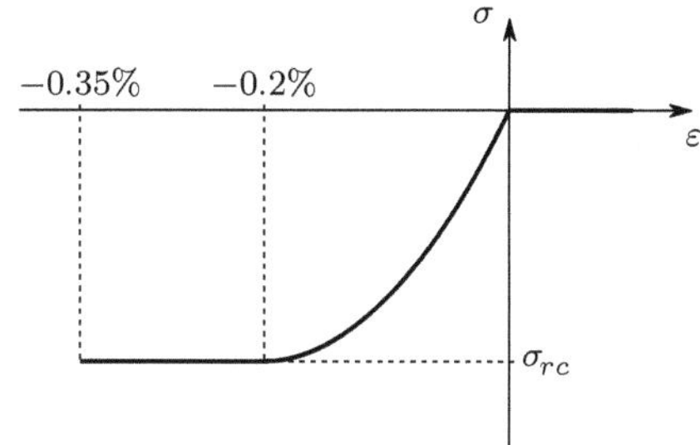

Fig. 99. Stress-strain diagram used in the computation of the ultimate strength of concrete elements

$$\overbrace{\left\{\begin{array}{l}\varepsilon_s = 0.01 \\ \varepsilon_c \le 0.002\end{array}\right. \quad \left\{\begin{array}{l}\varepsilon_s = 0.01 \\ 0.002 \le \varepsilon_c \le 0.0035\end{array}\right.}^{\text{steel failure}} \quad \overbrace{\left\{\begin{array}{l}\varepsilon_s \le \frac{\sigma_Y}{E_s} \\ \varepsilon_c = 0.0035\end{array}\right. \quad \left\{\begin{array}{l}\frac{\sigma_Y}{E_s} \le \varepsilon_s \le 0.01 \\ \varepsilon_c = 0.0035 \, .\end{array}\right.}^{\text{concrete failure}}$$

As an example, only the fourth possibility is analysed, i.e., the case of having the ultimate strain in the concrete and the steel in the yielding zone.[20] Under these conditions, the tensile force borne by the steel is $N_s = \Omega_s \sigma_Y$, where σ_Y is the nominal value of the yielding stress and Ω_s is the area of steel in the cross-section. Representing the position of the neutral axis by its distance from the upper side of the cross-section kh (with $0 \le k \le 1$), from the equilibrium condition of the stress resultants we get

$$\underbrace{\frac{2}{3}\frac{2}{3.5}bkh\sigma_{rc}}_{N_{c1}3(\text{parabola})} + \overbrace{\frac{1.5}{3.5}bkh\sigma_{rc}}^{N_{c2}3(\text{rectangle})} = \underbrace{\Omega_s\sigma_Y}_{N_s} \Rightarrow k = \frac{17}{21}\frac{\Omega_s}{bh}\frac{\sigma_Y}{\sigma_{rc}} \, .$$

The strain in the steel may then be computed from this value. The analysis will be valid if

$$\frac{\sigma_Y}{E_s} \le \varepsilon_s = 0.0035\frac{1-k}{k} \le 0.01 \, . \tag{183}$$

The ultimate bending moment may then be obtained by computing the moment of the stresses with respect to the neutral axis

$$M = N_s\,(1 - kh) + N_{c1}\frac{5}{14}kh + N_{c2}\frac{11}{14}kh \quad \text{with} \quad \left\{\begin{array}{l}N_s = \Omega_s\sigma_Y \\ N_{c1} = \frac{8}{21}bkh\sigma_{rc} \\ N_{c2} = \frac{3}{7}bkh\sigma_{rc} \, .\end{array}\right.$$

---

[20] In the case of mild reinforcing steel, the yielding zone is about $\frac{9}{10}$ of the total range of strains considered in the stress-strain diagram. For this reason, with the quantities of steel used in current members, the failure takes place under these conditions.

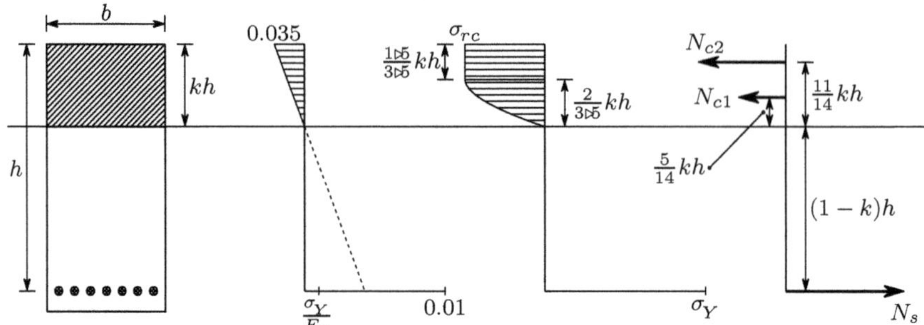

Fig. 100. Limit state of a reinforced concrete rectangular cross-section

The other possibilities of failure may be physically grouped in two. One occurs in members with a very low amount of reinforcing steel, where the steel reaches the maximum strain $\varepsilon_s = 0.01$ before the concrete ($\varepsilon_s > 0.01$ in (183)). The other possibility occurs in members with a very high amount of steel, where the maximum strain in the concrete is attained with the steel in the elastic phase ($\varepsilon_s < \frac{\sigma_Y}{E_s}$). Members with this kind of failure should be avoided, since the collapse is not preceded by plastic deformations, i.e., it is a brittle failure, which is not desirable from the point of view of the safety, as explained in Sect. VI.5. The other failure possibilities could be analysed in the same way as the situation analysed above, by means of the equilibrium conditions of the forces acting in the cross-section.

VII.11 Examples and Exercises

VII.1. Determine the bending strength of a bar with a square cross-section, when the action axis is parallel to:
(a) one of the sides;
(b) one of the diagonals.

Resolution

(a) The moment of inertia of a rectangular cross-section with height h, with respect to the symmetry axis which is parallel to the base b, is given by the expression

$$I = \frac{bh^3}{12}\,.$$

Since the maximum distance to the neutral axis is in this case $\frac{a}{2}$, the section modulus takes the value ($b = h = a$)

$$\frac{I}{v} = \frac{\frac{a^4}{12}}{\frac{a}{2}} = \frac{a^3}{6}\,.$$

Denoting by σ_{all} the nominal value of the material's resisting stress, the resisting bending moment is

$$M_{\text{all}}^{\text{a}} = \frac{a^3}{6}\sigma_{\text{all}} \ .$$

(b) If the action axis is parallel to one of the square's diagonals, the neutral axis is the other diagonal. The moment of inertia with respect to this axis is also $I = \frac{a^4}{12}$. In fact, in a square, the moment of inertia is the same with respect to any axis passing through the centroid, since the two principal moments of inertia are equal. The maximum distance to the neutral axis is, in this case, $v = \frac{a}{\sqrt{2}}$. Thus, the resisting bending moment takes the value

$$M_{\text{all}}^{\text{b}} = \frac{I}{v}\sigma_{\text{all}} = \frac{\frac{a^4}{12}}{\frac{a}{\sqrt{2}}}\sigma_{\text{all}} = \frac{a^3}{6\sqrt{2}}\sigma_{\text{all}} = \frac{1}{\sqrt{2}}M_{\text{all}}^{\text{a}} \approx 0.707 M_{\text{all}}^{\text{a}} \ .$$

VII.2. A tree trunk with circular cross-section of diameter d is to be cut to a rectangular cross-section of base b and height h. Determine the dimensions b and h, in order to maximize:

(a) the bending stiffness;
(b) the bending strength.

Resolution

(a) The geometrical parameter which enters into the definition of the bending stiffness is the moment of inertia. Since the diagonal of the rectangle cannot exceed the diameter of the trunk d, the moment of inertia may be expressed as a function of b, yielding $(h^2 + b^2 = d^2)$

$$I = \frac{bh^3}{12} = \frac{b\left(d^2 - b^2\right)^{\frac{3}{2}}}{12} \ .$$

The value of b which maximizes I may be obtained from the condition of a zero derivative in order to b, which gives

$$\frac{\mathrm{d}I}{\mathrm{d}b} = 0 \ \Rightarrow \ d^2 - 4b^2 = 0 \ \Rightarrow \ b = \frac{d}{2} \ \Rightarrow \ h = \frac{\sqrt{3}}{2}d \ \Rightarrow \ \frac{h}{b} = \sqrt{3} \approx 1.732 \ .$$

(b) To obtain the maximum bending strength, the section modulus must be maximized. The same procedure as before yields

$$\frac{I}{v} = \frac{bh^2}{6} = \frac{b\left(d^2 - b^2\right)}{6} \ ; \qquad \frac{\mathrm{d}}{\mathrm{d}b}\left(\frac{I}{v}\right) = 0 \ \Rightarrow \ d^2 - 3b^2 = 0$$

$$\Rightarrow \ b = \frac{d}{\sqrt{3}} \ \Rightarrow \ h = \frac{\sqrt{2}}{\sqrt{3}}d \ \Rightarrow \ \frac{h}{b} = \sqrt{2} \approx 1.414 \ .$$

VII.3. Compare the section moduli of the three following rectangular cross-sections with the same area $\Omega = bh$. The action axis is parallel to the height h.
(a) Base b and height h.
(b) Base $\frac{b}{2}$ and height $2h$.
(c) Base $2b$ and height $\frac{h}{2}$.

VII.4. Wire made of $S\,235$ steel [10] with a circular cross-section of diameter d is wound around a cylindrical drum for transportation. Determine the minimum diameter D of the winding needed to avoid residual deformation, when the wire is unwound.

Resolution

The wire will not have a residual curvature, if the yielding strain is not exceeded during the winding process, i.e, if the curvature does not exceed the value ((142), $y = \frac{d}{2}$)

$$\varepsilon_{\max} = \frac{1}{\rho}\frac{d}{2} \leq \varepsilon_Y \;\Rightarrow\; \frac{1}{\rho} \leq \frac{2\varepsilon_Y}{d}\ .$$

The yielding stress of this steel and its modulus of elasticity are $\sigma_Y = 235 MPa$ and $E = 206 GPa$ [10], respectively. Thus, the minimum diameter of the winding will be

$$D = 2\rho = \frac{d}{\varepsilon_Y} = \frac{Ed}{\sigma_Y} = \frac{206 \times 10^9}{235 \times 10^6} \approx 877d\ .$$

The exact minimum diameter of the drum would be $D_{\text{drum}} = D - \frac{d}{2}$.

VII.5. Express the section modulus and the moment of inertia as functions of the cross-section area Ω and the height h in the following cross-sections (the action axis is parallel to the height h):
(a) rectangle of base b and height h;
(b) circle of radius r;
(c) isosceles triangle of base b and height h;
(d) rhombus with a horizontal dimension b and height h;
(e) I-beam INP200 [9];
(f) I-beam HE200B [9].

Resolution

(a) Rectangle

$$I = \frac{bh^3}{12} = \frac{1}{12}bhh^2 = \frac{1}{12}\Omega h^2 \approx 0.0833\Omega h^2; \quad \frac{I}{v} = \frac{I}{\frac{h}{2}} = \frac{1}{6}\Omega h \approx 0.1667\Omega h\ .$$

(b) Circle

$$I = \frac{\pi r^4}{4} = \pi r^2 \frac{(2r)^2}{16} = \frac{1}{16}\Omega h^2 = 0.0625\Omega h^2;$$

$$\frac{I}{v} = \frac{I}{\frac{h}{2}} = 0.125\Omega h \ .$$

(c) Triangle

$$I = \frac{bh^3}{36} = \frac{1}{18}\frac{bh}{2}h^2 = \frac{1}{18}\Omega h^2 \approx 0.0556\Omega h^2;$$

$$\frac{I}{v} = \frac{I}{\frac{2}{3}h} = \frac{1}{12}\Omega h \approx 0.0833\Omega h \ .$$

(d) Rhombus:

$$I = \frac{bh^3}{48} = \frac{1}{24}\frac{bh}{2}h^2 = \frac{1}{24}\Omega h^2 \approx 0.0417\Omega h^2;$$

$$\frac{I}{v} = \frac{I}{\frac{h}{2}} = \frac{1}{12}\Omega h \approx 0.0833\Omega h \ .$$

(e) I-beam INP200 ([9], 7.1.1):

$$\begin{cases} I = 2140cm^4 \\[2mm] \Omega = 33.5cm^3 \\[2mm] h = 20cm \end{cases} \Rightarrow \begin{cases} I = \frac{2140}{33.5 \times 20^2}\Omega h^2 \approx 0.1597\Omega h^2 \\[2mm] \frac{I}{v} = \frac{I}{\frac{h}{2}} \approx 0.3194\Omega h \ . \end{cases}$$

(f) I-beam HE200B ([9], 7.1.3):

$$\begin{cases} I = 5696cm^4 \\[2mm] \Omega = 78.1cm^3 \\[2mm] h = 20cm \end{cases} \Rightarrow \begin{cases} I = \frac{5696}{78.1 \times 20^2}\Omega h^2 \approx 0.1823\Omega h^2 \\[2mm] \frac{I}{v} = \frac{I}{\frac{h}{2}} \approx 0.3647\Omega h \ . \end{cases}$$

From these examples, we conclude that, by choosing cross-sections with less material in the region around the neutral axis, like the I-beams, both the bending stiffness and the bending strength are substantially increased, without there being any need to increase the amount of material (represented by Ω) or the height of the cross-section.

VII.6. The bar with the square cross-section represented in Fig. VII.6 is made of a brittle material with linear elastic behaviour until rupture. Determine the increase that can be obtained in bending strength by cutting the bar as represented in the Figure.

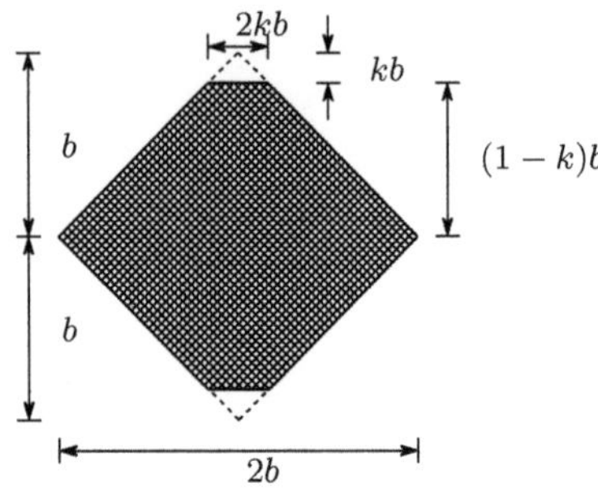

Fig. VII.6

Resolution

Since the material is brittle, rupture takes place for tensile stresses, when the rupture strain is attained, which causes a crack in a transversal direction to the fibres. As a consequence of the stress concentration at the tip of the crack (see Sect. VI.9), it propagates immediately to the whole cross-section causing the failure of the bar. For this reason, the bending strength can only be increased by improving the elastic loading capacity, i.e., by improving the section modulus. The two small symmetrical cuts shown in the Figure reduce the maximum distance to the neutral axis, v without a great reduction of the moment of inertia I, which may increase the section modulus $\frac{I}{v}$.

The moment of inertia of the cross-section with the cuts defined by kb, may be computed from the expressions for the moment of inertia of a rhombus $(I = \frac{bh^3}{48})$ and of a triangle $(I = \frac{bh^3}{36})$ and from the parallel-axis theorem, yielding (Fig. VII.6)

$$I = \frac{2b\,(2b)^3}{48} - 2\left[\frac{2kb\,(kb)^3}{36} + \frac{1}{2}2kb\,kb\left(1 - \frac{2}{3}k\right)^2 b^2\right]$$

$$= \left[\frac{1}{3} - \frac{k^4}{9} - 2k^2\left(1 - \frac{2}{3}k\right)^2\right]b^4\ .$$

The section modulus takes the value

$$\frac{I}{v} = \frac{I}{(1-k)\,b} = \frac{b^3}{1-k}\left[\frac{1}{3} - \frac{k^4}{9} - 2k^2\left(1 - \frac{2}{3}k\right)^2\right]\ .$$

The value of k which maximizes this quantity may be computed by analytical or numerical means, leading to the conclusion that, for $k = \frac{1}{9}$, the section modulus attains the maximum value $0.35117b^3$. Comparing this value with the section modulus of the original cross-section, we get

$$\begin{cases} k = 0 \ \Rightarrow\ \frac{I}{v} = \frac{b^3}{3} = 0.33333b^3 \\[2mm] k = \frac{1}{9} \ \Rightarrow\ \frac{I}{v} = \left(\frac{I}{v}\right)_{\max} = 0.35117b^3 \end{cases} \quad \Rightarrow \quad \frac{\left(\frac{I}{v}\right)_{\max}}{\left(\frac{I}{v}\right)} = 1.0535\ .$$

We conclude that the cuts can increase the bending strength of the bar by 5.35%.

VII.7. Figure VII.7 represents the cross-section of a bar supporting the indicated bending moment M. Justifying the procedure used, determine the variation of the failure bending moment, when the top and bottom small rectangles (top: $2a \times 4a$; bottom: $2 \times a \times 4a$) are removed, so that an I-shaped cross-section is obtained, for:

(a) a brittle material with linear elastic behaviour and rupture stress σ_r;
(b) a ductile material with elastic perfectly plastic behaviour with yielding stress σ_Y.

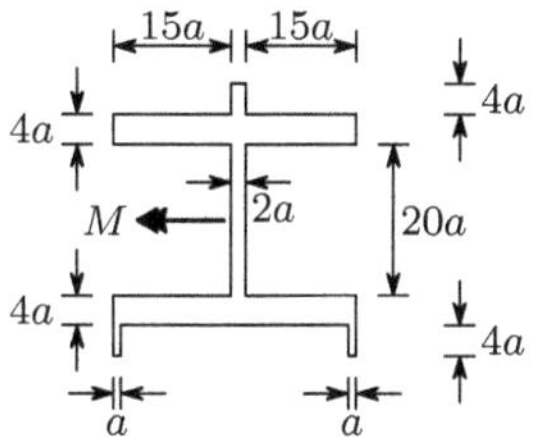

Fig. VII.7

Resolution

(a) In the case of a brittle material with linear elastic behaviour until rupture, the relation between the ultimate bending moments, with and without the small rectangles, coincides with the relation between the section moduli in the two situations. In the case of the original section (Fig. VII.7), the moment of inertia and the section modulus are given by the expressions

$$I_1 = \frac{2a(36a)^3}{12} + \frac{30a(28a)^3}{12} - \frac{30a(20a)^3}{12} = 42656a^4$$

$$\Rightarrow \left(\frac{I}{v}\right)_1 = \frac{42656a^4}{18a} = \frac{21328}{9}a^3 \approx 2369.78a^3 \ .$$

In the case of the cross-section without the small rectangles, the same quantities take the values

$$I_2 = \frac{32a(28a)^3}{12} - \frac{30a(20a)^3}{12} = \frac{115616a^4}{3} \approx 38538.67a^4$$

$$\Rightarrow \left(\frac{I}{v}\right)_2 = \frac{\frac{115616a^4}{3}}{14a} \approx 2752.76a^3 \ .$$

The relation between the section moduli in the two situations is then

$$\frac{\left(\frac{I}{v}\right)_2}{\left(\frac{I}{v}\right)_1} \approx 1.16161 \ .$$

We conclude that the removal of the small rectangles increases the ultimate bending strength by 16%

(b) In the case of a ductile material with elastic perfectly plastic behaviour, the relation between the ultimate bending moments is given by the relation between the plastic section moduli. In the original cross-section (Fig. VII.7) this quantity takes the value

$$Z_1 = 2 \times (18a \times 2a \times 9a + 30a \times 4a \times 12a) = 3528a^3 \ .$$

The removal of the small rectangles causes a fall in the plastic section modulus, which corresponds to the first moment of the removed area

$$Z_2 = Z_1 - 2 \times 2a \times 4a \times 16a = 3272a^3 \ .$$

The relation between the plastic section moduli in the two situations is then

$$\frac{Z_2}{Z_1} = \frac{3272a^3}{3528a^3} \approx 0.92744 \ ,$$

which represents a reduction in the ultimate strength of 7.256%.

VII.8. A bar with a rectangular cross-section with the dimensions $b \times 2b$ supports a bending moment whose action axis is vertical and makes an angle of $45°$ with the symmetry axis of the rectangle. Determine:

(a) the maximum stress in the cross-section;
(b) the position of the neutral axis;
(c) the curvature of the bar.
(d) Compare the answers to questions a) and c), with the corresponding quantities obtained when the bar is rotated so that the action axis becomes parallel to the larger sides of the rectangle.

Resolution

(a) Since we have a cross-section which has a rectangular convex contour with a symmetry axis, (152) may be used to compute the maximum stress. To this end, it is necessary to compute the following quantities (axis y is parallel to the largest side of the rectangle)

$$\left(\frac{I}{v}\right)_x = \frac{bh^2}{6} = \frac{b(2b)^2}{6} = \frac{2}{3}b^3 \qquad \left(\frac{I}{v}\right)_y = \frac{hb^2}{6} = \frac{2bb^2}{6} = \frac{1}{3}b^3$$

$$M_x = M\cos\alpha = \frac{\sqrt{2}}{2}M \qquad M_y = M\sin\alpha = \frac{\sqrt{2}}{2}M \ .$$

Substituting these values in (152), we get

$$\sigma_{\max} = \frac{\frac{\sqrt{2}}{2}M}{\frac{2}{3}b^3} + \frac{\frac{\sqrt{2}}{2}M}{\frac{1}{3}b^3} = \frac{9\sqrt{2}}{4}\frac{M}{b^3} \approx 3.182\frac{M}{b^3} \ .$$

(b) The position of the neutral axis is defined by angle β, which may be obtained by means of (151), yielding

$$\tan\beta = \frac{I_x}{I_y}\tan 45° = \frac{\frac{b(2b)^3}{12}}{\frac{2bb^3}{12}} = 4 \ \Rightarrow \ \beta = \arctan(4) \approx 75.96° \ .$$

We conclude that the deflection plane makes an angle of $75.96° - 45° = 30.96°$ with the vertical plane

(c) The curvature may be computed by means of (154), yielding

$$\frac{1}{\rho} = \frac{M}{E}\sqrt{\frac{\frac{1}{2}}{\left[\frac{b(2b)^3}{12}\right]^2} + \frac{\frac{1}{2}}{\left(\frac{2bb^3}{12}\right)^2}} \approx 4.373\frac{M}{b^4 E} \ .$$

(d) If the cross-section is rotated so that the larger sides of the rectangle become vertical, the maximum stress and the curvature take the values

$$\sigma_{\max} = \frac{M}{\frac{2}{3}b^3} = 1.5\frac{M}{b^3} \qquad \left(\text{instead of } 3.182\frac{M}{b^3}\right)$$

$$\frac{1}{\rho} = \frac{M}{E\frac{b(2b)^3}{12}} = 1.5\frac{M}{b^4 E} \qquad \left(\text{instead of } 4.373\frac{M}{b^4 E}\right) \ .$$

VII.9. Demonstrate the last equality of (154).

Resolution

The last equality of (154) is equivalent to the expression

$$\frac{\sin^2\theta}{I_n^2} = \frac{\cos^2\alpha}{I_x^2} + \frac{\sin^2\alpha}{I_y^2} \ , \tag{a}$$

since we have $I_\theta = \frac{I_n}{\sin\theta}$ (148). Equality (a) may be demonstrated on the basis of the following expressions

$$\tan\alpha = \frac{I_y}{I_x}\tan\beta \quad (151) \quad (b) \qquad \sin^2\alpha = \frac{\tan^2\alpha}{1 + \tan^2\alpha} \quad (e)$$

$$\theta = \frac{\pi}{2} - \beta + \alpha \quad (\text{Fig. 75}) \quad (c) \qquad \cos^2\alpha = \frac{1}{1 + \tan^2\alpha} \quad (f)$$

$$I_n = I_x\cos^2\beta + I_y\sin^2\beta \quad (d) \qquad \sin(a+b) = \sin a\cos b + \cos a\sin b \quad (g) \ .$$

Substituting $\sin^2 \alpha$ and $\cos^2 \alpha$ in the second term of (a) by (e) and (f), respectively, and using (b), we may establish the relation

$$\frac{\cos^2 \alpha}{I_x^2} + \frac{\sin^2 \alpha}{I_y^2} = \frac{1}{I_x^2 \cos^2 \beta + I_y^2 \sin^2 \beta} . \tag{h}$$

By means of (g), we may express $\sin\theta$ as a function of angles α and β, yielding

$$\sin\theta = \sin\left[\left(\frac{\pi}{2} - \beta\right) + \alpha\right] = \cos\alpha\,(\cos\beta + \sin\beta \tan\alpha)$$
$$= \cos\alpha\left(\cos\beta + \sin\beta\frac{I_y}{I_x}\tan\beta\right) .$$

Using this expression, the first term of (a) may be transformed, so that it is expressed in terms of I_x, I_y and β

$$\frac{\sin^2\theta}{I_n^2} = \cos^2\alpha\frac{\left(\cos\beta + \sin\beta\frac{I_y}{I_x}\tan\beta\right)^2}{I_n^2}$$

$$= \frac{\overbrace{\cos^2\alpha}^{1}}{1 + \frac{I_y^2}{I_x^2}\tan^2\beta}\;\frac{\cos^2\beta + 2\frac{I_y}{I_x}\sin^2\beta + \frac{I_y^2}{I_x^2}\frac{\sin^4\beta}{\cos^2\beta}}{I_n^2}$$

$$= \frac{1}{I_n^2}\frac{\overbrace{I_x^2\cos^4\beta + 2I_xI_y\sin^2\beta\cos^2\beta + I_y^2\sin^4\beta}^{I_n^2\,(\,\text{d}\,)}}{I_x^2\cos^2\beta + I_y^2\sin^2\beta} = \frac{1}{I_x^2\cos^2\beta + I_y^2\sin^2\beta} .$$

This result coincides with (h), which shows that (a) is correct.

VII.10. Determine the shape and dimensions of the core of the following cross-sections:
    (a) rectangle with base b and height h;
    (b) circle with radius r;
    (c) rhombus with symmetry axes b and h;
    (d) equilateral triangle with side length a;
    (e) ellipse with semi-axes lengths a and b.

VII.11. The cantilever beam represented in Fig. VII.11-a is made of two materials, a and b, with elasticity moduli $E_a = 2E$ and $E_b = 5E$. The beam supports a vertical loading p by surface unit and a horizontal concentrated force $P = 500pa^2$, as indicated in the Figure. Determine the maximum normal stress in each of the materials.

Resolution

Both the bending moments caused by the vertical and horizontal forces attain maximum values at the built-in end, so that the maximum values of the normal stress occur in that cross-section.

As the beam is made of two materials with linear elastic behaviour, it is necessary to compute the centroid's position, weighting the first area moments with the elasticity moduli of the two materials, as given by (167). Since the cross-section has a vertical axis of symmetry, the position of the centroid is completely defined by the distance d (Fig. VII.11-b)

$$d = \frac{\Omega_a d_a E_a + \Omega_b d_b E_b}{\Omega_a E_a + \Omega_b E_b} = \frac{241}{34}a \approx 7.08824a$$

$$\text{with } \begin{cases} d_a &=& 10a \\ d_b &=& 4.5a \end{cases} \text{ and } \begin{cases} \Omega_a &=& 40a^2 \\ \Omega_b &=& 18a^2 \,. \end{cases}$$

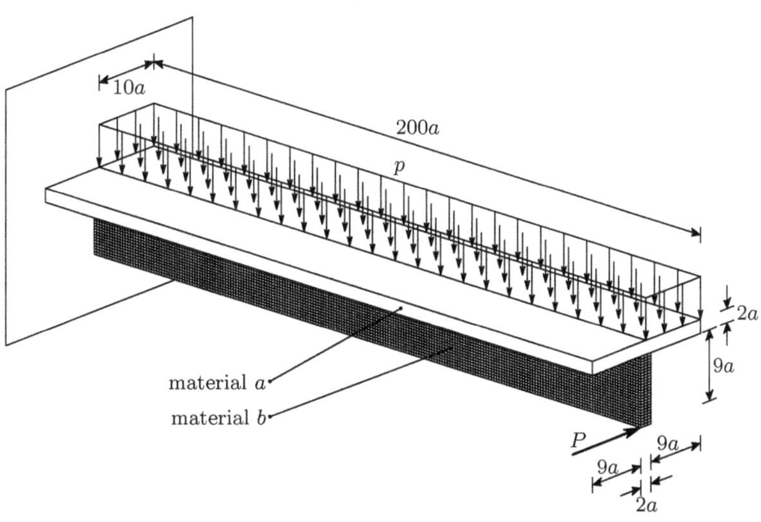

Fig. VII.11-a

The weighted moments of inertia of the cross-section with respect to the principal axes x and y (Fig. VII.11-b) take the values (172)

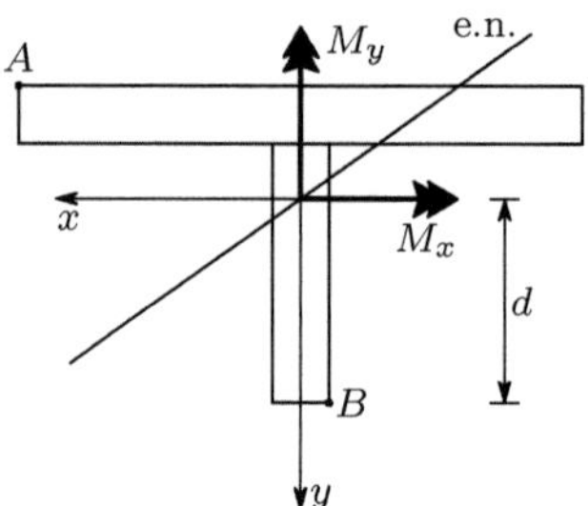

Fig. VII.11-b

$$J_x = E_a \int_{\Omega_a} y^2 \, \mathrm{d}\Omega_a + E_b \int_{\Omega_b} y^2 \, \mathrm{d}\Omega_b$$

$$= 2E \left[\frac{20a(2a)^3}{12} + 40a^2(2.91176a)^2 \right]$$

$$+ 5E \left[\frac{2a(9a)^3}{12} + 18a^2(2.58824a)^2 \right]$$

$$\approx 1915.34 Ea^4$$

$$J_y = E_a \int_{\Omega_a} x^2 \, \mathrm{d}\Omega_a + E_b \int_{\Omega_b} x^2 \, \mathrm{d}\Omega_b$$

$$= 2E \left[\frac{2a(20a)^3}{12} \right] + 5E \left[\frac{9a(2a)^3}{12} \right] \approx 2696.67 Ea^4 \ .$$

The bending moments at the left end cross-section (built-in end) take the values

$$M_x = -\frac{10pa\,(200a)^2}{2} = -2 \times 10^5 pa^3 \qquad M_y = -500pa^2(200a) = -1 \times 10^5 pa^3 \ .$$

Since the neutral axis must be in the same quadrant as the resultant bending moment (Fig. VII.11-b), points A and B are the farthest points from the neutral axes in materials a and b, respectively. The stresses in these points may be computed by means of (172), yielding

$$\sigma_{a-\mathrm{max}} = \frac{M_x E_a}{J_x} y - \frac{M_y E_a}{J_y} x$$

$$= \frac{-2 \times 10^5 pa^3 (2E)}{1915.34 a^4 E}(-3.91176a) - \frac{-1 \times 10^5 pa^3 (2E)}{2696.67 a^4 E}(10a) \approx 1558.59\,p \, ,$$

$$\sigma_{b-\mathrm{max}} = \frac{M_x E_b}{J_x} y - \frac{M_y E_b}{J_y} x$$

$$= \frac{-2 \times 10^5 pa^3 (5E)}{1915.34 a^4 E}(7.08824a) - \frac{-1 \times 10^5 pa^3 (5E)}{2696.67 a^4 E}(-a) \approx -3886.18\,p \ .$$

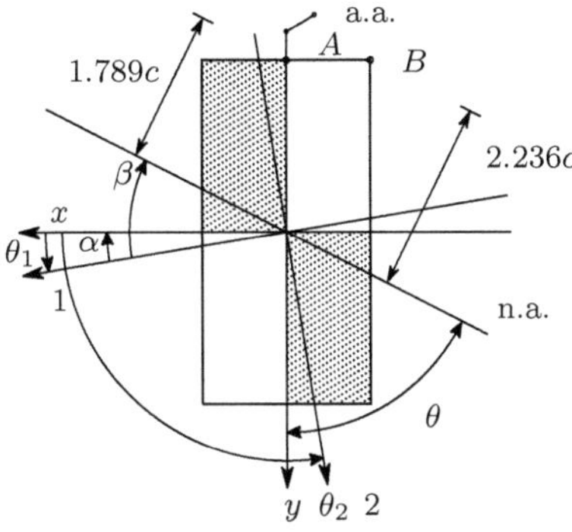

Fig. VII.12

VII.12. Consider the cross-section of a composite bar as depicted in Fig. VI.20. The moduli of elasticity of materials a and b are, respectively, $E_a = 2E$ and $E_b = E$.

(a) Determine the orientation of the principal axes of bending, i.e., the orientation of the action axes which cause plane bending.

(b) Determine the orientation of the neutral axes and the maximum stresses in the two materials, caused by a bending moment M with a vertical action axis.

Resolution

(a) The principal bending axes may be computed from the weighted moments and the product of inertia with respect to the axes x and y represented in Fig. VII.12. These quantities take the values (cf. (168) and (170))

$$J_x = 2E\frac{c(4c)^3}{12} + E\frac{c(4c)^3}{12} = 16Ec^4$$

$$J_y = 2E\frac{2c(2c)^3}{12} + E\frac{2c(2c)^3}{12} = 4Ec^4$$

$$J_{xy} = -2E\frac{c^2(2c)^2}{4} \times 2 + E\frac{c^2(2c)^2}{4} \times 2$$

$$= -2Ec^4 \ .$$

The principal directions of bending may be determined by means of the same expression that is used for the computation of the principal directions of inertia in homogeneous cross-sections, yielding

$$\tan 2\theta = -\frac{2J_{xy}}{J_x - J_y} = \frac{4}{16 - 4} \Rightarrow \begin{cases} \theta_1 = 9.22° \\ \\ \theta_2 = 99.22° \ . \end{cases}$$

(b) The orientation of the neutral axis may be found by means of the expression resulting from the condition $\sigma_a = 0$ in (172), which yields $\beta = \arctan\left(\frac{J_1}{J_2}\tan\alpha\right)$, where 1 and 2 are the weighted principal directions of inertia (the principal directions of bending), α is the angle between the positive directions of the moment vector and axis 1 and β is the angle between the neutral axis and the positive direction of axis 1. The weighted principal moments of inertia may be computed from the values of J_x, J_y and J_{xy} above, by means of the expressions of rotation of reference axes of inertia

$$J_1 = J_x \cos^2\theta_1 + J_y \sin^2\theta_1 - 2J_{xy}\sin\theta_1\cos\theta_1 = 16.32Ec^4$$

$$J_2 = J_x \cos^2\theta_2 + J_y \sin^2\theta_2 - 2J_{xy}\sin\theta_2\cos\theta_2 = 3.675Ec^4 \ .$$

Since the position of the moment vector (direction x, Fig. VII.12) is in this case given by $\alpha = -9.22°$, we get for angle β the value (Fig. VII.12)

$$\beta = \arctan\left[\frac{J_1}{J_2}\tan(-9.22°)\right] = -35.79° \ .$$

The stresses could be obtained from (172). As an alternative, (169) may be used. In order to use the second possibility, the moment of inertia with respect to the neutral axis is needed

$$J_n = J_1 \cos^2\beta + J_2 \sin^2\beta = 12.00Ec^4 \ .$$

From Fig. VII.12 we conclude that the angle between the action and neutral axes is

$$\theta = 180° - (35.79° - 9.22°) - 90° = 63.43° \ .$$

The maximum distances to the neutral axis are

$$\begin{cases} v_a = 2c\cos(35.79° - 9.22°) = 1.789c \\[2ex] v_b = 2c\cos(35.79° - 9.22°) + c\sin(35.79° - 9.22°) = 2.236c \ , \end{cases}$$

respectively for materials a and b (points A and B, Fig. VII.12). The maximum stresses in the two materials are then (169)

$$J_\theta = \frac{12.00Ec^4}{\sin 63.43°} = 13.42Ec^4 \Rightarrow \begin{cases} \sigma_{a-\max} = \frac{M2E}{13.42Ec^4}1.789c = 0.267\frac{M}{c^3} \\[2ex] \sigma_{b-\max} = \frac{ME}{13.42Ec^4}2.236c = 0.167\frac{M}{c^3} \ . \end{cases}$$

VII.13. Consider a prismatic bar that has a rectangular cross-section with a height h, made of a material with elastic perfectly plastic behaviour characterized by the elasticity modulus E and the yielding stress σ_Y. A bending moment with an action axis parallel to the height h, with the magnitude $M = 1.3M_e$ (M_e is the maximum bending moment in the elastic regime) is applied and subsequently removed. Determine:

(a) the curvature of the bar in the loading phase;
(b) the residual curvature after unloading;
(c) the residual stresses.

Resolution

(a) The curvature in the loading phase is given directly by (179), yielding

$$\frac{1}{\rho} = \frac{2\sigma_Y}{hE} \frac{1}{\sqrt{3 - 2 \times 1.3}} = 1.581 \frac{2\sigma_Y}{hE} = 1.581 \frac{1}{\rho_e} \,,$$

where ρ_e is the curvature radius for $M = M_e$.

(b) The deformation recovery in the unloading is elastic and proportional to the removed bending moment. Thus, we have

$$\frac{1}{\rho_{\text{unload}}} = -1.3 \frac{1}{\rho_e} \Rightarrow \frac{1}{\rho_{\text{residual}}} = (1.581 - 1.3) \frac{1}{\rho_e} = 0.281 \frac{1}{\rho_e} \,.$$

(c) To compute the residual stresses it is necessary to determine the height of the cross-section which remains in the elastic regime in the loading phase (h_e, Fig. 98). To this end, (178) may be used, yielding

$$M = M_e \left[\frac{3}{2} - \frac{1}{2} \left(\frac{h_e}{h} \right)^2 \right] \Rightarrow h_e = \sqrt{3 - 2 \frac{M}{M_e}} \, h = \sqrt{0.4} \, h \approx 0.632 h \,.$$

According to Fig. 98, the stresses caused by unloading are $1.3\sigma_Y$ and $\sqrt{0.4} \times 1.3\sigma_Y \approx 0.822\sigma_Y$, respectively in the farthest fibers and in the fibres at distance $\frac{h_e}{2}$ from the neutral axis. Thus, the residual stresses distribution takes the form represented in the last diagram of Fig. 98, with a residual stress in the farthest fibres of $0.3\sigma_Y$, while in the fibres at distance $\frac{h_e}{2}$ from the neutral axis the residual stress takes the value $\sigma_Y - 0.822\sigma_Y = 0.178\sigma_Y$. The residual curvature may also be computed from this last stress (181), yielding

$$\frac{1}{\rho_{\text{residual}}} = \frac{2 \times 0.178\sigma_Y}{E \times 0.632h} = 0.281 \frac{2\sigma_Y}{Eh} = 0.281 \frac{1}{\rho_e} \,.$$

VII.14. Compute the shape factors of the I-beams INP200 and IPE200 [9].

Resolution

INP200

The plastic section modulus ((180) and following text) may be expressed as a function of the first area moment of half cross-section with respect to the neutral axis (S_x, [9], 7.1.1), i.e.,

$$S_x = 125cm^3 \Rightarrow Z = 2S_x = 250cm^3 \,.$$

The shape factor may be given by the relation between the plastic (Z) and the elastic $\left(\frac{I}{v}\right)$ section moduli, yielding

$$\frac{I}{v} = 214cm^3 \Rightarrow \varphi = \frac{Z}{\frac{I}{v}} = \frac{250}{214} \approx 1.168\ .$$

IPE200

The same procedure gives the result ([9], 7.1.2)

$$\begin{cases} S_x = 110cm^3 \\[2em] \dfrac{I}{v} = 194cm^3 \end{cases} \Rightarrow \varphi = \frac{2 \times 110}{194} \approx 1.134\ .$$

VII.15. Consider the cantilever beam with a rectangular cross-section and variable height represented in Fig. VII.15. Compare the exact value of the stress in point A, obtained from the solutions of the Theory of Elasticity, with the approximate solution furnished by the bending theory for the same stress. The cross-section has a constant width b.

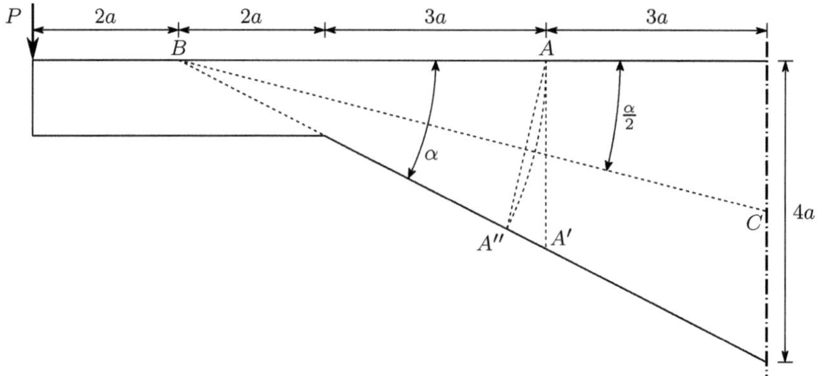

Fig. VII.15

Resolution

According to the theory of prismatic members, in cross-section AA' we have shear force and bending moment, which induces in point A the normal stress

$$\begin{cases} M = 7Pa \\[1em] \overline{AA'} = 2.5a \end{cases} \Rightarrow \sigma = \frac{7Pa}{\frac{b(2.5a)^2}{6}} = 6.72\frac{P}{ab}\ .$$

The solution of the Theory of Elasticity may be obtained by combining the solutions presented in Sects. VI.7.c and VII.8.b for the wedge shaped element.

To this end, it is necessary to consider the system of forces, which is statically equivalent to force P, but acting in point B. Force P is decomposed into two components, one (N) in the direction of the wedge axis (segment BC), and other (V) in the perpendicular direction. Thus, we have

$$M = 2Pa \qquad V = P\cos\frac{\alpha}{2} \qquad N = -P\sin\frac{\alpha}{2}\ .$$

In accordance with (165), the bending moment M causes the stress ($\alpha = \arccos\frac{1}{2}$, $r = 5a$)

$$\sigma_M = \frac{2Pa}{b\,(5a)^2}\,\frac{2}{1 - \frac{\alpha}{\tan\alpha}} \approx 2.2007\frac{P}{ab}\ .$$

The stress caused by the shear force V may be computed from (164), yielding

$$\sigma_V = \frac{2}{\alpha - \sin\alpha}\,\frac{P\cos\frac{\alpha}{2}\sin\frac{\alpha}{2}}{b5a} \approx 5.4425\frac{P}{ab}\ .$$

The axial force N induces the stress (cf. Subsect. VI.7.c)

$$\sigma_N = \frac{2}{\alpha + \sin\alpha}\,\frac{-P\sin\frac{\alpha}{2}\cos\frac{\alpha}{2}}{b5a} \approx -0.0982\frac{P}{ab}\ .$$

Thus, the total stress in point A is

$$\sigma = \sigma_M + \sigma_V + \sigma_N = 7.545\frac{P}{ab}\ .$$

This value is about 12% larger than the solution of the bending theory, which exceeds the predictions indicated in Subsect. VII.8.b. This is because there, a perpendicular section to the wedge axis was considered. In fact, if we consider, instead of section AA', the section AA'', the solution of the bending theory becomes substantially closer to the exact solution, exceeding it by about 5%.

$$\overline{AA''} = 2\times 5a\sin\frac{\alpha}{2} \ \Rightarrow\ \sigma = \frac{7Pa}{\dfrac{b\left(10\sin\frac{\alpha}{2}a\right)^2}{6}} = 7.9566\frac{P}{ab}\ .$$

VII.16. The prismatic bar with the cross-section depicted in Fig. VII.16-a is made of two materials, a and b, and undergoes a uniform temperature increase Δt. The materials have linear elastic behaviour defined by the parameters

$$E_a = E \qquad E_b = 2E \qquad \alpha_a = \alpha \qquad \alpha_b = 2\alpha\ .$$

(a) Determine the elongation and the curvature introduced by Δt (the bar has length l).

(b) Determine the distribution of stresses in the cross-section.

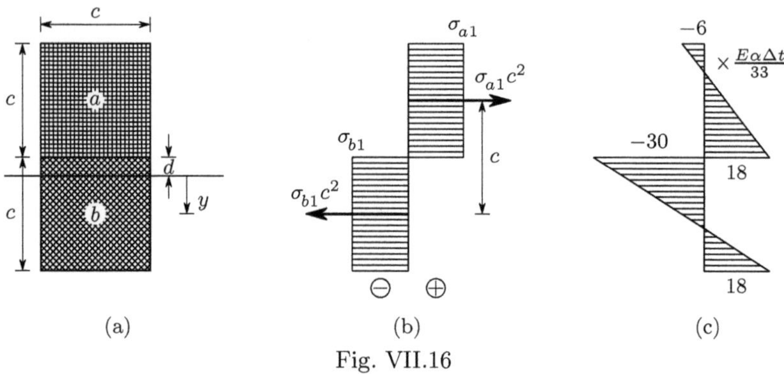

Fig. VII.16

Resolution

(a) The temperature variation causes bending in the bar, as explained in Sect.
VI.6.d (Fig. 65). In order to compute the corresponding curvature, let us
first suppose that the bending is prevented by applying adequate bending
moments at both ends of the bar. Under these conditions, the stresses
in the cross-section are given by (135), since the bar remains straight,
yielding $(\Omega_a = \Omega_b = c^2)$

$$\begin{cases} \sigma_{a1} = \dfrac{2E^2c^4}{Ec^2 + 2Ec^2}\Delta t \dfrac{\alpha}{c^2} = \dfrac{2}{3}E\alpha\Delta t \\[2ex] \sigma_{b1} = \dfrac{2E^2c^4}{Ec^2 + 2Ec^2}\Delta t \dfrac{-\alpha}{c^2} = -\dfrac{2}{3}E\alpha\Delta t \ . \end{cases}$$

The couple of forces corresponding to the stress distribution in the cross-
section (Fig. VII.16-b) is the bending moment needed to prevent bending.
This moment takes the value

$$M = \sigma_{a1}c^2 \times c = \frac{2}{3}c^3 E\alpha\Delta t \ .$$

The elongation of the bar is not affected by this bending moment, and
so it may be computed from the stress in one of the materials. Using, for
example, σ_{a1}, we get from (117)

$$\Delta l = l\left(\alpha\Delta t + \frac{\sigma_{a1}}{E_a}\right) = l\left(\alpha\Delta t + \frac{2}{3}\alpha\Delta t\right) = \frac{5}{3}\alpha\Delta t l \ .$$

If a bending moment $M' = -M$, is subsequently applied to the bar, the
total bending moment vanishes and only the temperature variation remains. Thus, the curvature acquired by the bar in this second loading
phase is the curvature caused by the temperature variation. This curvature may be obtained by the theory of bending of composite members

described in Sect. VII.9. To this end, the position of the neutral axis must be computed ((167) and Fig. VII.16-a)

$$c^2 \left(\frac{c}{2} + d\right) E = c^2 \left(\frac{c}{2} - d\right) 2E \Rightarrow d = \frac{c}{6} \ .$$

The weighted moment of inertia takes the value (168)

$$J_n = E\left[\frac{c^4}{12} + c^2 \left(\frac{c}{2} + \frac{c}{6}\right)^2\right] + 2E\left[\frac{c^4}{12} + c^2 \left(\frac{c}{2} - \frac{c}{6}\right)^2\right] = \frac{11}{12} E c^4 \ .$$

The curvature is then (169, $\theta = 90°$)

$$\frac{1}{\rho} = \frac{M'}{J_n} = \frac{\frac{2}{3} c^3 E \alpha \Delta t}{\frac{11}{12} E c^4} = \frac{8}{11} \frac{\alpha \Delta t}{c} \ .$$

(b) The stresses caused by the bending moment M' may by computed by means of (169), yielding

$$\begin{cases} \sigma_{a2} = \dfrac{ME_a}{J_n} y = \dfrac{\frac{2}{3} c^3 E \alpha \Delta t E}{\frac{11}{12} E c^4} y = \dfrac{8}{11} E \alpha \Delta t \dfrac{y}{c} \\[3mm] \sigma_{b2} = \dfrac{ME_b}{J_n} y = \dfrac{\frac{2}{3} c^3 E \alpha \Delta t 2E}{\frac{11}{12} E c^4} y = \dfrac{16}{11} E \alpha \Delta t \dfrac{y}{c} \ . \end{cases}$$

By superposing the stresses caused by the temperature variation in the straight bar to the bending stresses, the total stresses are obtained

$$\begin{cases} \sigma_a = \sigma_{a1} + \sigma_{a2} = \left(\dfrac{2}{3} + \dfrac{8}{11} \dfrac{y}{c}\right) E \alpha \Delta t \\[3mm] \sigma_b = \sigma_{b1} + \sigma_{b2} = \left(-\dfrac{2}{3} + \dfrac{16}{11} \dfrac{y}{c}\right) E \alpha \Delta t \ . \end{cases}$$

Particularizing these stresses for the upper fibres ($y = -(c + d) = -\frac{7}{6}c$), to the interface between the two materials ($y = -\frac{c}{6}$) and to the bottom fibres ($y = c - d = \frac{5}{6}c$), the stress distribution represented in Fig. VII.16-c is obtained.

VII.17. Figure VII.17-a represents the cross-section of a beam made of a ductile material with elastic perfectly plastic behaviour. The bending moment M acts in the cross-section. Compute the shape factor of this cross-section.

Resolution

The shape factor is the relation between the plastic moment M_p and the maximum moment in the elastic regime M_e, which is equivalent to the relation between the plastic and elastic section moduli, Z and $\frac{I}{v}$, respectively

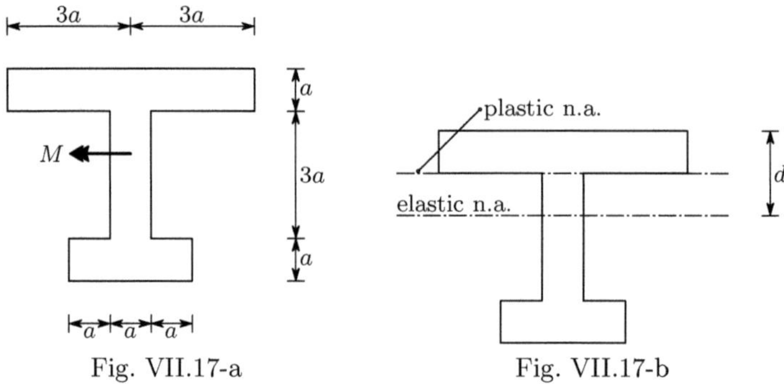

Fig. VII.17-a   Fig. VII.17-b

$$\varphi = \frac{M_p}{M_e} = \frac{Z}{\left(\frac{I}{v}\right)} \ .$$

Since the cross-section does not have a horizontal axis of symmetry, the neutral axis does not have the same position in the elastic and plastic phase, suffering a displacement during the elasto-plastic phase, from the centroid of the cross-section, to a position which divides the cross-section into two equal areas.

The centroid's position may be defined by the distance d (Fig. VII.17-b), which takes the value

$$d = \frac{6a^2 \frac{a}{2} + 3a^2(\frac{5}{2}a + \frac{9}{2}a)}{12a^2} = 2a \ .$$

The moment of inertia, with respect to the elastic neutral axis, is

$$I = \frac{6a^2}{12} + 6a^2 \left(\frac{3}{2}a\right)^2 + \frac{a\,(3a)^3}{12} + 3a^2 \left(\frac{1}{2}a\right)^2 + \frac{3a^4}{12} + 3a^2 \left(\frac{5}{2}a\right)^2 = 36a^4 \ .$$

The elastic section modulus is then

$$\frac{I}{v} = \frac{36a^4}{3a} = 12a^3 \ .$$

The plastic section modulus may be computed from (180), with the neutral axis in the position indicated in Fig. VII.17-b (plastic n.a.), yielding

$$Z = \frac{\Omega}{2}\,(y_{G1} + y_{G2}) = 6a^2 \left(\frac{a}{2}\right) + 3a^2 \left(\frac{3}{2}a + \frac{7}{2}a\right) = 18a^3 \ .$$

The shape factor is then

$$\varphi = \frac{Z}{\left(\frac{I}{v}\right)} = \frac{18a^3}{12a^3} = 1.5 \ .$$

VII.18. Figure VII.18 represents the cross-section of a bar made of a material with elastic perfectly plastic behaviour with a yielding stress σ_Y. Determine the maximum values of the bending moment which can be applied to this cross-section in elastic and in elasto-plastic regime. What is the shape factor of this cross-section?

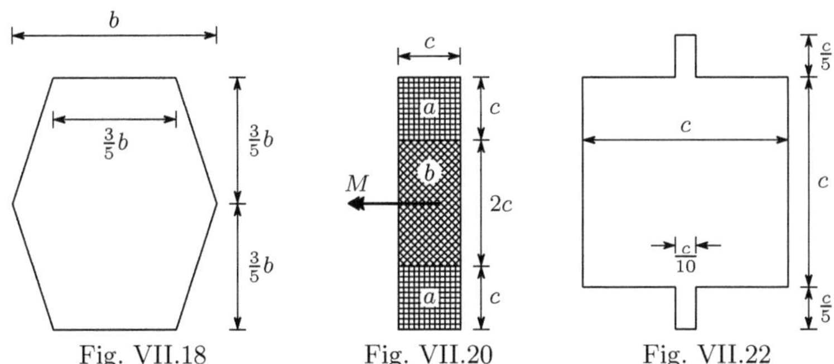

Fig. VII.18   Fig. VII.20   Fig. VII.22

VII.19. A steel wire with a yielding stress ε_Y and rectangular cross-section with dimensions $a \times 3a$ is wound around a cylindrical drum. Determine the minimum diameter the drum can have in order to avoid permanent deformations in the wire. enddescription

VII.20. Figure VII.20 represents the cross-section of a bar made of two materials, a and b, which have elasticity moduli $E_a = 2E$ and $E_b = 3E$ and coefficients of thermal expansion $\alpha_a = 3\alpha$ and $\alpha_b = 2\alpha$. The bar undergoes a uniform temperature increase Δt and supports the bending moment M. Determine:
(a) the elongation of the bar, knowing that it has an initial length l;
(b) the curvature of the bar;
(c) the distribution of stresses in the cross-section.

VII.21. To the bar considered in example VI.15 a bending moment M is applied, whose action axis makes a 30° angle with the vertical.
(a) Determine the curvature of the bar, if only elastic deformations occur.
(b) If M is gradually increased, which of the two materials yields at first? Justify the answer and determine the relation between the maximum stress in this material and moment M, considering only elastic deformations.

VII.22. Supposing that the bar with the cross-section represented in Fig. VII.22 is made of a brittle material with linear elastic behaviour until rupture, ascertain if it is possible to increase its bending strength by removing the two small rectangles, so that a square cross-section $c \times c$ is obtained. Justify the answer.

Answer the same question, supposing now that the material is ductile and that the bar does not undergo cyclic loading.

VII.23. The bar with the cross-section represented in Fig. VII.23 is made of a material with elastic perfectly plastic rheological behaviour, defined by the yielding stress σ_Y and by the elasticity modulus E. Determine:
   (a) the bending moment which is needed to plastify the flanges, while the web remains in the elastic regime;
   (b) the residual stresses, when this bending moment is removed;
   (c) the residual curvature of the bar.

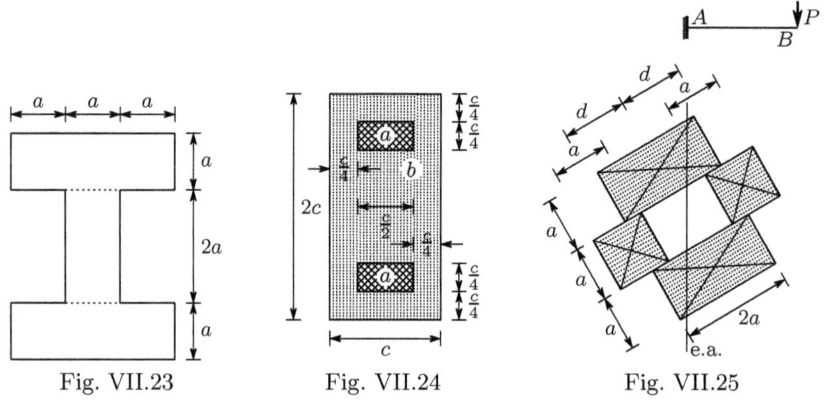

Fig. VII.23   Fig. VII.24   Fig. VII.25

VII.24. The bar whose cross-section is represented in Fig. VII.24 is made of two materials, a and b, with linear elastic behaviour defined by the elasticity moduli $E_a = 4E$ and $E_b = E$, respectively. Determine the maximum stresses in each material caused by a bending moment M with a vertical action axis.

VII.25. The cantilever beam $\overline{AB}$ (Fig. VII.25) is made of a material with linear elastic behaviour and is obtained by assembling four bars, so that the cross-section represented in Fig. VII.25 is obtained. Determine the distance d, so that the displacement of point B has the same direction as the line of action of the force P. Justify the procedure used.

VII.26. Determine and compare the plastic section moduli of the cross-section depicted in Fig. VII.23 and of a rectangular cross-section with the same area and the same height.

Resolution

The cross-section represented in Fig. VII.23 has a area $\Omega = 8a^2$. A rectangular cross-section with the same area and the same height has a width $2a$. The plastic moduli of the cross-section depicted in Fig. VII.23 (Z_1) and of the rectangular cross-section (Z_2) are, respectively

$$Z_1 = 2 \times \left(3a^2 \times \frac{3}{2}a + a^2 \times \frac{a}{2} \right) = 10a^3$$

and

$$Z_2 = 2 \times 4a^2 \times a = 8a^3 \ .$$

We confirm that, even in the case of constant tensile and compressive stresses, the cross-section with less material in the region around the neutral axis has a larger resisting moment, as mentioned in Sect. VII.3 (see Footnote 38).

VII.27. Show that the solution obtained for the pure bending of a prismatic bar made of a material with linear elastic behaviour obeys every condition of equilibrium and compatibility.

Resolution

Considering a reference system, where axis x coincides with the neutral axis and axis z is the centroidal axis of the bar, the solution of the problem may be described by the expressions

$$\sigma_z = \frac{Ey}{\rho}; \ \varepsilon_z = \frac{y}{\rho}; \ \sigma_x = \sigma_y = \tau_{xy} = \tau_{yz} = \tau_{xz} = \varepsilon_x = \varepsilon_y = \gamma_{xy} = \gamma_{yz} = \gamma_{xz} = 0 \ .$$

These expressions obey the constitutive law of the material ($\sigma = E\varepsilon$). Substituting them in the differential equations of equilibrium (5), we find at once that they are satisfied ($\frac{\partial \sigma_z}{\partial z} = 0$). The same happens with the conditions of equilibrium at the lateral boundary (8, $n = 0 \Rightarrow n\sigma_z = 0$). In the end cross-sections we have $n = 1$, i.e., $\overline{Z} = n\sigma_z = \sigma_z$, which means that the equilibrium conditions are satisfied only if the bending moments are applied by means of forces distributed as defined by the linear law defining σ_z. If this does not happen, the solution is only valid for points which are sufficiently far from the end cross-sections for Saint-Venant's principle to be considered valid.

The local conditions of strain compatibility (53) are automatically satisfied, since the only non-zero stress (σ_z) is a linear function of coordinate y. In the case of a multiply connected cross-section, the integral conditions of compatibility would have to be verified, which would require the analysis of the displacement functions corresponding to the above strain distribution. This analysis is, however, not presented here (see, e.g. [1] or [4]). We may also conclude that the integral conditions of compatibility are satisfied by the analysis explained in Sect. VII.6.

VIII

Shear Force

VIII.1 General Considerations

Pure bending is a very rare loading condition. In fact, slender members are very often under the action of shear forces caused by transversal loading or by end moments. The presence of the shear force V implies that the bending moment cannot be constant, since $V = \frac{\mathrm{d}M}{\mathrm{d}z}$ (non-uniform bending: $M \neq 0$ and $V \neq 0$). The shear force is balanced by shearing stresses τ_{zx} and τ_{zy}, acting on the cross-section of the bar. Denoting by V_x and V_y the components of the shear force in the reference axes x and y, the shearing stress distribution in the cross-section must obey the conditions

$$\int_\Omega \tau_{zx}\,\mathrm{d}\Omega = V_x \qquad \text{and} \qquad \int_\Omega \tau_{zy}\,\mathrm{d}\Omega = V_y \ . \tag{184}$$

A supplementary condition is furnished by the reciprocity of shearing stresses in perpendicular facets, which is also an equilibrium condition (see Subsect. II.3.a). According to this condition, if there are no shear forces with a component in the longitudinal direction, applied in the lateral surface of the bar, the shearing stress will be zero in that direction and, as a consequence, in the points of the cross-section which are close to the boundary, the component of the shearing stress which is perpendicular to it will also be zero (Fig. 101). Thus, in the points of the cross-section at an infinitesimal distance to its boundary, *the shearing stress will be tangent to the border line.*

It is obvious that there are infinite stress distributions which obey this condition and also satisfy (184). We have, therefore, a problem with an infinite degree of indeterminacy. The law of conservation of plane sections cannot be used to solve the problem, since, as explained in Sects. V.10 and VII.1, the shear force is not a symmetrical internal force. Besides, the superposition principle cannot be used to analyse the effects of the bending moment and of the shear force separately. In fact, this principle refers to distinct sets of external loads and it is not possible to find a system of transversal forces

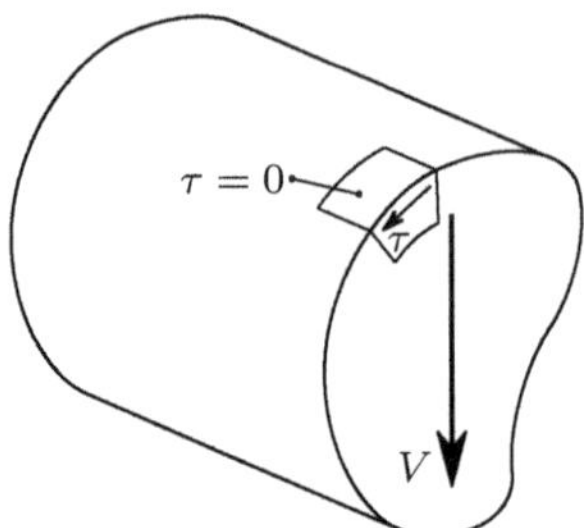

Fig. 101. Shearing stress at the boundary of the cross-section

which causes shear force without introducing a bending moment, since $M = \int V \, \mathrm{d}z + C$, although the opposite is possible, as seen in the analysis of the bending moment.

For these reasons, the analysis of the effect of the shear force expounded here is limited to prismatic bars made of materials with linear elastic behaviour. Furthermore, the following starting hypothesis must be considered (Saint-Venant's hypothesis): *the normal stresses caused by the bending moment in the case of non-uniform bending may be computed by the expressions developed for circular bending.* The validity of this hypothesis will be discussed later. First, it is used to develop the basic tool for the analysis of the effect of the shear force acting on the cross-section: the expression for the computation of the *longitudinal shear force*, i.e., the shear force acting on longitudinal cylindrical surfaces which are parallel to the bar's axis.

VIII.2 The Longitudinal Shear Force

In a prismatic bar under non-uniform bending let us consider the piece defined by two cross-sections at an infinitesimal distance $\mathrm{d}z$ from each other. In this piece let us consider a longitudinal cylindrical surface, defined by the fibres contained in a straight or curved line of the cross-section (Sect. VII.2), as represented in Fig. 102 (squared surface). That line divides the cross-section into two distinct parts, which means that the longitudinal surface divides the piece of bar into two distinct bodies. In order to simplify the development, we first analyse only the case of plane bending.

The equilibrium conditions of the piece of bar as a whole yield the well-known relations between the transversal load P, the shear force in the cross-section V and the bending moment M. Using the sign conventions represented by considering as positive the directions depicted in Fig. 102, we get

$$\begin{cases} \sum F_y = 0 \Rightarrow P = -\frac{\mathrm{d}V}{\mathrm{d}z} \\[2ex] \sum M_x = 0 \Rightarrow V = \frac{\mathrm{d}M}{\mathrm{d}z} \, . \end{cases} \tag{185}$$

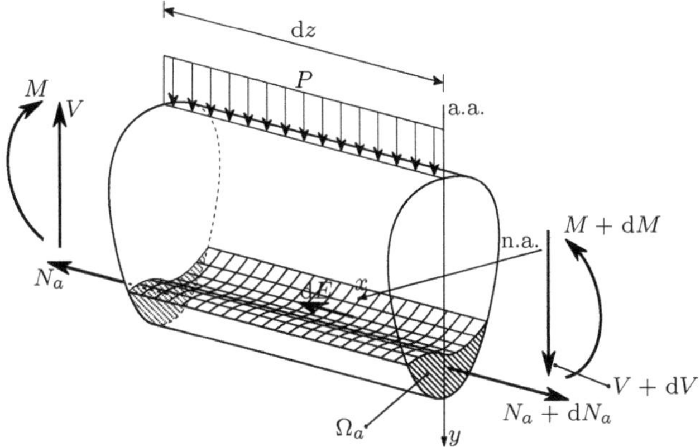

Fig. 102. Longitudinal shear force in a prismatic bar under non-uniform bending

Let us now consider the equilibrium condition of the longitudinal forces acting on the part of the bar defined by the hatched area Ω_a of the left and right cross-sections, which is separated from the remaining bar by the squared longitudinal surface. In the areas Ω_a of the left and right cross-sections, normal stresses caused by the bending moment are acting. According to the Saint-Venant's hypothesis, the forces resulting from these stresses in the left and right cross-sections are given by the expressions (Fig. 102)

$$\sigma = \frac{My}{I} \Rightarrow \begin{cases} N_a = \int_{\Omega_a} \sigma \, d\Omega_a = \frac{M}{I} \int_{\Omega_a} y \, d\Omega_a = \frac{MS}{I} \\[2ex] N_a + dN_a = \frac{M+dM}{I} \int_{\Omega_a} y \, d\Omega_a = \frac{MS}{I} + \frac{S \, dM}{I} \ . \end{cases} \tag{186}$$

In these expressions $S = \int_{\Omega_a} y \, d\Omega_a$ represents the first area moment of the area Ω_a with respect to the neutral axis. The resultant of these two opposite forces – dN_a – must be balanced by the *longitudinal shear force* dE, acting on the contact surface between the two bodies (the squared surface). Thus, this force takes the value ($dM = V \, dz$, (185))

$$dE = N_a + dN_a - N_a = \frac{S \, dM}{I} = \frac{V S}{I} \, dz \ . \tag{187}$$

If the equilibrium of the upper part were to be considered instead, an equal force with opposite direction would be obtained, since the unbalanced force dN_a would have the opposite direction. The first area moment would be $-S$, since the area moment of the whole cross-section in relation to the neutral axis is zero. From (187) we can see that, of all possible longitudinal surfaces, the neutral surface has the maximum longitudinal shear force, because the maximum absolute value of the first area moment S corresponds the whole tensioned area (or to the whole compressed area) of the cross-section.[1]

The longitudinal shear force per unit length is called the *longitudinal shear flow* and is given by the expression

$$f = \frac{\mathrm{d}E}{\mathrm{d}z} = \frac{VS}{I} \, . \tag{188}$$

In the case of inclined bending, the longitudinal shear force may be computed by superposing the forces corresponding to the decomposition of the bending moment and the shear force in the principal axes of inertia, which leads to the expression (cf.(150), $\mathrm{d}M_x = V_y\,\mathrm{d}z$ and $\mathrm{d}M_y = -V_x\,\mathrm{d}z$)

$$\mathrm{d}E = \left(\frac{V_y S_x}{I_x} + \frac{V_x S_y}{I_y} \right) \mathrm{d}z \, ,$$

where $S_x = \int_{\Omega_a} y\,\mathrm{d}\Omega_a$ and $S_y = \int_{\Omega_a} x\,\mathrm{d}\Omega_a$ are the first area moments of Ω_a with respect to the principal axes x and y, respectively. An alternative expression for inclined bending is presented in Subsect. VIII.3.f.

In order to illustrate the importance of this internal force caused by the shear force V, let us consider the cantilever beam depicted in Fig. 103, which is made of two bars with square cross-section $b \times b$.

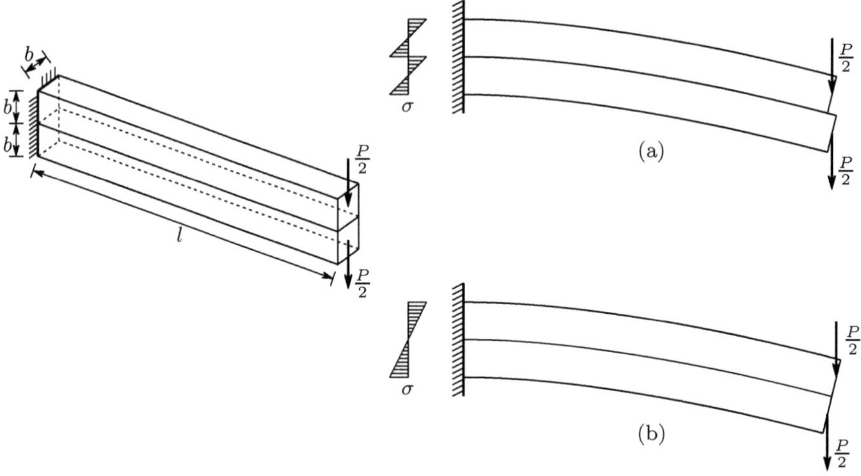

Fig. 103. Non-uniform bending of a built-up beam: **(a)** without friction in the contact surface; **(b)** bars perfectly connected together

If the contact surface between the two bars is lubricated, so that the friction force between the bars is eliminated, each bar will bend independently and a

---

[1]The same holds in the case of inclined bending, since the maximum value of $\mathrm{d}E$ corresponds to the difference between the resultants of the normal stresses acting on the whole tensioned area (or on the whole compressed area) of the cross-section.

relative sliding in the contact surface of the bars takes place, leading to the deformation and stress distribution represented in Fig. 103-a. The maximum stress caused by the bending moment, which occurs in the left end cross-section, may be computed considering the force $\frac{P}{2}$ acting on one beam with square cross-section $b \times b$, yielding

$$M = M_{\max} = \frac{Pl}{2} \Rightarrow \sigma^a_{\max} = \frac{M_{\max}}{\frac{I}{v}} = \frac{\frac{Pl}{2}}{\frac{b^3}{6}} = 3\frac{Pl}{b^3} \,.$$

In the same cross-section the curvature takes the value

$$\frac{1}{\rho_a} = \frac{M_{\max}}{EI} = \frac{\frac{Pl}{2}}{E\frac{b^4}{12}} = 6\frac{Pl}{Eb^4} \,.$$

If the two bars are perfectly connected together, so that the above-mentioned sliding is prevented, the two bars behave as a single unit with a cross-section $b \times 2b$. Thus, the deformation and the stress distribution take the forms represented in Fig. 103-b. The maximum stress and curvature are then given by

$$M = M_{\max} = Pl \Rightarrow \begin{cases} \sigma^b_{\max} = \dfrac{M_{\max}}{\frac{I}{v}} = \dfrac{Pl}{\frac{b(2b)^2}{6}} = \dfrac{3}{2}\dfrac{Pl}{b^3} = \dfrac{1}{2}\sigma^a_{\max} \\[2ex] \dfrac{1}{\rho_b} = \dfrac{M_{\max}}{EI} = \dfrac{Pl}{E\frac{b(2b)^3}{12}} = \dfrac{6}{4}\dfrac{Pl}{Eb^4} = \dfrac{1}{4}\dfrac{1}{\rho_a} \,. \end{cases}$$

We conclude that, by preventing the sliding in the contact surface, the bending stiffness is multiplied by four and the loading capacity of the beam duplicates, since the maximum stress caused by a given load P is divided by two, i.e., twice the load may be applied for the same maximum stress. In this case, the connection between the two bars must resist the shear flow (188)

$$f = \frac{\mathrm{d}E}{\mathrm{d}z} = \frac{VS}{I} = \frac{Pb^2\frac{b}{2}}{\frac{b(2b)^3}{12}} = \frac{3}{4}\frac{P}{b} \,.$$

In order to see how the cross-section deforms in the presence of a shear force, let us consider a piece with infinitesimal length $\mathrm{d}z$, of a bar with a rectangular cross-section. The bar is under non-uniform plane bending with the action axis parallel to height h, as represented in Fig. 104. The width b of the cross-section is very small, compared with the height h, so the shearing stresses in the cross-section may be considered as constant and parallel to the sides of the cross-section in the whole width.

In the horizontal surface $abcd$ the same shearing stress τ as in the cross-section is acting, as a consequence of the reciprocity of the shearing stresses. In this surface, the stress distribution may be admitted as uniform, since the dimension $\mathrm{d}z$ is infinitesimal. The resultant of this shearing stress is the

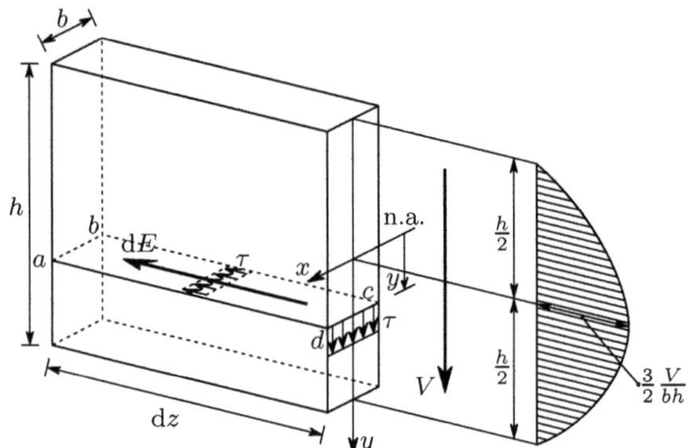

Fig. 104. Shearing stresses caused by the shear force V in a rectangular cross-section with small width

longitudinal shear force given by (187). Thus, the shearing stress takes the value

$$\tau b\,\mathrm{d}z = \mathrm{d}E = \frac{VS}{I}\,\mathrm{d}z \;\Rightarrow\; \tau(y) = \frac{VS(y)}{Ib} \;\Rightarrow\; \tau(y) = \frac{V}{I}\frac{1}{2}\left(\frac{h^2}{4}-y^2\right)\;. \quad (189)$$

This expression defines a parabolic stress distribution, as represented in Fig. 104. The maximum value of the shearing stress occurs on the neutral axis ($y=0$) and takes the value $\tau_{\mathrm{max}} = \frac{Vh^2}{8I} = \frac{3}{2}\frac{V}{bh}$.

Since the shearing strain is proportional to the sharing stress $\left(\gamma = \frac{\tau}{G}\right)$, the cross-section must deform in such a way, that the shearing stress vanishes in the fibres farthest from the neutral axis ($y = \frac{h}{2} \Rightarrow \tau = 0$) and attains a maximum value on the neutral axis ($y = 0 \Rightarrow \tau = \tau_{\mathrm{max}}$). If the cross-section were to remain plane, the shearing strain would be constant in the cross-section (Fig. 105-a) and the distribution of shearing stresses would not be as represented by (189). Thus, we conclude that, either the starting hypothesis for the analysis of the effect of the shear force is wrong (the Saint-Venant hypothesis), or the cross-section must deform as represented in Fig. 105-b.

However, by considering all pieces of infinitesimal length $\mathrm{d}z$ separately, we verify that, provided that the shear force is constant, the same warping in all cross-sections takes place. This means that the deformations of the different pieces are *compatible*, i.e., that the deformed infinitesimal pieces fit perfectly together. Thus, no additional normal stresses are needed to make deformations compatible, which means that the strain distribution resulting from Saint-Venant's hypothesis obeys all conditions of compatibility.

This example shows that the cross-section may warp without the need to change the length of the fibres ($\overline{aa} = \overline{a'a'}$, Fig. 105), provided that the shear force does not vary along the axis of the bar. Since the deformation caused by the shear force does not require changes in the fibres' length, this force may be resisted without altering the distribution of the normal stresses corresponding

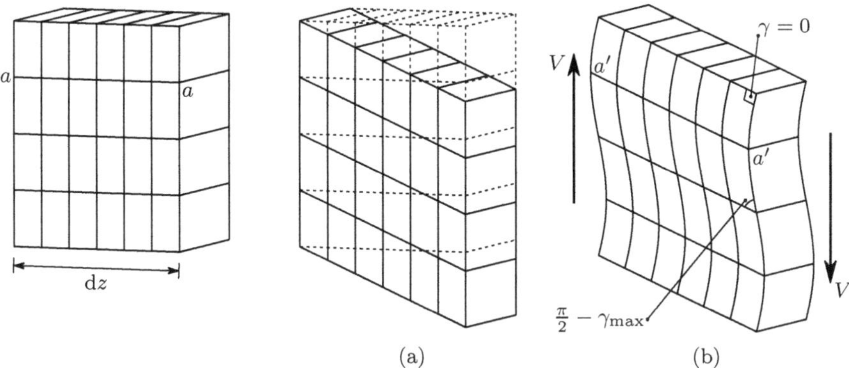

Fig. 105. Warping of a rectangular cross-section caused by the shear force V

to circular bending, i.e., there is no objection to the validity of the Saint-Venant hypothesis. This conclusion may be generalized to a cross-section of any shape, since the shearing strains corresponding to any distribution of shearing stresses may occur without the need to change the length of the fibres, provided that the warping is the same in all cross-sections.

These considerations are not a complete proof of the validity of the Saint-Venant hypothesis in the case of constant shear force. However, they do show that this possibility exists and the solutions of the Theory of Elasticity for particular problems confirm that, *if the shear force is constant, the distribution of normal stresses caused by the bending moment is the same as in circular bending, i.e., it is the same as when the cross-sections remain plane and perpendicular to the bar's axis.* This means that the law of conservation of plane sections is a sufficient condition for a linear distribution of the longitudinal strains in the cross-section, although it may not be necessary, as we conclude from the above considerations.

In the case of a non-constant shear force, this is no longer valid. However, as discussed in Sect. VII.7, the error affecting the computation of the normal stresses and, as a consequence, the computation of the longitudinal shear force by means of (187), is very small and may even vanish (see Sect. VIII.6).

From a practical point of view, (187) may thus be considered exact. However, the computation of the shearing stress from the longitudinal shear force always requires simplifying hypotheses, which introduce errors, whose importance depends on the shape of the cross-section. Thus, good approximations for the shearing stress distribution are obtained for symmetrical cross-sections, if the action axis of the shear force coincides with the symmetry axis and in the cases of *thin-walled* cross-sections. In other cases it is generally not possible to compute the shearing stresses by means of the elementary theory presented in this book. These cases, as well as the errors introduced by the simplifying hypotheses used are discussed below.

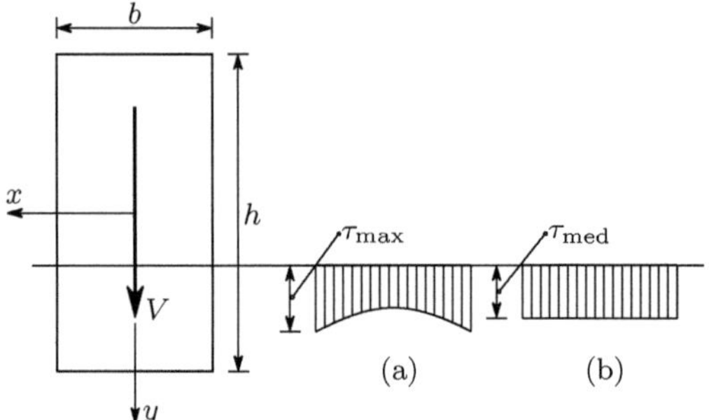

Fig. 106. Shearing stress τ_{zy} in a rectangular cross-section: (**a**) real distribution; (**b**) admitted distribution

VIII.3 Shearing Stresses Caused by the Shear Force

VIII.3.a Rectangular Cross-Sections

In rectangular cross-sections under plane bending the simplifying hypothesis which consists of considering the shearing strain as constant in the width of the cross-section is usually considered: that is, the stress varies only in the direction parallel to the action axis of the shear force. This corresponds to the generalization to rectangular sections with any width/height ratio of the assumptions used in previous section for the small width case. In the case of inclined non-uniform bending, the shear force is decomposed in the symmetry axes. Thus, in a point defined by its coordinates x and y, the two components of the shearing stress are ((189) and Fig. 104)

$$\begin{cases} \tau_{zy} = \frac{V_y}{I_x} \frac{1}{2} \left(\frac{h^2}{4} - y^2 \right) = \frac{V_y}{bh} \left[\frac{3}{2} - 6 \left(\frac{y}{h} \right)^2 \right] \\[2mm] \tau_{zx} = \frac{V_x}{I_y} \frac{1}{2} \left(\frac{b^2}{4} - x^2 \right) = \frac{V_x}{bh} \left[\frac{3}{2} - 6 \left(\frac{x}{b} \right)^2 \right] . \end{cases} \tag{190}$$

The Theory of Elasticity provides a solution for this problem, which is obtained without the simplifying hypothesis above. This solution indicates that the shearing stress is not constant in the direction perpendicular to the action axis of the shear force unless the Poisson's coefficient vanishes, but it has a maximum in the points close to the lateral sides, as represented in Fig. 106-a.

The maximum value of the shearing stress, which occurs for $x = \pm \frac{b}{2}$ and $y = 0$, may be computed by the expression (cf. e.g. [4])

$$\tau_{\max} = \alpha \frac{3}{2} \frac{T}{\Omega}$$

$$\text{with} \quad \alpha = 1 + \frac{\nu}{1 + \nu} \left(\frac{h}{b} \right)^2 \left[\frac{2}{3} - \frac{4}{\pi^2} \sum_{n=1,2,3,\ldots}^{\infty} \frac{1}{n^2 \cosh \left(n\pi \frac{h}{b} \right)} \right] . \tag{191}$$

The coefficient α represents the correction to be applied to the maximum stress obtained from (190), in the case of plane bending, $\tau_{\max} = \frac{3}{2}\frac{V}{\Omega}$. This coefficient depends on the height/width ratio (h/b) and on the Poisson coefficient of the material, ν. The following table gives values of α, computed from (191), for some cases.

α	$\nu = 0$	0.05	0.1	0.15	0.2	0.25	0.3	0.4	0.5
$h/b = 0.25$	1.0000	1.2352	1.4491	1.6443	1.8233	1.9879	2.1399	2.4113	2.6466
0.50	1.0000	1.0944	1.1802	1.2585	1.3303	1.3964	1.4574	1.5663	1.6606
0.75	1.0000	1.0498	1.0951	1.1365	1.1744	1.2093	1.2415	1.2990	1.3488
1.00	1.0000	1.0301	1.0574	1.0823	1.1052	1.1263	1.1457	1.1804	1.2104
1.25	1.0000	1.0198	1.0379	1.0543	1.0694	1.0833	1.0961	1.1190	1.1388
1.50	1.0000	1.0140	1.0266	1.0382	1.0488	1.0586	1.0676	1.0837	1.0977
2.00	1.0000	1.0079	1.0151	1.0217	1.0277	1.0333	1.0384	1.0475	1.0554
4.00	1.0000	1.0020	1.0038	1.0054	1.0069	1.0083	1.0096	1.0119	1.0139

This table shows that the error of the solution furnished by (190) increases with the value of the Poisson coefficient and decreases as the height/width ratio increases. The dependence of the error on the relation $\frac{h}{b}$ has greater practical relevance, since structural materials with a Poisson coefficient smaller than 0.05 are not common, while rectangular cross-sections with height/width ratios superior to 2 are widely used.

VIII.3.b Symmetrical Cross-Sections

In practical applications cross-sections that are symmetrical with respect to the action axis of the shear force are common. In these cases, the computation of the shearing stresses may be carried out by considering two simplifying hypotheses: the vertical component of the shearing stress τ_{zy} is constant in the direction perpendicular to the symmetry axis; the total stress vectors τ in a line perpendicular to the symmetry axis have directions converging to the point defined by the two tangents to the cross-section's contour on that line, as represented in Fig. 107.

The vertical component of the shearing stress may then be computed in the same way as in the rectangular cross-section, taking the value

$$\tau_{zy} = \frac{VS(y)}{Ib(y)} \, . \tag{192}$$

The horizontal component and the resultant stress may then be obtained from this value and angle ψ, yielding

$$\tau_{zx} = \tau_{zy} \tan \psi \quad \Leftrightarrow \quad \tau = \sqrt{\tau_{zx}^2 + \tau_{zy}^2} = \frac{\tau_{zy}}{\cos \psi} = \frac{VS}{Ib \cos \psi} \, . \tag{193}$$

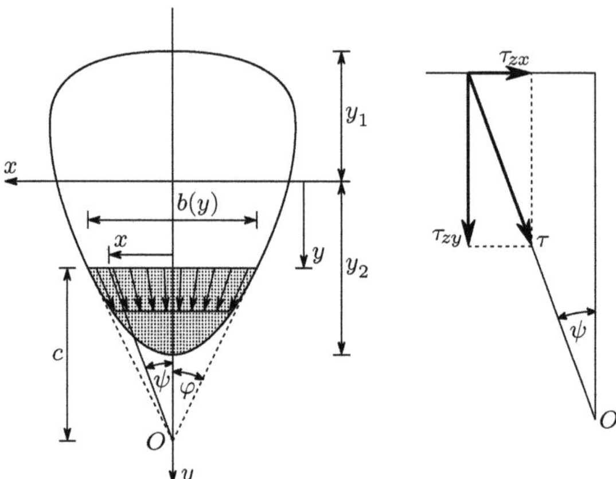

Fig. 107. Simplifying hypotheses for the computation of the shear force in a symmetrical cross-section

The maximum stress for a given value of y occurs clearly on the contour of the cross-section, taking the value $\tau_{\max} = \dfrac{VS}{Ib\cos\varphi}$.

As an applied example let us consider a circular cross-section (Fig. 108). The first area moment of the surface element defined by the central angle β is given by the expression (Fig. 108)

$$\mathrm{d}S = \overbrace{r\sin\beta}^{x}\ \overbrace{r\,\mathrm{d}\beta\,\sin\beta}^{\mathrm{d}y'}\ \overbrace{r\cos\beta}^{y'} = r^3\sin^2\beta\cos\beta\,\mathrm{d}\beta \ .$$

Integrating to the whole area defined by angle α (Fig. 108), we get

$$S = \int_{-\alpha}^{\alpha} r^3\sin^2\beta\cos\beta\,\mathrm{d}\beta = \frac{2}{3}r^3\sin^3\alpha \ . \tag{194}$$

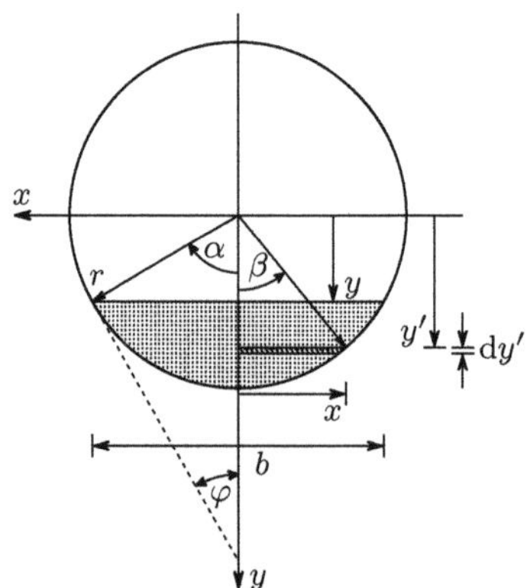

Fig. 108. Computation of the shearing stress in a circular cross-section

The shearing stress τ_{zy} corresponding to the area moment S (194) is then $(b = 2r \sin \alpha)$

$$\tau_{zy}(\alpha) = \frac{VS}{Ib} = \frac{V\frac{2}{3}r^3 \sin^3 \alpha}{\frac{\pi r^4}{4} 2r \sin \alpha} = \frac{4}{3} \frac{V}{\pi r^2} \sin^2 \alpha = \frac{4}{3} \frac{V}{\Omega} \sin^2 \alpha \ .$$

For a given value α, the maximum stress occurs at the boundary. From (193) we get

$$\tau = \frac{\tau_{zy}}{\cos \varphi} = \frac{\tau_{zy}}{\sin \alpha} = \frac{4}{3} \frac{V}{\Omega} \sin \alpha \ .$$

This expression attains a maximum for $\alpha = \frac{\pi}{2}$ (neutral axis), which means that the maximum shear stress in the cross-section takes the value

$$\alpha = \frac{\pi}{2} \ \Rightarrow \ \tau = \tau_{\max} = \frac{4}{3} \frac{V}{\Omega} \ .$$

The solution given by the Theory of Elasticity for this problem indicates that, unless the Poisson coefficient takes the value $\nu = 0.5$ (incompressible material), the stress distribution is not uniform in the neutral axis. The maximum value occurs in the centre of the circle and takes the value [4]

$$\tau_{\max} = \gamma \frac{4}{3} \frac{V}{\Omega} \quad \text{with} \quad \gamma = \frac{9 + 6\nu}{8\,(1 + \nu)} \ .$$

The error for the approximate solution vanishes for $\nu = 0.5$ ($\gamma = 1$) and takes the maximum value for a vanishing Poisson's coefficient ($\gamma = 1.125$). For the mean value $\nu = 0.25$, we get $\gamma = 1.05$. In the case of steel ($\nu = 0.3$) the error is 3.8% ($\gamma = 1.038$). We conclude that the error introduced by the simplifying hypotheses is relatively small.

VIII.3.c Open Thin-Walled Cross-Sections

Many of the slender members currently used in structural engineering, especially in metallic constructions, have thin-walled cross-sections, i.e., cross-sections made of straight or curved elements with small thickness, in comparison with the cross-section dimensions. Usual *profile sections*, such as I-beams, channel beams, angle sections, Z-sections, T-beams, circular or rectangular tubes, etc., are examples of this kind of member. In this Sub-section, we will deal with open thin-walled cross-sections, i.e., simply-connected thin-walled cross-sections.

As seen in the study of the shearing stresses in rectangular cross-sections, if the width is small compared with the height, the simplifying hypothesis of considering constant stresses in the thickness b is very close to the actual distribution. The same happens in thin-walled cross-sections, like that represented in Fig. 109. Thus, by considering the longitudinal surface which is perpendicular to the centre line of the cross-section wall and contains the point where the shearing stress is to be computed, the shearing stress may be obtained from the longitudinal shear force $\mathrm{d}E$. From (187) we get

$$\mathrm{d}E = \tau e \,\mathrm{d}z = \frac{VS}{I} \Rightarrow \tau = \frac{\mathrm{d}E}{e \,\mathrm{d}z} = \frac{VS}{Ie} \,, \tag{195}$$

where e represents the wall thickness in the point where τ is computed. The computation of the area moment S of thin walls may be simplified if the area is considered as concentrated on the centre line. Denoting by s a coordinate which follows that line (Fig. 109), we get for the first area moment needed to compute the shearing stress in the point defined by s[2]

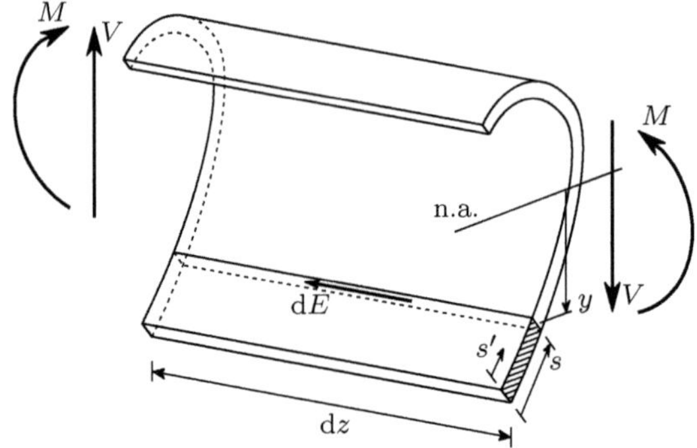

Fig. 109. Longitudinal shear force in a thin-walled cross-section

$$S\,(s) = \int_0^s e(s')y(s')\,\mathrm{d}s' \ .$$

In order to illustrate these considerations, the shearing stress distribution in the cross-section represented in Fig. 110, caused by a vertical shear force V is analysed.

In the flange element $\overline{AB}$ the area moment corresponding to the point of the centre line defined by the coordinate s_1 may be expressed by

$$S(s_1) = s_1 e \left(\frac{h}{4} + \frac{s_1}{2} \right) \ .$$

The shearing stress in this point is then

$$\tau(s_1) = \frac{V S(s_1)}{Ie} = \frac{V}{I} \left(\frac{h s_1}{4} + \frac{s_1^2}{2} \right) \ .$$

---

[2]If the same approximation is made for the moment of inertia, a completely consistent theory for thin-walled cross-sections with infinitesimal wall thickness is obtained, in the sense that the computed resultant of the shearing stress exactly balances the applied shear force. Otherwise, a discrepancy will appear, which is introduced by the wall curvature or by angle points in the centre line.

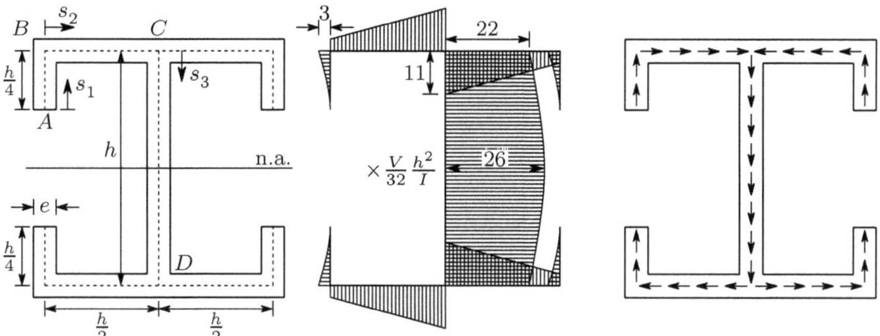

Fig. 110. Shearing stresses caused by a vertical positive shear force in a symmetrical open thin-walled cross-section

The maximum stress occurs for the maximum value of s_1 (point B), taking the value

$$s_1 = \frac{h}{4} \quad \Rightarrow \quad \tau = \tau_{\max}^{AB} = \frac{3}{32}h^2\frac{V}{I} \ .$$

In the flange element $\overline{BC}$ the area moment and the shearing stress may be expressed in terms of coordinate s_2, yielding

$$S\left(s_2\right) = \frac{3h^2e}{32} + s_2 e\frac{h}{2} \quad \Rightarrow \quad \tau = \frac{V}{I}\left(\frac{h}{2}s_2 + \frac{3h^2}{32}\right) \ .$$

In this wall segment the stress is a linear function of s_2 and takes the maximum value in point C

$$s_2 = \frac{h}{2} \quad \Rightarrow \quad \tau = \tau_{\max}^{BC} = \frac{11}{32}h^2\frac{V}{I} \ .$$

Finally, in the web (wall segment $\overline{CD}$) the area moment may be expressed as a function of coordinate s_3, yielding

$$S(s_3) = \frac{22}{32}h^2e + s_3 e\left(\frac{h}{2} - \frac{s_3}{2}\right) \quad \Rightarrow \quad \tau = \frac{V}{I}\left(\frac{22}{32}h^2 + \frac{s_3 h}{2} - \frac{s_3^2}{2}\right) \ .$$

This expression represents a parabolic stress distribution. The maximum value occurs on the neutral axis and takes the value

$$s_3 = \frac{h}{2} \quad \Rightarrow \quad \tau = \tau_{\max}^{CD} = \frac{26}{32}h^2\frac{V}{I} \ .$$

The direction of the shearing stresses may be obtained from the direction of the longitudinal shear force. For example, in order to get the stress direction in the flange element $\overline{AB}$, let us consider the balance of the longitudinal forces acting on a piece of this flange element, as represented in Fig. 111.

Let us assume a positive shear force (downward direction). As the flange element $\overline{AB}$ is above the neutral axis, it is in the compressed zone, if the

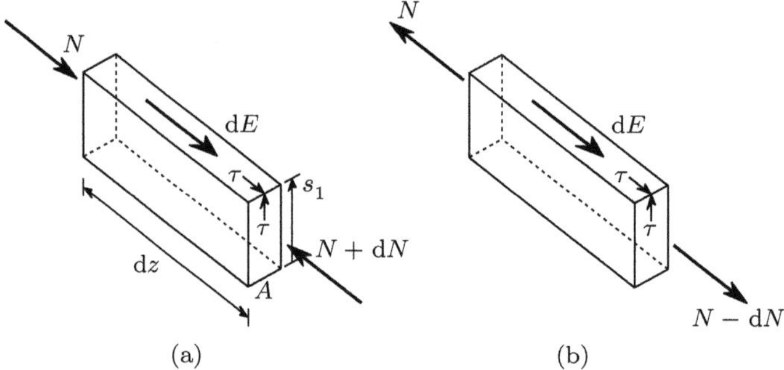

Fig. 111. Determination of the direction of the shearing stresses in the flange element $\overline{AB}$ (Fig. 110): (**a**) positive bending moment; (**b**) negative bending moment

bending moment is positive. A positive shear force will cause an increase in the bending moment, as $\mathrm{d}M = V\,\mathrm{d}z$, which will cause an increase $\mathrm{d}N$ in the compressive stress resultant N (Fig. 111-a). In the case of a negative bending moment, the flange element $\overline{AB}$ will be in the tensioned zone. However, a positive shear force will cause a decrease in the absolute value of the bending moment ($\mathrm{d}M > 0$ and $M < 0$) and, therefore, a decrease in the tensile stress resultant N, as represented in Fig. 111-b. In both cases, the same direction is obtained for the shearing stress τ, as expected, since this stress is caused by the shear force, which is the same in the two cases.

The direction of the shearing stresses in the segments $\overline{BC}$ and $\overline{CD}$ could be obtained in the same way. The symmetry of the cross-section leads to the directions of the shearing stresses represented in Fig. 110.

An additional tool to obtain the direction of the shearing stresses is furnished by the condition of constant shear flow in a point of convergence of two or more centre lines of the cross-section walls, as points B and C (Fig. 110). This condition may be obtained from the balance equation of the longitudinal

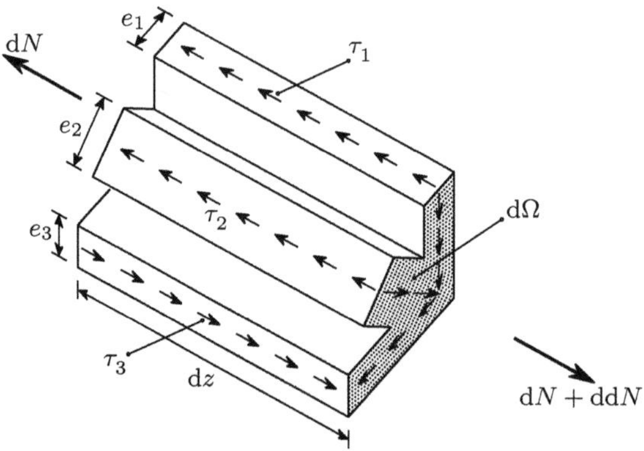

Fig. 112. Shear flow in a nodal point of a thin-walled cross-section

forces acting on an infinitesimal neighbourhood of one of these points (nodal points). In the case represented in Fig. 112, this equation takes the form

$$\overbrace{\left(-\tau_1 e_1 - \tau_2 e_2 + \tau_3 e_3\right) \mathrm{d}z}^{\text{infinitesimal quantity of second order}} + \underbrace{\mathrm{d}\sigma\, \mathrm{d}\Omega}_{\text{infinite simal quantity of third order }(\mathrm{dd}N)} = 0 \, .$$

The product $e\,\mathrm{d}z$ is an infinitesimal quantity of second order, since the thickness e is infinitesimal (cf. Footnote 55). Because $\mathrm{d}\Omega$ is also a second order infinitesimal quantity, $\mathrm{d}\sigma\mathrm{d}\Omega$ will be an infinitesimal quantity of third order. Thus, $\mathrm{d}\sigma\mathrm{d}\Omega$ is an infinitesimal quantity of higher order, which may be neglected, yielding

$$\overbrace{\tau_1 e_1 + \tau_2 e_2}^{\text{ingoing shear flow}} = \underbrace{\tau_3 e_3}_{\text{outgoing shear flow}} \, . \tag{196}$$

Generalizing (196) to a number n of centre lines converging to a nodal point, we get

$$\sum_{i=1}^{n} \tau_i e_i = 0 \, .$$

Taking the reciprocity of shearing stresses into consideration, this expression means that the sum of the products τe heading into the nodal point is equal to the sum of the products τe heading out. In other words, *the shear flow entering the node is equal to the shear flow leaving the node.* For example, in point C (Fig. 110) the shear flow entering the node is $2 \times \frac{11}{32}\frac{Vh^2 e}{I}$ and the outgoing flow is $\frac{22}{32}\frac{Vh^2 e}{I}$.

VIII.3.d Closed Thin-Walled Cross-Sections

If the cross-section is doubly-connected, i.e., if the centre line of the wall is a closed line, a longitudinal cut, like the one represented in Fig. 109, is not enough to separate the cross-section into two distinct parts. This means that two cuts must be made and that the longitudinal shear force $\mathrm{d}E$, given by (187), is the sum of the resultants of two different longitudinal shearing stresses, τ_1 and τ_2. The value of the shearing stress cannot be computed, therefore, unless an additional relation between τ_1 and τ_2 is found. However, in the case of a symmetrical cross-section, with respect to the action axis of the shear force, these stresses will be equal, provided that the two cuts are made in symmetrical points of the centre line, as represented in Fig. 118. In this case, the shearing stress may be computed by the expression

$$2\tau e\,\mathrm{d}z = \mathrm{d}E = \frac{VS}{I}\,\mathrm{d}z \;\Rightarrow\; \tau = \frac{VS}{2Ie}\,, \tag{197}$$

where S is the first area moment of the shaded area in Fig. 113.

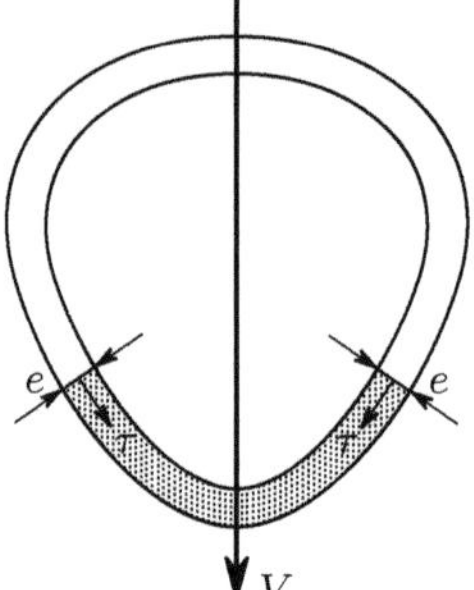

Fig. 113. Computation of the shearing stress in a closed symmetrical thin-walled cross section

If the cross-section is not symmetrical with respect to the action axis of the shear force, the problem becomes a statically indeterminate one, whose solution may be computed by means of the force method. As seen in Sect. VI.4, this method consists of releasing a sufficient number of connections to get a statically determinate problem, followed by the computation of the forces needed to avoid displacements in the released connections. In the present problem, the longitudinal connection in a point of the cross-section wall is released, so that an open cross-section is obtained. Under the action of the shear force, the two sides of the cut suffer a longitudinal relative displacement, as represented in Fig. 114-a. This displacement must then be eliminated, by applying a pair of shear forces dE to both sides of the cut (Fig. 114-b). The resulting stress in any point of the cross-section may be obtained by the superposition principle, by adding the stresses corresponding to the two situations (Fig. 114-c).

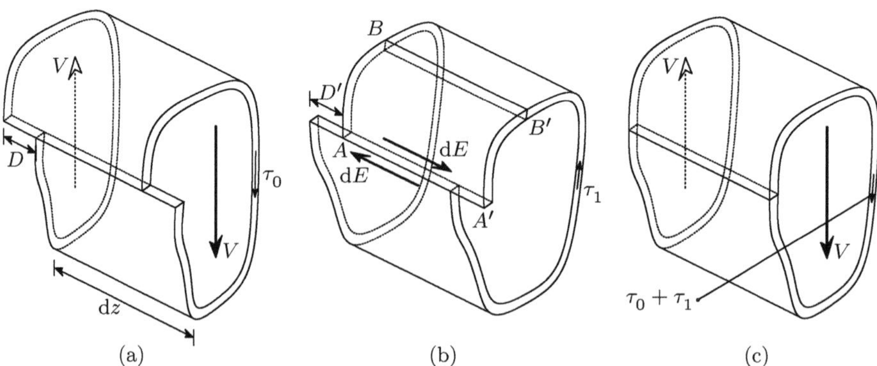

Fig. 114. Computation of the shear stresses in a non-symmetrical closed thin-walled cross-section

The relative displacement in direction z of two points of the centre line, located at an infinitesimal distance $\mathrm{d}s$ of each other, is $\mathrm{d}D = \gamma_0\,\mathrm{d}s$.[3] Thus, in the open cross-section, the relative displacement D of both sides of the cut, caused by the shear force V (Fig. 114-a), may be computed by integrating the shear strain γ_0 along the complete centre line of the wall, which yields $(\tau_0 = G\gamma_0)$

$$D = \oint \gamma_0\,\mathrm{d}s = \frac{1}{G}\oint \tau_0\,\mathrm{d}s = \frac{V}{IG}\oint \frac{S}{e}\,\mathrm{d}s \ . \tag{198}$$

In the situation depicted in Fig. 114-b, the shear flow $f = \tau_1(s)e(s)$ is constant along the whole centre line of the wall,[4] since there are no other forces applied to the bar apart from the pair of forces $\mathrm{d}E$. This conclusion is easily drawn by establishing the balance condition of the longitudinal forces acting on the piece defined by the longitudinal cut $\overline{AA'}$ and by any other longitudinal surface $\overline{BB'}$ (Fig. 114-b). This condition immediately means that the product $\tau_1 e = \frac{\mathrm{d}E}{\mathrm{d}z} = f$ is constant, even if e varies along the centre line. The longitudinal relative displacement D' caused by the pair of forces $\mathrm{d}E$, is then $(\tau_1 = \frac{f}{e})$

$$D' = \oint \gamma_1\,\mathrm{d}s = \oint \frac{\tau_1}{G}\,\mathrm{d}s = \frac{f}{G}\oint \frac{\mathrm{d}s}{e} \ . \tag{199}$$

The condition of compatibility requires that the displacement D' eliminates displacement D, which allows the computation of the shear flow f

$$D + D' = 0 \ \Rightarrow \ f = -\frac{V}{I}\frac{\oint \frac{S}{e}\,\mathrm{d}s}{\oint \frac{\mathrm{d}s}{e}} \ \Rightarrow \ \tau_1(s) = \frac{f}{e(s)} \ . \tag{200}$$

The shearing stress in the closed cross-section (Fig. 114-c) may then be computed by adding the stresses τ_0 and τ_1.

The closed line integrals appearing in the expressions above obviously refer to the line limiting the closed part of the cross-section, that is, they do not include simply-connected walls, as in the cross-section represented in Fig. 115.

The expressions above are valid for doubly-connected cross-sections, i.e., closed cross-sections with only one channel. In cross-sections with higher degrees of connection a number of longitudinal cuts equal to the degree of connection minus one is necessary to get a statically determinate problem, i.e., an open cross-section. As a consequence, the conditions of compatibility of the deformations in all the longitudinal cuts yield, instead of (200), a system

---

[3]This simple relation requires that the fibres remain parallel to each other in the deformation caused by the shear force. This condition is satisfied if there is no rotation of the cross-sections around a longitudinal axis of the prismatic bar, i.e., if torsion does not take place (see example XII.8).

[4]This shear flow defines a torsional moment (twisting moment or torque, see Chap. X.3). This moment corresponds to the translation of the shear force, from the shear centre of the open cross-section to the shear centre of the closed cross-section (see Sect. VIII.4 and example VIII.12).

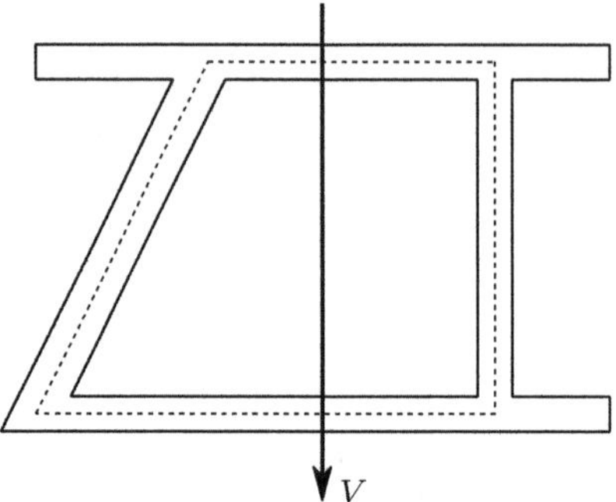

Fig. 115. Line, to which the closed line integrals in (198), (199) and (200) are referred (*dashed line*)

with a number of equations equal to the degree of connection minus one (see example VIII.7).

VIII.3.e Composite Members

In composite members the longitudinal shear force may be determined in the same way as in the case of homogeneous bars (187). The normal stress is in this case given by (169). Assuming, for simplicity, plane bending, as in the case represented in Fig. 116, we get the following expression for the longitudinal shear force (Fig. 116-b)

$$\begin{cases} d\sigma_a = \dfrac{dM\,E_a}{J_n}y \\[2ex] d\sigma_b = \dfrac{dM\,E_b}{J_n}y \end{cases} \Rightarrow \left| \begin{aligned} dE &= \int_{\Omega_{a1}} d\sigma_a\,d\Omega_a + \int_{\Omega_{b1}} d\sigma_b\,d\Omega_b \\ &= \frac{V}{J_n}\left(E_a \int_{\Omega_{a1}} y\,d\Omega_a + E_b \int_{\Omega_{b1}} y\,d\Omega_b \right) dz \ . \end{aligned} \right.$$

$$(201)$$

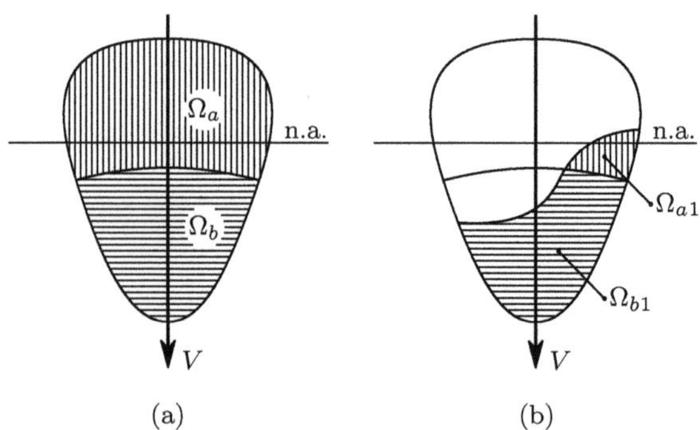

Fig. 116. Determination of the longitudinal shear force in composite members

In composite members, the longitudinal shear force in the contact surface between the two materials must usually be computed. In this particular case (201) takes a simpler form and the longitudinal shear force may be computed by any of the following expressions

$$
\begin{cases}
\Omega_{a1} = \Omega_a \\
\quad \Rightarrow \dfrac{\mathrm{d}E}{\mathrm{d}z} = \dfrac{V E_a S_a}{J_n} = \dfrac{V S_a}{I_{ha}} \quad \text{with} \quad S_a = \displaystyle\int_{\Omega_a} y\,\mathrm{d}\Omega_a \\
\Omega_{b1} = 0 \\[2mm]
\Omega_{a1} = 0 \\
\quad \Rightarrow \dfrac{\mathrm{d}E}{\mathrm{d}z} = \dfrac{V E_b S_b}{J_n} = \dfrac{V S_b}{I_{hb}} \quad \text{with} \quad S_b = \displaystyle\int_{\Omega_b} y\,\mathrm{d}\Omega_b \;. \\
\Omega_{b1} = \Omega_b
\end{cases}
\tag{202}
$$

VIII.3.f Non-Principal Reference Axes

In some cross-sections it is easy to compute the moments and product of inertia with respect to non-principal central axes, as well as distances and area moments. In Fig. 117 two examples of this kind of cross-section are represented.

In these cases it may be useful to compute the normal and shearing stresses directly from these axes, especially if one of them is parallel to the action axis.

The normal stresses may by computed by means of (140). From this equation an expression for the computation of the longitudinal shear force may then be developed. If the bending moment has only the M_x component and the axial force vanishes, the normal stress may be computed by the expression

$$
\sigma = \frac{I_y y - I_{xy} x}{I_x I_y - I_{xy}^2} M_x \;.
$$

The same line of reasoning used to develop (186), leads to the following expression for the longitudinal shear force (cf. Figs. 102 and 117)

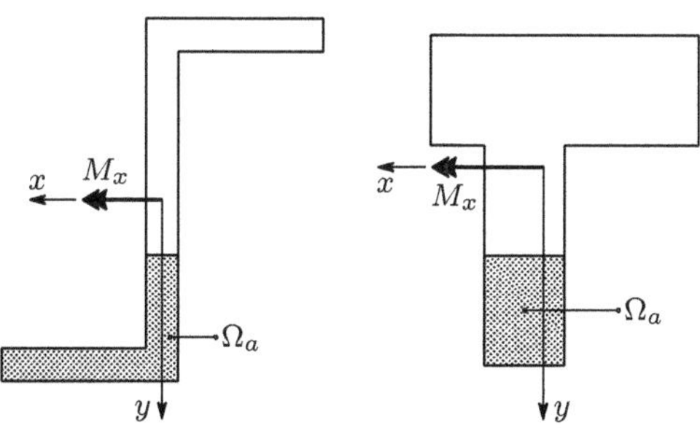

Fig. 117. Computation of the longitudinal shear force with non-principal reference axes

$$\mathrm{d}E = \int_{\Omega_a} \frac{I_y y - I_{xy} x}{I_x I_y - I_{xy}^2} \, \mathrm{d}M_x \, \mathrm{d}\Omega_a = V_y \frac{I_y S_x - I_{xy} S_y}{I_x I_y - I_{xy}^2} \, \mathrm{d}z$$

$$\text{with} \quad V_y = \frac{\mathrm{d}M_x}{\mathrm{d}z}, \quad S_x = \int_{\Omega_a} y \, \mathrm{d}\Omega_a \quad \text{and} \quad S_y = \int_{\Omega_a} x \, \mathrm{d}\Omega_a \; . \tag{203}$$

The shearing stresses may be computed from this expression, in the same way as was done on the basis of (187) (see example VIII.10).

VIII.4 The Shear Centre

When inclined circular bending was analysed (Sect. VII.4), we showed that a parallel displacement of the action axis does not change the normal stresses induced by the bending moment in the cross-section. However, if a shear force is acting (non-uniform bending), the equilibrium condition requires that the action axis of the shear force has a position which coincides with the line of action of the resultant of the shearing stresses. The position of the action axis of the shear force is therefore not arbitrary. There are two internal forces introducing shearing stresses in the cross-section: the shear force and the torsional moment. The expressions presented for the shearing stresses in this Chapter only take the shear force into consideration, since they are all based on the relation $\mathrm{d}M = V \, \mathrm{d}z$ (185). It is therefore assumed that the torsional moment is zero. If it is not, additional shearing stresses will appear. These stresses will be analysed in Chap. X.

Thus, to avoid torsion, the action axis of the shear stress must coincide with the line of action of the resultant of the shearing stresses computed by means of the expressions which were developed from (187) (longitudinal shear force caused by the cross-sectional shear force). By considering two shear forces with the directions of the principal central axes of inertia, and computing the position of the line of action of the resultant of the shearing stresses in each case, a point is defined by the intersection of these two lines, which has the following property: *if the line of action of the shear force passes through this point, it will not induce torsion of the bar.* This point is the *shear centre* of the cross-section.

The shear centre plays the same role in relation to the transversal forces, as the centroid in relation to the longitudinal (axial) forces: if the resultant axial force passes through the centroid of the cross-section, it will not induce bending; otherwise, composed bending will take place, with a bending moment given by the product of the axial force and the distance of its line of action to the centroid. In the same way, if the resultant of the forces acting on the cross-section plane (the shear force) does not pass through the shear centre, it will introduce a torsional moment, with a value given by the product of the shear force and the distance of its line of action to the shear centre.

The computation of the torsional moment must thus be made in relation to the shear centre, while the bending moment is computed with respect to the centroid. In the case of a cross-section with a symmetry axis, the shear centre is on this axis, since, for an action axis of the shear force coinciding with the symmetry axis, the shearing stress distribution will also be symmetric, which means that the line of action of its resultant coincides with the symmetry axis. Thus, if the cross-section has two symmetry axes the centroid and the shear centre will coincide. In other cases, these two points usually occupy different positions in the cross-section's plane.

We will demonstrate later (Chap. XII) that in prismatic bars made of materials with linear elastic behaviour, the shear centre coincides with the *torsion centre*, which is defined as the point around which the cross-section rotates in the twisting deformation induced by the torsional moment. For this reason, these two designations are sometimes indistinctly used.

While it is very easy to compute the position of the line of action of the resultant of the normal stresses in the case of pure axial force, since these stresses are constant in the cross-section, the computation of the line of action of the resultant of the shearing stresses is often complex, since the distribution of the stresses caused by the shear force is required. As seen in the previous sections, good approximations for these stresses are obtained only in the cases of symmetrical cross-sections with respect to the action axis of the shear force and in thin-walled cross-sections. In the first case, the position of the resultant is known. In the case of non-symmetrical cross-sections which cannot be considered as thin-walled, the problem of computing the shear centre's position cannot be solved by the approach used in the Strength of Materials. But the knowledge of the position of the shear centre is most important in the case of open thin walled cross-sections, since this kind of member is very weak in torsion, as will be seen in Chap. X. Fortunately, the stresses caused by the shear force in these cross-sections are easily computed with good approximation, as seen in Subsect. VIII.3.c.

In order to illustrate these considerations, the position of the shear centre of the channel cross-section represented in Fig. 118 is computed. As this cross-section has a symmetry axis, the shear centre will be located on this axis.

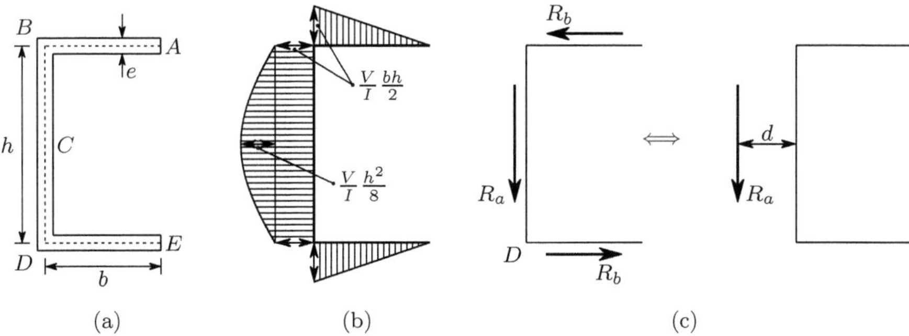

Fig. 118. Computation of the position of the shear centre in a thin-walled cross-section

Thus, in order to determine its position, it is enough to compute the distance d from the line of action of the resultant of the shearing stresses, introduced by a shear force perpendicular to the symmetry axis, to the centre line of the web (Fig. 118-c).

As the example of (Fig. 110) shows, the shearing stress has a linear distribution in the wall segments which are parallel to the neutral axis, and a parabolic distribution in the others. Besides, we know that the maximum stress occurs on the neutral axis. For these reasons, in example of (Fig. 118) the stress distribution is completely defined by the values in points B and C. For point B we get from (195)

$$S = be\frac{h}{2} \;\Rightarrow\; \tau = \frac{V}{I}\frac{bh}{2} \;.$$

For point C the same expression yields the value

$$S = \frac{beh}{2} + \frac{h}{2}e\frac{h}{4} \;\Rightarrow\; \tau = \frac{V}{I}\left(\frac{bh}{2} + \frac{h^2}{8}\right) \;.$$

The resultants of the shearing stress in the web (R_a) and in the flanges (R_b) may be computed from the diagram areas in Fig. 118-b, multiplied by the thickness e, yielding

$$R_a = \frac{V}{I}\left(\frac{bh}{2}he + \frac{2}{3}\frac{h^2}{8}he\right) = \frac{V}{I}\left[\underbrace{\frac{eh^3}{12}}_{I_a} + 2 \times \underbrace{be\left(\frac{h}{2}\right)^2}_{I_b - \frac{be^3}{12}}\right] \approx V \tag{204}$$

$$R_b \; \frac{1}{2}\frac{V}{I}\frac{bh}{2}be = \frac{V}{I}\frac{b^2 he}{4} \approx \frac{3}{\left(\frac{h}{b}\right)^2 + 6\frac{h}{b}}V \;.$$

It must be remarked here that, as mentioned in Footnote 55, an exact balance between the shear force and the resultant of the shearing stresses is only achieved if the moment of inertia of the flange, with respect to is centre line $\left(\frac{be^3}{12}\right)$, is neglected.[5] The condition of equivalence of moments with respect to point D (Fig. 118-c) allows the computation of the distance d, which defines the position of the shear centre

$$R_b h = R_a d \;\Rightarrow\; d = \frac{3}{\left(\frac{h}{b}\right)^2 + 6\frac{h}{b}}h \;. \tag{205}$$

---

[5] From a mathematical point of view, the theory expounded for thin-walled cross-sections is only valid if the thickness of the walls is infinitesimal, in comparison with the cross-section dimensions. In this case, the moment of inertia of the flange with respect to its centre line, $\frac{be^3}{12}$, is an infinitesimal quantity of third order, which may be neglected in presence of the infinitesimal quantity of first order resulting from the parallel-axis theorem, $\frac{beh^2}{4}$.

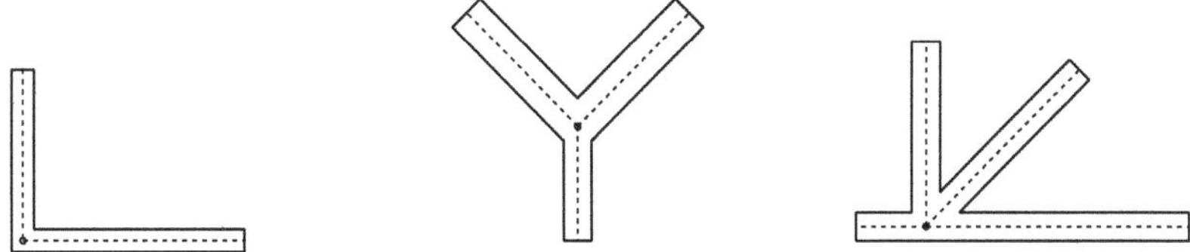

Fig. 119. Shear centre in thin-walled cross-sections with concurrent and straight wall elements

The thin-walled cross-sections with concurrent and straight wall elements, like those represented in Fig. 119, are a particularly simple case of determination of the shear centre. In fact, as the resultants of the shearing stresses in the different wall elements pass through the intersection of the centre lines, the moment of the shearing stress in relation to this point vanishes, which means that it is the shear centre.

VIII.5 Non-Prismatic Members

VIII.5.a Introduction

The basic equation for the analysis of the effect of the shear force (187) has been deduced for prismatic bars. So when the above expressions for the computation of shearing stresses are applied to non-prismatic members, errors are introduced. In order to get an idea of the importance of these errors, two examples of non-prismatic members, which are simple enough for an exact solution to be given by the Theory of Elasticity, are analysed.

VIII.5.b Slender Members with Curved Axis

As explained in Sect. VIII.2, the expression obtained for the shearing stress in a rectangular cross-section with a small thickness (189) coincides with the exact solution of the Theory of Elasticity. Thus, in a bar with the same cross-section, but with a curved axis, the discrepancies between the exact solution and the results obtained using (187) may be attributed to the fact that the bar's axis is not a straight line.

The bar represented in Fig. 120 has a circular axis and a rectangular cross-section with the dimensions $b \times h$ ($b \ll h$). The shear force in the cross-section B defined by the angle θ takes the value $V = -P\cos\theta$.

The shearing stress in that cross-section may be expressed as a function of the dimensionless coordinate η, which, multiplied by the height of the cross-section h, defines the distance to the centre line ($-\frac{1}{2} \leq \eta \leq \frac{1}{2}$, Fig. 120). The exact solution obtained by the Theory of Elasticity for the shearing stress on the cross-section defined by the angle θ may be defined by the expression [4]

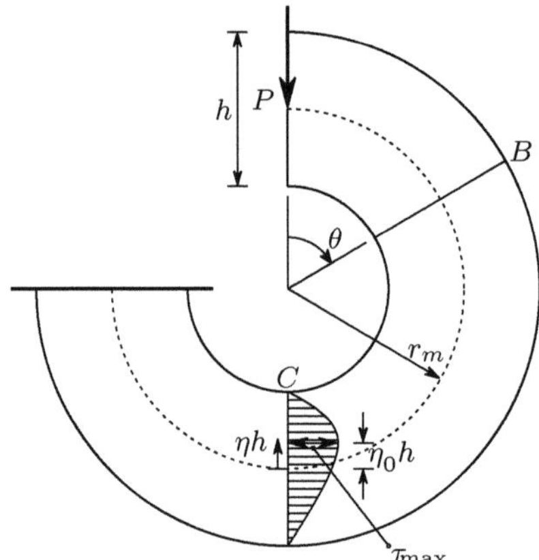

Fig. 120. Shearing stresses induced by the shear force in a bar with a curved axis

$$\tau = -\frac{3}{2}\frac{P\cos\theta}{bh}\underbrace{\frac{1+\eta\alpha+\frac{\left(1-\frac{\alpha^2}{4}\right)^2}{(1+\eta\alpha)^3}-\frac{2\left(1+\frac{\alpha^2}{4}\right)}{1+\eta\alpha}}{3-\frac{3}{\alpha}\left(1+\frac{\alpha^2}{4}\right)\ln\frac{2+\alpha}{2-\alpha}}}_{\gamma}\quad\text{with}\quad\alpha=\frac{h}{r_m}.\qquad(206)$$

In the limit case of a prismatic bar ($\alpha = 0$, $\theta = \pi$) this solution yields the same value as (190) $\left(\tau_{\max}=\frac{3}{2}\frac{V}{bh}\right)$.

When the relation α between the height of the cross-section and the curvature radius of the centre line r_m increases, the difference between the distributions of shearing stresses given by (206) and by the expression developed for prismatic bars increases also. This difference remains small, however, even for larger curvatures, as may be easily confirmed by computing the values of η and γ corresponding to the maximum shearing stress ($\eta = \eta_0 \Rightarrow \gamma = \gamma_{\max}$) for some values of α

α	0.0000	0.1000	0.2500	0.5000	0.7500	1.0000	1.5000
η_0	0.0000	0.0250	0.0626	0.1259	0.1905	0.2565	0.3885
$\gamma_{\max}$	1.0000	1.0009	1.0056	1.0233	1.0573	1.1166	1.4402

VIII.5.c Slender Members with Variable Cross-Section

In bars with variable cross-section the expressions developed on the basis of (187) may lead to completely erroneous results, at least in relation to the location of the maximum stress in the cross-section. For example, in the problem represented in Fig. 86, the exact solution shows that the shearing stress vanishes in the neutral axis and attains the maximum value in the farthest points from the neutral axis, as may be easily ascertained by a two-dimensional analysis of the stress state in those points, which totally contradicts the solution developed for prismatic bars.

Regarding the value of the maximum shearing stress in the cross-section, significant errors may also be introduced by the theory of prismatic bars, as may be easily verified by computing the maximum shearing stress in cross-section $\overline{AA'}$ (Fig. 86). From (164) we find that the maximum radial stress occurs in point A and takes the value

$$\varphi = \frac{\alpha}{2} \Rightarrow \sigma_r = \sigma_{r-\text{max}} = \frac{2}{\alpha - \sin\alpha} \frac{P}{br} \sin\frac{\alpha}{2} .$$

A two-dimensional analysis of the stress state shows that the shearing stress in a vertical facet takes the value

$$\tau_{\text{max}} = \frac{1}{2}\sigma_{r-\text{max}} \sin\alpha = \sin\frac{\alpha}{2} \cos\frac{\alpha}{2}\sigma_{r-\text{max}} = \frac{2\sin^2\frac{\alpha}{2}\cos\frac{\alpha}{2}}{\alpha - \sin\alpha} \frac{P}{br} .$$

The theory of prismatic bars yields the following value for the maximum shearing stress in the same cross-section, $\tau_{\text{max}-\text{p}}$

$$\begin{cases} h = 2r \sin\frac{\alpha}{2} \\ \\ V = P \end{cases} \Rightarrow \tau_{\text{max}-\text{p}} = \frac{3}{2}\frac{V}{bh} = \frac{3}{4}\frac{1}{\sin\frac{\alpha}{2}}\frac{P}{br} .$$

The relation between the exact value τ_{max} and the value yield by the theory of prismatic bars, $\tau_{\text{max}-\text{p}}$, depends only on angle α and may expressed by parameter β

$$\beta = \frac{\tau_{\text{max}}}{\tau_{\text{max}-\text{p}}} = \frac{8}{3}\frac{\sin^3\frac{\alpha}{2}\cos\frac{\alpha}{2}}{\alpha - \sin\alpha} .$$

The following Table gives the values of β corresponding to some values of angle α.

α	1°	10°	20°	30°	45°	60°
β	1.999	1.988	1.952	1.892	1.764	1.593

This example shows that the actual value of the maximum shearing stress in a slender member with a variable cross-section may be substantially higher than the value given by the theory of prismatic bars.

VIII.6 Influence of a Non-Constant Shear Force

The solution of the Theory of Elasticity for the shearing stresses in the example depicted in Fig. 85 (162) shows that (189) is exact ($V = p\left(\frac{l}{2} - z\right)$), although the expression defining the normal stresses σ_z (162) is different from the expression developed for the case of pure bending, on the basis of the law of conservation of plane sections. The variation of the shear force thus affects the distribution of normal stresses, but does not change the distribution of shearing stresses. This is due to the fact that the second element of the expression of σ_z (which represents the correction to be added to (146), to take the variation of the shearing stress into account) is independent of z, i.e., it is constant in all cross-sections, so it does not introduce a longitudinal shear force.

Also in the case of the example depicted in Fig. 120 the shear force is not constant. However, the distribution of shearing stresses in the cross-section is not altered by the variation of the shear force, since the exact solution (206) shows that the shearing stress is proportional to the shear force $V = -P\cos\theta$.

Considering these examples and the fact that the normal stress computed by means of the expressions developed on the basis of the Saint Venant hypothesis are very close to the exact solution (Sect. VII.7), we may conclude that the variation of the shear force does not affect the validity of the fundamental expression for studying the effect of the shear force (187).

VIII.7 Stress State in Slender Members

Generally, in slender members, the stresses that act on perpendicular facets to the cross-section plane and are parallel to it – σ_x, σ_y and τ_{xy} – either vanish, as happens if there are no forces applied on the bar element under consideration and the Poisson coefficient is constant, or are sufficiently small to be neglected (see example VIII.16). We thus have a plane stress state. Obviously, this does not apply to the regions in the vicinity of sudden changes in the cross-section dimensions, angle points of the bar's axis or strongly concentrated loads. However, in these cases the theory of prismatic bars is not valid.

According to these considerations, the stress state in a slender member under non-uniform bending may be analysed in the plane perpendicular to the cross-section which contains the shear force vector in the point under consideration.[6] In thin-walled cross-sections this is the longitudinal plane containing the wall centre line. In this kind of cross-section and also in rectangular sections under plane bending, the plane stress state may be visualized by means of the *principal stress trajectories*. These lines represent, in each point, the principal directions of the stress state. As the stress tensor only has a normal and a shearing component, the maximum principal stress is always a tensile one and the minimum principal stress is always compressive, as may be easily verified by drawing the Mohr circles corresponding to tensile and compressive

---

[6]This conclusion remains valid if a torsional moment is also acting, since this internal force only causes shearing stresses in the cross-section and not σ_x, σ_y or τ_{xy}, as will be seen in Chap. X.

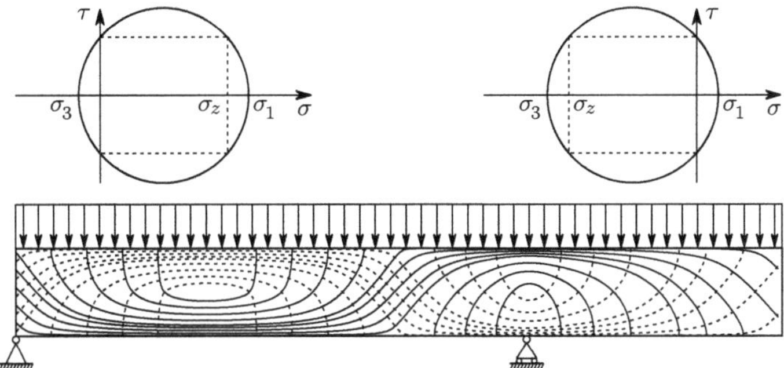

Fig. 121. Principal stress trajectories in a simply supported beam: —— tensile trajectories; ------ compressive trajectories

normal stresses σ_z, as represented in Fig. 121. The same Figure also shows the principal stress trajectories in a simply supported beam under a uniformly distributed load.

In the points on the neutral surface a purely deviatoric stress state is installed, since $\sigma = 0$. The principal direction are thus at $45°$ angles with the cross-section plane. If there are no shearing loads applied on the surface of the bar, one of the principal directions is perpendicular and the other is tangent to the surface. However, in the right end cross-section the principal directions are indeterminate, since there are no stresses in theses points. The principal stress trajectories, with an inclination of $45°$, appearing in the left end cross-section result from the fact that the principal directions were computed by means of the theory of prismatic bars, assuming that the left reaction force is applied as a shear force acting on that cross-section. In the same way, the perturbation introduced by the concentrated load corresponding to the reaction force on the right support was not considered. Actually, in this region the compressive trajectories converge to the support. Furthermore, this reaction force was considered to be distributed over a small length, in order to avoid discontinuities in the shearing stress distribution, which would introduce corners into the principal stress trajectories.

The safety evaluation in bars under non-uniform bending usually includes three points:

- verification of the maximum normal stress in the fibres farthest from the neutral axis;
- verification of the maximum shearing stress, which usually occurs on the neutral axis;
- verification of the two-dimensional stress state in the points where the shearing and normal stresses simultaneously reach higher values. In these points, a yielding or a rupture criterion must be used. In ductile materials the von Mises criterion is generally used (see example VIII.17).

The third verification is especially important in I-beams and channels, in the points where web and flange connect, in the case of cross-sections where both the bending moment and the shear force attain higher values, as in the cross-sections which are close to the right support in the beam represented in Fig. 121. In these points, the normal stress is not much lower than the maximum value, since they are close to the fibres farthest from the neutral axis. The shearing stress is also close to the maximum value appearing on the neutral axis, because the area moment of the flanges is not substantially smaller than the area moment of half section. These considerations are summarized in Fig. 122.

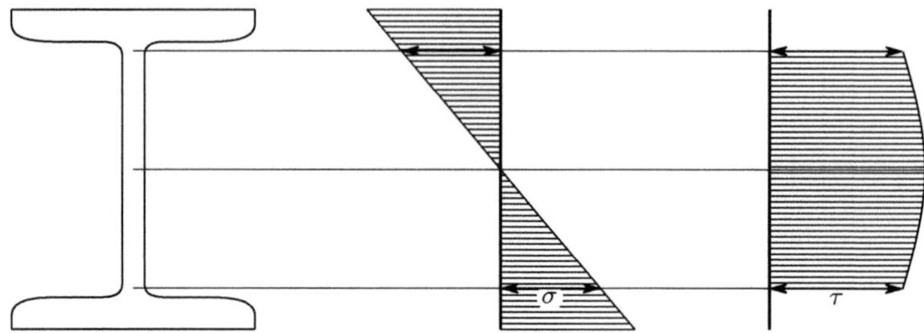

Fig. 122. Stress state (σ, τ) in connection points between web and flange in a I-beam

VIII.8 Examples and Exercises

VIII.1. Figure VIII.1 shows the cross-section of a simply supported beam with a span $100a$, under a uniformly distributed load p. The beam is made by connecting a bar with rectangular cross-section $a \times 3a$ and four bars with square cross-section $a \times a$. Determine the longitudinal shear force acting in each connection.

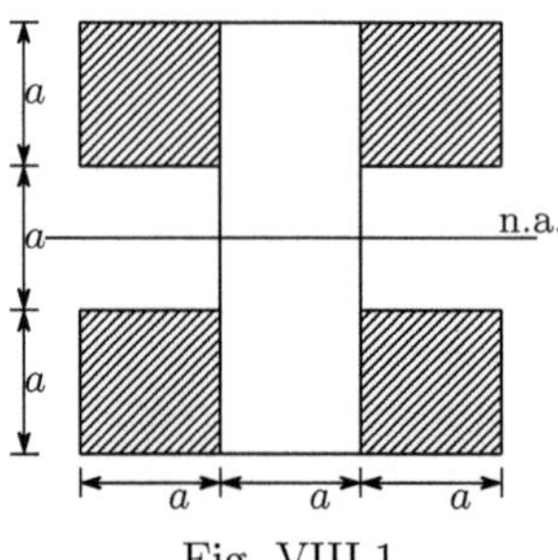

Fig. VIII.1

Resolution

The problem may be solved by directly applying (187). To this end, it is necessary to compute the moment of inertia of the cross-section and the first area moment S of one of the hatched areas in Fig. VIII.1 with respect to the neutral axis. These quantities take the values

$$I = \frac{(3a)^4}{12} - \frac{2a \times a^3}{12} = \frac{79}{12}a^4 \quad \text{and} \quad S = a^3 ,$$

respectively. The longitudinal shear force per unit length in each connection is then

$$\frac{\mathrm{d}E}{\mathrm{d}z} = \frac{VS}{I} = V\frac{a^3}{\frac{79}{12}a^4} = \frac{12}{79}\frac{V}{a} .$$

Since the longitudinal shear force is proportional to the cross-sectional shear force V, we conclude that it varies linearly between $-\frac{12}{79}\frac{50pa}{a} = -\frac{600}{79}p$ and $\frac{600}{79}p$ ($V_{\max} = 50pa$).

VIII.2. Determine the maximum shearing stress in a cross-section with the shape of an isosceles triangle of base b and height h, caused by a shear force acting on the symmetry axis (Fig. VIII.2-a).

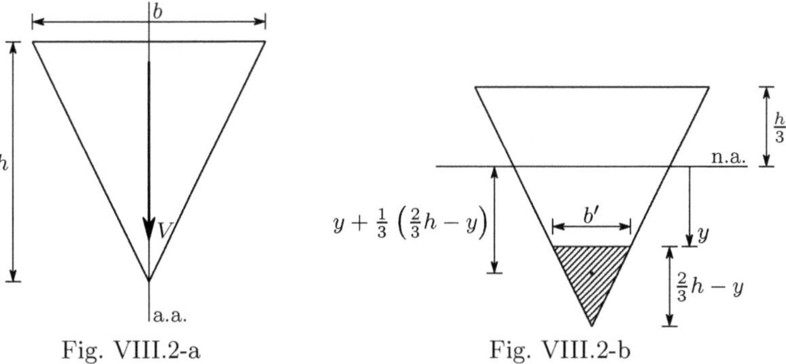

Fig. VIII.2-a   Fig. VIII.2-b

Resolution

Since the cross-section is symmetrical with respect to the action axis of the shear force, the problem may be solved by means of the theory expounded in Subsect. VIII.3.b. The moment of inertia of the cross-section takes the value $I = \frac{bh^3}{36}$ (Fig. VIII.2-a). The first area moment of the area defined by the distance y (shaded area in Fig. VIII.2-b) is given by the expression

$$S = \frac{1}{2}b'\left(\frac{2}{3}h - y\right)\left[y + \frac{1}{3}\left(\frac{2}{3}h - y\right)\right] \quad \text{with} \quad b' = \frac{b}{h}\left(\frac{2}{3}h - y\right) .$$

The vertical component of the shearing stress may obtained from (192), yielding

$$\tau_{xy} = \frac{VS}{Ib'} = \frac{V}{I}\left(\frac{1}{9}hy + \frac{2}{27}h^2 - \frac{1}{3}y^2\right).$$

Differentiating this expression in relation to y and equating to zero, the value of y corresponding to the maximum shearing stress is obtained

$$\frac{\mathrm{d}\tau_{xy}}{\mathrm{d}y} = 0 \;\Rightarrow\; \frac{1}{9}h - \frac{2}{3}y = 0 \;\Rightarrow\; y = \frac{h}{6}.$$

The maximum value of the vertical component of the shearing stress is then

$$y = \frac{h}{6} \;\Rightarrow\; \tau_{xy} = \tau_{xy}^{\max} = \frac{1}{12}\frac{Vh^2}{I} = \frac{1}{12}\frac{36}{bh^3}Vh^2 = 3\frac{V}{bh} = \frac{3}{2}\frac{V}{\Omega}.$$

Since angle φ (Fig. 107) is constant, we conclude that the maximum shearing stress occurs at the sides of the cross-section at the distance $y = \frac{h}{6}$ from the neutral axis and takes the value (cf. (193))

$$\tau_{\max} = \frac{\tau_{xy}^{\max}}{\cos\varphi} = \frac{3}{\cos\left(\arctan\frac{b}{2h}\right)}\frac{V}{bh}.$$

As a rule, the maximum shearing stress occurs on the neutral axis. In this case it does not take place, which is because the cross-section width is not constant in the region around the neutral axis.

VIII.3. Determine the distribution of shearing stresses induced by a vertical shear force V in the open thin-walled cross-section depicted in Fig. VIII.3-a.

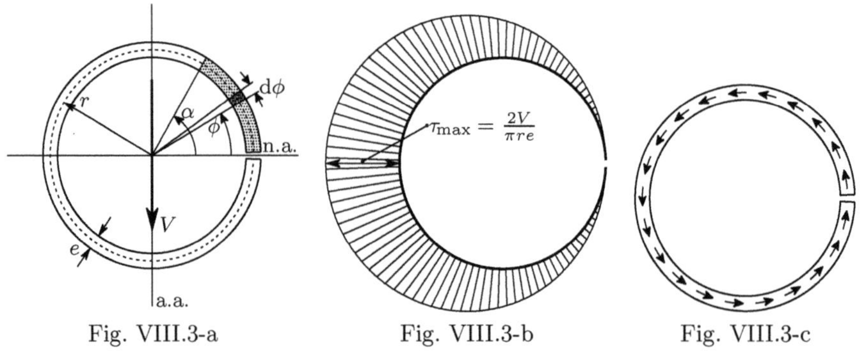

Fig. VIII.3-a   Fig. VIII.3-b   Fig. VIII.3-c

Resolution

Using polar coordinates, we may define the position of a point on the centre line by means of angle α (Fig. VIII.3-a). Denoting the integration variable by ϕ, the first area moment of the shaded area defined by angle α takes the value

$$\mathrm{d}S = er\,\mathrm{d}\phi \times r\sin\phi \ \Rightarrow\ S\left(\alpha\right) = er^2 \int_0^\alpha \sin\phi\,\mathrm{d}\phi = \left(1 - \cos\alpha\right) er^2 \ .$$

The moment of inertia of the cross-section with respect to the neutral axis may also be computed by integration along the centre line, yielding

$$I = \int_0^{2\pi} \left(r\sin\phi\right)^2 \times er\,\mathrm{d}\phi = er^3 \int_0^{2\pi} \sin^2\phi\,\mathrm{d}\phi = \pi er^3 \ .$$

The shearing stress is then defined by the expression

$$\tau = \frac{VS}{Ie} = \frac{V}{\pi er}\left(1 - \cos\alpha\right) \ .$$

This stress distribution defines the diagram represented in Fig. VIII.3-b. The direction of the shearing strain may be found, as described in Subsect. VIII.3.c (Fig. 111), which leads to the directions represented in Fig. VIII.3-c.

VIII.4. Determine the distribution of shearing stresses in a thin-walled circular tube with a wall-thickness e and a radius of the center line r.

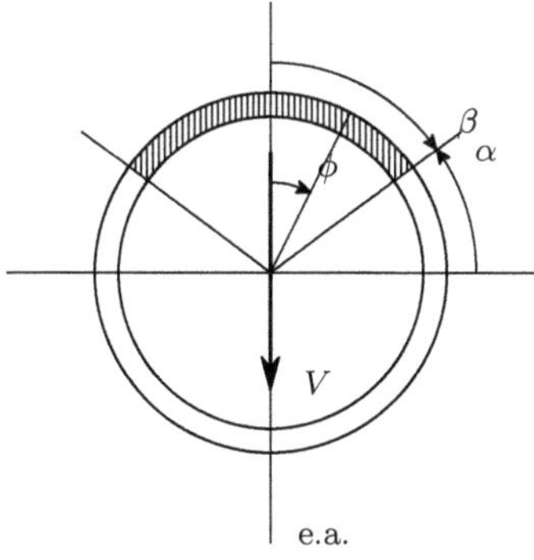

Fig. VIII.4

Resolution

Since we have a symmetrical cross-section, (197) may be used to compute the shearing stresses. Defining the position of the two symmetrical longitudinal cuts by angle β (Fig. VIII.4), we get, for the first area moment of the hatched area,

$$S\left(\beta\right) = 2\int_0^\beta r\cos\phi \times er\,\mathrm{d}\phi = 2er^2\sin\beta \ .$$

The shearing stress then takes the value

$$\tau = \frac{VS}{2Ie} = \frac{V}{I}r^2\sin\beta \ .$$

VIII.5. Determine the distribution of shearing stresses in the cross-section considered in example VIII.4 without using symmetry considerations.

Resolution

The open cross-section considered in example VIII.3 may be used as the statically determinate base problem. Denoting the moment of inertia of the cross-section by I, the shearing stress in the open cross-section is given by the expression

$$\tau_0 = \frac{V}{I} r^2 \left(1 - \cos\alpha\right) \, .$$

The closed line integrals contained in (200) are given by the expressions

$$\oint \frac{S}{e}\,\mathrm{d}s = \int_0^{2\pi} \underbrace{r^2\left(1 - \cos\alpha\right)}_{\frac{S}{e}}\underbrace{r\,\mathrm{d}\alpha}_{\mathrm{d}s} = 2\pi r^3 \qquad \text{and} \qquad \oint \frac{\mathrm{d}s}{e} = \frac{2\pi r}{e} \, .$$

Substituting these expressions into (200), we get

$$\tau_1 = -\frac{V}{Ie}\frac{\oint \frac{S}{e}\,\mathrm{d}s}{\oint \frac{\mathrm{d}s}{e}} = -\frac{V}{Ie}\frac{2\pi r^3}{\frac{2\pi r}{e}} = -\frac{V}{I} r^2 \, .$$

The shearing stress in the closed cross-section is then

$$\tau = \tau_0 + \tau_1 = -\frac{V}{I} r^2 \cos\alpha \, .$$

This value coincides with the solution obtained in example VIII.4, since $\sin\beta = \cos\alpha$. The difference in the sign results from the fact that in example VIII.4 the shearing stress is considered positive when it has the direction of progression of angle β, while in example VIII.3 the direction of progression of angle α is adopted as positive.

VIII.6. Determine the distribution of shearing stresses in the thin-walled cross-section represented in Fig. VIII.6-a. The cross-section wall has a constant thickness e.

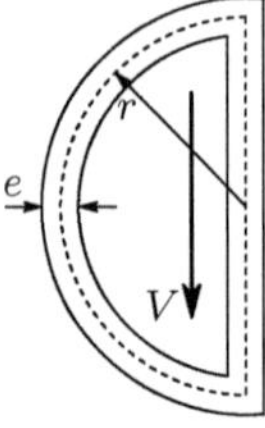

Fig. VIII.6-a

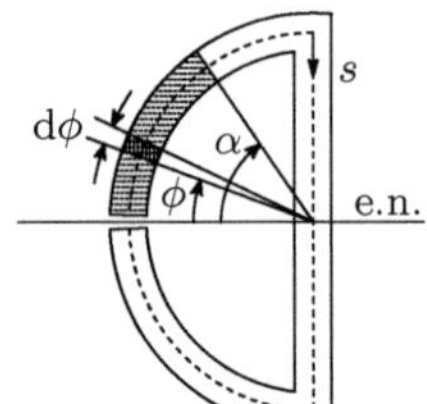

Fig. VIII.6-b

Resolution

Since the cross-section is doubly-connected and the action axis of shear force is not coincident with, or parallel to a symmetry axis, the shearing stress must be computed by means of (198)–(200). As a statically determinate base problem the open thin-walled cross-section represented in Fig. VIII.6-b may be used.

As in example VIII.3, the coordinate α may be considered to define the position of a point in the centre line of the curved part of the wall (Fig. VIII.6-b). Thus, the first area moment of the shaded area defined by angle α may be computed by means of the expression

$$S(\alpha) = \int_0^\alpha r \sin\phi \times er \, \mathrm{d}\phi = er^2 \int_0^\alpha \sin\phi \, \mathrm{d}\phi = (1 - \cos\alpha)\, er^2 \,.$$

The shearing stress corresponding to this area moment is, then,

$$\tau_0(\alpha) = \frac{VS}{Ie} = \frac{V(1 - \cos\alpha)\, er^2}{Ie} = \frac{Vr^2}{I}(1 - \cos\alpha) \,.$$

In the straight wall, the position of a point on the centre line may be defined by coordinate s (Fig. VIII.6-b). The first area moment and the corresponding shearing stresses are then

$$S(s) = S\left(\alpha = \frac{\pi}{2}\right) + se\left(r - \frac{s}{2}\right) = er^2 + ers - e\frac{s^2}{2} \quad \tau_0(s) = \frac{V}{I}\left(r^2 + rs - \frac{s^2}{2}\right) \,.$$

Evaluating the integrals contained in (200), we get

$$\oint \frac{\mathrm{d}s}{e} = \frac{1}{e}(\pi r + 2r)$$

$$\oint \frac{S}{e}\,\mathrm{d}s = 2\int_0^{\frac{\pi}{2}} \frac{(1 - \cos\alpha)\, er^2}{e}\underbrace{r\,\mathrm{d}\alpha}_{\mathrm{d}s} + \int_0^{2r}\left(r^2 + rs - \frac{s^2}{2}\right)\mathrm{d}s = \left(\pi + \frac{2}{3}\right)r^3 \,.$$

From (200) we get the shear flow

$$f = -\frac{V \oint \frac{S}{e}\,\mathrm{d}s}{I \oint \frac{\mathrm{d}s}{e}} = -\frac{V\left(\pi + \frac{2}{3}\right)r^3}{I\,\frac{1}{e}(\pi r + 2r)} = -\frac{Ver^2}{I}\frac{3\pi + 2}{3\pi + 6} \,.$$

Since the wall thickness is constant, the shearing stress corresponding to this shear flow is also constant and takes the value

$$\tau_1 = \frac{f}{e} = -\frac{Vr^2}{I}\frac{3\pi + 2}{3\pi + 6} \,.$$

The total shearing stress in the close thin-walled cross-section may be obtained by adding the stresses in the statically determinate base problem (τ_0) to the stress τ_1, which yields, respectively for the curved and straight walls, the expressions

$$\tau(\alpha) = \tau_0(\alpha) + \tau_1 = \frac{Vr^2}{I}\left(1 - \cos\alpha\right) - \frac{Vr^2}{I}\frac{3\pi+2}{3\pi+6} = \frac{Vr^2}{I}\left(\frac{4}{3\pi+6} - \cos\alpha\right)$$

$$\tau(s) = \tau_0(s) + \tau_1 = \frac{V}{I}\left(r^2 + rs - \frac{s^2}{2} - \frac{3\pi+2}{3\pi+6}\right).$$

In Figs. VIII.6-c a diagram showing the distribution of the shearing stress values in the cross-section is presented. Figure VIII.6-d shows the direction of the shearing stresses.

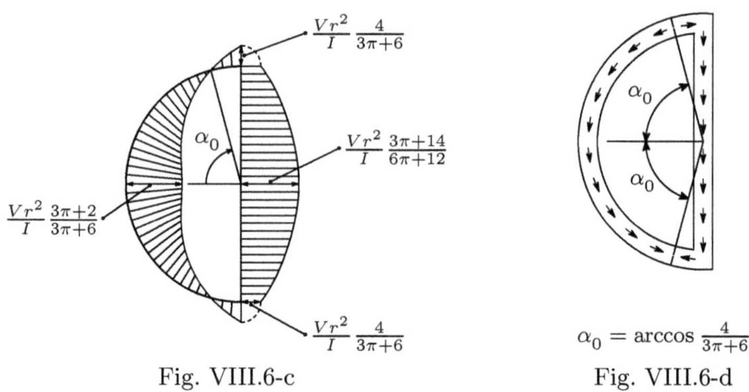

Fig. VIII.6-c   Fig. VIII.6-d

VIII.7. Develop expressions allowing the computation of the shearing stress caused by the shear force in a triply-connected thin-walled cross-section.

Resolution

The degree of static indeterminacy is two, since it is necessary to cut the cross-section wall in two points to get an open thin-walled cross-section. Let us assume that the two cuts are made in the points c_1 and c_2 (Fig. VIII.7). Denoting the coordinates along the centre line in the three walls by s_1, s_2 and s_3, the relative displacement D_1 in cut c_1 may be obtained by applying (198) to the channel defined by points a, b and c_1, which yields, considering as positive the coordinates which define a clockwise rotation around the channel

$$D_1 = \frac{V}{IG}\oint\frac{S}{e}\,\mathrm{d}s = \frac{V}{IG}\left(\int_{c_1}^{b}\frac{S}{e}\,\mathrm{d}s_1 - \int_{a}^{b}\frac{S}{e}\,\mathrm{d}s_3 + \int_{a}^{c_1}\frac{S}{e}\,\mathrm{d}s_1\right).$$

In the same way, we get for the relative displacement D_2 in cut c_2

$$D_2 = \frac{V}{IG}\left(\int_{c_2}^{a}\frac{S}{e}\,\mathrm{d}s_2 + \int_{a}^{b}\frac{S}{e}\,\mathrm{d}s_3 + \int_{b}^{c_2}\frac{S}{e}\,\mathrm{d}s_2\right).$$

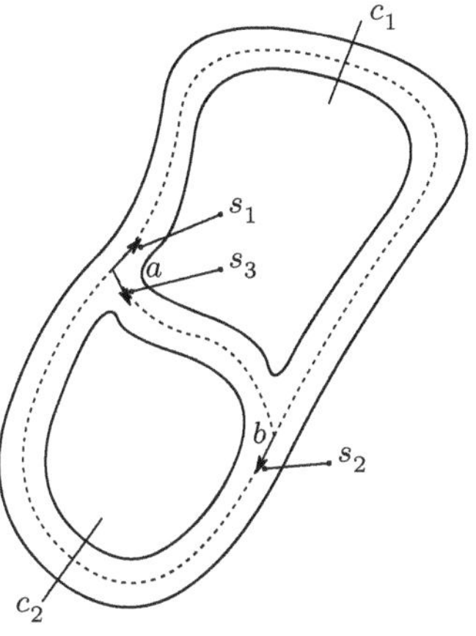

Fig. VIII.7

The displacements D_1 and D_2 are eliminated by the shear flows f_1, f_2 and f_3, corresponding to coordinates s_1, s_2 and s_3, respectively (Fig. VIII.7). Applying (199) to the two channels of the cross-section, we get

$$D_1' = \frac{f}{G} \oint \frac{\mathrm{d}s}{e} = \frac{f_1}{G} \int_{c_1}^{b} \frac{\mathrm{d}s_1}{e_1} - \frac{f_3}{G} \int_{a}^{b} \frac{\mathrm{d}s_3}{e_3} + \frac{f_1}{G} \int_{a}^{c_1} \frac{\mathrm{d}s_1}{e_1}$$

$$D_2' = \frac{f}{G} \oint \frac{\mathrm{d}s}{e} = \frac{f_2}{G} \int_{c_2}^{a} \frac{\mathrm{d}s_2}{e_2} + \frac{f_3}{G} \int_{a}^{b} \frac{\mathrm{d}s_3}{e_3} + \frac{f_2}{G} \int_{b}^{c_2} \frac{\mathrm{d}s_2}{e_2} \ .$$

The shear flows f_1, f_2 and f_3 are the unknowns of the problem, which may be computed by solving the system of equations represented by the compatibility conditions $D_1 + D_1' = 0$ and $D_2 + D_2' = 0$ and the condition of equilibrium of the flows in node a or in node b, $f_1 + f_3 = f_2$ (Fig. 112).

VIII.8. The beam represented in Fig. VIII.8 is made of concrete and reinforced with two steel plates, as shown. Assuming that the concrete does not crack in the tensioned zone, determine the longitudinal shear force in each steel-concrete connection. Consider $E_{\text{steel}} = 10 E_{\text{concrete}}$.

Resolution

The weighted moment of inertia of the cross-section takes the value ($E_s = E_{\text{steel}}$ and $E_c = E_{\text{concrete}}$)

$$J_n = 2 \left[\frac{b \left(\frac{b}{10} \right)^3}{12} + \frac{b^2}{10} \left(b + \frac{b}{20} \right)^2 \right] E_s + E_c \frac{b \left(2b \right)^3}{12}$$

$$= b^4 E_s \left[0.2207 + \frac{E_c}{E_s} \frac{2}{3} \right] = 0.2873 b^4 E_s \ .$$

The first moment of the area occupied by a steel plate in the cross-section is

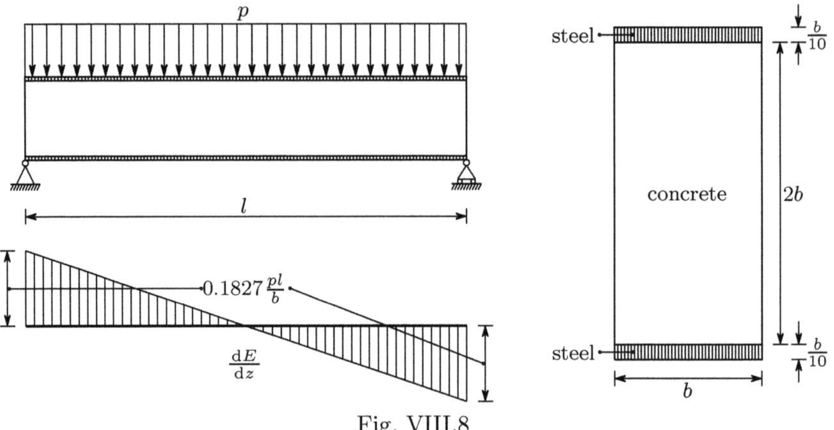

Fig. VIII.8

$$S_a = b\frac{b}{10}\left(b + \frac{b}{20}\right) = 0.105b^3 \ .$$

Substituting the values of J_n and S_a in the first of (202), we get the longitudinal shear force per unit length

$$\frac{\mathrm{d}E}{\mathrm{d}z} = \frac{VE_sS_a}{J_n} = \frac{VE_s0.105b^3}{0.2873b^4E_s} = 0.3654\frac{V}{b} \ .$$

This force attains the maximum value in the cross-sections over the supports, taking the value (Fig. VIII.8)

$$V_{\mathrm{max}} = \frac{pl}{2} \Rightarrow \left(\frac{\mathrm{d}E}{\mathrm{d}z}\right)_{\mathrm{max}} = 0.1827\frac{pl}{b} \ .$$

VIII.9. In the cantilever beam considered in example VII.11 (Fig. VII.11-a), determine the distribution of the longitudinal shear force per unit length in the connection between the two materials.

Resolution

The horizontal shear force V_x does not cause a longitudinal shear force in the surface between the two materials, since axis y (Fig. VII.11-b) is a symmetry axis and that surface is perpendicular to this axis. The longitudinal shear force introduced by the cross-sectional shear force V_y may be computed by means of (202).

The weighted moment of inertia of the cross-section takes the value $J_x = 1915.34Ea^4$ (cf. example VII.11). The moment of the area occupied by material a, with respect to the neutral axis, is

$$S_a = 40a^2 \times 2.91176a \approx 116.470\,a^3 \ .$$

Considering a coordinate z with origin in the free end and pointing leftwards, the shear force V_y is defined by the expression

$$V_y(z) = 10paz .$$

The longitudinal shear force per unit length, as function of coordinate z, takes then the value (202)

$$E'(z) = \frac{\mathrm{d}E}{\mathrm{d}z} = \frac{V_y S_a E_a}{J_x} = \frac{10paz \times 116.470\,a^3 \times 2E}{1915.34 E a^4} \approx 1.21618\,pz .$$

This force attains the maximum value in the built-in end, yielding

$$z = z_{\max} = 200a \;\Rightarrow\; E' = E'_{\max} = 243.236\,pa .$$

VIII.10. Determine the shearing stress in the point of connection between the web and a flange in the cross-section represented in Fig. VIII.10.

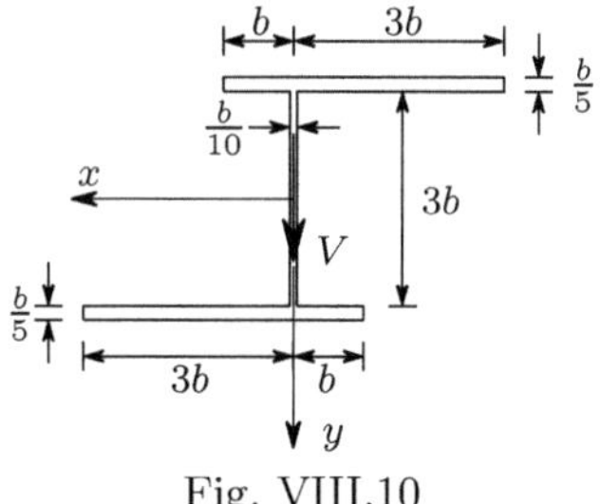

Fig. VIII.10

Resolution

The problem may be solved by means of (203). To this end, it is necessary to compute the moments and the product of inertia in relation to axes x and y. These quantities take the values

$$I_x = \frac{\frac{b}{10}\,(3b)^3}{12} + 2\left[\frac{4b\left(\frac{b}{5}\right)^3}{12} + 4b\frac{b}{5}\left(1.5b + \frac{b}{10}\right)^2\right] = 4.32633 b^4$$

$$I_y = 2\left(\frac{\frac{b}{5}\,(3b)^3}{3} + \frac{\frac{b}{5}b^3}{3}\right) + \frac{3b\left(\frac{b}{10}\right)^3}{12} = 3.73358 b^4$$

$$I_{xy} = 4b \times \frac{b}{5} \times b \times \left(1.5b + \frac{b}{10}\right) \times 2 = 2.56b^4 .$$

In order to get the shearing stress in the connection point between the web and a flange, the first area moments of a flange, with respect to axes x and y

must be computed. Considering the bottom flange, these quantities are given by the expressions

$$S_x = 4b \times \frac{b}{5} \times \left(\frac{3}{2}b + \frac{b}{10}\right) = 1.28b^3 \qquad \text{and} \qquad S_y = 4b \times \frac{b}{5} \times b = 0.8b^3 \ .$$

The longitudinal shear force per unit length in this point is, then, (203)

$$\frac{\mathrm{d}E}{\mathrm{d}z} = \frac{3.73358b^4 \times 1.28b^3 - 2.56b^4 \times 0.8b^3}{4.32633b^4 \times 3.73358b^4 - \left(2.56b^4\right)^2} V = 0.28450\frac{V}{b} \ .$$

Thus, the shearing stress in that point takes the value

$$\tau = \frac{\mathrm{d}E}{\frac{b}{10}\,\mathrm{d}z} = \frac{0.2845}{0.1b}\frac{V}{b} = 2.845\frac{V}{b^2} \ .$$

As an alternative, this stress could be determined by decomposing the shear force in the two principal directions of inertia. However the volume of computation would be substantially larger, since it would be necessary to compute the principal moments and directions of inertia and two shearing stresses (one for each plane of bending).

VIII.11. Determine the position of the shear centre in the cross-section represented in Fig. VIII.3-a.

Resolution

As seen in example VIII.3, the stresses caused by the shear force in the cross-section are given by the expression

$$\tau = \frac{V}{\pi e r}\left(1 - \cos\alpha\right) \ .$$

The position of the shear centre on the symmetry axis may be obtained by the equivalence condition of the moments of the shearing stresses and of the shear force with respect to any point of the cross-section's plane. The point which leads to the simplest expressions is the centre of the cross-section. In this case, we have

$$M = \int_0^{2\pi} r \times \tau e r \,\mathrm{d}\alpha = \frac{Vr}{\pi}\int_0^{2\pi} (1 - \cos\alpha)\,\mathrm{d}\alpha = 2Vr \ .$$

The shearing stress resultant obviously takes the value V (this may easily be confirmed by evaluating the integral $\int_0^{2\pi} -\tau e r \cos\alpha\,\mathrm{d}\alpha$). The moment of the stresses will be equal to the moment of the shear force if the action axis of V is at a distance d from the cross-section centre given by the expression

$$V \times d = 2Vr \ \Rightarrow\ d = 2r \ .$$

Since the cross-section is symmetrical in relation to the horizontal axis passing through the centre, the shear centre will be on this axis, at a distance $2r$ from the centroid, to the left.

VIII.12. Considering the cross-section defined in example VIII.6, determine the position of the shear centre:
(a) in the statically determinate base problem (bar with the longitudinal cut, Fig. VIII.6-b);
(b) in the closed cross-section.

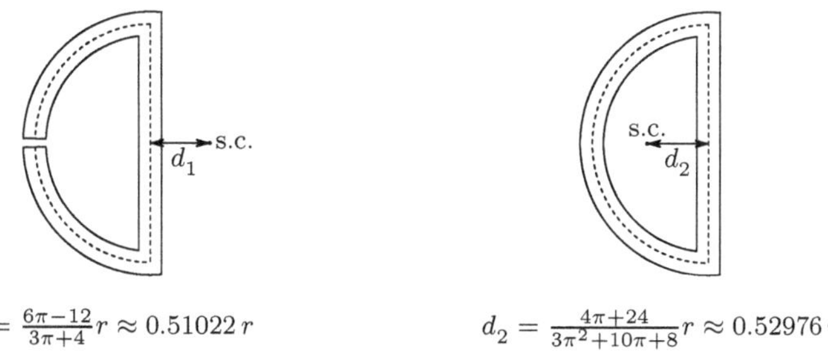

$$d_1 = \frac{6\pi-12}{3\pi+4}r \approx 0.51022\,r \qquad d_2 = \frac{4\pi+24}{3\pi^2+10\pi+8}r \approx 0.52976\,r$$

Fig. VIII.12

Resolution

In order to determine the position of the shear centre, the moment of inertia must be expressed as a function of the given geometrical data. Considering that the cross-section is thin-walled, this quantity may be given by the expression (Fig. VIII.6-b)

$$I = 2 \int_0^{\frac{\pi}{2}} (r \sin \phi)^2\, re\, d\phi + \frac{e\,(2r)^3}{12} = \left(\frac{\pi}{2} + \frac{2}{3}\right) er^3 . \qquad \text{(VIII.12-a)}$$

(a) The simplest expression for the moment of the shearing stresses is found if the centre of the half-circumference defining the centre line of the curved wall is used as reference point. This moment has a clockwise direction and takes the value

$$\tau_0 = \frac{Vr^2}{I}(1 - \cos \alpha) \Rightarrow \left| \begin{array}{l} M = 2 \int_0^{\frac{\pi}{2}} r \times \tau er\, d\alpha \\[2mm] = \frac{2Ver^4}{I} \int_0^{\frac{\pi}{2}} (1 - \cos \alpha)\, d\alpha = \frac{Ver^4}{I}(\pi - 2) . \end{array} \right.$$

This moment will be equivalent to the moment of the shear force with respect to the same point if its line of action is at a distance d to the right of this point, so the following condition is satisfied

$$Vd = M \Rightarrow Vd = \frac{Ver^4}{I}(\pi - 2) \Rightarrow d = \frac{er^4}{I}(\pi - 2) = \frac{6\pi - 12}{3\pi + 4}r .$$

where I has been substituted by (VIII.12-a).

(b) In the case of the closed cross-section, the procedure is similar, the only difference being, that the shearing stress in the curved wall is given by the expression

$$\tau(\alpha) = \frac{Vr^2}{I}\left(\frac{4}{3\pi+6} - \cos\alpha\right) .$$

The moment of the stresses with respect to the centre of the half-circumference, considered as positive in the clockwise direction, is then given by the expression

$$M = \frac{2Ver^4}{I}\int_0^{\frac{\pi}{2}}\left(\frac{4}{3\pi+6} - \cos\alpha\right) = -\frac{Ver^4}{I}\frac{2\pi+12}{3\pi+6} .$$

The equivalence condition of the moments of the stresses and the shear force yields the position of the shear centre

$$Vd = M \Rightarrow Vd = -\frac{Ver^4}{I}\frac{2\pi+12}{3\pi+6} \Rightarrow d = -\frac{4\pi+24}{3\pi^2+10\pi+8}r .$$

The minus sign means that the shear centre is located to the left of the reference point.

Figure VIII.12 shows these two cross-sections with their corresponding shear centres. As mentioned in Footnote 57, the shear flow f computed in example VIII.6 corresponds to a torsional moment. The relation between the shear flow and the torsional moment will be studied in Sect. X.3. This relation is given in (240) and may be written in the form $T = 2Af$, where A represents the area limited by the wall's centre line in the closed cross-section. In the present case, we have

$$\begin{cases} f = \frac{Ver^3}{I}\frac{3\pi+2}{3\pi+6} \\[2ex] A = \frac{\pi r^2}{2} \\[2ex] I = \left(\frac{\pi}{2}+\frac{2}{3}\right)er^3 \end{cases} \Rightarrow T = 2\times\frac{\pi r^2}{2}\times\frac{Ver^3}{\left(\frac{\pi}{2}+\frac{2}{3}\right)er^3}\frac{3\pi+2}{3\pi+6} \approx 1.03398\,Vr .$$

This torsional moment is equal to $V(d_1 + d_2)$, that is, it corresponds to the translation of the shear force from the shear centre of the open cross-section to the shear center of the closed section.

VIII.13. The cantilever beam represented in Fig. VIII.13-a has a closed cross-section from A to B and an open cross-section from B to C. The wall has constant thickness e. The plane of application of the distributed load p contains the centre line of the web in segment BC of the beam. Draw the diagram representing the torsional moment of the beam.

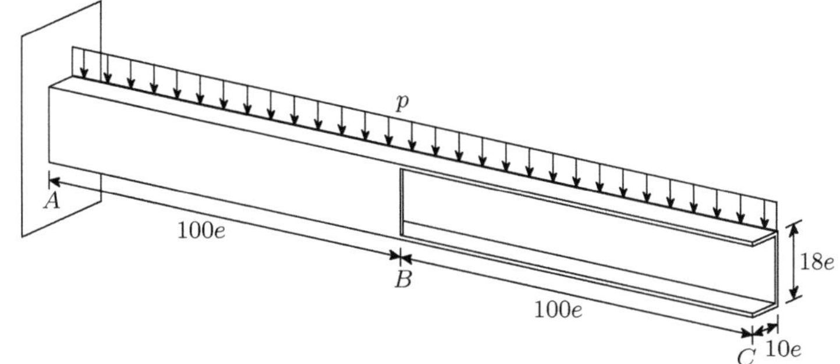

Fig. VIII.13-a

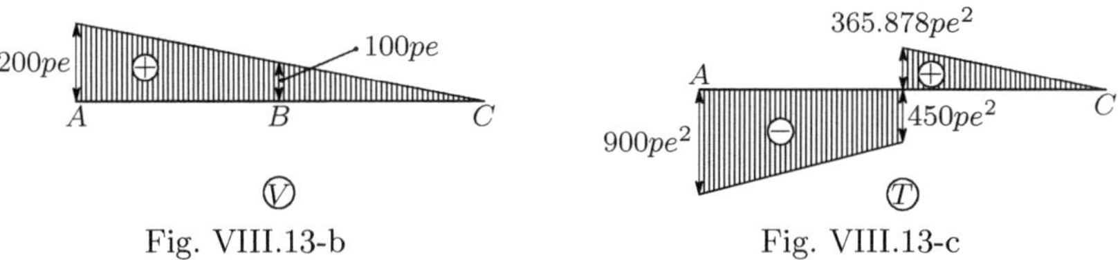

Fig. VIII.13-b

Fig. VIII.13-c

Resolution

The shear centre of the cross-section in segment AB coincides with the centroid, since the cross-section is doubly symmetric. The distance from the action axis of the shear force to the shear centre is $4.5\,e$. In the case of segment BC, the position of the shear centre may be computed from (205). In this case, we have

$$\begin{cases} h = 17e \\ b = 9e \end{cases} \Rightarrow d = \frac{3}{\left(\frac{h}{b}\right)^2 + 6\frac{h}{b}}h = \frac{3}{\left(\frac{17e}{9.5e}\right)^2 + 6\frac{17e}{9.5e}}17e \approx 3.65878\,e\,.$$

The diagram representing the distribution of the shear force is given in Fig. VIII.13-b. By multiplying the shear force by the distance of its line of action to the shear centre, a negative torsional moment is obtained for segment AB, while in part BC it takes a positive value, as represented in Fig. VIII.13-c.

VIII.14. Draw a diagram representing the distribution of torsional moment in the cantilever beam considered in example VII.11 and represented in Fig. VII.11-a.

Resolution

The beam has a thin-walled cross-section with straight centre lines converging in a point, so that this point – the centroid of the rectangle made of material a – coincides with the shear centre of the cross-section. In fact, the non-homogeneity of the cross-section does not affect the validity of the line of reasoning developed for this kind of cross-sections (Fig. 119).

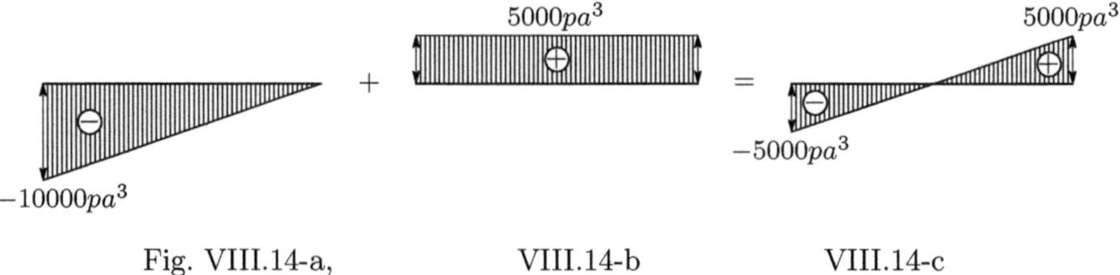

Fig. VIII.14-a,   VIII.14-b   VIII.14-c

The vertical load p causes a negative torsional moment with the value $V_y \times 5a$, as represented in Fig. VIII.14-a. The horizontal load $P = 500pa^2$, induces a constant positive torsional moment with the value $P \times 10a$, which is shown in Fig. VIII.14-b. By superposing these two diagrams, we get the total moment given in Fig. VIII.14-c.

VIII.15 Figure VIII.15 represents a thin-walled cross-section made of two materials a (hatched areas) and b, with the moduli of elasticity $E_a = 9E$ and $E_b = 3E$. Determine the position of the shear centre of this cross-section.

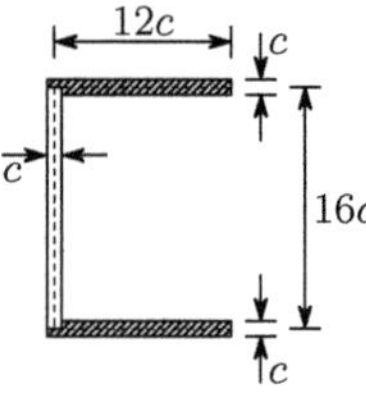

Fig. VIII.15

Resolution

The longitudinal shear force in composite bars under plane bending may be computed by means of (201). Representing by S_a and S_b the first area moments contained in (201), we may develop the following expression for the computation of the shearing stress

$$\tau = \frac{\mathrm{d}E}{e\,\mathrm{d}z} = \frac{V\left(E_a S_a + E_b S_b\right)}{J_n e} .$$

Since the cross-section has a symmetry axis, the shear centre is on this axis and its position may be obtained by considering a shear force with the direction of the other principal axis of inertia (a vertical axis in this case). The weighted moment of inertia, with respect to the horizontal axis takes the value

$$J_n = \frac{c\left(16c\right)^3}{12}3E + 2 \times 12c^2 \times \left(8c\right)^2 \times 9E = 14848Ec^2 .$$

The procedure leading to the computation of the position of the shearing stress resultant from the above indicated expressions for the shearing stress τ and for the moment of inertia I is exactly the same as that used in the example described in Sect. VIII.4. The distribution of shearing stress takes the same form, as indicated in Fig. 118-b. The stress in the angle point between the web and a flange takes the value $(S_a = 12c^2 \times 8c$ and $S_b = 0)$

$$\tau_1 = \frac{V \times 9E \times 12c^2 \times 8c}{14848Ec^4 \times c} = \frac{27}{464}\frac{V}{c^2} \approx 58.190 \times 10^{-3}\frac{V}{c^2}\ .$$

As in the case of Fig. 118-b, we compute the difference between the maximum stress in the web and τ_1. This difference is defined by the first area moment of the half web with respect to the neutral axis, taking the value

$$\tau_2 = \tau_{\max} - \tau_1 = \frac{V}{14848Ec^4 \times c}8c^2 \times 4c \times 3E = \frac{3}{464}\frac{V}{c^2}\ .$$

The resultant of the shearing stresses in a flange is equal to the area of the stress diagram (triangular diagram, Fig. 118-b), multiplied by the thickness $e = c$

$$R_b = \frac{1}{2}\tau_1 \times 12c \times c = \frac{81}{232}V\ .$$

The moment of the shearing stresses with respect to one of the angle points of the wall's centre line may be computed directly from R_b, yielding

$$M = R_b \times 16c = \frac{162}{29}Vc\ .$$

The resultant of the shearing stresses is equal to the shear force. This fact may easily be confirmed by computing the stress resultant in the web, R_a, which is simultaneously the total stress resultant, since the stresses in the flanges are perpendicular to the shear force

$$R_a = \tau_1 \times 16c^2 + \frac{2}{3}\tau_2 \times 16c^2 = \left(\frac{27}{464} \times 16 + \frac{2}{3}\frac{3}{464} \times 16\right)V = V\ .$$

By equating the moment of the stresses to the moment of the shear force, we get the distance d of the action axis of this force to the centre line of the web, such that it does not induce torsion. This distance defines the position of the shear centre (Fig. 118-c)

$$M = V \times d \Rightarrow \frac{162}{29}Vc = Vd \Rightarrow d = \frac{162}{29}c \approx 5.5862\,c\ .$$

VIII.16. In the beam represented in Fig. 85, find the relation between the maximum values of the stresses σ_y and σ_z (162), as a function of the relation between the height h of the cross-section and the span l.

Resolution

From the second of (162) we can see that the maximum value of σ_y occurs in the points belonging to the upper fibres and that it takes the value $-p$. Dividing this value by $\sigma_{z-\max}$ (162), we get the relation

$$\frac{\sigma_{y-\max}}{\sigma_{z-\max}} = \frac{-p}{\frac{ph^3}{\frac{h^3}{12}}\left[\frac{1}{16}\left(\frac{l}{h}\right)^2 + \frac{1}{60}\right]} = -\frac{5}{\frac{15}{4}\left(\frac{l}{h}\right)^2 + 1} \, .$$

In slender members we generally have $l \geq 10h$. For $l = 10h$, $\sigma_{y-\max}$ is only about 1% of $\sigma_{z-\max}$.

VIII.17. Develop expressions for the direct application of the Tresca's and von Mises' yielding criteria to the safety evaluation of slender members.

Resolution

The principal stresses of the stress state in a point of a slender member are given by the expressions (cf. (38), with $\sigma_x = \sigma$, $\sigma_y = 0$ and $\tau_{xy} = \tau$)

$$\sigma_1 = \frac{\sigma}{2} + \sqrt{\left(\frac{\sigma}{2}\right)^2 + \tau^2} \qquad \text{and} \qquad \sigma_3 = \frac{\sigma}{2} - \sqrt{\left(\frac{\sigma}{2}\right)^2 + \tau^2} \, ,$$

where σ and τ are the normal and shearing stresses in a point of the cross-section. Substituting these values in the expression of Tresca's yielding criterion (104), we get

$$\sigma_1 - \sigma_3 \leq \sigma_{\text{all}} \Rightarrow \sqrt{\sigma^2 + 4\tau^2} \leq \sigma_{\text{all}} \, .$$

Using the same procedure in relation to von Mises' yielding criterion (105), we get ($\sigma_2 = 0$)

$$\sqrt{\sigma_1^2 - \sigma_1\sigma_3 + \sigma_3^2} \leq \sigma_{\text{all}} \Rightarrow \sqrt{\sigma^2 + 3\tau^2} \leq \sigma_{\text{all}} \, .$$

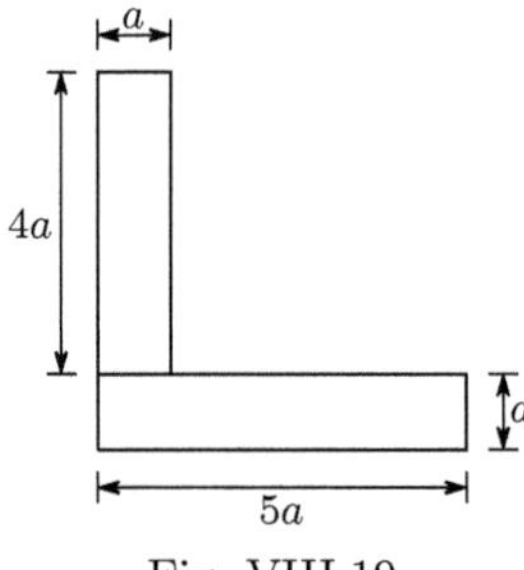

Fig. VIII.19

VIII.18. A cantilever beam of length l, carrying a uniformly distributed load p, is made of two materials, a and b, as depicted in Fig. VII.24. The two materials have linear elastic behaviour, with elasticity moduli $E_a = 4E$ and $E_b = E$. Determine the longitudinal shear force in the connection between one rectangle of material a and the material b around it.

VIII.19. Figure VIII.19 represents the cross-section of a cantilever beam of length $100a$. The beam is composed of two bars with rectangular cross-section, connected as shown in the figure. The beam supports a vertical, uniformly distributed load p.

(a) What is the longitudinal shear force in the connection between the two bars?

(b) What should be the position of the plane containing the loading, so that no torsion takes place?

IX

Bending Deflections

IX.1 Deflections Caused by the Bending Moment

IX.1.a Introduction

When the effect of the bending moment was studied (Chap. VII), several expressions were developed relating the bending moment to the curvature (145, 148, 154, 169, 173, 179, 181 and 182). As remarked a few times already, the validity of these expressions is not limited by the size of the deformations and rotations. In this section, three methods are described for computing cross-section displacement and rotation, for a given curvature distribution. Other methods, based on energy considerations, will be seen in Chap. XII.

To this end, let us consider that the axis of the bar after the deformation (the deflection curve) has the shape represented in Fig. 123.

The figure shows that $\rho\,\mathrm{d}\varphi = \mathrm{d}s$. Since a positive bending moment causes a negative curvature in the reference system (y, z) used in Fig. 123, we get, in the case of linear elastic plane bending (145)

$$\frac{1}{\rho} = \frac{\mathrm{d}\varphi}{\mathrm{d}s} = -\frac{M}{EI} \,.$$

(207)

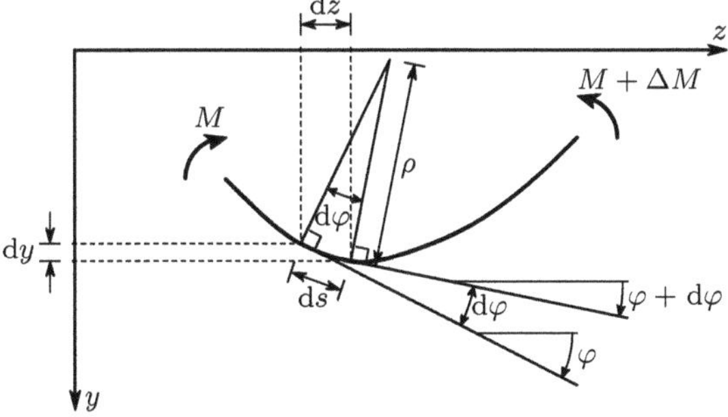

Fig. 123. Deformation caused by a positive bending moment

Angle φ may be related to the coordinates of the deflection curve, yielding

$$\varphi = \arctan \frac{\mathrm{d}y}{\mathrm{d}z} \;\Rightarrow\; \frac{1}{\rho} = \frac{\mathrm{d}\varphi}{\mathrm{d}z}\frac{\mathrm{d}z}{\mathrm{d}s} = \frac{\mathrm{d}}{\mathrm{d}z}\left(\arctan\frac{\mathrm{d}y}{\mathrm{d}z}\right)\frac{\mathrm{d}z}{\mathrm{d}s}$$

$$= \frac{\frac{\mathrm{d}^2 y}{\mathrm{d}z^2}}{1+\left(\frac{\mathrm{d}y}{\mathrm{d}z}\right)^2}\frac{\mathrm{d}z}{\mathrm{d}s} = \frac{\frac{\mathrm{d}^2 y}{\mathrm{d}z^2}}{1+\left(\frac{\mathrm{d}y}{\mathrm{d}z}\right)^2}\frac{\mathrm{d}z}{\sqrt{\mathrm{d}y^2 + \mathrm{d}z^2}} = \frac{\frac{\mathrm{d}^2 y}{\mathrm{d}z^2}}{\left[1+\left(\frac{\mathrm{d}y}{\mathrm{d}z}\right)^2\right]^{\frac{3}{2}}} \,. \tag{208}$$

In the case of linear elastic plane bending, the second equality of (207) is valid. In this case, the relation between the bending moment M and the coordinates of the deflection curve is defined by the differential equation

$$\frac{M}{EI} = -\frac{\frac{\mathrm{d}^2 y}{\mathrm{d}z^2}}{\left[1+\left(\frac{\mathrm{d}y}{\mathrm{d}z}\right)^2\right]^{\frac{3}{2}}} \,. \tag{209}$$

The resolution of this equation is quite complex. However, in the usual applications in the small deformation domain, the rotations are almost always very small. If axis z has the direction of the bar's axis before the deformation, the tangent of the rotation angle $\frac{\mathrm{d}y}{\mathrm{d}z}$ (the slope of the deflection curve) is also very small. Furthermore, as only the square of this quantity appears in (208) and (209), it may be neglected in the addition to 1, yielding

$$\frac{1}{\rho} = \frac{\mathrm{d}^2 y}{\mathrm{d}z^2} \quad \text{(any material)}$$

$$-\frac{M}{EI} = \frac{\mathrm{d}^2 y}{\mathrm{d}z^2} \quad \text{(linear elastic material)} \,. \tag{210}$$

IX.1.b Method of Integration of the Curvature Equation

This method consists of the integration of (210) in order to get the slope of the deflection curve $\frac{\mathrm{d}y}{\mathrm{d}z}$ and the displacements of the cross-sections, which are here denoted by y. Thus, we get (since φ is an infinitesimal quantity, we have $\varphi \approx \tan\varphi$)[1]

$$\varphi = \frac{\mathrm{d}y}{\mathrm{d}z} = \int \frac{1}{\rho}\,\mathrm{d}z + \varphi_0$$

$$\varphi = \frac{\mathrm{d}y}{\mathrm{d}z} = -\int \frac{M}{J_\theta}\,\mathrm{d}z + \varphi_0 \tag{211}$$

$$\varphi = \frac{\mathrm{d}y}{\mathrm{d}z} = -\frac{1}{EI}\int M\,\mathrm{d}z + \varphi_0 \,,$$

---

[1] In these expressions and in those that follow the integral symbols ($\int$) represent the primitives, since the integration constants are given explicitly (φ_0 in the case of (211)).

respectively for slender members made of any material, for members made of linear elastic materials (169) and for homogeneous bars made of linear elastic materials, with constant cross-section, under plane bending. Integrating once again, we get the coordinate y of the deflection line, i.e., the displacement of the bar in the direction perpendicular to its axis. For the same cases as in (211), we get

$$y = \int \frac{\mathrm{d}y}{\mathrm{d}z}\,\mathrm{d}z = \iint \frac{1}{\rho}\,\mathrm{d}z^2 + \varphi_0 z + y_0$$

$$y = -\iint \frac{M}{J_\theta}\,\mathrm{d}z^2 + \varphi_0 z + y_0 \tag{212}$$

$$y = -\frac{1}{EI} \iint M\,\mathrm{d}z^2 + \varphi_0 z + y_0 \, .$$

The integration constants φ_0 and y_0 are obtained from the support and continuity conditions. The following examples are given: in a built-in end the rotation and displacement will be zero; in a symmetry point, with respect to geometry and loading, the rotation will vanish; in a pin connection between two bars (hinge), the displacement must be the same in the two joined ends; in a connection between two bars which resists bending, but does not resist shear, the rotation must be the same in the two joined ends; in a point of discontinuity of the curvature equation, with continuity of the bar, the rotation and the deflection given by the left and right curvature equations must be the same; etc.

Now, let us note that, provided that the distance z is measured in the undeformed configuration, the simplification introduced in the development of (210) does not introduce any error in the computation of the rotation by means of (211), even for larger rotations. In fact, as no axial deformation is being considered, the coordinate s in (207), after the deformation, takes the same value as the coordinate z in the undeformed configuration. In order to clarify this point, expressions for the computation of rotations and deflections as functions of coordinate s are developed below.

From (207) we get, without any simplifying hypothesis

$$\varphi = \int \frac{1}{\rho}\,\mathrm{d}s + \varphi_0 \, . \tag{213}$$

From Fig. 123 we conclude easily that an infinitesimal increment $\mathrm{d}s$ of coordinate s corresponds to increments $\mathrm{d}y = \mathrm{d}s \sin\varphi$ and $\mathrm{d}z = \mathrm{d}s \cos\varphi$ in the vertical and horizontal coordinates, respectively. Integrating and substituting (213) in these expressions, we get the following expressions for the coordinates of the deflection curve

$$y = \int \sin\varphi\,\mathrm{d}s + y_0 = \int \sin\left(\int \frac{1}{\rho}\,\mathrm{d}s + \varphi_0\right)\mathrm{d}s + y_0$$

$$z = \int \cos\varphi\,\mathrm{d}s + z_0 = \int \cos\left(\int \frac{1}{\rho}\,\mathrm{d}s + \varphi_0\right)\mathrm{d}s + z_0 \, . \tag{214}$$

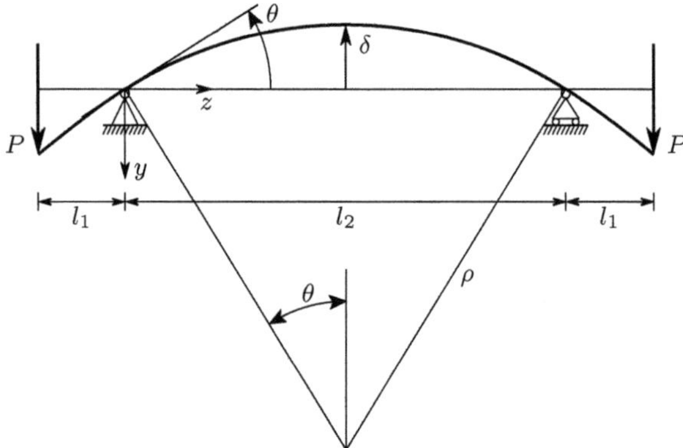

Fig. 124. Deformation in circular bending

Since there is no axial deformation, the coordinate s after the deformation takes the same value as the coordinate z before the deformation. From (213) and (214) we get therefore, without any restriction in the size of the rotations,

$$\begin{cases} \varphi = \int \frac{1}{\rho}\,\mathrm{d}z + \varphi_0 \\[2mm] y = \int \sin\left(\int \frac{1}{\rho}\,\mathrm{d}z + \varphi_0\right)\mathrm{d}z + y_0\,. \end{cases} \tag{215}$$

The first of these expressions coincides with the first of (211) and has been obtained without the need of any simplifying hypothesis. With respect to the second one, it corresponds to the last of (208), with the difference that there (208) the coordinates y and z refer to the deformed configuration, while in (215) coordinate z refers to the undeformed bar. When the rotations φ are small, we may admit that $\sin\varphi \approx \varphi$ and that $\cos\varphi \approx 1$. As a consequence, coordinate z coincides in the undeformed and deformed configurations and the second of (215) coincides with the first of (212).[2]

These considerations may be illustrated by a simple example. Let us consider the beam represented in Fig. 124.

---

[2]Problems of Mechanics of Materials, where the deformations are too large to be considered as infinitesimal, may be formulated by describing the geometry of the body, either in the deformed, or in the undeformed configuration. In the first case we have a *Eulerian formulation*, while the second is called a *Lagrangian formulation*. Generally, in Solid Mechanics, the latter leads to simpler expressions, while in liquids the opposite is true. In the present case, (215) represent a Lagrangian description of the relation between the curvature and the corresponding rotations and displacements, since the coordinate z refers to the undeformed configuration and y represents the displacement. Equations (208) and (209) are a Eulerian description of the same relations, since y and z are the coordinates of the axis of the deformed bar. In this case, too, we conclude that the Lagrangian formulation leads to simpler expressions.

As the beam is under circular bending between the supports ($M_0 = Pl_1$),[3] its deflection curve is an arc of circumference, so that the displacement δ and the rotation over the supports, θ, may be computed exactly, i.e., without the simplifying hypothesis of small rotations. The absolute value of the rotation θ is then given by

$$\theta\rho = \frac{l_2}{2} \;\Rightarrow\; \theta = \frac{1}{\rho}\frac{l_2}{2} = \frac{M_0 l_2}{2EI} \; . \tag{216}$$

The exact value of the displacement δ is ($\rho = \frac{EI}{M_0}$, (145))

$$\delta = \rho - \rho\cos\theta = \frac{EI}{M_0}\left(1 - \cos\frac{M_0 l_2}{2EI}\right) \; . \tag{217}$$

The method of integration of the curvature equation yields ($M = -M_0$)

$$\varphi(z) = \frac{1}{EI}\int M_0\,\mathrm{d}z + \varphi_0 = \frac{M_0}{EI}z + \varphi_0$$

$$z = \frac{l_2}{2} \;\Rightarrow\; \varphi = 0 \;\Rightarrow\; \varphi_0 = -\frac{M_0 l_2}{2EI} \;\Rightarrow\; \varphi(z) = \frac{M_0}{EI}\left(z - \frac{l_2}{2}\right) \tag{218}$$

$$z = 0 \;\Rightarrow\; \varphi = \theta = -\frac{M_0 l_2}{2EI} \; .$$

This rotation has the same absolute value as given by (216), which confirms that no error is introduced by the assumption of small rotations. In the case of the displacement δ, we get from the last of (212) and from (218) ($z = 0 \Rightarrow y = 0 \Rightarrow y_0 = 0$)

$$y(z) = \int \varphi(z)\,\mathrm{d}z = \frac{M_0}{EI}\left(\frac{z^2}{2} - \frac{l_2}{2}z\right); \qquad z = \frac{l}{2} \;\Rightarrow\; y = \delta = -\frac{M_0 l_2^2}{8EI} \; . \tag{219}$$

This expression is different from (217), which reflects the approximation introduced by the assumption of small rotations. If (215) is used instead, an expression with the same absolute value, as yielded by geometrical considerations (217), is obtained

$$y = \int \sin\varphi\,\mathrm{d}z + y_0 = \int \sin\left[\frac{M_0}{EI}\left(z - \frac{l_2}{2}\right)\right]\mathrm{d}z + y_0$$

$$= -\frac{EI}{M_0}\cos\left[\frac{M_0}{EI}\left(z - \frac{l_2}{2}\right)\right] + y_0 \; ;$$

---

[3]In fact, the bending moment does not remain exactly equal to Pl_1 during deformation, since the distance l_1 between the point of application of P and the support varies during the process. However, in this case only the beam segment between the supports needs to be considered, so that only the value of M_0 needs to be considered, and not l_1.

$$z = 0 \Rightarrow y = 0 \Rightarrow y_0 = \frac{EI}{M_0} \cos \frac{M_0 l_2}{2EI}$$

$$\Rightarrow y = \frac{EI}{M_0} \left\{ \cos \frac{M_0 l_2}{2EI} - \cos \left[\frac{M_0}{EI} \left(z - \frac{l_2}{2} \right) \right] \right\} ;$$

$$z = \frac{l_2}{2} \Rightarrow y = -\frac{EI}{M_0} \left(1 - \cos \frac{M_0 l_2}{2EI} \right) . \tag{220}$$

In order to get an idea about the error introduced by the assumption of infinitesimal rotations, the exact value of the displacement – δ_e – given by (217) or (220) is compared with the approximate value – δ_a – obtained by means of the method of integration of the curvature equation (219). This error depends only on angle θ, as it is easily confirmed by substituting $\frac{M_0}{EI}$ by $\frac{2\theta}{l_2}$ in the above-mentioned expressions (216), yielding (θ in radians)

$$\frac{M_0}{EI} = \frac{2\theta}{l_2} \Rightarrow \begin{cases} \delta_e = \frac{l_2}{2\theta} (1 - \cos \theta) \\ \delta_a = \frac{\theta l_2}{4} \end{cases} \Rightarrow \frac{\delta_a}{\delta_e} = \frac{1}{2} \frac{\theta^2}{1 - \cos \theta} .$$

From this expression, we conclude that a relative error of 5% in the approximate solution ($\frac{\delta_a}{\delta_e} = 1.05$) requires the rotation θ to attain the value of 0.7645 radians (43.8°!!!). This example shows that large rotations are needed to get a significant error from the introduction of the assumption of infinitesimal rotations, in the computation of the deflections from a given curvature equation. This is because the sine of an angle remains close to the measure of the angle up to appreciable angle magnitudes (215). For example, the angle which exceeds the value of its sine by 5% has the magnitude of 0.541 radians (31°).

IX.1.c The Conjugate Beam Method

The conjugate beam method is based on an analogy between the determination of the shear force and bending moment equations, $V(z)$ and $M(z)$, respectively, from the function defining the transversal loading $P(z)$, and the determination of the rotation and deflection equations, $\varphi(z)$ and $y(z)$, respectively, from the curvature equation $\frac{1}{\rho}(z)$ (211) and (212). The usual sign conventions lead to the relations (185)

$$P(z) \Rightarrow V(z) = -\int P \, dz + V_0 \quad \Rightarrow M(z) = -\iint P \, dz^2 + V_0 z + M_0$$

$$\frac{1}{\rho} = -\frac{M}{EI} \Rightarrow \varphi(z) = -\int \frac{M}{EI} \, dz + \varphi_0 \Rightarrow y(z) = -\iint \frac{M}{EI} \, dz^2 + \varphi_0 z + y_0 .$$

$$\tag{221}$$

The integration constants are in both cases defined by the support conditions. But these may be different in the two cases. In the first we have, for example, vanishing shear force and bending moment in a free end of a beam, while in the second case the rotation and the deflection vanish in a built-in end.

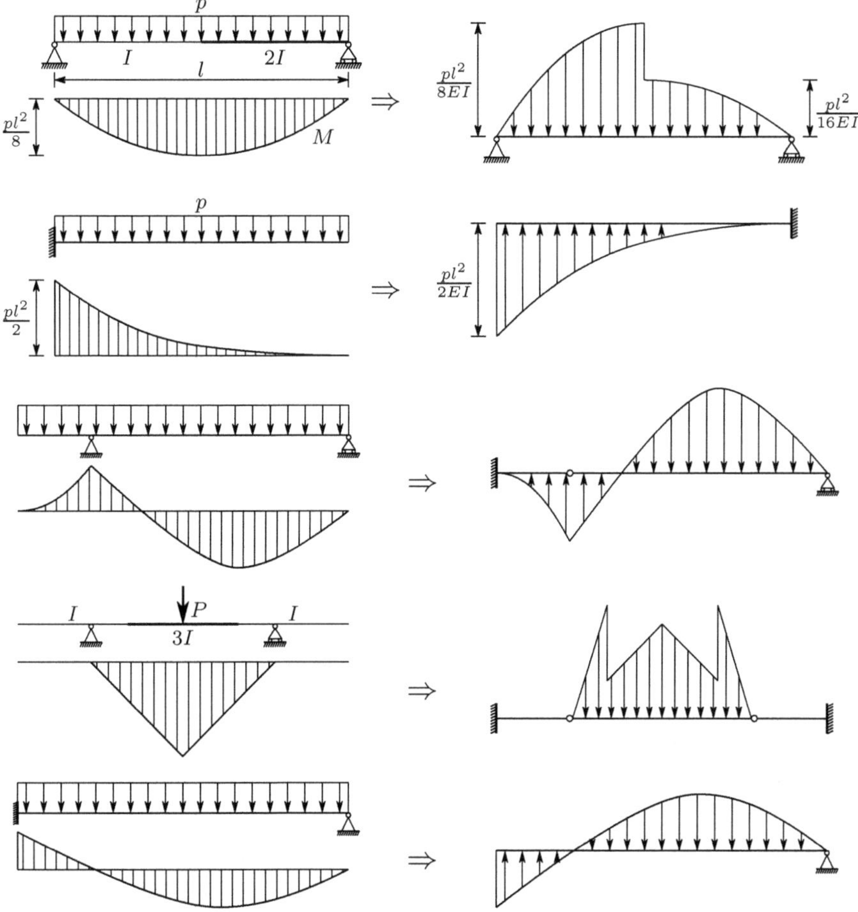

Fig. 125. Examples of conjugate beams

The conjugate beam method consists of computing the rotations and deflections by means of a fictitious beam – the conjugate beam – loaded with a transversal distributed force corresponding to the curvature equation ($\frac{M}{EI} = -\frac{1}{\rho}$). The shear force and the bending moment in the conjugate beam take values corresponding to the rotation and deflection of the actual beam, respectively, as (221) clearly shows. The two beams – the actual and the conjugate – have the same geometry, but may have different supports, since the conditions determining the value of the integration constants may be different, as mentioned above. In Fig. 125 some examples of beams and the corresponding conjugates (on the right) are presented.

The conjugate of a statically determinate beam is also statically determinate. In fact, a statically determinate beam is freely deformable, which means that its deformation is compatible with the support conditions, for any distribution of bending moments, i.e., for any curvature equation. Furthermore, rigid-body motions are prevented, which means that the deformation is determinate. Transposed to the conjugate beam, this means that, irrespective of the distribution of transversal loads $P' = \frac{M}{EI}$, the equilibrium must be guaranteed. Thus, the conjugate beam must have enough supports to be at least statically determinate. But it cannot be statically indeterminate, since in this case the shear forces and bending moments could not be obtained by equilibrium considerations, which would mean indeterminate deflections in the actual beam.

In statically indeterminate beams the only curvature distributions possible are those that lead to deflections and rotations which are compatible with the support conditions. This is reflected in the fact that the conjugate of a statically indeterminate beam has fewer external connections than are necessary to guarantee the equilibrium for any loading. This means that the loading $P' = \frac{M}{EI}$ acting on the conjugate beam must satisfy certain conditions. These are, in fact, the deformation compatibility conditions in the actual beam. For example, in the last beam represented in Fig. 125, equilibrium only takes place if the moment of the transversal load $P' = \frac{M}{EI}$ with respect to the support vanishes. This corresponds exactly to the condition of zero displacement in the right support of the actual beam.

IX.1.d Moment-Area Method

This method allows the direct determination of the relative rotation of two cross-sections in a bar (first moment-area theorem) and of the distance between the tangent to the deflection line in a point and any other point of this line (second moment-area theorem). In order to demonstrate these theorems let us consider a piece of the deflection line defined by two cross-sections, a and b, at a distance l of each other, as represented in Fig. 126.

The relative rotation of two cross-sections at an infinitesimal distance $\mathrm{d}s$ from each other is $\mathrm{d}\varphi = \frac{1}{\rho}\mathrm{d}s = \frac{M}{EI}\mathrm{d}s$. Adding the relative rotations of all infinitesimal segments $\mathrm{d}s$ contained between the cross-sections a and b, i.e., integrating between a and b, we get the relative rotation of these two cross-sections

$$\varphi = \int_a^b \frac{M}{EI}\,\mathrm{d}s = \int_0^l \frac{M}{EI}\,\mathrm{d}z \ .$$

This integral represents the area defined by the bending moment diagram divided by the bending stiffness EI – the curvature diagram – between points a and b of the deflection line. This conclusion describes the *first moment-area theorem*, which may stated as follows:

> *the relative rotation of two cross-sections is equal to the area defined by the curvature diagram between these two cross-sections.*

The vertical distance between the tangents to the deflection line in two points at an infinitesimal distance $\mathrm{d}s$ from each other may be given by

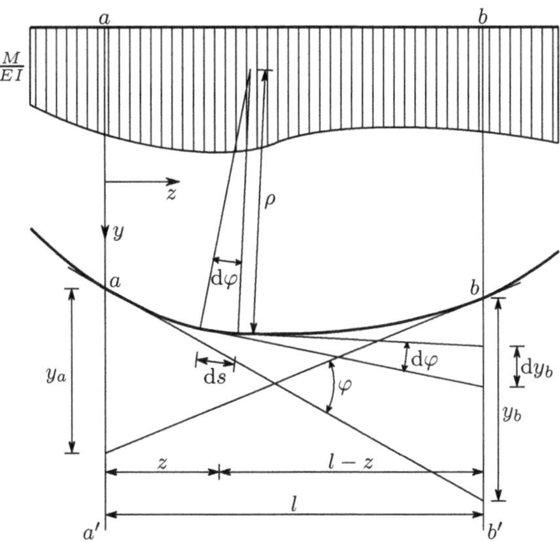

Fig. 126. Deformation of a piece of a prismatic bar under a positive bending moment

$\mathrm{d}y_b \approx (l - z)\,\mathrm{d}\varphi$ (Fig. 126), *provided that the rotation angle φ is small enough to be considered as infinitesimal.* Under these conditions, coordinates s and z take the same values and the distance y_b (Fig. 126) may be computed by integrating $\mathrm{d}y_b$ between a and b, yielding

$$y_b \approx \int_a^b (l - z)\,\mathrm{d}\varphi = \int_a^b (l - z)\,\frac{1}{\rho}\,\mathrm{d}s \approx \int_0^l \frac{M}{EI}\,(l - z)\,\mathrm{d}z \ .$$

This integral describes the *second moment-area theorem*, which states that:

> *the distance between the tangent to a point a of the deflection line and another point b of the same line is given by the first area moment of the curvature diagram $\frac{1}{\rho} = \frac{M}{EI}$ between points a and b with respect to a line perpendicular to the undeformed bar axis passing through point b.*

Similarly, the distance y_a is given by the first area moment of the same surface with respect to line aa' (Fig. 126).

The moment-area method may be directly applied to the computation of rotations and deflections in bars where the position of a cross-section with a zero rotation is known, since, by considering point a on that cross-section, the first and the second moment-area theorems immediately give the rotation and the displacement in relation to the cross-section with zero rotation, respectively. Where the position of a cross-section with a vanishing rotation is not known a priori, this method may also be applied, although not directly.

The three methods described for the computation of deflections caused by the bending moment yield exactly the same results, since they are all based on the curvature function, and the same simplifying hypothesis is used in all them. In order to illustrate their use, the rotation and the deflection of cross-section C of the beam represented in Fig. 127 are computed using the three methods.

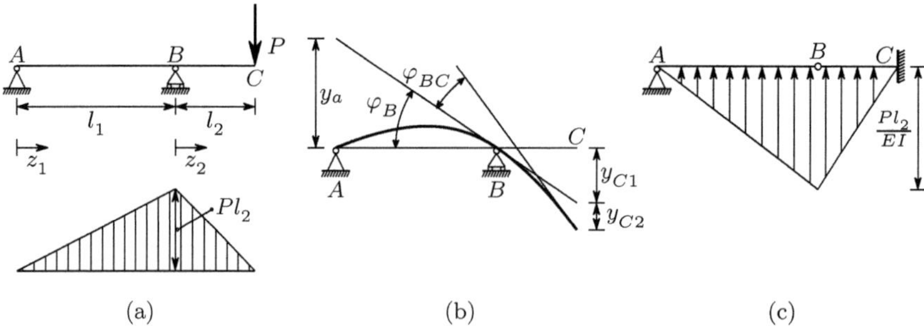

Fig. 127. Computation of rotations and deflections in a beam

The method of integration of the curvature equation gives the following expression, for the rotation in the segment $\overline{AB}$ as a function of z_1 (Fig. 127-a)

$$\varphi(z_1) = -\int \frac{M}{EI}\,dz_1 + \varphi_0 = \frac{1}{EI}\int \frac{Pl_2}{l_1} z_1\,dz_1 + \varphi_0 = \frac{1}{EI}\frac{Pl_2}{l_1}\frac{z_1^2}{2} + \varphi_0\,.$$

The integration constant φ_0 cannot be determined yet, since we do not know the rotation of any cross-section in the segment $\overline{AB}$. Integrating once more, we get the deflection equation in $\overline{AB}$

$$y(z_1) = -\iint \frac{M}{EI}\,dz_1^2 + \varphi_0 z_1 + y_0 = \frac{1}{EI}\frac{Pl_2}{l_1}\frac{z_1^3}{6} + \varphi_0 z_1 + y_0\,.$$

Since the deflection in point A ($z_1 = 0$) must vanish we conclude that $y_0 = 0$. The integration constant φ_0 may now be obtained from the condition of zero displacement in point B, yielding

$$z_1 = l_1 \;\Rightarrow\; y = 0 \;\Rightarrow\; \varphi_0 = -\frac{Pl_1 l_2}{6EI} \;\Rightarrow\; \begin{cases} \varphi(z_1) = \dfrac{P}{EI}\left(\dfrac{l_2}{l_1}\dfrac{z_1^2}{2} - \dfrac{l_1 l_2}{6} \right) \\[2ex] y(z_1) = \dfrac{P}{EI}\left(\dfrac{l_2}{l_1}\dfrac{z_1^3}{6} - \dfrac{l_1 l_2}{6}z_1 \right)\,. \end{cases}$$

The maximum displacement in the beam segment $\overline{AB}$ may be obtained from the condition of zero rotation, yielding

$$\varphi = 0 \;\Rightarrow\; z_1 = \frac{l_1}{\sqrt{3}} \;\Rightarrow\; y_{\max} = -\frac{1}{9\sqrt{3}}\frac{Pl_1^2 l_2}{EI}\,.$$

In order to analyse the deformation of beam segment $\overline{BC}$, the rotation of cross-section B is needed. This rotation takes the value

$$z_1 = l_1 \;\Rightarrow\; \varphi = \varphi_B = \frac{Pl_1 l_2}{3EI} \;. \tag{222}$$

In beam segment $\overline{BC}$, the bending moment, expressed as a function of co-ordinate z_2, is given by $M(z_2) = -P(l_2 - z_2)$. Integrating once, we get the rotation

$$\varphi(z_2) = -\int \frac{M}{EI}\,\mathrm{d}z_2 + \varphi_0 = \frac{P}{EI}\left(l_2 z_2 - \frac{z_2^2}{2}\right) + \varphi_0 \;.$$

The condition of continuity of the beam in cross-section B, $\varphi_B^{\text{left}} = \varphi_B^{\text{right}}$ leads us to conclude that the integration constant φ_0 is the above-computed rotation of section B, φ_B. The integration constant y_0 must vanish, since the displacement in B is zero ($z_2 = 0 \Rightarrow y = 0$). The rotation and the deflection in the beam segment $\overline{BC}$ are then

$$\begin{cases} \varphi(z_2) = \frac{P}{EI}\left(\frac{l_1 l_2}{3} + l_2 z_2 - \frac{z_2^2}{2}\right) \\ y(z_2) = \frac{P}{EI}\left(\frac{l_1 l_2}{3} z_2 + \frac{l_2}{2} z_2^2 - \frac{1}{6} z_2^3\right) \;. \end{cases}$$

By particularizing these expressions to $z_2 = l_2$, the rotation and the deflection of point C are obtained, yielding

$$z_2 = l_2 \;\Rightarrow\; \begin{cases} \varphi = \varphi_C = \frac{Pl_2}{EI}\left(\frac{l_1}{3} + \frac{l_2}{2}\right) \\ y = y_C = \frac{Pl_2^2}{3EI}\left(l_1 + l_2\right) \;. \end{cases} \tag{223}$$

The application of the moment-area method to this problem is not direct, since the position of the cross-section with zero rotation is not known a priori. This problem may be circumvented by computing the rotation in B from the distance y_A (Fig. 127-b), which can be obtained by means of the second moment-area theorem. According to this theorem, distance y_A is equal to the first area moment of the curvature diagram between points A and B, with respect to the vertical line passing through point A (Fig. 127-a), which yields

$$y_A = \frac{1}{2}\frac{Pl_2}{EI}l_1\frac{2}{3}l_1 = \frac{Pl_1^2 l_2}{3EI} \;\Rightarrow\; \varphi_B = \frac{y_A}{l_1} = \frac{Pl_1 l_2}{3EI} \;.$$

This is the same value as obtained by the method of integration of the curvature equation (222). The rotation of cross-section C may then be obtained by adding to this value the relative rotations of cross-sections B and C, φ_{BC}, which may be computed with the first moment-area theorem, yielding (Fig. 127-b)

$$\varphi_{BC} = \frac{1}{2}\frac{Pl_2}{EI}l_2 = \frac{Pl_2^2}{2EI} \;\Rightarrow\; \varphi_C = \varphi_B + \varphi_{BC} = \frac{Pl_2}{EI}\left(\frac{l_1}{3} + \frac{l_2}{2}\right) \;.$$

This is the same value as given in (223). The distance y_{C1} is given by the product of the rotation φ_B by the distance l_2. The second moment-area theorem yields the distance y_{C2}. The deflection of cross-section C is then

$$\begin{cases} y_{C1} = \varphi_B l_2 = \dfrac{Pl_1 l_2^2}{3EI} \\[2mm] y_{C2} = \dfrac{1}{2}\dfrac{Pl_2}{EI}l_2\dfrac{2}{3}l_2 = \dfrac{Pl_2^3}{3EI} \end{cases} \Rightarrow y_C = y_{C1} + y_{C2} = \dfrac{Pl_2^2}{3EI}(l_1 + l_2) \ .$$

This value coincides with that given in (223), as expected.

The conjugate beam is represented in Fig. 127-c. Note that the intermediate support is transformed into an intermediate hinge, since the displacement in this point vanishes for any curvature distribution. This means that the bending moment in the conjugate beam must vanish for any loading (i.e., for any curvature distribution in the actual beam). The free end of the actual beam (point C) is transformed into a built-in support, since reaction forces introducing a shear force (a rotation in the actual beam) and a bending moment (a deflection in the actual beam) must appear in cross-section C. Thus, the rotation and deflection of cross-section C are given by the shear force and the bending moment in the built-in end C of the conjugate beam (Fig. 127-c). These values are easily computed and confirm (223).

Generally, the method of integration of the curvature equation is advantageous if the equation of the deflection line is needed, as, for example, if the position of the cross-section with the maximum deflection is to be computed, or in the case of the computation of influence lines (see Sect. XII.3.d). For the computation of localized rotations or deflections, it is usually simpler to use the moment-area method of the conjugate beam method. When these methods are used and the curvature diagram does not have a simple geometrical form, it is useful to decompose it in simple shapes such as rectangles, triangles or parabolas, since in this case the corresponding area and centroid positions do not need to be computed by integration (see, e.g., example IX.9).

IX.2 Deflections Caused by the Shear Force

IX.2.a Introduction

The methods analysed in the previous Section do not include deformation caused by shear force, since they are based in the relation between bending moment and curvature. The shear force causes a transversal relative displacement of two cross-sections (shear displacement), as represented in Fig. 105-b. This displacement corresponds to the relative motion of the resultants of the shearing stresses caused by the shear force ($\gamma_m\,dz$, Fig. 128). However, this displacement cannot be obtained directly from the corresponding shearing strains, because these strains are not constant in the cross-section: if the

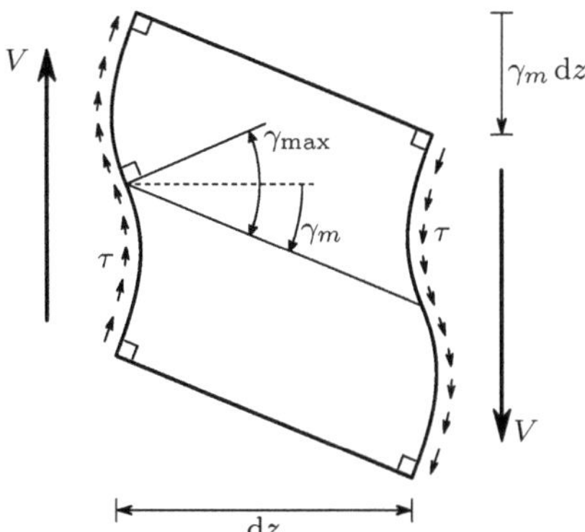

Fig. 128. Relative displacement of two cross sections caused by the shear force V.
------ initial orientation of the bar's axis

shearing strain in the fibres farthest from the neutral axis were to be used, a vanishing shear displacement would be obtained, while the shearing strain at the neutral axis would lead to a maximum value of shear displacement ($\gamma_{\max}\,\mathrm{d}z$, Fig. 128). These two extreme cases differ only by a rigid body rotation. As a consequence of this rotation the problem of computing the shear displacement corresponding to a given distribution of shearing strains becomes indeterminate.

However, this rigid body motion may be eliminated by means of the principle of energy conservation. In fact, because in a rigid body rotation, the stresses inside a piece of bar like that represented in Fig. 128, do not change, the work done by the pair of forces V in that rotation is not offset by the variation of elastic potential energy stored by the piece of bar. Considering a piece of bar of infinitesimal length $\mathrm{d}z$ under the action of a bending moment M and a shear force V, we can easily verify that the work done by the bending moment, $\mathrm{d}W_M$, in the relative rotation of the end cross-sections, is equal to the elastic potential energy stored in the deformation caused by the normal stresses σ, $\mathrm{d}U_\sigma$. For the simplest case of plane bending, we have

$$\mathrm{d}W_M = \frac{1}{2}M\,\mathrm{d}\varphi = \frac{1}{2}M\frac{1}{\rho}\mathrm{d}z = \frac{1}{2}\frac{M^2}{EI}\mathrm{d}z$$

$$\mathrm{d}U_\sigma = \frac{1}{2}\int_\Omega \sigma\varepsilon\,\mathrm{d}\Omega\,\mathrm{d}z = \frac{1}{2}\int_\Omega \frac{My}{I}\frac{My}{EI}\,\mathrm{d}\Omega\,\mathrm{d}z = \frac{1}{2}\frac{M^2}{EI^2}\int_\Omega y^2\,\mathrm{d}\Omega\,\mathrm{d}z = \frac{1}{2}\frac{M^2}{EI}\mathrm{d}z\ .$$

The principle of energy conservation states that the work done by the external forces is equal to the elastic potential energy stored by the bar. In this analysis V and M are taken to be external forces applied to the piece of bar. Since, as seen above, the work done by the bending moment is balanced by the energy corresponding to the normal stresses, the work $\mathrm{d}W_V$, done by the pair of shear forces V in the relative displacement $\gamma_m\,\mathrm{d}z$ (Fig. 128), must also be balanced by the elastic potential energy corresponding to the shear strains, $\mathrm{d}U_\tau$. This condition yields the relation

$$\begin{cases} \mathrm{d}W_V = \frac{1}{2}V\gamma_m\,\mathrm{d}z \\[2mm] \mathrm{d}U_\tau = \frac{1}{2}\int_\Omega \tau\gamma\,\mathrm{d}\Omega\,\mathrm{d}z \end{cases} \quad ; \quad \mathrm{d}W_V = \mathrm{d}U_\tau \Rightarrow V\gamma_m = \int_\Omega \tau\gamma\,\mathrm{d}\Omega \ . \quad (224)$$

We may also arrive at this conclusion by observing that, in the deformation caused by the shear force, the relative rotation of the end cross-sections and the elongation of the fibres are zero, which means that in this deformation we have $\mathrm{d}W_M = \mathrm{d}U_\sigma = 0$. Equation (224) may also be obtained by means of equilibrium considerations (see example IX.4). The relative shear displacement per unit length may be represented by the *mean shearing strain* γ_m (Fig. 128). This quantity may be obtained from (224).

The elastic potential energy stored per unit length of the bar, in the deformation caused by the shear force may then be given by the expression (224)

$$\frac{\mathrm{d}U_\tau}{\mathrm{d}z} = \frac{\mathrm{d}W_V}{\mathrm{d}z} = \frac{1}{2}V\gamma_m = \frac{1}{2G}\int_\Omega \tau^2\,\mathrm{d}\Omega \ . \quad (225)$$

As seen in Chap. VIII, the shearing stress is obtained from the longitudinal shear force. Although different expressions have been obtained for the various types of cross-section, the shearing stress may be represented by the general expression

$$\tau = \frac{V}{I}R\left(x,y\right) \Rightarrow \frac{\mathrm{d}U_\tau}{\mathrm{d}z} = \frac{V^2}{2GI^2}\int_\Omega \left[R\left(x,y\right)\right]^2\,\mathrm{d}\Omega \ . \quad (226)$$

$R\left(x,y\right)$ is a function of the coordinates of the point in the cross-section. This function includes the first area moment S defined in (187) and has the dimensions of an area. For example, in a symmetrical cross-section with respect to the action axis of the shear force, we have $R = \frac{S}{b\cos\psi}$, as it is immediately seen from (193). With this function, another form may be given to the expression of the elastic potential energy stored per unit length, in the deformation caused by the shear force. From (226) we get

$$\frac{\mathrm{d}U_\tau}{\mathrm{d}z} = \frac{1}{2}\frac{V^2}{G\Omega_r} \quad \text{with} \quad \Omega_r = \frac{I^2}{\int_\Omega R^2\,\mathrm{d}\Omega} \ . \quad (227)$$

The quantity Ω_r has the dimensions of an area and is usually called the *reduced area* of the cross-section. Ω_r depends only on the geometry of the cross-section and on the orientation of the action axis of the shear force. It represents the area that the cross-section would need to store the same elastic potential energy with a constant shear force. From (225) and (227) we get

$$\gamma_m = \frac{V}{G\Omega_r} \ . \quad (228)$$

IX.2.b Rectangular Cross-Sections

In a rectangular cross-section we have $R = \frac{S}{b} = \frac{1}{2}\left(\frac{h^2}{4} - y^2\right)$, as we may conclude from (189). The reduced area is then

$$\Omega_r = \frac{\left(\frac{bh^3}{12}\right)^2}{\int_{-\frac{h}{2}}^{\frac{h}{2}} \frac{1}{4}\left(\frac{h^2}{4} - y^2\right)^2 b\,dy} = \frac{bh}{1.2} = \frac{\Omega}{1.2} \, . \tag{229}$$

As an example, let us consider a simply-supported beam with a rectangular cross-section of width b and height h, under a concentrated load at mid-span, as represented in Fig. 129.

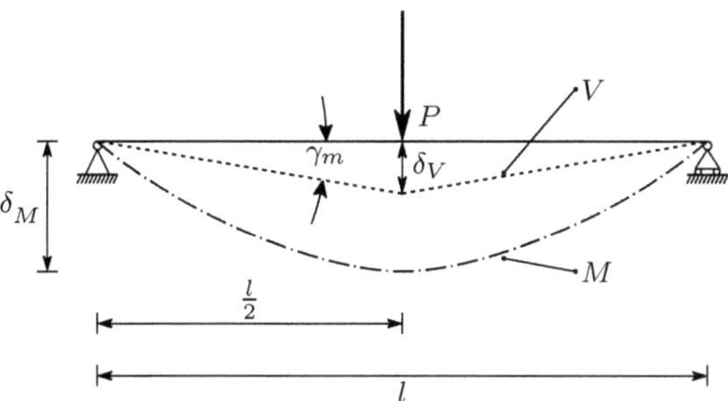

Fig. 129. Deformation of a simply supported beam caused by the bending moment M and by the shear force V

The contribution of the shear force to the displacement of the middle cross-section, δ_V, may be computed from (228) and (229), yielding $\left(V = \frac{P}{2}\right)$

$$\delta_V = \frac{l}{2}\gamma_m = \frac{l}{2}\frac{\frac{P}{2}}{\frac{E}{2(1+\nu)}\frac{bh}{1.2}} = \frac{0.6Pl\,(1+\nu)}{bhE} \, .$$

The contribution of the curvature caused by the bending moment to the same displacement, δ_M, may be obtained by means of the moment-area method. Applying the second moment-area theorem to the piece of beam between one of the supports and the middle cross-section, we get, since in this cross-section the rotation vanishes $\left(M_{\max} = \frac{Pl}{4}\right)$,

$$\delta_M = \frac{Pl}{4E\frac{bh^3}{12}}\frac{l}{2}\frac{1}{2}\frac{2}{3}\frac{l}{2} = \frac{Pl^3}{4Ebh^3} \, .$$

The ratio between the contributions of the shear force and of the bending moment to the total displacement is then

$$\frac{\delta_V}{\delta_M} = 2.4\,(1+\nu)\left(\frac{h}{l}\right)^2 \, .$$

From this expression we conclude immediately that the contribution of the shear force to the displacement is small, since, in a slender member, we usually have $l \gg h$. For example, for $l = 10h$ and $\nu = 0.25$, the deformation caused by the shear force represents only 3% of the total deformation. For this reason, the contribution of the shear force to the total deformation is often disregarded, specially when the h/l ratio is small.

IX.2.c Symmetrical Cross-Sections

In the case of symmetrical cross-sections with respect to the action axis of the shear force, the shearing stress is given by (193). In this case we have, since $x = c \tan \psi$ (Fig. 107),

$$R^2 = \frac{S^2}{b^2 \cos^2 \psi} = \frac{S^2}{b^2} \left(1 + \tan^2 \psi \right)$$

$$\Rightarrow \int_\Omega R^2 \, d\Omega = \int_{-y_1}^{y_2} \frac{S^2}{b^2} \int_{-\varphi}^{\varphi} \left(1 + \tan^2 \psi \right) \underbrace{\frac{c}{\cos^2 \psi} \, d\psi}_{dx} \, dy$$

$$= \int_{-y_1}^{y_2} \frac{S^2}{b^2} \underbrace{2c \tan \varphi}_{b} \left(1 + \frac{1}{3} \tan^2 \varphi \right) dy$$

$$= \int_{-y_1}^{y_2} \frac{S^2}{b} \left(1 + \frac{1}{3} \tan^2 \varphi \right) dy \ .$$

Particularizing this expression for a circular cross-section, we get from (194) and Fig. 108

$$\begin{cases} y & = r \cos \alpha \\ dy & = -r \sin \alpha \, d\alpha \\ b & = 2r \sin \alpha \\ S & = \frac{2}{3} r^3 \sin^3 \alpha \\ \tan \varphi = \frac{1}{\tan \alpha} \end{cases} \Rightarrow \left| \begin{array}{l} \int_\Omega R^2 \, d\Omega \\ = -\frac{2}{9} r^6 \int_\pi^0 \sin^6 \alpha \left(1 + \frac{1}{3} \frac{\cos^2 \alpha}{\sin^2 \alpha} \right) d\alpha \\ = \frac{2}{27} \pi r^6 \ . \end{array} \right.$$

The reduced area of a circular cross-section takes then the value

$$\Omega_r = \frac{I^2}{\int_\Omega R^2 \, d\Omega} = \frac{\left(\frac{\pi r^4}{4} \right)^2}{\frac{2}{27} \pi r^6} = \frac{27}{32} \pi r^2 \approx \frac{\Omega}{1.185} \ .$$

IX.2.d Thin-Walled Cross-Sections

According to the simplifying hypothesis used in the analysis of thin-walled cross-sections – a constant shearing stress in the wall thickness e – the integration of the function R^2 may be performed along the centre line of the wall. Thus, in the case of an open cross-section, we get from (195) and (227) ($\mathrm{d}\Omega = e\,\mathrm{d}s$)

$$R = \frac{S}{e} \quad \Rightarrow \quad \Omega_r = \frac{I^2}{\int_\Omega \frac{S^2}{e^2}\,\mathrm{d}\Omega} = \frac{I^2}{\int_L \frac{S^2}{e}\,\mathrm{d}s} \, , \tag{230}$$

where s is a coordinate following the wall's centre line and L represents this line. In the case of a closed thin-walled cross-section, function R has a different form (see example IX.8).

As an example, we compute the reduced area of the cross-section represented in Fig. 118 for the special case of having $b = h$. In the flange (segment AB), the first area moment S may be expressed as a function of coordinate s_1 with origin in point A (Fig. 118-a). In the web (segment BD), we consider a coordinate s_2 with origin in point C. Since the thickness is constant and $b = h$, the integration of R^2 along the whole centre lines yields

$$\begin{cases} S_1\left(s_1\right) = \frac{h}{2}s_1 e \\[2mm] S_2\left(s_2\right) = \left(\frac{5}{8}h^2 - \frac{s_2^2}{2}\right)e \end{cases}$$

$$\Rightarrow \quad \int_L \frac{S^2}{e}\,\mathrm{d}s = \underbrace{2e\int_0^h \left(\frac{h}{2}s_1\right)^2\,\mathrm{d}s_1}_{\text{flanges}} + \underbrace{e\int_{-\frac{h}{2}}^{\frac{h}{2}} \left(\frac{5}{8}h^2 - \frac{s_2^2}{2}\right)^2\,\mathrm{d}s_2}_{\text{web}} = \frac{61}{120}eh^5 \, .$$

The moment of inertia takes the value $I = \frac{7}{12}eh^3$, ((204), $b = h$), which gives the reduced area

$$\Omega_r = \frac{I^2}{\int_L \frac{S^2}{e}\,\mathrm{d}s} = \frac{\left(\frac{7}{12}eh^3\right)^2}{\frac{61}{120}eh^5} \approx \frac{eh}{1.494} \, . \tag{231}$$

It must be noted that there is a significant discrepancy between this value and the value resulting from a common assumption in structural analysis, which consists of considering the web area, eh, as the reduced area in the case of profile sections like I-beams and C-channels. This simplifying assumption is based on geometrical considerations in the web's plane and does not take the flange deformation caused by the shearing stresses into consideration, and so it does not satisfy the principle of energy conservation[4] (see example IX.7).

Another method of computing the deformation caused by the shear force is sometimes mentioned and used. It is based on the assumption that, during the deformation caused by the shear force, the cross-section remains perpendicular to the bar axis in the region around the cross-section centroid. It is, however, possible to show that this assumption is not valid. This is due either to the action of the perturbations on the normal stress distribution, caused by the

---

[4] A three-dimensional analysis of the deformation caused by shearing stresses shows that the mean shearing strain of the cross-section is superior to the mean shearing strain in the web.

fact that the cross-sections can not deform freely at the supports and on points of discontinuity of the shear force function (concentrated loads), or to the additional normal stresses which appear when the shear force is not constant (second element of the first of (162)).

In order to clarify this point, we analyse the first of these cases, for the example of a cantilever beam with rectangular cross-section, under a concentrated load P, which is applied in the form of shearing stresses, whose distribution satisfies the parabolic law represented by (189), (Fig. 130).

The normal stresses represented by (146) and the shearing stresses given by (189) coincide with the solution of the Theory of Elasticity for this problem, provided that the cross-section's width is small compared with the height. However, this solution is not valid for the region which is close to the built-in support, since this cross-section cannot deform freely. Let us assume that it must remain plane. The warping of the cross-section is avoided by additional stresses which must have a *vanishing resultant,* since the shear force and the bending moment are statically determinate (P and Pl, respectively). Thus, the plane cross-section must have the orientation defined qualitatively by line AA'. If the cross-section were to remain perpendicular to the bar axis in the region around the centroid (lines BB' perpendicular to the axis of the undeformed bar) the deflections caused by the shear force would be larger. However, in this case, the balance conditions would not be satisfied, since the stresses required to force the cross-section to remain plane with the orientation defined by line BB' would not have a zero resulting moment.

Later we shall use the Saint Venant principle and the theorem of Castigliano (Sect. XII.3.b) to show that the deformation caused by the shear force, computed by energy considerations, corresponds to considering the line AA' as perpendicular to the undeformed position of the bar axis.

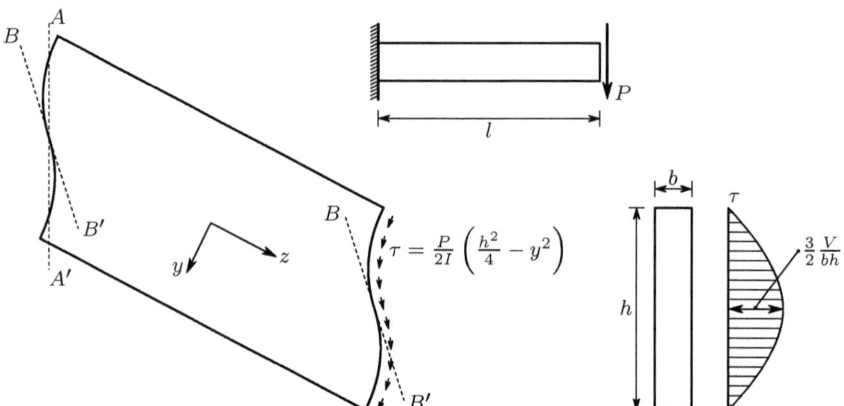

Fig. 130. Deformation caused by the shear force in a cantilever beam with rectangular cross-section

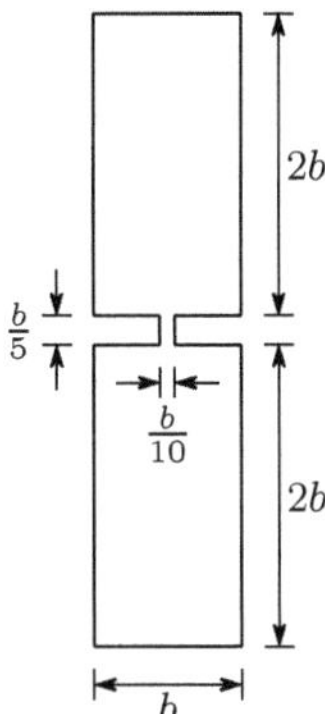

Fig. 131. Deformations caused by a vertical shear force (admitting the simplifying hypotheses for symmetrical cross-sections): $\gamma_m = 0.514366\frac{V}{Ga^2}$; $\gamma_{\max} = 3.56449\frac{V}{Ga^2}$; $\frac{\gamma_{\max}}{\gamma_m} = 6.93$

In the case of rectangular cross-sections the above-mentioned hypothesis leads to an error which is given by the ratio (189 and 229)

$$\frac{\gamma_{\max}}{\gamma_m} = \frac{\frac{3}{2}\frac{V}{G\Omega}}{1.2\frac{V}{G\Omega}} = 1.25 \ .$$

The deflection caused by shear would be overestimated by 25% in this case. In other types of cross-section it may be considerably more, as, for example, in the cross-section represented in Fig. 131, where a high shearing stress occurs in the centroid. The relative error is in this case almost 600%. The computation of the shear deflections by means of geometrical considerations may also lead to an underestimation of the shear deflections, as in the case of (231), where the usual assumptions lead to $\gamma_{\max} \approx \frac{V}{Geh}$, and the reduced area concept (228) gives $\gamma_m = 1.494\frac{V}{Geh}$.

IX.3 Statically Indeterminate Frames Under Bending

IX.3.a Introduction

In Sect. VI.4, the force and displacement methods for the computation of internal forces in statically indeterminate structures were introduced. In this Section, only the force method will be used to determine reaction and internal forces in statically indeterminate frames under bending, since the use of the displacement method in these structures requires the prior solution of simple, statically indeterminate, problems, such as the bending moments introduced into a bar by a unit rotation of one of its ends.

As an introductory example, the internal forces in the beam depicted in Fig. 132 are computed by means of the force method. In order to simplify the explanation, the deformations caused by the shear force are disregarded, so that only the deflections introduced by the bending moment are considered.[5] As a statically determinate base structure, the cantilever beam obtained by removing the left support is considered. Because no horizontal forces or temperature variations are acting, the horizontal reaction force in this support vanishes, so that the only hyperstatic unknown is the vertical reaction force R_A. Both the deflection y_A, caused in the released connection by the applied force P, and the displacement y'_A caused by the hyperstatic unknown R_A in the statically determinate base structure are easily obtained by means of the second moment-area theorem, yielding

$$\begin{cases} y_A = \frac{P(l-a)^2}{6EI}\,(2l+a) \\[2ex] y'_A = \frac{l^3 R_A}{3EI} \end{cases} \quad ; \quad y_A = y'_A \Rightarrow R_A = \frac{P\,(l-a)^2\,(2l+a)}{2l^3}\,.$$

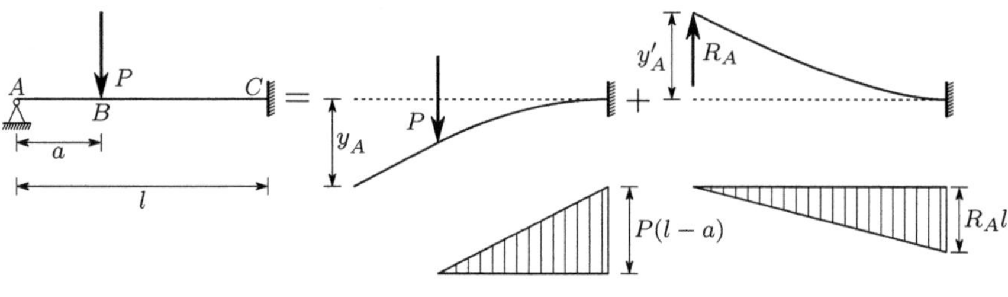

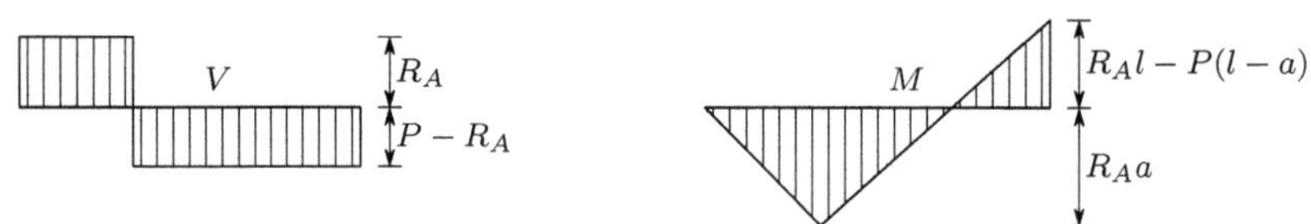

Fig. 132. Example of analysis of a statically indeterminate beam

The internal forces (shear force and bending moment) may be computed, either directly from the equilibrium conditions, or by means of the superposition principle, yielding the diagrams depicted in Fig. 132.

In the case of a beam with a degree of static indeterminacy greater than one, a system with a number of equations equal to the degree of indeterminacy is obtained. In Fig. 133 the application of the force method to a beam with degree two of indeterminacy is schematically represented.

The hyperstatic unknowns in this case are the reaction moments at the built-in ends, M_A and M_B. These moments may be computed by solving the

---

[5]This simplification is commonly used in engineering practice, as has been justified in the example described in Fig. 129.

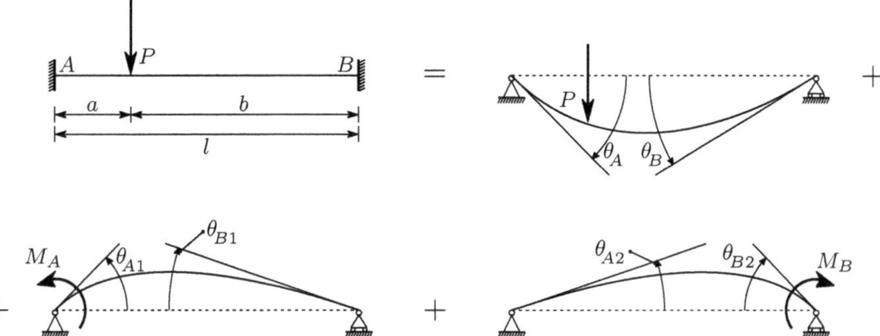

Fig. 133. Analysis of a beam with degree two of static indeterminacy, by means of the force method

system of two equations which results from the conditions of zero rotation in the supports A and B

$$\begin{cases} \theta_A = \theta_{A1} + \theta_{A2} \\ \theta_B = \theta_{B1} + \theta_{B2} \,. \end{cases}$$

IX.3.b Equation of Two Moments

In the analysis of framed structures under bending, the equation usually called the *equation of two moments* is often used when the diagrams describing the distribution of bending moments are drawn. This equation results from the decomposition of the system of forces acting in a piece of bar into two self-equilibrated sub-systems. One of these has vanishing bending moments at the ends of the piece of bar, and the other has no applied loads. The internal forces introduced by the first one are the same as would appear in a simply supported bar with the applied loads. The second sub-system corresponds to the same simply supported beam under the action of end moments. These considerations are summarized in Fig. 134. From this figure, we can see that the total internal forces may be expressed in terms of the internal forces in the simply-supported beam and the end moments, M_a and M_b, as given by the expressions

$$\begin{cases} V(z) = V_1 + \dfrac{M_b - M_a}{l} \\ M(z) = M_1 + M_a \dfrac{l-z}{l} + M_b \dfrac{z}{l} \,. \end{cases}$$

IX.3.c Equation of Three Moments

In the computation of internal forces in continuous beams of constant cross-section the so-called *equation of three moments* is often used. This equation results from applying the force method to this kind of structure, considering as a statically determinate base problem the series of simply-supported

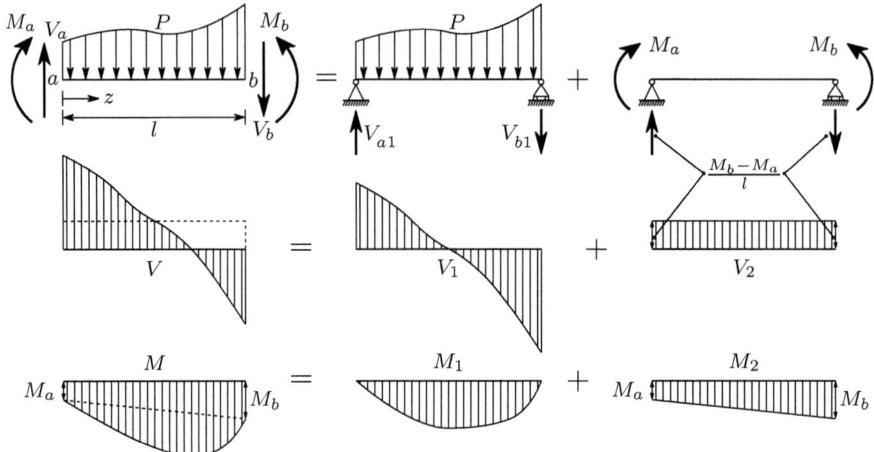

Fig. 134. Deduction of the equation of two moments

beams obtained when hinges are introduced into the cross-sections that lie over the supports, i.e.. when the bending connection in these cross-sections is released. The condition of compatibility is given by the rotation continuity condition, i.e., by the statement that the two bar ends on a support have the same rotation. Figure 135 shows a continuous beam with three released bending connections. The corresponding hyperstatic unknowns are M_{n-1}, M_n and M_{n+1}. The equation of three moments defines a relation between these quantities.

The right end of beam segment l_n undergoes rotations caused by the loading in the statically determinate (simply-supported) base beam and by the moments acting on the extremities of the beam segment, M_{n-1} and M_n. These three rotations may be computed by means of the second moment-area

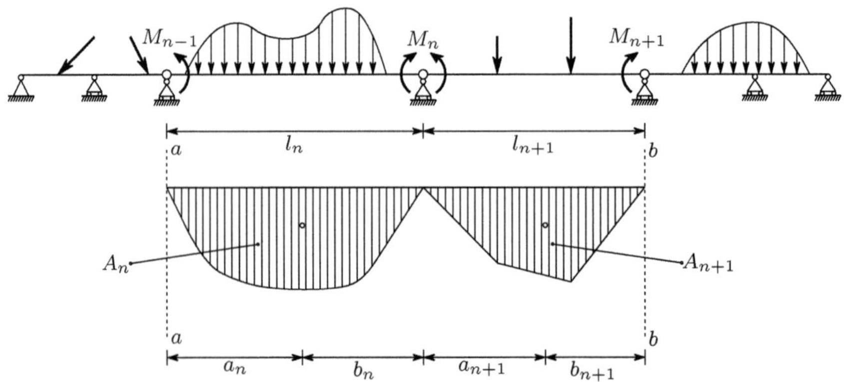

Fig. 135. Deduction of the equation of three moments

theorem, using the same technique as in the case of rotation φ_B in Fig. 127-b (first area moment of the curvature diagram with respect to line aa, divided by l_n). Adding these three components we get, taking the rotations in the counterclockwise direction to be positive,

$$\theta_{ne} = \frac{A_n a_n}{E I l_n} - \frac{M_{n-1} l_n}{6EI} - \frac{M_n l_n}{3EI} \, .$$

In this expression A_n represents the area defined by the bending moment diagram in the simply-supported beam l_n and a_n is the distance of its centroid to line aa (Fig. 135).

In the same way, the rotation of the left end of the beam segment l_{n+1} may be obtained, yielding

$$\theta_{nd} = -\frac{A_{n+1} b_{n+1}}{E I l_{n+1}} + \frac{M_n l_{n+1}}{3EI} + \frac{M_{n+1} l_{n+1}}{6EI} \, .$$

The condition of compatibility, $\theta_{ne} = \theta_{nd}$, then yields a relation between the bending moments M_{n-1}, M_n and M_{n+1} which represents the equation of three moments

$$M_{n-1} l_n + 2 M_n \left(l_n + l_{n+1} \right) + M_{n+1} l_{n+1} = 6 \left(\frac{A_n a_n}{l_n} + \frac{A_{n+1} b_{n+1}}{l_{n+1}} \right) \, . \tag{232}$$

An equation of this type may be written for each of the intermediate supports. Since the degree of static indeterminacy is equal to the number of intermediate supports, we get a system of equations whose solution gives the values of the hyperstatic unknowns, $M_1, M_2, \ldots, M_n$.

As an example of application, let us consider the beam represented in Fig. 136.

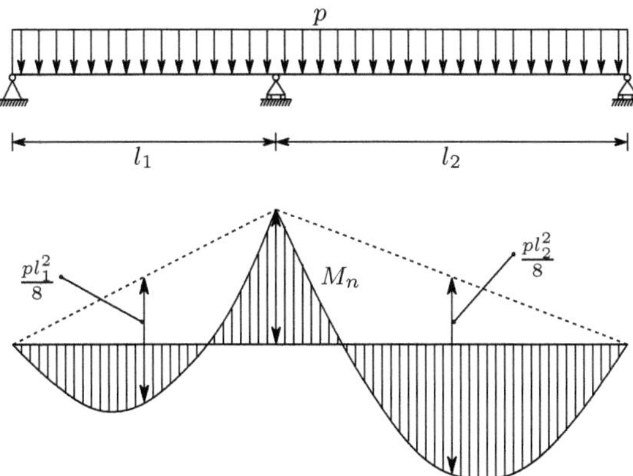

Fig. 136. Application of the equation of three moments to a continuous beam with two spans

In this case, we have

$$l_n = l_1 \qquad\qquad A_n = \tfrac{p l_1^3}{12} \quad A_{n+1} = \tfrac{p l_2^3}{12}$$
$$l_{n+1} = l_2$$
$$M_{n-1} = M_{n+1} = 0 \quad a_n = \tfrac{l_1}{2} \qquad b_n = \tfrac{l_2}{2} \; .$$

Introducing these values into (232), we get

$$M_n = \frac{p}{8} \frac{l_1^3 + l_2^3}{l_1 + l_2}; \qquad l_1 = l_2 = l \Rightarrow M_n = \frac{p l^2}{8} \; .$$

The bending moment diagram may be obtained by means of the theorem of two moments (Fig. 136).

It may happen that the bending moment areas, A_n or A_{n+1}, are zero, without the first area moment of the bending moment diagram vanishing. In this case (232) is not directly applicable (see example IX.11).

IX.4 Elasto-Plastic Analysis Under Bending

As described in Sect. VII.10 (Fig. 97), if the bending moment acting in a bar made of a ductile material with elastic perfectly plastic behaviour exceeds the value corresponding to the yield strain in the fibres farthest from the neutral axis ($M_e = \frac{I}{v}\sigma_Y$), the cross-section enters into the elasto-plastic regime until a plastic hinge forms and bending yielding takes place. The transition from the elastic phase to the yielding of the cross-section is a gradual process which, from a theoretical point of view, is only complete for an infinite curvature, as mentioned in Sect. VII.10. However, the bending moment gets close to the ultimate value for a relatively small curvature, especially in the case of cross-sections which are optimized for the bending strength, like I-beams and C-channels. In these cross-sections the shape factor is only slightly higher than 1, which means that the relation between the bending moment and the curvature gets very close to an elastic perfectly plastic behaviour, as represented in Fig. 137 (the dashed curves represent the actual behaviour). This is because, after yielding of the flanges, the bending strength of these cross-sections is practically exhausted. By assuming that the plastic hinge forms instantaneously, an elastic perfectly plastic behaviour in bending is obtained, which allows a simple elasto-plastic analysis, similar to that performed in the study of the axial force (Sect. VI.4). This analysis may also be carried out in the case of cross-sections with larger shape factors, but it must be born in mind that it is less accurate in this case.

The main objective of the elasto-plastic analysis is the computation of the ultimate loading capacity of the structure, or, more precisely, of the factor by which the load must be multiplied so that a *collapse mechanism* is formed. Because the formation of a plastic hinge corresponds to the loss of an internal

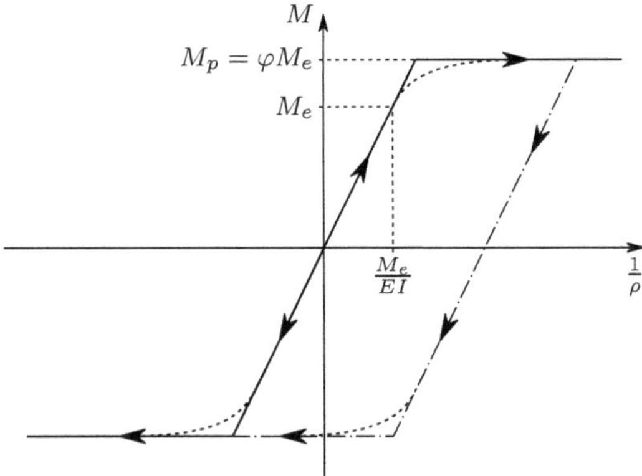

Fig. 137. Elastic perfectly plastic behaviour in bending

connection, this happens when yielding takes place in a number of cross-sections equal to the degree of static indeterminacy plus one.

Since statically determinate structures have the exact minimum number of connections required to guarantee the static equilibrium, a plastic hinge is enough for the formation of the collapse mechanism. In this kind of structure, the internal forces are independent of the material behaviour. The plastic hinge which leads to the collapse will form in the cross-section with the largest ratio between the acting bending moment and the ultimate plastic moment (see example IX.12).

In the case of statically indeterminate structures, the ultimate loading capacity of the structure may be obtained by two methods.

– The first consists of the sequential analysis of the bending moment distribution in the structure, starting with the initial elastic phase, and tracking the formation of each plastic hinge until the collapse mechanism is formed. This procedure is called the *static method.*
– The second consists of the direct analysis of all the possible collapse mechanisms, computing the load corresponding to each of them. The actual collapse mechanism will be that with the lowest collapse load. This procedure is called the *kinematic method.*

The second of these methods is the direct application of a plastic limit analysis theorem called the *kinematic limit load theorem*, which may be stated as follows (cf. e.g. [13]):

a structure will not have the capacity to resist a given loading if there is a mechanism of purely plastic deformation, in which the corresponding internal forces are not able to equilibrate the given loads.

There is another plastic limit analysis theorem called *static limit load theorem* (this theorem has no relation to the static method described above). It may be stated as follows [13]:

a structure will have the capacity to resist a given loading if there is a distribution of internal forces, which is able to equilibrate the given external forces, without exceeding the yielding stress of the material in any point of the structure.

None of the usual methods of plastic analysis is directly based on the static limit load theorem, although it is possible to use it for this purpose. However, it is especially useful in the kinematic method, as a complement to the kinematic limit load theorem. In fact, the correctness of the value obtained for the collapse load by means of a guessed mechanism, may be verified using the static limit load theorem: if, in that mechanism, with the corresponding loading, the acting bending moment does not exceed the ultimate plastic moment in any cross-section, the static theorem ensures that the actual collapse load is not lower than the applied loading. On the other hand, the kinematic theorem ensures that the actual collapse load is no higher than the value corresponding to the analysed collapse mechanism. The combination of the two theorems thus leads to the conclusion that the guessed mechanism is the actual collapse mechanism and the corresponding load the actual collapse load.

In order to illustrate these considerations, the plastic collapse load of the beam represented in Fig. 138 is computed, using both the static and the kinematic methods.

In the static method, the bending moment distribution in the elastic phase is at first computed (Fig. 138-a-*i*). From this distribution, we conclude that the first plastic hinge forms at the built-in support, when the load reaches the value $P = \frac{256}{108}\frac{M_p}{l} \approx 2.370\frac{M_p}{l}$. After the formation of this hinge, the bending moment in cross-section A retains the constant value M_p for increasing values of P, since the curvature is in the yielding zone (Fig. 137). In this phase, the

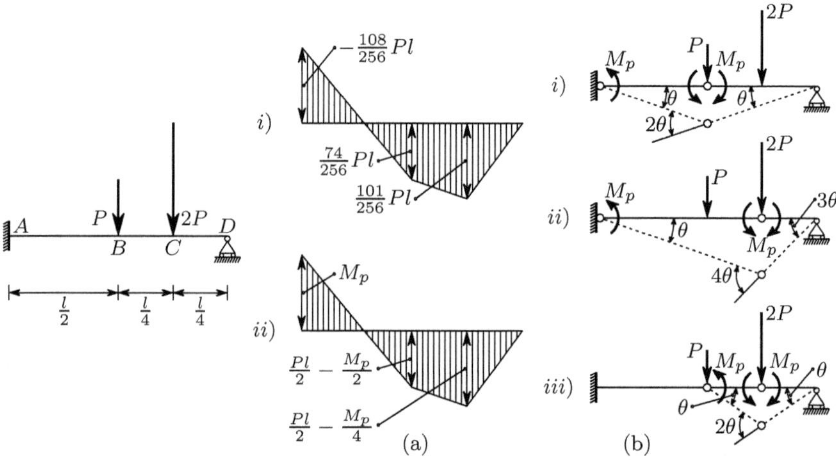

Fig. 138. Computation of the plastic collapse load of a statically indeterminate beam: **(a)** static method; **(b)** kinematic method

bending moment diagram takes the form given in Fig. 138-a-*ii*. From it, we conclude that the second plastic hinge forms in cross-section C, when the load reaches the value $P = 2.5\frac{M_p}{l}\left(\frac{Pl}{2} - \frac{M_p}{4} = M_p\right)$. Since the degree of static indeterminacy is one, the second plastic hinge induces the collapse of the structure.

In the kinematic method, all the three possible collapse mechanisms with two plastic hinges are analysed and the corresponding values of P are computed. This task may be carried out either by means of equilibrium conditions or using the Theorem of Virtual Displacements. This theorem, which is explained in Sect. XII.4, usually leads to a simpler computation. The values of P corresponding to the collapse mechanisms represented in Fig. 138-b are $P = 3\frac{M_p}{l}$, $P = 2.5\frac{M_p}{l}$ and $P = 6\frac{M_p}{l}$, respectively for the mechanisms *i*, *ii* and *iii*. The lowest value of P corresponds to mechanism *ii*, which is therefore the actual collapse mechanism. Both the position of the plastic hinges and the value of P coincide with the result obtained by means of the static method. If the bending moment diagrams of the three mechanisms are drawn, it is found that only in mechanism *ii* is the value M_p not exceeded in any cross-section, as required by the static limit load theorem.

It is better to use the kinematic method when the number of possible collapse configurations is small and the degree of static indeterminacy is high, since the computation effort corresponding to each collapse configuration is very small compared with that required for the analysis of a statically indeterminate structure. In the case of structures under distributed loads the location of the maximum bending moments may change during the elasto-plastic deformation, which makes the identification of the possible collapse mechanisms more difficult (see example IX.14).

As the kinematic method is based on a limit load theorem, it cannot take the order of application of the external forces into consideration. In fact, the collapse load obtained is only valid for a proportional variation of the applied loads. Since the problems where plastic deformations take place are not conservative, it is essential that the actual order of application of the loads is taken into consideration when computing the structural response. This can only be done by using a sequential method of analysis as the static method.

IX.5 Examples and Exercises

IX.1. In the beam represented in Fig. IX.1 determine the displacement of cross-section B, the rotations of the left and right cross-sections of hinge B and the rotation of cross-section C, using the method of integration of the curvature equation, the conjugate beam method and the moment-area method.

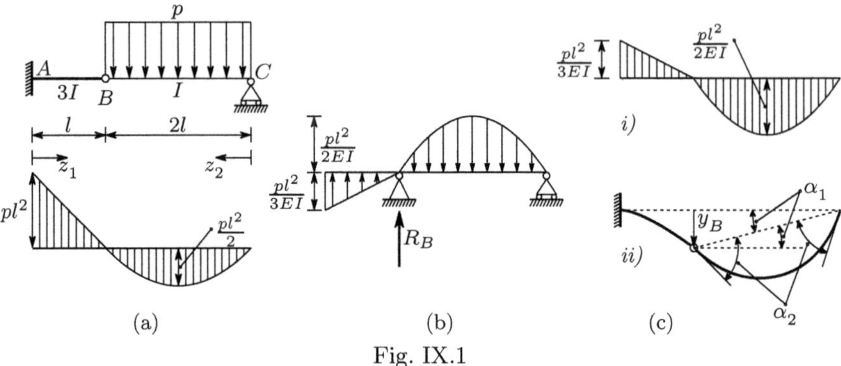

Fig. IX.1

Resolution

Method of integration of the curvature equation

The displacement of cross-section B, y_B, and the rotation of the left cross-section of hinge B may be obtained from the deformation of the beam segment AB. Expressing the bending moment as a function of coordinate z_1 (Fig. IX.1-a) and integrating, we get the rotations in this beam segment ($3EI$ is the bending stiffness of beam AB)

$$M(z_1) = -pl^2 + plz_1 \Rightarrow \varphi(z_1) = -\frac{1}{3EI} \int M\,dz_1 = \frac{pl}{3EI}\left(-\frac{z_1^2}{2} + lz_1\right) + \varphi_0 .$$

The integration constant φ_0 vanishes, since for $z_1 = 0$ we have $\varphi = 0$. The rotation of the left cross-section of hinge B then takes the value

$$\begin{cases} \varphi(z_1) = \dfrac{pl}{3EI}\left(-\dfrac{z_1^2}{2} + lz_1\right) \\ z_1 \quad\;\; = l \end{cases} \Rightarrow \varphi = \varphi_B^l = \frac{pl^3}{6EI} .$$

Since z_1 is positive from left to right and a downwards displacement y is positive, the positive value of φ_B^l means that this rotation is in the clockwise direction. Integrating the expression of the rotations, the displacement function is obtained

$$y(z_1) = \int \varphi\,dz_1 = \frac{pl}{3EI}\left(-\frac{z_1^3}{6} + l\frac{z_1^2}{2}\right) + y_0 .$$

As the displacement of cross-section A is zero, the integration constant y_0 must vanish. The displacement of cross-section B is then

$$z_1 = l \Rightarrow y = y_B = \frac{pl^4}{9EI} .$$

This value is positive, which means that it is a downward displacement.

The knowledge of the displacement y_b and of the deformation of beam segment BC allow the computation of the two remaining rotations. Expressing the bending moment as a function of coordinate z_2 (Fig. IX.1-a) and integrating, we get the rotation function of beam segment BC

$$M(z_2) = plz_2 - \frac{pz_2^2}{2} \Rightarrow \varphi(z_2) = -\frac{1}{EI} \int M \, dz_2 = \frac{p}{EI} \left(\frac{z_2^3}{6} - l\frac{z_2^2}{2} + C \right) .$$

The integration constant C cannot be computed yet, since we do not know the rotation of any cross-section of beam segment BC. Integrating once more, we get the displacement function

$$y(z_2) = \int \varphi \, dz_2 = \frac{p}{EI} \left(\frac{z_2^4}{24} - l\frac{z_2^3}{6} + Cz_2 \right) + y_0 .$$

The integration constant y_0 vanishes, since in the cross-section defined by $z_2 = 0$ the displacement is prevented by the support C. The integration constant C may now be computed from the displacement of section B, yielding

$$z_2 = 2l \Rightarrow y = y_B = \frac{pl^4}{9EI} \Rightarrow \frac{(2l)^4}{24} - \frac{l(2l)^3}{6} + C2l = \frac{l^4}{9} \Rightarrow C = \frac{7}{18}l^3 .$$

The rotation function in beam segment BC and the rotations of the right cross-section of hinge B and of cross-section C are then

$$\varphi(z_2) = \frac{p}{EI} \left(\frac{z_2^3}{6} - l\frac{z_2^2}{2} + \frac{7}{18}l^3 \right) \Rightarrow \begin{cases} z_2 = 2l \Rightarrow \varphi = \varphi_B^r = -\frac{5}{18}\frac{pl^3}{EI} \\[2ex] z_2 = 0 \Rightarrow \varphi = \varphi_C = \frac{7}{18}\frac{pl^3}{EI} . \end{cases}$$

Since coordinate z_2 is positive from right to left, we conclude that the rotation φ_B^r is in the clockwise direction, while section C rotates in the counterclockwise direction.

Conjugate beam method

The conjugate of the given beam is represented in Fig. IX.1-b. The built-in end transforms into a free end, since no reaction forces may appear in this point of the conjugate beam. The intermediate hinge transforms into an intermediate support, since the discontinuity introduced by the hinge in the rotation function of the actual beam implies a discontinuity in the shear force function of the conjugate beam, which is introduced by the concentrated force corresponding to the reaction force. The shear force in the left cross-section of support B of the conjugate beam gives the rotation φ_B^l in the left cross-section of hinge B. Computing the resultant of the vertical forces acting on the left of cross-section B, we get

$$V_B^l = \varphi_B^l = \frac{pl^2}{3EI} \times \frac{l}{2} = \frac{pl^3}{6EI} .$$

The bending moment in the same cross-section corresponds to its displacement and is given by

$$M'_B = y_B = \frac{pl^3}{6EI} \times \frac{2}{3}l = \frac{pl^4}{9EI} \, .$$

The reaction force in support B may be computed from the condition of zero moment with respect to point C, yielding

$$\frac{pl^3}{6EI} \times \left(2l + \frac{2}{3}l \right) + R_B \times 2l = \frac{2}{3}\frac{pl^2}{2EI}2l \times l \;\Rightarrow\; R_B = \frac{pl^3}{9EI} \, .$$

The shear force in the right cross-section of support B is then

$$V_B^{fr} = \varphi_B^r = \frac{pl^3}{6EI} + R_B = \frac{5}{18}\frac{pl^3}{EI} \, .$$

This is a positive rotation, i.e., a rotation in the clockwise direction. The shear force in cross-section C is the resultant of the vertical forces acting to the left of this cross-section, yielding

$$V_C^l = \varphi_C = V_B^{fr} - \frac{2}{3}\frac{pl^2}{2EI}2l = -\frac{7}{18}\frac{pl^3}{EI} \, .$$

The negative value means that the cross-section rotates in the counterclockwise direction.

Moment-area method

Since the moment-area method has been expounded without defining positive values for the rotations and displacements, it is helpful to draw the qualitative shape of the deflection line, in accordance with the curvature diagram, as represented in Fig. IX.1-c-*ii*. The moment-area theorems may be directly applied to beam segment AB, since it is a cantilever. The area of the curvature diagram between A and B (Fig. IX.1-c-*i*) gives the rotation φ_B^l of the left cross-section of hinge B, and the first area moment of the same area with respect to a vertical line passing through point B gives the displacement y_B. These quantities take then the values

$$\varphi_B^l = \frac{pl^2}{3EI} \times \frac{l}{2} = \frac{pl^3}{6EI} \quad ; \qquad y_B = \frac{pl^3}{6EI} \times \frac{2}{3}l = \frac{pl^4}{9EI} \, .$$

Angle α_1 (Fig. IX.1-c-*ii*) may be computed from y_B. Angle α_2 corresponds to the relative rotations of the middle cross-section of beam segment BC and one of the end cross-sections of the same beam segment, i.e., it is half the area of the curvature diagram in segment BC. Thus, we get the following values for the rotations φ_B^r and φ_C

$$\begin{cases} \alpha_1 = \frac{y_B}{2l} = \frac{pl^3}{18EI} \\[2mm] \alpha_2 = \frac{2}{3}\frac{pl^2}{2EI}l = \frac{pl^3}{3EI} \end{cases} \Rightarrow \begin{cases} \varphi_B^r = \alpha_2 - \alpha_1 = \frac{5}{18}\frac{pl^3}{EI} \\[2mm] \varphi_C = \alpha_1 + \alpha_2 = \frac{7}{18}\frac{pl^3}{EI} \, . \end{cases}$$

IX.2. In the beam represented in Fig. IX.2-a the segment AB has a flexural stiffness $2EI$, while the remaining beam has stiffness EI. Using the conjugate beam method, determine the relative displacement of cross-sections B and C.

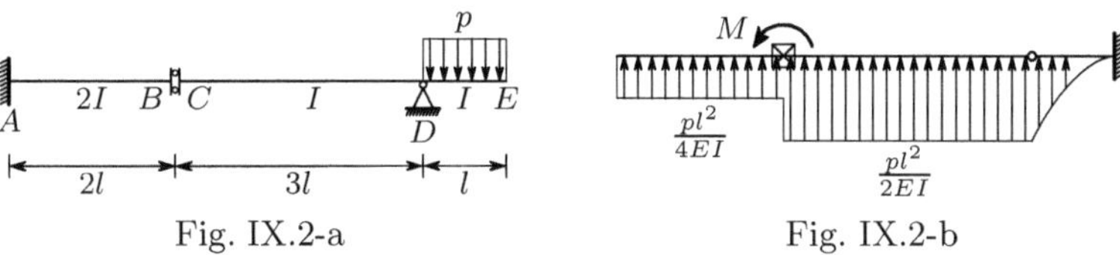

Fig. IX.2-a   Fig. IX.2-b

Resolution

The conjugate beam must have a free end in point A, since there, neither displacement, nor rotation take place (V and M must vanish in the conjugate beam). In point D it must have a hinge, since the displacement in this point is prevented by the support, which means that the bending moment must vanish in the conjugate beam. The free end E transforms into a built-in end, since the rotation and displacement are not zero, which means that the shear force and the bending moment cannot vanish in cross-section E of the conjugate beam. With regard to the connection between cross-sections B and C, the conjugate beam must have a support which introduces a discontinuity in the bending moment diagram, since the displacement function has a discontinuity between B and C. On the other hand, a vertical reaction force must not appear, since it would cause a discontinuity in the shear force diagram, which would mean different rotations in cross-sections B and C. Thus, the conjugate beam has a support in point B which prevents the rotation, but not the displacement. Figure IX.2-b illustrates the conjugate beam.

The reaction M in the support B of the conjugate beam takes the value

$$M = \frac{1}{EI}\left(\frac{pl^2}{4} \times 2l \times 4l + \frac{pl^2}{2} \times 3l \times \frac{3}{2}l\right) = \frac{17}{4}\frac{pl^4}{EI}\,.$$

This moment corresponds to the discontinuity in the bending moment diagram of the conjugate beam, i.e., to the discontinuity on the displacement of the actual beam between cross-sections B and C, which is the relative displacement of these cross-sections.

IX.3. The bending moment in cross-section A of the beam represented in Fig. IX.3-a takes the value $M_A = -\frac{pl^2}{8}$. The beam has a flexural stiffness EI.

    (a) Determine the rotation of cross-section B using the conjugate beam method.

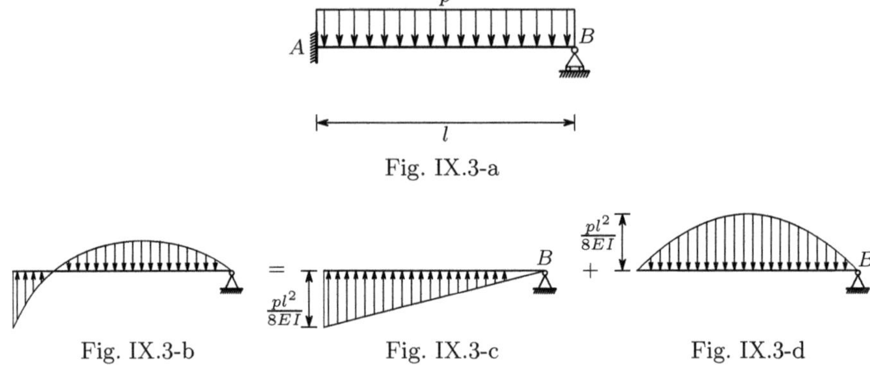

Fig. IX.3-a

Fig. IX.3-b   Fig. IX.3-c   Fig. IX.3-d

(b) Show that the equilibrium conditions are satisfied in the conjugate
    beam.

Resolution

(a) The conjugate of this hyperstatic beam is the beam represented in
    Fig. IX.3-b. In order to simplify the computation, it is best to decom-
    pose the loading acting on the beam into the two components represented
    in Figs. IX.3-c and IX.3-d.
    The rotation of cross-section B is given by the shear force in the same
    cross-section of the conjugate beam which takes the value

$$\theta_B = -\frac{1}{2} \times \frac{pl^2}{8EI} \times l + \frac{2}{3} \times \frac{pl^2}{8EI} \times l = -\frac{1}{16}\frac{pl^3}{EI} + \frac{1}{12}\frac{pl^3}{EI} = \frac{pl^3}{48EI} \,.$$

(b) The conjugate beam will be in equilibrium if the moment of its load-
    ing with respect to point B vanishes. This condition corresponds to the
    condition of zero displacement in support B of the actual beam, so that
    it is necessarily satisfied if the internal forces in this beam are correctly
    computed. Considering the moments in relation to point B of the loads
    represented in Figs. IX.3-c and IX.3-d, we see at once that they are equal
    and have opposite directions

$$\frac{1}{16}\frac{pl^3}{EI} \times \frac{2}{3}l - \frac{1}{12}\frac{pl^3}{EI} \times \frac{1}{2}l = \left(\frac{1}{24} - \frac{1}{24}\right)\frac{pl^4}{EI} = 0 \,.$$

IX.4. Deduce (224) without using the principle of energy conservation.

Resolution

Equation (224) may also be deduced by means of static equivalence considerations, between the applied shear force and the shearing stresses in the warped

cross-section. As a consequence of this equivalence, the resultant of the shearing stresses must have the same direction as the shear force. This happens only if the component of the resultant of the shearing stresses in the fibres' direction (Fig. 128) is equal to the component of the shear force in the same direction. The angle between the shearing stress in a point and the normal to the fibres' direction is equal to the double shearing strain γ (distortion) in that point. The angle between the shear force and that normal is γ_m (Fig. 128). Since angle γ_m is small, the projection of the shear force V in the fibres' direction is $V \sin \gamma_m \approx V \gamma_m$. The component of the resultant of the shearing stresses acting in an infinitesimal area $d\Omega$ of the cross-section in the fibres' direction is $\tau\, d\Omega\, \gamma$. Thus, the components in the fibres' direction of the shear force and of the resultant of the shearing stresses will be equal, if the following relation is satisfied

$$\int_\Omega \tau\gamma\, d\Omega = V\gamma_m \ .$$

IX.5. Determine and compare the contributions of the bending moment and of the shear force to the displacement of cross-section B in the cantilever beam represented in Fig. IX.5-a. The beam has a rectangular cross-section with width b and height h.

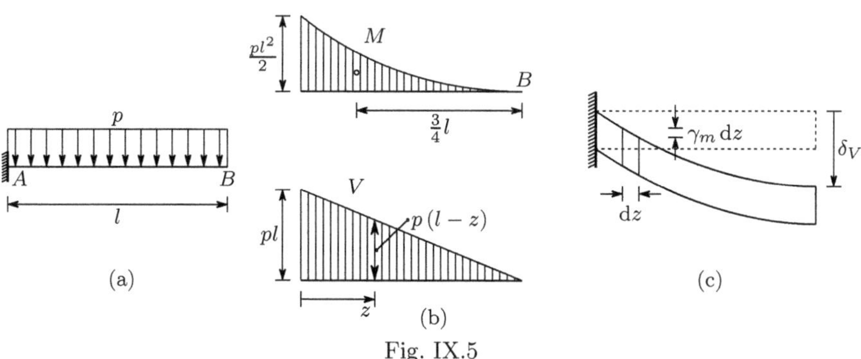

Fig. IX.5

Resolution

The contribution of the bending moment may be computed by means of the second moment-area theorem. Thus, the first moment of the curvature diagram (Fig. IX.5-b) with respect to a vertical line passing through point B, yields

$$\delta_M = \frac{1}{EI}\frac{1}{3}\frac{pl^2}{2}l \times \frac{3}{4}l = \frac{pl^4}{8EI} = \frac{3}{2}\frac{pl^4}{bh^3 E} \ .$$

The relative transversal displacement of two cross-sections at an infinitesimal distance dz from each other, caused by the shear force, is $\gamma_m\, dz$ (Fig. IX.5-c). The relative transversal displacement of cross-sections A and B is then

$$\delta_V = \int_0^l \gamma_m \, \mathrm{d}z = \int_0^l \frac{V}{G\Omega_r} \, \mathrm{d}z \ .$$

Expressing V as a function of coordinate z (Fig. IX.5-b), we get

$$\begin{cases} \Omega_r = \dfrac{bh}{1.2} \\[2mm] V = p\,(l - z) \end{cases} \Rightarrow \delta_V = \frac{1.2}{Gbh} \int_0^l p\,(l - z)\, \mathrm{d}z = \frac{0.6pl^2}{Gbh} = 1.2\,(1 + \nu)\,\frac{pl^2}{Ebh} \ .$$

The relation between the contributions of the bending moment and of the shear force is then

$$\frac{\delta_V}{\delta_M} = 0.8\,(1 + \nu)\left(\frac{h}{l}\right)^2 \ .$$

As in the example presented in Sect. IX.2.b, we conclude that the contribution of the shear force to this deflection is very small. For example, for $\nu = 0.25$ and $\frac{h}{l} = 0.1$, we have $\frac{\delta_V}{\delta_M} = 1\%$.

IX.6. Figure IX.6 shows the thin-walled cross-section of a simply-supported beam with a span $l = 10h$, which carries a uniformly distributed load p. Determine and compare the contributions of the bending moment and of the shear force to the displacement of the mid-span cross-section. Consider $\nu = 0.3$.

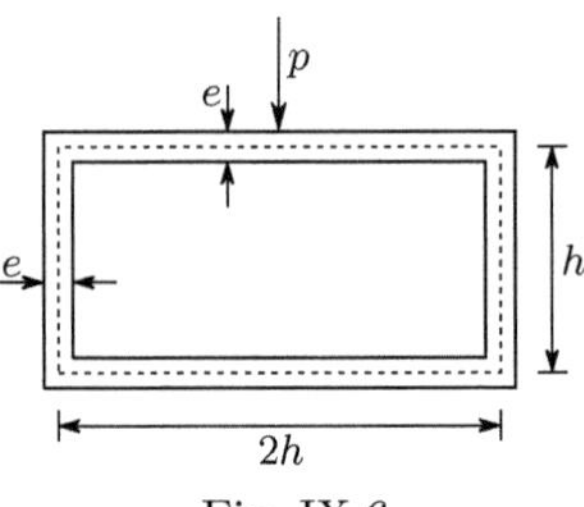

Fig. IX.6

Resolution

Since the cross-section is symmetric with respect to the action axis of the shear force, the distribution of shearing stresses is also symmetric. As a consequence, the reduced area of the cross-section is twice that of the half cross-section, which has been determined in Sect. IX.2.d (231). Thus, we have $\Omega_r = 1.339\,eh$. The moment of inertia of the cross-section is also twice the value included in (231), i.e., $I = \frac{7}{6}eh^3$.

The contribution of the bending moment may be computed by means of the second moment-area theorem. According to this theorem, the displacement of the mid-span cross-section is given by the first moment of the area of the curvature diagram between the mid-span cross-section and one of the supports, with respect to the vertical line passing through the support. Since the diagram is parabolic, its area is two thirds of the area of the rectangle containing the half parabola, and its centroid is at distance $\frac{5}{8}\frac{10h}{2}=\frac{25}{8}h$ from the support. Thus, we have

$$\delta_M=\frac{1}{E\frac{7}{6}eh^3}\times\frac{2}{3}\frac{p\,(10h)^2}{8}5h\times\frac{25}{8}h\approx 111.61\frac{ph}{eE}\ .$$

The contribution of the shear force corresponds to the relative transversal displacement of one support cross-section and the mid-span cross-section. Considering a coordinate z with origin in the mid-span cross-section, we have

$$V=pz\ \Rightarrow\ \delta_V=\int_0^{\frac{l}{2}}\frac{V}{G\Omega_r}\,\mathrm{d}z=\frac{p}{\frac{E}{2(1+0.3)}1.339eh}\int_0^{5h}z\,\mathrm{d}z=24.276\frac{ph}{Ee}\ .$$

In this example, the shear force makes an important contribution, since we have $\delta_V=0.2175\delta_M$.

IX.7. Figure IX.7 represents the cross-section of a simply supported steel beam with a span of $1.5m$, which carries a uniformly distributed load of $30kN/m$.
    (a) Determine the reduced area of this cross-section and compare it with the web area.
    (b) Determine the displacement of the mid-span cross-section, considering the contributions of the bending moment and of the shear force. Consider $E=206\times 10^9\,Pa$ and $G=80\times 10^9\,Pa$.

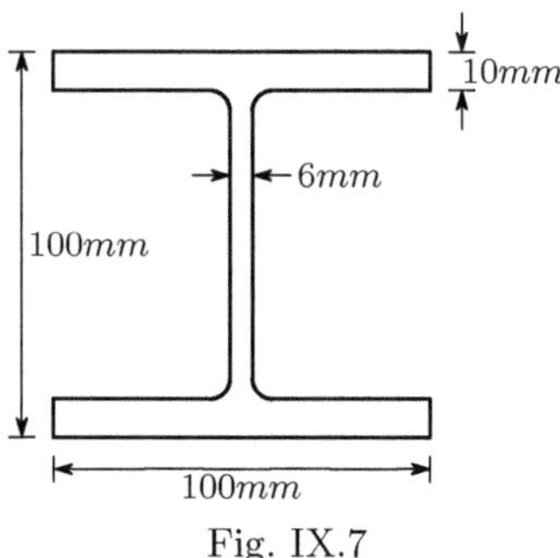

Fig. IX.7

Resolution

(a) The reduced area may be computed by means of (230). The moment of inertia of the cross-section takes the value

$$I=\frac{100^4}{12}-\frac{94\times 80^3}{12}=4.3227\times 10^6\ mm^4\ .$$

The integral $\int_L \frac{S^2}{e}\,\mathrm{d}s$ in the flanges of the cross-section takes the value (s is a coordinate with origin in the free end of a flange)

$$\int_{\text{flanges}} \frac{S^2}{e}\,\mathrm{d}s = 4\int_0^{50} \frac{1}{10}\left(10s \times 45\right)^2\,\mathrm{d}s = 3.375 \times 10^9 \ .$$

In the web, the first area moment S, expressed as a function of a coordinate s with origin in the cross-section centroid, is given by

$$S = 100 \times 10 \times 45 + 6 \times 45 \times \frac{45}{2} - 6s\frac{s}{2} = 51075 - 3s^2 \ .$$

With this area moment we get

$$\int_{\text{web}} \frac{S^2}{e}\,\mathrm{d}s = \int_{-45}^{45} \frac{1}{6}\left(51075 - 3s^2\right)^2\,\mathrm{d}s = 3.6138 \times 10^{10} \ .$$

Equation (230) yields then

$$\Omega_r = \frac{I^2}{\int_L \frac{S^2}{e}\,\mathrm{d}s} = \frac{\left(4.3227 \times 10^6\right)^2}{3.375 \times 10^9 + 3.6138 \times 10^{10}} = 472.9\,mm^2 \ .$$

The web's area takes the value

$$\Omega_{\text{web}} = 6 \times 100 = 600\,mm^2 \ .$$

(b) Considering a coordinate z with origin in the mid-span cross-section, the shear force is defined by $V = pz$. The contribution of this force to the displacement of the mid-span cross-section is then

$$\delta_V = \int_0^{\frac{l}{2}} \frac{pz}{G\Omega_r}\,\mathrm{d}z = \frac{pl^2}{8G\Omega_r}$$

$$= \frac{30000 \times 1.5^2}{8 \times 80 \times 10^9 \times 472.9 \times 10^{-6}} = 2.2303 \times 10^{-4}\,m = 0.223\,mm \ .$$

The contribution of the curvature to the same displacement may be computed by means of the second moment-area theorem, yielding (see example IX.6)

$$\delta_M = \frac{5}{384}\frac{pl^4}{EI} = \frac{5}{384}\frac{30000 \times 1.5^4}{206 \times 10^9 \times 4.3227 \times 10^{-6}}$$
$$= 2.2208 \times 10^{-3}\,m = 2.22mm \ .$$

The total displacement is then

$$\delta = \delta_M + \delta_V = 2.22 + 0.223 = 2.44\,mm \ .$$

IX.8. Determine the reduced area of a thin-walled tubular cross-section. The wall centre line has a radius r. The wall has a thickness e.

Resolution

Since we have a closed cross-section, the reduced area cannot be directly computed from (230). Function R may be easily obtained, however, from the expression of the shearing stress caused by the shear force (see example VIII.4)

$$\tau = \frac{V}{I} r^2 \sin\beta \Rightarrow R = r^2 \sin\beta \ .$$

From this expression we get

$$\int_\Omega R^2 \, \mathrm{d}\Omega = 2 \int_0^\pi \left(r^2 \sin\beta \right)^2 \overbrace{er\,\mathrm{d}\beta}^{\mathrm{d}\Omega} = 2er^5 \int_0^\pi \sin^2\beta\,\mathrm{d}\beta = \pi er^5 \ .$$

The moment of inertia of the cross-section is $I = \pi er^3$ (see example VIII.3). Thus, the reduced area takes the value

$$\Omega_r = \frac{I^2}{\int_\Omega R^2\,\mathrm{d}\Omega} = \frac{\left(\pi er^3\right)^2}{\pi er^5} = \pi re = \frac{\Omega}{2} \ .$$

The reduced area of the open thin-walled cross-section analysed in example VIII.3 may be obtained in the same way, yielding $\Omega_r = \frac{\Omega}{6}$.

IX.9. Determine the reaction forces in the supports A and B of the beam depicted in Fig. IX.9.

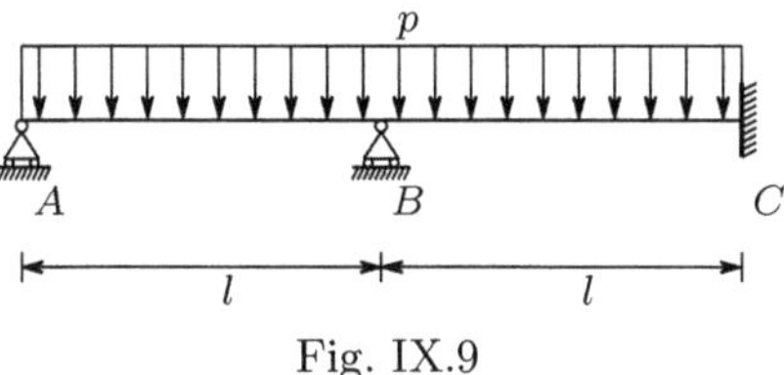

Fig. IX.9

Resolution

The problem may be solved by means of the force method, using, for example, the statically determinate base structure depicted in Fig. IX.9-a. The displacements of cross-sections A and B are easily computed using the second moment-area theorem, since cross-section C does not rotate. The displacement of cross-section A takes the value (Fig. IX.9-a)

$$y_A = \frac{1}{3} \frac{p\,(2l)^2}{2EI} 2l \times \frac{3}{4} 2l = \frac{p\,(2l)^4}{8EI} = 2\frac{pl^4}{EI} \ .$$

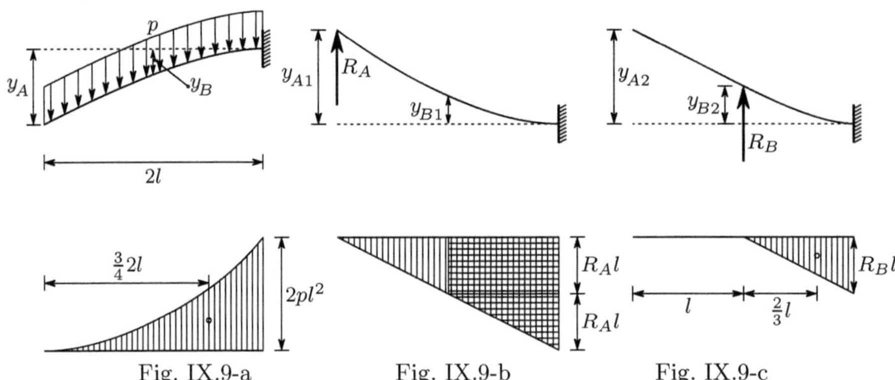

Fig. IX.9-a   Fig. IX.9-b   Fig. IX.9-c

In the case of cross-section B, the explicit evaluation of the integral representing the first area moment of the curvature diagram between B and C may be
avoided by decomposing the bending moment diagram in beam segment BC
into three components, so that known areas are obtained: the moment caused
by the distributed load p in BC (the tympanum of a half parabola), the shear
force in B ($V_B = pl \rightarrow$ triangular diagram) and by the bending moment in B
($M_B = \frac{pl^2}{2} \rightarrow$ rectangular diagram). Adding these three components, we get

$$
y_B = \frac{1}{EI} \left(\underbrace{\frac{1}{3}\frac{pl^2}{2} \times l \times \frac{3}{4}l}_{\substack{\text{tympanum of} \\ \text{half parabola}}} + \underbrace{\frac{1}{2}pl^2 \times l \times \frac{2}{3}l}_{\text{triangle}} + \underbrace{\frac{pl^2}{2} \times l \times \frac{l}{2}}_{\text{rectangle}} \right) = \frac{17}{24}\frac{pl^4}{EI} \; .
$$

The displacements y_A and y_B are offset by the deflections caused by the
reaction forces in the supports, R_A and R_B. These displacements may also be
computed by means of the second moment-area theorem. Thus, we get from
the bending moment diagram caused by R_A (Fig. IX.9-b)

$$
y_{A1} = \frac{1}{2}2lR_A \times 2l \times \frac{2}{3}2l \times \frac{1}{EI} = R_A\frac{8}{3}\frac{l^3}{EI}
$$

$$
y_{B1} = \frac{1}{EI} \left(lR_A \times l \times \frac{l}{2} + \frac{1}{2}R_A \times l \times l \times \frac{2}{3}l \right) = R_A\frac{5}{6}\frac{l^3}{EI} \; .
$$

In the same way, the displacements caused by R_B are given by the expressions
(Fig. IX.9-c)

$$
y_{A2} = \frac{1}{2}lR_B \times l \times \left(l + \frac{2}{3}l \right) \times \frac{1}{EI} = R_B\frac{5}{6}\frac{l^3}{EI}
$$

$$
y_{B2} = \frac{1}{2}lR_B \times l \times \frac{2}{3}l \times \frac{1}{EI} = R_B\frac{1}{3}\frac{l^3}{EI} \; .
$$

The reaction forces R_A and R_B are then computed by solving the system of
equations defined by the deformation compatibility conditions

$$\begin{cases} y_A = y_{A1} + y_{A2} \\ y_B = y_{B1} + y_{B2} \end{cases} \Leftrightarrow \begin{cases} 2pl = \frac{8}{3}R_A + \frac{5}{6}R_B \\ \frac{17}{24}pl = \frac{5}{6}R_A + \frac{1}{3}R_B \ . \end{cases} \Rightarrow \begin{cases} R_A = \frac{11}{28}pl \\ R_B = \frac{32}{28}pl \ . \end{cases}$$

IX.10. Using the equation of three moments, determine the bending moments in the cross-sections over the supports B and C of the beam represented in Fig. IX.10.

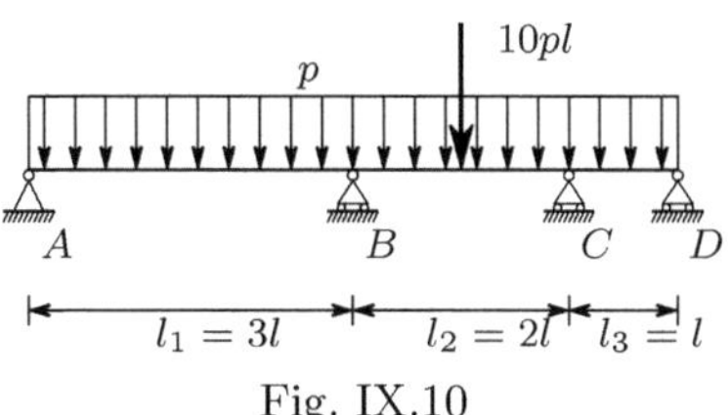

Fig. IX.10

Resolution

The problem may be solved by direct application of (232) to the intermediate supports B and C. The areas defined by the bending moment diagrams in the statically determinate base system, for the beam segments $\overline{AB}$, $\overline{BC}$ and $\overline{CD}$, respectively, take the values

$$A_1 = \frac{2}{3}\frac{pl_1^2}{8} \times l_1 = \frac{pl_1^3}{12} = \frac{9}{4}pl^3 \quad A_2 = \frac{pl_2^3}{12} + \frac{1}{2}\frac{10pl \times l_2}{4} \times l_2 = \frac{17}{3}pl^3 \quad A_3 = \frac{pl^3}{12} \ .$$

The distances from the centroids of these areas to the supports are respectively

$$a_1 = b_1 = \frac{3}{2}l \qquad a_2 = b_2 = l \qquad a_3 = b_3 = \frac{1}{2}l \ .$$

Denoting by M_1 and M_2 the bending moments in the cross-sections over the supports B and C, respectively, the deformation compatibility condition in node B is (232)

$$\begin{cases} M_{n-1} = 0 \\ M_n = M_1 \\ M_{n+1} = M_2 \end{cases} \Rightarrow \begin{array}{l} 2M_1 5l + M_2 2l = 6\left(\frac{\frac{9}{4}pl^3 \times \frac{3}{2}l}{3l} + \frac{\frac{17}{3}pl^3 \times l}{2l}\right) \\ \Rightarrow \quad 5M_1 + M_2 = \frac{95}{8}pl^2 \ . \end{array}$$

In the same way, we get for node C

$$\begin{cases} M_{n-1} = M_1 \\ M_n = M_2 \\ M_{n+1} = 0 \end{cases} \Rightarrow \begin{array}{l} M_1 2l + 2M_2 3l = 6\left(\frac{17}{3}pl^3 \times \frac{1}{2} + \frac{pl^3}{12} \times \frac{1}{2}\right) \\ \Rightarrow \quad M_1 + 3M_2 = \frac{69}{8}pl^2 \ . \end{array}$$

Solving the system formed by the two compatibility conditions, the bending moments M_1 and M_2 are obtained

$$\begin{cases} 5M_1 + M_2 = \frac{95}{8}pl^2 \\ M_1 + 3M_2 = \frac{69}{8}pl^2 \end{cases} \Rightarrow \begin{cases} M_1 = 1.929pl^2 \\ M_2 = 2.232pl^2 \ . \end{cases}$$

IX.11. Using the equation of three moments, determine the bending moment in the cross-section over the intermediate support of the beam represented in Fig. IX.11-a.

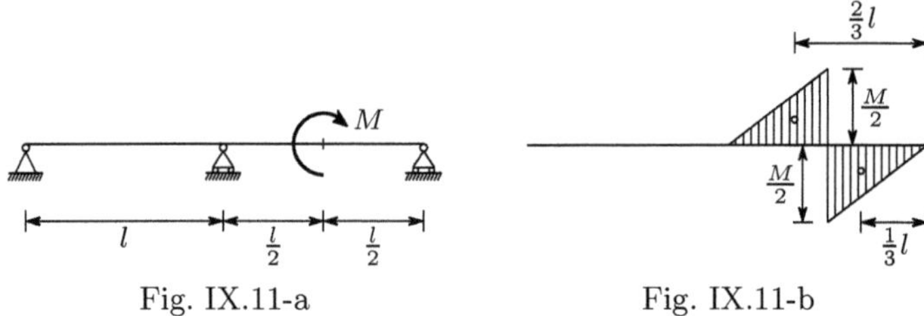

Fig. IX.11-a   Fig. IX.11-b

Resolution

Since in this case the area A_{n+1} vanishes, but the first area moment is not zero, we cannot apply (232) directly. The problem is however easily solved by explicitly computing the first area moment of the bending moment diagram (which is represented by $A_{n+1}b_{n+1}$ in (232)), yielding (Fig. IX.11-b)

$$2M_n 2l = 6 \left(-\frac{1}{2}\frac{M}{2}\frac{l}{2} \times \frac{2}{3}l + \frac{Ml}{8} \times \frac{1}{3}l \right) \frac{1}{l} \Rightarrow M_n = -\frac{M}{16} \ .$$

IX.12. Figure IX.12-a represents a beam whose central zone (AB) has a cross-section with a greater bending strength. The ultimate plastic moment in this zone is $2M_p$, while in the remaining beam it takes the value M_p. Determine the value of the distributed load p which causes the collapse of the beam.

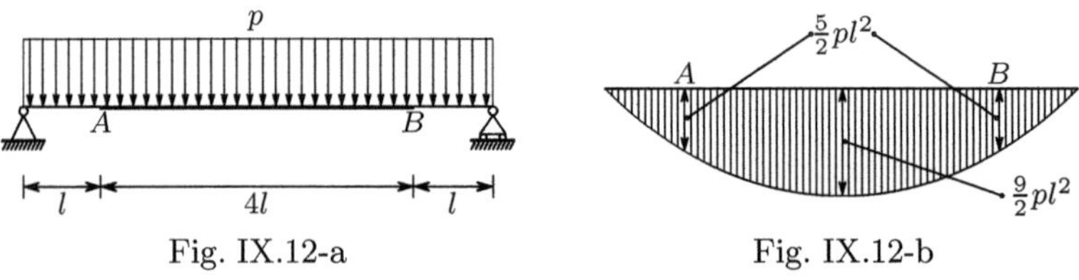

Fig. IX.12-a   Fig. IX.12-b

Resolution

Since the beam is statically determinate, collapse will occur when the first plastic hinge forms. This will happen in the cross-section where the relation between the acting bending moment and the ultimate plastic moment is the largest. In this beam there are two possibilities: formation of a plastic hinge in the cross-section with the largest bending moment of the reinforced zone – the middle cross-section – when the bending moment reaches the value $2M_p$ in that cross-section, or simultaneous formation of two plastic hinges in the two cross-sections with the largest bending moments in the parts of the beam with ultimate plastic moment M_p (cross-sections A and B). The values of p corresponding to each of these situations are (Fig. IX.12-b)

$$\frac{9}{2}pl^2 = 2M_p \Rightarrow p = \frac{4}{9}\frac{M_p}{l^2} \approx 0.4444\frac{M_p}{l^2}$$

$$\frac{5}{2}pl^2 = M_p \Rightarrow p = \frac{2}{5}\frac{M_p}{l^2} = 0.4\frac{M_p}{l^2} \; .$$

The collapse load is the smallest of these two values. Thus, the beam collapses when two plastic hinges form in cross-sections A and B, which happens when the load reaches the value $p = 0.4\frac{M_p}{l^2}$.

IX.13. The beam represented in Fig. IX.13 has a cross-section with an ultimate plastic moment M_p. Determine the value of P which induces the collapse of the beam, using the static and the kinematic methods.

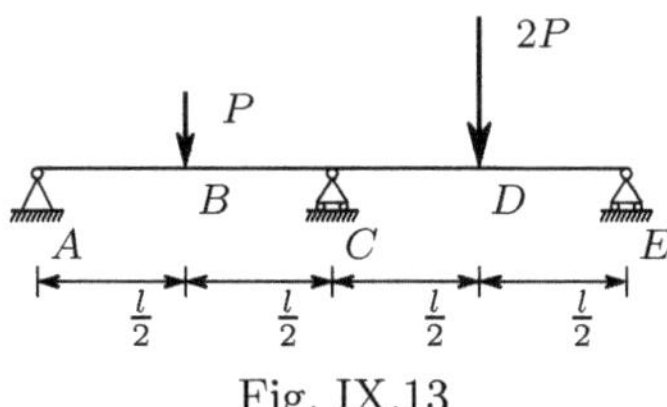

Fig. IX.13

Resolution

Static method

The diagram of bending moments in the elastic phase may be easily obtained by means of the equation of three moments. The bending moment in cross-section C is (232)

$$\begin{cases} A_n = \frac{Pl^2}{8}; & A_{n+1} = \frac{Pl^2}{4} \\ a_n = b_{n+1} = \frac{l}{2} \end{cases} \Rightarrow M_n = \frac{9}{32}Pl \; .$$

The complete diagram of bending moments is easily drawn using the equation of two moments (Fig. IX.13-a).

As the maximum bending moment occurs in cross-section D, the first plastic hinge will form there, when the bending moment reaches the value

$$\frac{23}{64}Pl = M_p \;\Rightarrow\; P = \frac{64}{23}\frac{M_p}{l} \; .$$

For higher values of P, the bending moment in cross-section D will remain constant with the value M_p. Since the original beam has a degree of static indeterminacy of one, the formation of the plastic hinge renders the beam statically determinate. The reaction force in support E may be obtained by the condition of a bending moment M_p in cross-section D, yielding

$$R_E\frac{l}{2} = M_p \;\Rightarrow\; R_E = 2\frac{M_p}{l} \; .$$

With this value, the bending moment diagram for $P > \frac{64}{23}\frac{M_p}{l}$ is immediately obtained (Fig. IX.13-b).

The formation of a second plastic hinge induces the collapse of the beam. From the values of the bending moments in cross-sections B and C shown in the diagram in Fig. IX.13-b, it is not clear which of them has the maximum absolute value. However, determining the value of P corresponding to the formation of a plastic hinge in cross-section C,

$$-Pl + 2M_p = -M_p \;\Rightarrow\; P = 3\frac{M_p}{l} \; ,$$

we verify that, for this value of P, the bending moment in cross-section B is still inferior to M_p

$$M_B = M_p - \frac{3\frac{M_p}{l}l}{4} = \frac{M_p}{4} < M_p \; .$$

Thus, the second plastic hinge will form in cross-section C, for $P = 3\frac{M_p}{l}$.

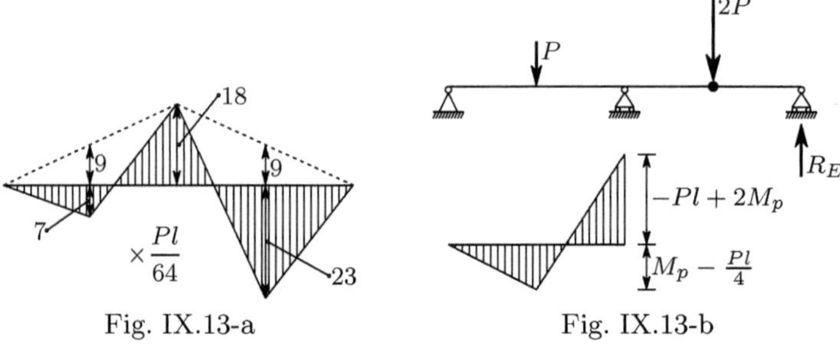

Fig. IX.13-a   Fig. IX.13-b

Kinematic Method

Since the plastic hinges may form in the cross-sections where the bending moment attains a local maximum, there are three possibilities for the positions of the two plastic hinges needed to induce collapse (Fig. IX.13-c). The computation of the value of P corresponding to each configuration may be obtained by equilibrium conditions, or by means of the Theorem of Virtual Displacements (see Sect. XII.4).

In configuration 1, the conditions of static equilibrium yield

$$M_D = M_p \Rightarrow R_E = 2\frac{M_p}{l}\,;$$

$$M_C = -M_p \Rightarrow R_E l - 2P\frac{l}{2} = -M_p$$

$$\Rightarrow 2\frac{M_p}{l}l - 2P\frac{l}{2} = -M_p$$

$$\Rightarrow P = P_1 = 3\frac{M_p}{l}\,.$$

In the same way, the value of P which balances the internal forces with $M_B = M_p$ and $M_C = -M_p$ (configuration 2) may be obtained, yielding $P = P_2 = 6\frac{M_p}{l}$.

In the case of configuration 3, the reaction forces in the supports A and E may be expressed as functions of M_p yielding

$$M_B = -M_p \Rightarrow R_A = -2\frac{M_p}{l}\,; \qquad M_D = M_p \Rightarrow R_E = 2\frac{M_p}{l}\,.$$

The condition stating that the bending moment in cross-section C must have the same value, computed either by means of the forces to the left, or using the forces to the right of the cross-section, yields

$$M_C = -2\frac{M_p}{l}l - P\frac{l}{2} = 2\frac{M_p}{l}l - 2P\frac{l}{2} \Rightarrow P = 8\frac{M_p}{l}\,.$$

Since the smallest value of P corresponds to configuration 1, this is the real collapse configuration, which confirms the result obtained by means of the static method.

IX.14. The beam represented in Fig. IX.14 has a cross-section with an ultimate plastic moment M_p. Determine the value of the load p which induces the collapse of the beam using the static and kinematic methods.

Resolution

Static Method

The bending moment diagram in the elastic phase may be easily obtained by means of the force method, taking the bending moment in cross-section A as the hyperstatic unknown, which takes the value $M = \frac{pl^2}{8}$. The equation of

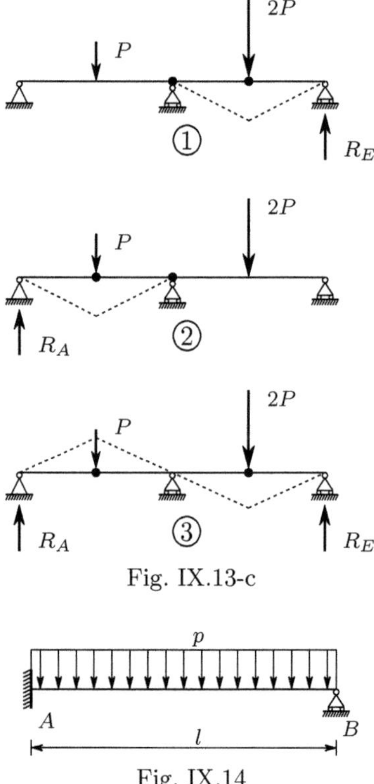

Fig. IX.13-c

Fig. IX.14

two moments leads to the diagram depicted in Fig. IX.14-a. The maximum value of the bending moment occurs clearly in cross-section A. Thus, the first plastic hinge forms there, when p reaches the value

$$\frac{pl^2}{8} = M_p \;\Rightarrow\; p = \frac{8M_p}{l^2}\,.$$

When p exceeds this value, the bending moment in cross-section A remains constant and equal to M_p. Thus, for $p \geq \frac{8M_p}{l^2}$, the forces acting on the beam are those represented in Fig. IX.14-b. The corresponding bending moments are represented in Fig. IX.14-c.

As the bending moment M_A remains constant, while p increases, the bending moments are not proportional to the load p. This means that the position of the cross-section where the maximum bending moment occurs does not remain constant, but suffers a displacement to the left, as p increases. For this reason, the computation of the value of P, which corresponds to the formation of the second plastic hinge, is more complex than in the case of structures where only concentrated loads are acting.

Using the equation of two moments, the bending moment and the shear force may be expressed as functions of coordinate z (Fig. IX.14-c), without

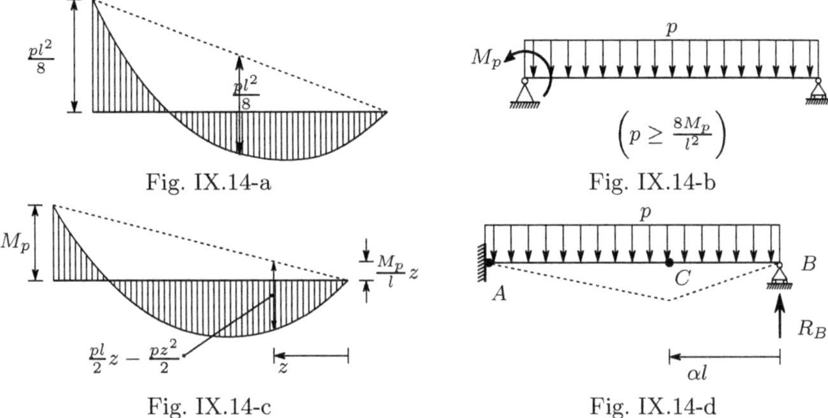

Fig. IX.14-a   Fig. IX.14-b

Fig. IX.14-c   Fig. IX.14-d

needing to determine the value of the reaction forces, yielding

$$M\left(z\right) = -\frac{M_p}{l}z + \frac{pl}{2}z - \frac{pz^2}{2} \Rightarrow V\left(z\right) = \frac{pl}{2} - \frac{M_p}{l} - pz \ .$$

The location of the maximum value of the bending moment may be obtained from the condition $V = 0$. Denoting by z_0 the distance from the cross-section where the maximum bending moment occurs to support B, we get the maximum value of the bending moment as a function of p

$$V = 0 \Rightarrow z = z_0 = \frac{l}{2} - \frac{M_p}{pl}$$

$$\Rightarrow M_{\max} = \left(\frac{pl}{2} - \frac{M_p}{l}\right)\left(\frac{l}{2} - \frac{M_p}{pl}\right) - \frac{p}{2}\left(\frac{l}{2} - \frac{M_p}{pl}\right)^2 = \frac{pl^2}{8} - \frac{M_p}{2} + \frac{M_p^2}{2pl^2} \ .$$

Collapse takes place when this moment reaches the value M_p, i.e.,

$$M_{\max} = M_p \Rightarrow l^4 p^2 - 12l^2 M_p p + 4M_p^2 = 0 \ .$$

This equation has two solutions, $p = 0.343\frac{M_p}{l^2}$ and $p = 11.657\frac{M_p}{l^2}$, of which only the second has physical meaning, since this analysis is only valid for $p \geq 8\frac{M_p}{l^2}$. Thus, collapse occurs for $p = 11.657\frac{M_p}{l^2}$, when a plastic hinge forms at the distance from the right support

$$p = p_{\mathrm{col}} = 11.657\frac{M_p}{l^2} \Rightarrow z_0 = \frac{l}{2} - \frac{M_p}{11.657\frac{M_p}{l^2}l} = 0.4142l \ .$$

Kinematic Method

Since the diagram of bending moments has only two maximum values and only two plastic hinges are necessary to induce the collapse, only the configuration represented in Fig. IX.14-d has to be analysed. However, as in the static method, the position of the cross-section where the maximum positive bending moment occurs is not known. Among the infinity of possible positions, the actual one is that corresponding to the minimum value of the load p.

Denoting by αl the distance from the plastic hinge C to support B (Fig. IX.14-d), the reaction force on the right support, expressed as a function of this distance, takes the value

$$M_C = M_p \Rightarrow R_B \alpha l - p\frac{\alpha^2 l^2}{2} = M_p \Rightarrow R_B = \frac{M_p}{\alpha l} + p\frac{\alpha l}{2} \; .$$

Since in the built-in end we have $M_A = -M_p$, the value of load p corresponding to a given value of α is

$$M_A = R_B l - \frac{pl^2}{2} = -M_p \Rightarrow \left(\frac{M_p}{\alpha l} + p\frac{\alpha l}{2}\right) l - \frac{pl^2}{2} = -M_p$$

$$\Rightarrow p = \beta\frac{M_p}{l^2} \quad \text{with} \quad \beta = \frac{2\alpha + 2}{\alpha - \alpha^2} \; .$$

The value of α which minimizes the value of p is an algebraic extremum of function β, i.e., one of the values

$$\frac{\mathrm{d}\beta}{\mathrm{d}\alpha} = 0 \Rightarrow \frac{2\left(\alpha - \alpha^2\right) - (1 - 2\alpha)\left(2\alpha + 2\right)}{\left(\alpha - \alpha^2\right)^2} = 0 \Rightarrow \alpha = 0.4142 \; \vee \; \alpha = -2.4142 \; .$$

Only the first of these solutions has a physical meaning, since $0 \leq \alpha \leq 1$. The collapse load is then the one corresponding to $\alpha = 0.4142$, which coincides with the value obtained by means of the static method

$$\alpha = 0.4142 \Rightarrow \beta = \frac{2 \times 0.4142 + 2}{0.4142 - 0.4142^2} = 11.657 \Rightarrow p_{\mathrm{col}} = 11.657\frac{M_p}{l^2} \; .$$

IX.15. The beam represented in Fig. IX.15-a has a cross-section with an ultimate plastic moment M_p. Determine the value of p which causes the collapse of the beam.

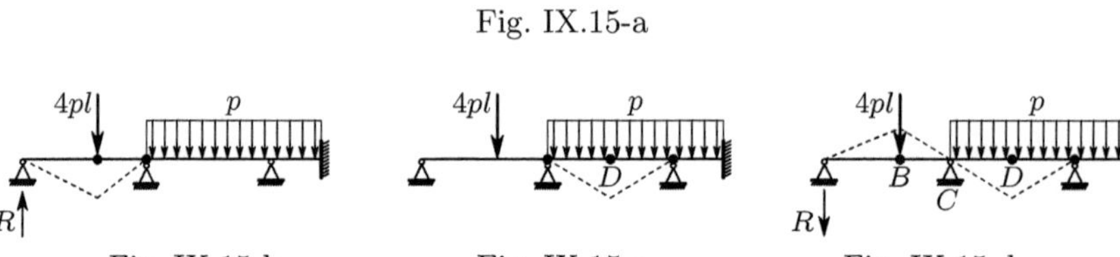

Resolution

As the beam is statically indeterminate of degree 3, an analysis by means of the static method would be very lengthy. Thus, the kinematic method is used. Although the beam has distributed loads and no symmetry conditions, the two most probable collapse mechanisms may be easily analysed. The first consists of the collapse of the first beam segment, as represented in Fig. IX.15-b. The corresponding value of p is

$$R \times 3l = M_p \Rightarrow R = \frac{M_p}{3l} \; ; \qquad R \times 5l - 4pl \times 2l = -M_p \Rightarrow p = \frac{1}{3}\frac{M_p}{l^2} \; .$$

The load corresponding to the second mechanism (Fig. IX.12-c) is also easily obtained, since the equation of two moments leads to the conclusion that the intermediate plastic hinge is at the same distance from the two supports of this beam segment (although the maximum bending moment is not in this position prior to collapse) and that the corresponding load is

$$\frac{p\,(5l)^2}{8} = 2M_p \Rightarrow p = \frac{16}{25}\frac{M_p}{l^2} = 0.64\frac{M_p}{l^2} > \frac{1}{3}\frac{M_p}{l^2} \; .$$

The analysis of the third mechanism (Fig. IX.12-d) would be more lengthy, since it would be necessary to determine the position of the cross-section where the intermediate plastic hinge forms (point D). However, it is easy to verify that this mechanism cannot form. In fact, computing the absolute value of the bending moment in cross-section C, we get

$$R \times 3l = M_p \Rightarrow R = \frac{M_p}{3l} \; ; \qquad M_C = R \times 5l + 4pl \times 2l = \frac{5}{3}M_p + 8pl^2 > M_p \; .$$

This result shows that a plastic hinge forms in cross-section C, before it forms in cross-section B.

Thus, the collapse load is that corresponding to the first mechanism, $p = \frac{1}{3}\frac{M_p}{l^2}$. The confirmation of this value by means of the static theorem is a little lengthy, since we have a partial collapse and the part of the beam not affected by it is still statically indeterminate of degree two. This verification is left as an exercise for the reader.

IX.16. The beam represented in Fig. IX.16 is made of material with elastic perfectly plastic behaviour. The beam segment AB has a cross-section with an ultimate plastic moment $3M_p$, while segment BC yields when the bending moment reaches the value M_p. Determine the value of p which induces the collapse of the beam.

Resolution

The beam has basically two possibilities of plastic collapse: one with two plastic hinges with negative bending moment in cross-sections B and C and one

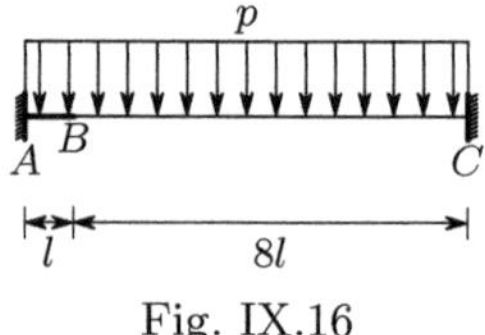

Fig. IX.16

plastic hinge with a positive moment in an intermediate position of beam segment BC. The second mechanism is similar to the first one, with the difference that the left plastic hinge occurs in cross-section A (instead of B).

The analysis of the first mechanism is very easy, since the bending moment diagram in beam segment BC is symmetric when the bending moments in cross-sections B and C are equal, as happens in this collapse configuration ($M_B = M_C = -M_p$). The corresponding load may be easily determined by means of the equation of two moments, yielding

$$\frac{p\,(8l)^2}{8} = 2M_p \;\Rightarrow\; p = \frac{M_p}{4l^2}\,.$$

The analysis of the second mechanism requires an investigation of the position of the hinge with a positive bending moment, since in this case there is no symmetry of the bending moment diagram. But it is easier to use the static theorem to confirm the collapse load corresponding to the first mechanism. The bending moment in cross-section A, in the first configuration, takes the value

$$M_A = -M_p - \frac{p \times 8l}{2} \times l - \frac{pl^2}{2} = -M_p - \frac{9}{2}pl^2 = -M_p - \frac{9}{2}\frac{M_p}{4l^2}l^2 = -\frac{17}{8}M_p\,.$$

This bending moment is inferior to the ultimate plastic moment of the cross-section ($3M_p$), which means the plastic moment is not exceeded in any cross-section. Thus the static theorem guarantees that the loading capacity of the beam is not less than $\frac{M_p}{4l^2}$. The combined static and kinematic theorems lead to the conclusion that this is the actual collapse load.

IX.17. The beams represented in Fig IX.17 are made of materials with linear elastic behaviour and have cross-sections with the given moments of inertia. Using the method of the conjugate beam, determine:
  (a) the rotation of cross-section D of the beam represented in Fig. IX.17-a;
  (b) the deflection of the point of application of load P in the beam depicted in Fig. IX.17-b;
  (c) the deflection of the point of application of load P in the beam depicted in Fig. IX.17-c;
  (d) the deflection of point B of the beam represented in Fig. IX.17-d.
IX.18. Using the equation of three moments, determine the diagrams of bending moments in the beams represented in Figs. IX.18-a and IX.18-b.

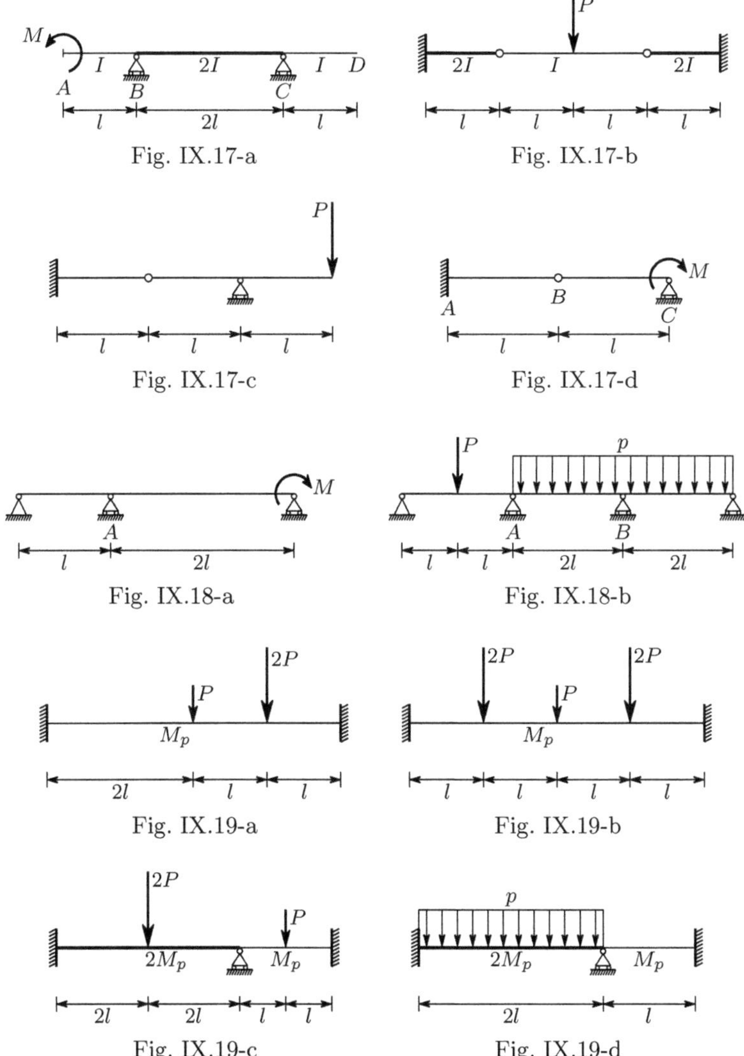

Fig. IX.17-a

Fig. IX.17-b

Fig. IX.17-c

Fig. IX.17-d

Fig. IX.18-a

Fig. IX.18-b

Fig. IX.19-a

Fig. IX.19-b

Fig. IX.19-c

Fig. IX.19-d

IX.19. The beams represented in Fig. IX.19 are made of materials with elastic perfectly plastic behaviour and have cross-sections with the indicated ultimate plastic moments. Determine the corresponding collapse loads.

X

Torsion

X.1 Introduction

In a slender member under the action of a torsional moment (also called twisting moment or torque) shearing stresses appear, whose moment about the bar axis is equal to the applied torque. In the same way as the shearing stresses caused by the shear force, these stresses must be tangent to the contour in the points lying close the boundary of the cross-section, as depicted in Fig. 101.

These two conditions are not sufficient to determine the distribution of shearing stresses in the cross-section. Furthermore, the twisting moment is not a symmetrical loading with respect to the middle cross-section of a piece of bar, so that the law of conservation of plane cross-sections, in the form demonstrated in Sect. V.10.c, is not valid. For these reasons, the elementary theories of the Strength of Materials cannot solve the problem of torsion of slender members, except in two cases where the shape of the cross-section creates supplementary conditions. In the first – circular cross-sections – the symmetry of the cross-section with respect to the bar axis allows the complete definition of the deformation caused by the torque. In the second – closed thin-walled cross-sections – a small simplifying hypothesis and the above mentioned conditions of equilibrium allow the computation of the shearing stresses caused by the torque.

In this chapter these two cases are analysed and some general considerations about the distribution of the stresses induced in the cross-section by the twisting moment are made. We also give and discuss some useful expressions, obtained by means of the Theory of Elasticity for the computation of shearing stresses and angles of twist in rectangular and open thin-walled cross-sections.

X.2 Circular Cross-Sections

A prismatic bar with a circular cross-section has a symmetrical geometry with respect to any plane passing through the bar axis. If, in addition, the material also has symmetrical rheological properties with respect to these planes, which happens if the material is isotropic or monotropic with the monotropy direction parallel to the bar axis, the bar is totally symmetric with respect to the bar axis, i.e., it is axisymmetric. As a consequence of this type of symmetry, all the points of a cross-section lying on a circumference with the centre in the bar axis, are in the same conditions with respect to the centre of the cross-section. If we consider a vector applied at the centre of the cross-section, representing the torque acting on the bar, all the points of that circumference are also in the same conditions with respect to that vector. As a consequence, all the points will undergo the same displacement in relation to the bar axis, i.e., the radial, circumferential and longitudinal components of the displacement will be the same in all points of the circumference. This means that the circumference will remain on a plane perpendicular to the bar axis and that its centre will remain on that axis.

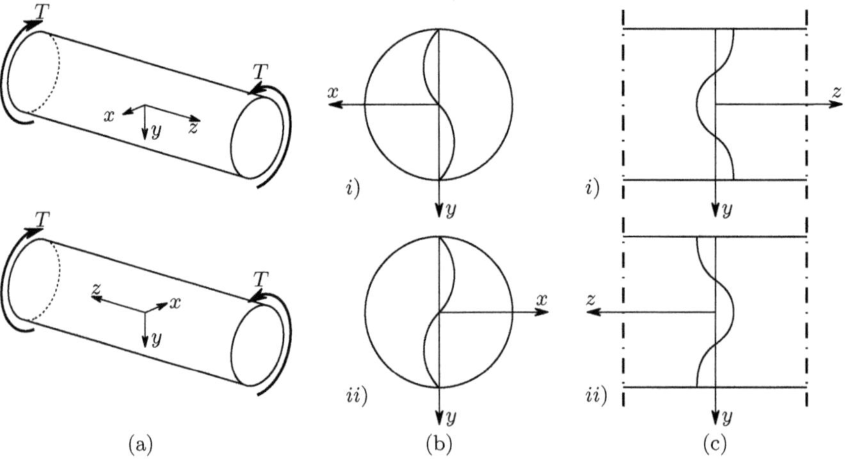

Fig. 139. Prismatic bar with circular cross-section under a positive torsional moment T

Let us consider now all the concentric circumferences in the cross-section of the bar. The considerations above do not imply that all circumferences undergo the same rotation and that they remain in the same plane. This means that a diameter could get curvature, both in the cross-section plane (Fig. 139-b) and in a longitudinal plane (Fig. 139-c). In order to demonstrate that this is not possible, let us consider two situations:

1- a torque T, is applied to the bar and causes in diameter y of the cross-section the deformations represented in Figs. 139-b-i and 139-c-i.

2- The same twisting moment T is applied and, subsequently, the bar is rotated 180° about diameter y of the middle cross-section. As a consequence, the deformed diameter acquires the shape represented in Figs. 139-b-*ii* and 139-c-*ii*.

As a consequence of the antisymmetry of the loading and of the symmetry of the bar, the 180°-rotation changes neither the geometry of the bar nor the direction of the applied torque, i.e., the geometry of the bar and the applied loading are exactly the same in the two situations. Therefore, the deformed shape must also be the same in the two situations, which only happens if the diameter y remains a straight line.

Although this argument has been made for the middle cross-section of the bar, it is valid for any cross-section which is sufficiently far from the extremities of the bar, or from the points where forces are applied, to accept the validity of the Saint-Venant principle. In fact, for any cross-section, a piece of bar may be defined such that its middle cross-section coincides with the cross-section under consideration, since that principle allows the substitution of any system of forces by its resultant (the torque, in this case).

The above considerations lead to the conclusion that the cross-sections remain plane, circular, parallel to each other and with their centers in the bar axis during torsional deformation, and that all diameters of a cross-section suffer the same rotation, i.e., the angle between any two diameters remains unchanged. This means that the only possible deformation of the cross-section is a diameter change, which, if it occurs, must take the same value in all cross-sections, since the torque is the same in all them. An axial deformation may also occur, since it does not conflict with the symmetry considerations stated above. The rotation of the cross-sections about the bar axis must take place, because otherwise the torque would not do work. Thus, it represents the main torsional deformation.

We may therefore conclude that, in a prismatic bar with circular cross-section, under a constant twisting moment (uniform torsion), the shearing strain between the radial and the circumferential directions must vanish, since the diameters must remain straight (cross-section's plane) and between the radial and the longitudinal directions (planes containing the axis of the bar), i.e., *among the radial, circumferential and longitudinal directions, shearing strains are only possible between the circumferential and the longitudinal directions.* The radial direction is therefore a principal direction of the deformation state in any point of the bar.

These conclusions have been drawn without any reference to the rheological properties of the material, except the symmetry condition with respect to the bar axis. Thus, they are valid provided that the material is isotropic or monotropic with the monotropy direction parallel to the bar axis, irrespective of any other considerations about the material behaviour, such as linearity of the stress-strain relation, kind of deformation (elastic, plastic or viscous), and the size of the deformations.

Once the way the bar deforms has been defined a relation may be established between the torsional deformation and the strains in a point. To this end, let us consider a bar under uniform torsion, as represented in Fig. 140-a. The relative rotation of two cross-sections is proportional to the distance between them. Denoting by θ the relative rotation of two cross-sections at a unit distance from each other (unit rotation, $\theta = \frac{\varphi}{l}$, Fig. 140-a), the relative rotation between two cross-sections at an infinitesimal distance $\mathrm{d}z$ from each other is $\theta\,\mathrm{d}z$. This rotation causes the deformation depicted in Fig. 140-b. Considering the upper cross-section to be fixed, the torsional deformation causes the displacement of points b and B to the positions b' and B', respectively. In the case of small deformations, the distances $\overline{AB}$ and $\overline{AB'}$ may be considered as equal, as are $\overline{ab}$ and $\overline{ab'}$. The distortion (double shear strain) in the points of the circumference with a radius r is given by the angle γ between the line segments $\overline{ab}$ and $\overline{ab'}$. This distortion may be obtained from the distance $\overline{bb'}$, which may be related to the relative rotation of the cross-sections $\theta\,\mathrm{d}z$, yielding

$$bb' = r\,\theta\,\mathrm{d}z = \gamma\,\mathrm{d}z \;\Rightarrow\; \gamma = r\theta\,. \tag{233}$$

This expression is still independent of the rheological properties of the material the bar is made of. But the development of the relation between the torque and the shearing stress requires the consideration of the constitutive law of the material, in addition to equilibrium conditions. In the case of a linear elastic behaviour, the shearing stress is proportional to the distortion, i.e., $\tau = G\gamma = Gr\theta$ (233).

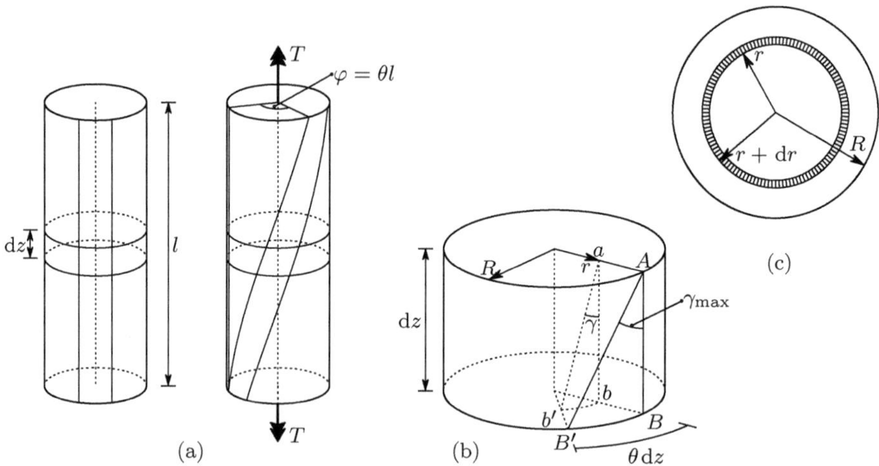

Fig. 140. Deformation of a bar with circular cross-section under the action of a positive torque T

The stresses acting in the region defined by the radii r and $r + \mathrm{d}r$ (Fig. 140-c) may be considered as constant, since $\mathrm{d}r$ is an infinitesimal quantity. Computing the moment of these stresses with respect to the bar axis and integrating to the whole cross-section, we get, using (233), the relation between the torque T and the unit rotation θ

$$\mathrm{d}T = (\tau 2\pi r\,\mathrm{d}r)\,r = G\gamma 2\pi r^2\,\mathrm{d}r = G\theta 2\pi r^3\,\mathrm{d}r$$

$$\Rightarrow \quad T = 2G\theta\pi \int_0^R r^3\,\mathrm{d}r = G\theta I_p \Rightarrow \theta = \frac{T}{GI_p} \quad \text{with} \quad I_p = \frac{\pi R^4}{2}\ . \tag{234}$$

I_p represents the polar moment of inertia of the cross-section with respect to the center. Clearly, the quantity GI_p represents the *torsional stiffness* of the bar, since the larger the value of this quantity, the smaller the deformation θ induced by the torsional moment T. Substituting (234) in (233) and multiplying by the shear modulus G, we get the relation between the torque T and the shearing stress in a point at distance r from the centre

$$\gamma = r\frac{T}{GI_p} \Rightarrow \tau = G\gamma = \frac{Tr}{I_p}\ . \tag{235}$$

The maximum stress occurs at the contour $(r = r_{\max} = R)$, taking the value

$$\tau_{\max} = \frac{T}{\frac{I_p}{R}}\ . \tag{236}$$

The parameter $\frac{I_p}{R} = \frac{\pi R^3}{2}$ is called the *torsion modulus* of the cross-section.

There is a clear analogy between (233), (234), (235) and (236) and (142), (145), (146) and (147) obtained for plane bending. This is because, in both cases, the cross-sections remain plane and rotate about an axis, which makes the strains proportional to a distance (r and y, respectively).

In the case of a circular tube, all the symmetry considerations remain valid. The only difference arises in the computation of the polar moment of inertia and of the torsion modulus, which are given by the expressions (R_i and R_e are the internal and external radii of the cross-section, respectively)

$$I_p = \frac{\pi}{2}\left(R_e^4 - R_i^4\right) \Rightarrow \frac{I_p}{R} = \frac{I_p}{R_e} = \frac{\pi}{2R_e}\left(R_e^4 - R_i^4\right)\ .$$

Tubes lead to greater material economy in the resistance to the torque than solid circular cross-sections, since the material around the centre makes a small contribution to the torsional strength. In fact, in this region the stress is always low $(\tau = \frac{T}{I_p}r)$ and the moment arm of these stresses in relation to the bar axis is small. Also in relation to this, there is a clear analogy with the case of plane bending (see the considerations about the rational shape of cross-sections in bending, described in the final part of Sect. VII.3).

As a consequence of the reciprocity of shearing stresses, the same shearing stress acts in facets belonging to planes which contain the bar axis as in the cross-section, as represented in Fig. 141-a. Since the other components of the stress tensor vanish,[1] we have in every point a purely deviatoric plane state of stress, whose principal directions make 45°-angles with the fibres and are in the plane perpendicular to the cross-section radius passing through the point under consideration. As a consequence, the principal stress trajectories on the surface of the bar, or in interior concentric cylindrical surfaces, have the shape of a helix at 45°, as represented in Fig. 141.

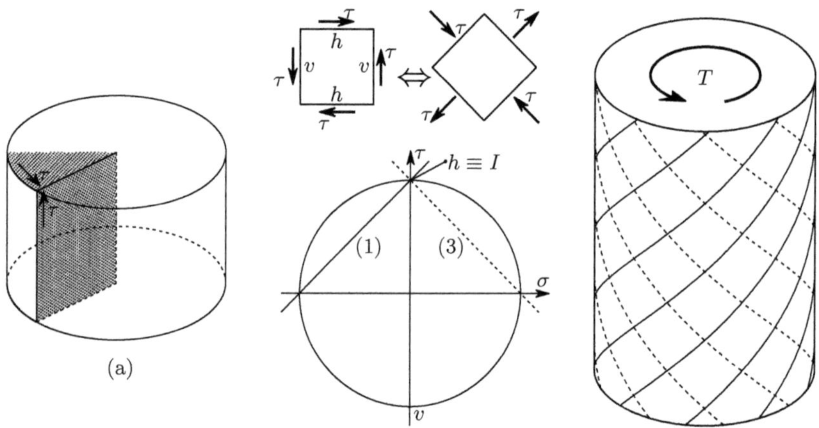

Fig. 141. Principal stress trajectories on the surface of a cylindric bar under a positive torsional moment: —— tensile trajectories; ······ compressive trajectories

In structural members made of brittle materials, if one of the principal stresses is positive, failure is usually caused by the tensile principal stress and the rupture lines coincide with the compressive trajectories. This is easily

---

[1]This is only valid if the shearing strains corresponding to the cross-section rotation are sufficiently small to be considered infinitesimal. In this case, if we consider the deformations caused by the shearing stress in the tubes with infinitesimal thickness dr separately (Fig. 140-c), we conclude that their deformations are compatible, since the shearing stress does not cause changes in the length or diameter of the tubes.

On the contrary, in the case of finite rotations, the inclination of the fibres with respect to the bar axis, caused by the finite shearing strains, leads to shortening of the tubes. This shortening vanishes in the center and reaches a maximum in the contour of the cross-section. Since the cross-section must remain plane, as required by the symmetry considerations set forth above, additional normal stresses must appear in the cross-section, and these are tensile in the outer region and compressive in the inner region. These stresses will also cause diameter and thickness changes in the different cylinders, so that other normal stresses will appear in the radial and circumferential directions, in order to make the transversal deformations compatible. Thus, we conclude that, in the case of non-infinitesimal rotations of the cross-sections, a shortening of the bar and a diameter variation of its cross-section take place.

confirmed using, for example, a piece of classroom chalk, where the torsional failure surface caused by a positive torque follows the dashed lines represented in Fig. 141.

The expressions developed above are valid for materials with linear elastic behaviour and with symmetric rheological properties with respect to the bar's axis, even in the case of non-isotropic materials, since the only element of the constitutive law which is used is the relation between shearing stress and distortion. Thus, they are also valid for a monotropic material, provided that the monotropy direction is parallel to the bar axis. These conditions occur in a wood cylinder. This material has a lower shearing strength in the longitudinal planes, so that torsion failure occurs under the action of the shearing stresses acting in the fibres' direction in longitudinal planes containing the bar's axis.

In a composite bar where two materials are arranged concentrically, as in the case depicted in Fig. 142, the axial symmetry conditions still hold. Thus, the theory of cylindrical bars is easily generalized to the torsion of these bars. The computation of the stresses caused by torsion is, however, a statically indeterminate problem, since the torque carried by each material is proportional to the corresponding torsional stiffness. Both the force and the displacement methods may be used. In the case of the displacement method, the unit rotation θ may be used as the kinematic unknown. Since this rotation is the same in the two parts, we have, denoting by T_a and T_b the torque carried by the materials a and b, respectively

$$\begin{cases} \theta = \dfrac{T_a}{G_a I_{pa}} = \dfrac{T_b}{G_b I_{pb}} \\[2ex] T = T_a + T_b \end{cases} \Rightarrow \begin{cases} T_a = \dfrac{G_a I_{pa}}{G_a I_{pa} + G_b I_{pb}} T & \Rightarrow & \tau_a = \dfrac{T_a}{I_{pa}} r \\[2ex] T_b = \dfrac{G_b I_{pb}}{G_a I_{pa} + G_b I_{pb}} T & \Rightarrow & \tau_b = \dfrac{T_b}{I_{pb}} r \\[2ex] \theta = \dfrac{T}{G_a I_{pa} + G_b I_{pb}} \, . \end{cases}$$

In slender members with a non-constant torsional moment and/or non-constant cross-section diameter the condition of antisymmetry (i.e., symmetry of rotation around axis y, cf. Fig. 139) is no longer valid, so that we may not conclude that the diameters must remain straight. However, if the variation is slow, the above theory of torsion in circular cross-sections may be considered as approximately valid. In this case, the relative rotation φ of two cross-sections located at a distance l from each other may be computed by the expression

$$\mathrm{d}\varphi = \theta\, \mathrm{d}z = \frac{T}{G I_p}\, \mathrm{d}z \Rightarrow \varphi = \frac{1}{G} \int_0^l \frac{T}{I_p}\, \mathrm{d}z \, . \tag{237}$$

X.2.a Torsion in the Elasto-Plastic Regime

In prismatic bars with a circular cross-section under uniform torsion it is easy to analyse the stresses and strains in the case of non-linear material

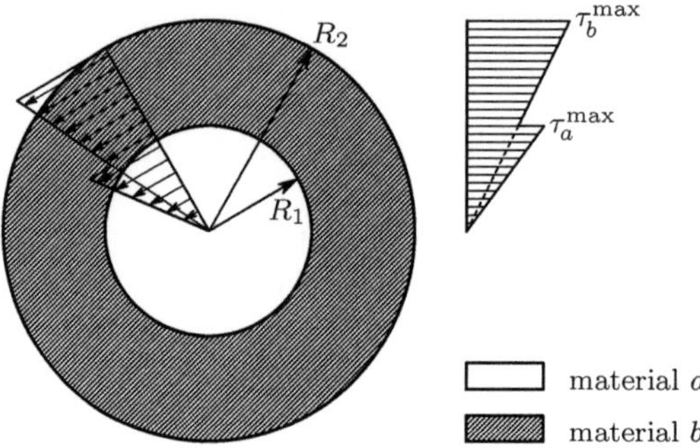

Fig. 142. Stresses caused by torsion in a composite circular cross-section $(G_a > G_b)$: $I_{pa} = \frac{\pi R_1^4}{2} \Rightarrow \tau_a^{\max} = \frac{2T_a}{\pi R_1^3}$; $I_{pb} = \frac{\pi}{2}\left(R_2^4 - R_1^4\right) \Rightarrow \tau_b^{\max} = \frac{2R_2 T_b}{\pi\left(R_2^4 - R_1^4\right)}$

behaviour, such as when the yielding shearing strain is exceeded in the points farthest from the centre of the cross-section. Considering the most simple elasto-plastic behaviour – elastic perfectly plastic behaviour – the relation between the shearing stress and the shearing strain takes the same form as the relation between the normal stress and the longitudinal strain, defined in Fig. 61.

The linear relation between the unit rotation θ and the distortion γ (233) is not affected by the yielding, since it is a purely kinematic relation. Thus, the distribution of shearing stresses takes the form given in Fig. 143-a.

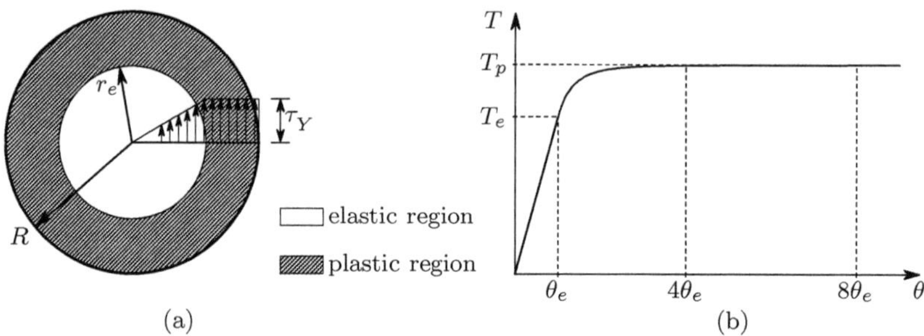

Fig. 143. Torsion in the elasto-plastic regime: **(a)** shearing stress distribution; **(b)** torque-rotation diagram

The twisting moment corresponding to the stress distribution represented in Fig. 143-a may be expressed as a function of the radius of the elastic region, r_e, by computing the moment of the shearing stresses with respect to the bar axis. The component of the twisting moment which corresponds to the elastic region may be obtained from (236), if R is substituted by r_e and $\tau_{\max}$ by τ_Y. Since, in the plastic zone, the shearing stress is constant and equal to the material's yielding stress τ_Y, the total torque is

$$
T = \tau_Y \frac{\pi r_e^3}{2} + \int_{r_e}^{R} r \times \tau_Y 2\pi r\,\mathrm{d}r = \tau_Y \left[\frac{\pi r_e^3}{2} + \frac{2\pi}{3}\left(R^3 - r_e^3 \right) \right]
$$
$$
= T_e \left[\frac{4}{3} - \frac{1}{3}\left(\frac{r_e}{R} \right)^3 \right] \quad \text{with} \quad T_e = \frac{\pi R^3}{2}\tau_Y .
$$
(238)

T_e represents the maximum torque that can be resisted by the cross-section in the elastic regime. From this expression we conclude immediately that, when r_e goes to zero (total plastification of the cross-section), the torque takes the value

$$
r_e = 0 \Rightarrow T = T_p = \frac{4}{3}T_e .
$$

T_p represents the yielding torque of the cross-section. By means of (233) r_e may be expressed as a function of the unit rotation θ, yielding (γ_Y is the yielding distortion)

$$
r = r_e \Rightarrow \gamma = \gamma_Y \Rightarrow \gamma_Y = \theta r_e \Rightarrow r_e = \frac{\gamma_Y}{\theta} .
$$

The maximum unit rotation in the elastic regime may also be expressed as a function of γ_Y. From (233) we get

$$
r = R \Rightarrow \gamma = \gamma_Y \Rightarrow \theta = \theta_e = \frac{\gamma_Y}{R} .
$$

By eliminating γ_Y from these two expressions and substituting r_e in (238), we get a relation between the torque T and the unit rotation θ

$$
\begin{cases} r_e = \dfrac{\gamma_Y}{\theta} \\[2mm] \theta_e = \dfrac{\gamma_Y}{R} \end{cases} \Rightarrow r_e = R\frac{\theta_e}{\theta} \Rightarrow T = T_e \left[\frac{4}{3} - \frac{1}{3}\left(\frac{\theta_e}{\theta} \right)^3 \right] .
$$

Figure 143-b shows this relation. From it, we conclude that the transition from the elastic phase to yielding is fast, since for $\theta = 4\theta_e$, T is already practically coincident with T_p. Solving this equation in order to θ, we get ($\theta_e = \frac{\gamma_Y}{R} = \frac{\tau_Y}{GR}$; $T_e = \frac{\pi R^3}{2}\tau_Y$)

$$
\theta = \frac{\tau_Y}{GR}\left(4 - \frac{6T}{\pi R^3 \tau_Y} \right)^{-\frac{1}{3}} .
$$
(239)

If the bar is unloaded after torsion in the elasto-plastic regime, residual deformations remain as a consequence of the plastic distortions. In the same way as in the case of bending (see Sect. VII.10.c), the unloading is accompanied by an elastic deformation recovery. The superposition of the stresses corresponding to loading (Fig. 144-a) and unloading (Fig. 144-b) yields the residual shearing stresses depicted in Fig. 144-c.

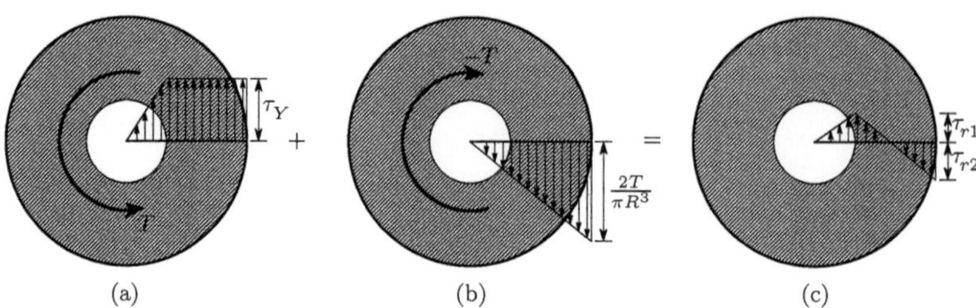

Fig. 144. Residual stresses after torsion in the elasto-plastic regime

The extreme values of the residual stresses, τ_{r1} and τ_{r2} (Fig. 144-c), are then

$$\tau_{r1} = \tau_Y - \frac{2T}{\pi R^3} \times \frac{r_e}{R} = \tau_Y - \frac{2T}{\pi R^3}\left(4 - \frac{6T}{\pi R^3 \tau_Y}\right)^{\frac{1}{3}}$$

$$\tau_{r2} = \frac{2T}{\pi R^3} - \tau_Y \, .$$

The last equality in the expression of τ_{r1} has been obtained by eliminating $\frac{r_e}{R}$, using (238) for the purpose.

The residual unit rotation θ_r may be obtained by subtracting the rotation recovery in the unloading ($\frac{T}{GI_p}$) from the rotation corresponding to the loading phase θ. From (239) we get

$$\theta_r = \theta - \frac{T}{GI_p} = \frac{\tau_Y}{GR}\left(4 - \frac{6T}{\pi R^3 \tau_Y}\right)^{-\frac{1}{3}} - \frac{2T}{G\pi R^4} \, .$$

This expression can also be obtained from the residual stress τ_{r1} (Fig. 144-c). In fact, we have $\tau_{r1} = G\theta_r r_e$, since the plastic distortion is zero, in the points where τ_{r1} occurs.

X.3 Closed Thin-Walled Cross-Sections

The shearing stresses at the cross-section's boundary must take the direction of the border line, as described in Sect. X.1. As the point under consideration moves to the interior of the cross-section, the magnitude and the direction of the shearing stress change, but they must do so continuously, unless some material discontinuity arises, such as the interface of two materials in the case of composite bars.[2] However, it is obvious that, if the stress function has a zero value at a point, the stress may have different directions at infinitesimal distances on either side of that point.

----

[2] A discontinuity in the shearing stress in a homogeneous cross-section would cause a discontinuity in the distortion, which would mean that the material has a crack in this point.

In the case of thin-walled cross-sections, if the wall thickness is small, it seems reasonable to admit that *the shearing stress remains nearly constant in the thickness direction*, since it does not have prescribed values at the boundary, in the same way as in the case of the shearing stresses caused by the shear force (cf. Subsect. VIII.3.c). In order to investigate the conditions in which this possibility exists, let us consider the two thin-walled cross-sections depicted in Fig. 145 under the action of a twisting moment.

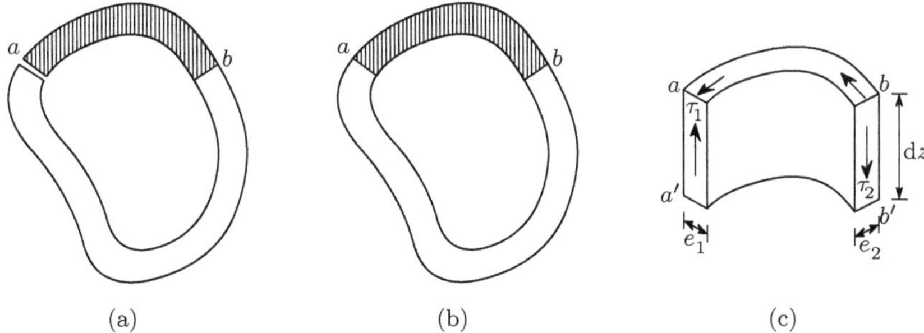

Fig. 145. Shearing stresses caused by a twisting moment in open and closed thin-walled cross-sections

The two cross-sections have the same geometry, but one of them is open (Fig. 145-a) and the other is closed (Fig. 145-b). Let us consider now the equilibrium condition in the longitudinal direction of the bar of the forces acting in the prism defined by the hatched area of the cross-section, ab, with the infinitesimal height dz (Fig. 145-c). In the longitudinal surfaces aa' and bb' shearing stresses with the same magnitude as in the cross-section are acting, as required by the condition of reciprocity of shearing stresses. The equilibrium condition in direction z requires that the resultants of the stresses acting in the surfaces aa' and bb' are equal and have opposite directions.[3]

In the case of the open cross-section, there are no stresses acting on surface aa'. Thus, the equilibrium condition in direction z implies that the stresses acting on surface bb' have a vanishing resultant force. If this force is zero, the shearing stress cannot be constant in the thickness, unless the stress vanishes in every point of the cross-section (b is a generic point of the wall centre line),

___

[3]It can be shown that, for infinitesimal deformations, the normal stresses in the cross-section vanish in the case of uniform torsion. We can also reach this conclusion by means of the argument set forth in Footnote 65. In fact, as infinitesimal distortions do not cause a change in the length of the fibres, and the normal stresses are not necessary to balance the torsional moment, they will vanish, since there is no need to make longitudinal deformations compatible. However, even if such normal stresses were to exist, they would be equal in all cross-sections (uniform torsion), so that their resultant in the prism considered would vanish.

which would imply a vanishing torque. Therefore, the above approximation cannot be used for the case of open thin-walled cross-sections under torsion.

In the case of the closed cross-section, however, the stresses acting on surface aa' may have a non-zero resultant and the equilibrium condition in direction z leads to the conclusion that the stress resultants in surfaces aa' and bb' are equal. Therefore, there is no objection to the validity of the simplifying hypothesis that the shearing stress may be accepted as constant in the wall thickness. This assumption is confirmed by the solutions obtained by means of the Theory of Elasticity and also by the membrane analogy (Subsect. X.4.c).

According to these considerations, in the case of closed thin-walled cross-section, we may define a *shear flow*, $f = \tau e$, which is constant along the whole centre line of the cross-section wall ($\tau_1 e_1 \, dz = \tau_2 e_2 \, dz$, Fig. 145-c). On the basis of the hypothesis above, by the end of the 19th century Bredt had developed a theory for the computation of stresses and deformations in bars with closed thin-walled cross-section under torsion [12].

The relation between the torque and the shearing stress may be obtained by simple equilibrium considerations. To this end, let us consider an infinitesimal length ds of the centre line (Fig. 146). The moment of the force acting in the rectangle of length ds and thickness e, with respect to an arbitrary point O of the surface limited by the centre line of the wall is given by the expression

$$dT = \tau e \, ds \, r = f \, ds \, r \, .$$

As the infinitesimal moments dT all take the same direction, the relation between the torque T and the shearing stress may be obtained by the integration

$$T = \oint f r \, ds = f 2A \;\Rightarrow\; f = \frac{T}{2A} \;\Rightarrow\; \tau = \frac{T}{2Ae}$$
$$\text{with} \quad A = \frac{1}{2} \oint r \, ds \, . \tag{240}$$

The quantity A represents the area of the surface limited by the wall's centre line, since $\frac{1}{2} r \, ds$ is the area of the vertically hatched triangle in Fig. 146. It can be easily shown that, even where point O is not in the interior of the surface limited by the wall's centre line, or in the case of large concavities where the infinitesimal moments dT do not take all the same direction, the equality $\oint r \, ds = 2A$ remains valid.

Equation (240) is known as *Bredt's first formula*. Is has been developed without needing any reference to the material's rheological properties. Thus, it is valid for linear or non-linear material behaviour, elastic or plastic deformation, isotropic or anisotropic materials, material homogeneity or heterogeneity, etc.[4] This means that the assumption of constant shearing stresses

---

[4] In the case of a composite wall (two materials in the thickness) Bredt's first formula is only valid in terms of the shear flow, $f = \frac{T}{2A}$, since the shearing stress cannot be constant in the thickness, if the shear modulus of the two materials is not the same (see example X.6).

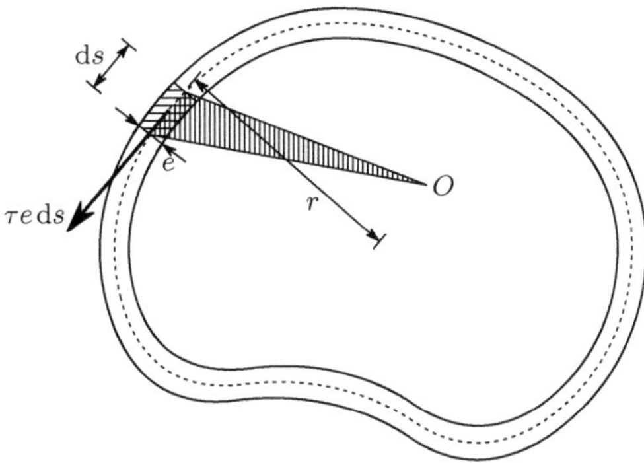

Fig. 146. Determination of the shearing stress caused by the twisting moment

in the thickness renders the problem statically determinate, since the stress distribution does not depend on the rheological behaviour of the material or on the way the bar deforms.

To determine the deformation caused by torsion the material behaviour must obviously be taken into consideration. Furthermore as the way the bar deforms under torsion is not defined *a priori* and thus we do not know what part of the distortion causes the bar to rotate around its axis (torsional deformation) and what part causes warping of the cross-section, the unit torsional rotation θ cannot be obtained by means of geometrical considerations. This is a similar problem to that of computing the deflections caused by the shear force (Sect. IX.2) and may be solved in the same way, i.e., by means of the principle of energy conservation. If the material has elastic behaviour, the work done by the torque in the torsional deformation is equal to the elastic potential energy stored by the distortions. If the material behaviour is linear, these two quantities are $W = \frac{1}{2}T\theta$ and $U = \frac{1}{2}\int_\Omega \tau\gamma\,d\Omega$, per unit length of the bar, respectively for the work and the energy. Thus, the principle of energy conservation leads to the conclusion ($d\Omega = e\,ds$)

$$W = U \Rightarrow T\theta = \int_\Omega \frac{\tau^2}{G}\,d\Omega = \oint \frac{\tau^2 e}{G}\,ds \Rightarrow \theta = \frac{1}{GT}\oint \tau^2 e\,ds \ .$$

Substituting the shearing stress given by (240) in this expression, we get the unit rotation θ as a function of the torque, the shear modulus G and the geometrical properties of the cross-section, which are represented by the quantity J,

$$\theta = \frac{T}{4GA^2}\oint \frac{ds}{e} = \frac{T}{GJ} \quad \text{with} \quad J = \frac{4A^2}{\oint \frac{ds}{e}} \ . \tag{241}$$

This expression is known as *Bredt's second formula*. The product GJ represents clearly the torsional stiffness. The integral $\oint \frac{ds}{e}$ is easily obtained considering wall segments with constant thickness. In the particular case of a constant thickness, we have

$$J = \frac{4A^2 e}{p} \, .$$
(242)

where p represents the perimeter of the wall's centre line. The unit rotation θ, in fact, depends only on the distribution of shearing strains in the cross-section. The relation between the distortions γ and θ may be obtained from (240) and (241), yielding

$$\theta = \underbrace{\frac{T}{2A}}_{f} \frac{1}{2GA} \oint \frac{\mathrm{d}s}{e} = \frac{1}{2GA} \oint \frac{f}{e} \, \mathrm{d}s = \frac{1}{2A} \oint \frac{\tau}{G} \, \mathrm{d}s = \frac{\oint \gamma \, \mathrm{d}s}{2A} \, .$$
(243)

Although this relation has been obtained by considering a linear elastic material behaviour ($\tau = G\gamma$), it is valid for any rheological behaviour, since it is a purely kinematic relation, i.e., it relates only kinematic quantities: the distortions γ and the rotation θ. The principle of energy conservation, assuming that the material has a linear elastic behaviour, is only a means to achieve this relation. Since this relation is a purely kinematic one, it may be used, regardless of any considerations about the forces which cause the distortions.[5] It may therefore be used in tubular thin-walled bars made of materials which do not obey Hooke's law: once the stresses are computed by means of the Bredt's first formula, the distortions may be computed by means of the constitutive law of the material. The unit rotation may then be obtained using (243). It can also be used to compute the rotation of closed parts (channels) of thin-walled cross-sections with a degree of connection superior to two, although the shearing stresses acting in the walls of a channel may not obey the equilibrium conditions (see example X.8).

If the closed thin-walled cross-section has open parts, i.e., simply connected wall elements, as in the cross-section depicted in Fig. 115, the integral in (241) obviously refers to the closed part of the cross-section. Thus, the area A is the area limited by the closed part of the centre line. In this kind of cross-section the contribution of the simply-connected wall elements to torsional strength is much smaller than the contribution of the closed part of the wall, as will be seen in Sect. X.4 by means of the membrane analogy (see example X.11).

The longitudinal connections of closed thin-walled built-up bars, as in the cases represented in Fig. 147, must resist a force per unit length which is equal to the shear flow $f = \tau e = \frac{T}{2A}$, as may easily be concluded from the longitudinal condition of equilibrium in Fig. 145-c.

---

[5]This conclusion is a particular form of the theorem of virtual forces which is widely used in structural analysis in the systematization of the force method. According to this theorem, the forces and stresses used to relate strains to displacements (in this case γ and θ), using energy considerations, do not need to be the actual ones, but only to satisfy the equilibrium conditions (see Subsect. XII.4.b). The relation between strains and displacements is purely kinematic, since, in a supported body, once the strain distribution is known, the corresponding displacements may be obtained by integrating the strain functions along the body.

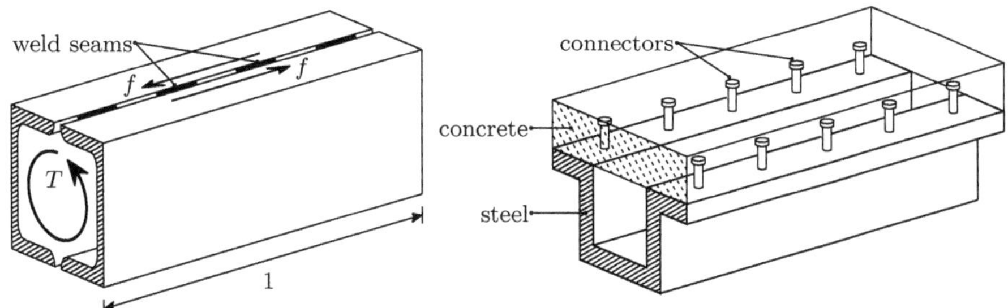

Fig. 147. Connection of closed thin-walled built-up bars under torsion

X.3.a Applicability of the Bredt Formulas

Bredt's theory has been developed on the basis of a simplifying assumption for the stress distribution, which gets closer to the actual distribution, when the wall becomes less thick. Thus, in the practical use of these formulas, this question must be raised: until what values of the relation between the wall thickness and cross-section dimensions is the error introduced by the simplifying hypothesis small enough to be disregarded ? A generally accurate answer to this question is hard to find, since the distribution of the shearing stresses caused by torsion strongly depends on the shape of the cross-section. However, the analysis of a particular case, where an exact solution is known, may give an idea about the limits of applicability of Bredt's theory.

For a circular tube the theory of circular bars explained in Sect. X.2 yields an exact solution for the stresses and the unit rotation induced by torsion. This solution may then be compared with the approximate solution furnished by Bredt's theory. The inner and outer radius of the cross-section, R_i and R_e, respectively, may be expressed by

$$R_i = r - \frac{e}{2} \qquad \text{and} \qquad R_e = r + \frac{e}{2} \,,$$

where r and e represent the mean radius and the wall's thickness, respectively. The polar moment of inertia is then

$$I_p = \frac{\pi}{2} \left(R_e^4 - R_i^4 \right) = \frac{\pi}{2} \left[\left(r + \frac{e}{2} \right)^4 - \left(r - \frac{e}{2} \right)^4 \right] = 2\pi r^3 e + \frac{\pi}{2} r e^3 \,.$$

The exact values of the maximum shearing stress and of the unit rotation are then given by the expressions

$$\tau_{\max}^C = \frac{T R_e}{I_p} = T \frac{r + \frac{e}{2}}{2\pi r^3 e + \frac{\pi}{2} r e^3} = \frac{T}{\pi} \frac{2r + e}{r e \left(4r^2 + e^2 \right)}$$

and

$$\theta^C = \frac{T}{G I_p} = \frac{T}{G \left(2\pi r^3 e + \frac{\pi}{2} r e^3 \right)} = \frac{T}{G\pi} \frac{2}{r e \left(4r^2 + e^2 \right)} \,.$$

Bredt's theory gives, for the same quantities, the expressions

$$\tau_{\max}^{B} = \frac{T}{2Ae} = \frac{T}{2\pi r^2 e} = \frac{T}{\pi}\frac{1}{2r^2 e}$$

and

$$\theta^{B} = \frac{Tp}{4GA^2 e} = \frac{T2\pi r}{4G\pi^2 r^4 e} = \frac{T}{G\pi}\frac{1}{2r^3 e}.$$

Defining a parameter α as the relation between the thickness e and the mean radius r, the relations between the exact and the approximate solutions are given by

$$\beta = \frac{\tau_{\max}^{C}}{\tau_{\max}^{B}} = \frac{2r^2 e\,(2r + e)}{re\,(4r^2 + e^2)} = \frac{4 + 2\alpha}{4 + \alpha^2}$$

and

$$\gamma = \frac{\theta^{C}}{\theta^{B}} = \frac{4r^3 e}{re\,(4r^2 + e^2)} = \frac{4}{4 + \alpha^2}.$$

The error depends only on parameter α. Computing the values of β and γ for some values of α, we get

$\alpha = e/r$	0	0.05	0.1	0.15	0.25	0.5	1
$\beta = \tau_{\max}^{C}/\tau_{\max}^{B}$	1.0000	1.0244	1.0474	1.0690	1.1077	1.1765	1.2000
$\gamma = \theta^{C}/\theta^{B}$	1.0000	0.9994	0.9975	0.9944	0.9846	0.9412	0.8000

We conclude that the error of the maximum stress starts to be important for relatively low values of parameter α ($\alpha > 0.1$), while the error associated to the rotation remains small until larger values of parameter α ($\alpha > 0.5$).

As mentioned at the beginning of this section, it is not easy to generalize these conclusions to other cross-section shapes. However, they show that, even for relatively small ratios between the wall thickness and the cross-section dimensions, considerable errors may be introduced when the maximum stress is evaluated by means of the first Bredt's formula. The error may be even larger if the cross-section has sharp concavities due to a stress concentration in those points, as will be seen in Sect. X.4. However, if the bar is made of a ductile material and there is no fatigue risk, the stress concentration may be disregarded.

X.4 General Case

X.4.a Introduction

In the general case of a non-circular cross-section warping takes place, i.e., the cross-section does not remain plane. This is easily confirmed by taking, for example, a rubber prism with a rectangular cross-section, where vertical and horizontal lines were drawn before a torsion deformation, as represented in Fig. 148. The warping occurs because the shearing stress acting in points which are close to the boundary must be parallel to the border line. In fact, if the cross-section were to remain plane and perpendicular to the bar axis, the shearing stress would be perpendicular to the radius joining the torsion centre with the point under consideration. Only in the case of circular cross-sections is this vector tangent to the boundary (see exercise X.14).[6] However, Saint-Venant has shown that, for infinitesimal deformations, the projection of the warped cross-section in the plane perpendicular to the bar's axis rotates as a rigid surface in the torsion deformation. On the basis of this statement he developed a general torsion theory for prismatic bars made of isotropic materials with linear elastic behaviour under infinitesimal deformations. The explanation of this theory is beyond the scope of this book, however (for a brief introduction, see, e.g., [2])

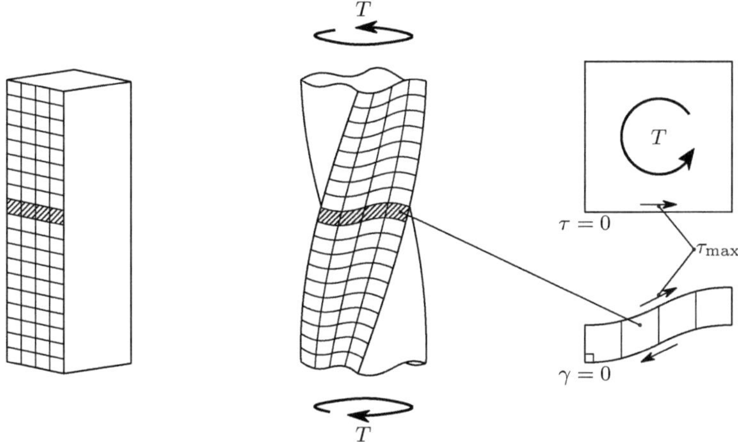

Fig. 148. Deformation of a rubber prism under a positive twisting moment

The conditions of symmetry and equilibrium allow some conclusions to be drawn. For example, if the cross-section has convex corners, the shearing stress will vanish there, since the stress vector cannot be parallel to two different directions.[7] In the example of Fig. 148, the symmetry of the bar with respect to a plane containing its axis and perpendicular to the lateral faces, and the

---

[6]It may be shown that only circular cross-sections can remain plane, even where the condition of perpendicularity with respect to the bar axis is not stated.

[7]Apparently, this line of reasoning could also be applied to concave corners. However, instead of that, the theoretical solution of the problem indicates a singularity which consists of an infinite value for the stress and a sudden change in the direction of the stress vector. These stress concentrations are described below.

antisymmetry of the deformation with respect to the same plane, lead to the conclusion that the maximum distortion occurs at the mid-point of the faces.

Conclusions like these are however more easily drawn with the help of *physical analogies*. We have such an analogy between two physical phenomena, when the mathematical expressions which describe them have the same form, while containing different physical entities.

The distribution of stresses in the interior of a bar under torsion is not an intuitive phenomenon, since stresses can only be observed indirectly, either by means of the deformations they cause on the surface of the bar, other by more or less sophisticated experimental methods. Other phenomena, described by equations with the same form as the differential equations of torsion, as some liquid flows, or the transversal deformation of a membrane under some conditions, are much easier to visualize, which allows qualitative conclusions about the distribution of shearing stresses caused by torsion to be drawn by analogy with those phenomena. Two of these analogies are briefly described below.

X.4.b Hydrodynamical Analogy

There are several analogies between liquid flows and the distribution of shearing stresses in the cross-section of a prismatic bar under torsion. The most widely used is the Greenhill's hydrodynamical analogy of [4].

Let us consider a vessel with a horizontal bottom and vertical walls, whose cross-section is the same as that under consideration, which contains a perfect liquid, (i.e., incompressible and without viscosity). The liquid has a uniform rotation motion (i.e., with the same rotation in every point).

Greenhill has found that the differential equations describing the motion of the liquid have the same shape as the equations which describe the torsion problem. In the torsion equations stress plays the same role as the velocity vector in the equations corresponding to the liquid motion. The boundary conditions are also the same in both problems, since both the velocity vector and the shearing stress vector must be tangent to the border line. This analogy yields some useful conclusions about the distribution of shearing stresses. For example, we can easily see that the maximum shearing stress in the torsion of a bar with rectangular cross-section occurs at the mid-point of the larger sides, since there the distance between the flow lines is smaller, which means that the velocity of the flow is larger (Fig. 149-a). In the cross-section in Fig. 149-b it is obvious that the flow lines get very close to each other in the regions around the concavities, especially if they have sharp corners, which means that there is a large stress concentration at these points.[8]

---

[8]As mentioned in Footnote 71, the mathematical equations indicate that in concave angle points the stress becomes infinite, so that this kind of concavity should be avoided, especially in the case of brittle materials or cyclic loading.

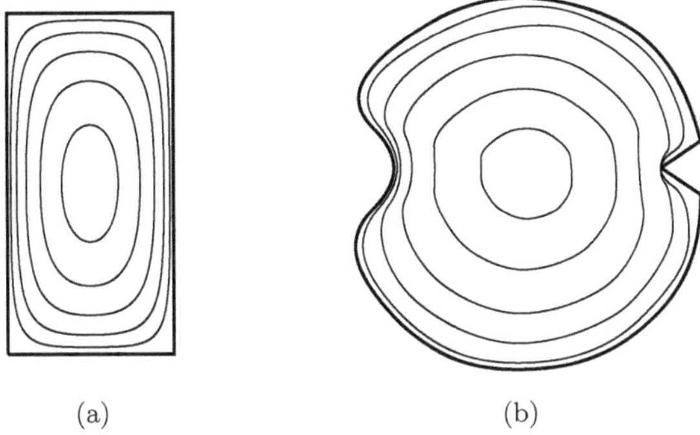

(a)   (b)

Fig. 149. Flow lines in the uniform rotation of a liquid

X.4.c Membrane Analogy

The membrane analogy was introduced by Prandtl (1903), [2],[4],[7]. He found that the equations for torsion on a prismatic bar with a simply-connected cross-section have the same shape as the differential equations which describe the deflections of an initially plane, elastic, homogeneous membrane[9] fixed to a plane frame with the shape of the cross-section, under a uniform pressure, provided that the deflections are small, or, more precisely, that its slope is small enough for the curvature to be described as the second derivative of the deflection (Fig. 150).

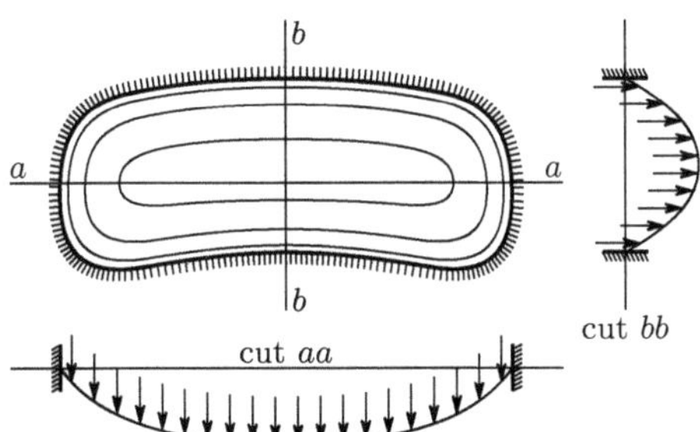

Fig. 150. Level curves of a membrane, deformed by a uniform pressure

The shearing stress at a point takes the direction of the membrane's level curves and is proportional to the steepest slope at that point. The closer the level curves are, the larger is the steepest slope. The flow lines represented in Fig. 149 may also represent the level curves in the membrane defined above

---

[9]The membrane has a constant thickness and is made of a material which does not resist shearing stresses. This means that the stress state is isotropic at any point of the membrane. These conditions are found in a soap film, for example.

for those cross-sections. Thus, by means of the membrane analogy, the stress distribution, and particularly the stress concentrations, become even more evident than in the case of the hydrodynamical analogy. Furthermore, the membrane analogy also gives information about the magnitude of the torque, since it is proportional to the volume limited by the undeformed (plane) and deformed membrane configurations.

In the case of a closed thin-walled cross-section with degree of connection two, the analogous membrane is that defined by the vertically hatched surface (1) of the cut represented in Fig. 151-b, since the analogous membrane may be obtained by removing from the membrane corresponding to the full cross-section, the membrane corresponding to the inner channel (hatched surface 2). In fact, if the wall thickness is small, we may accept that the inner contour of the thin-walled cross-section coincides with a level curve of the membrane corresponding to the full cross-section. Since the level curve takes the direction of the shearing stress, the stress vanishes in the facets perpendicular to the cross-section which take the direction of level curves. Therefore, the cross-section corresponding to the empty space and the shearing stresses acting on it may be removed without altering the equilibrium conditions in the remaining cross-section. In terms of membrane analogy, this is equivalent to removing the membrane corresponding to the empty space in the cross-section (hatched surface 2, Fig. 151-b).

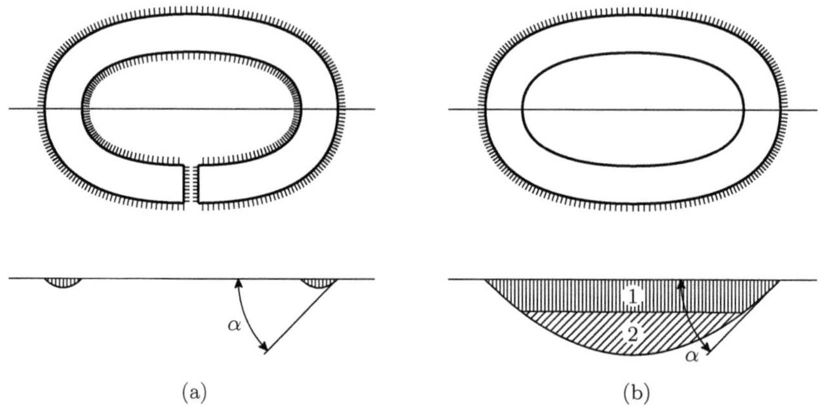

Fig. 151. Membrane analogy in a thin-walled cross-section: **(a)** open cross-section **(b)** closed cross-section

We confirm that in a closed thin-walled cross-section the stress may be considered constant in the thickness, since the slope of the membrane is approximately constant, while in the open cross-section the slope varies and changes sign in the thickness, vanishing in the middle region of the cross-section wall. Furthermore, we can easily ascertain that, for the same maximum stress, $\tau_{\max} = k \tan \alpha$ (Fig. 151), the torque corresponding to the open section is much smaller than that corresponding to the closed section, as we conclude from comparing the two hatched areas in Fig. 151-a with the hatched area 1 in Fig. 151-b.

X.4.d Rectangular Cross-Sections

The problem of torsion in prismatic bars with a rectangular cross-section has been solved by means of the Theory of Elasticity. The solution is given in the form of infinite series. The unit rotation θ may be computed by the expression [4]

$$\theta = \frac{T}{c_1 G h b^3} \quad \text{with} \quad c_1 = \frac{1}{3} - \frac{64}{\pi^5} \frac{b}{h} \sum_{n=1,3,5,\ldots}^{\infty} \frac{1}{n^5} \tanh\left(\frac{n\pi}{2}\frac{h}{b}\right) .$$

The maximum shearing stress which, as mentioned above, occurs in the mid-points of the larger sides (length h), is given by the expression

$$\tau_{\max} = \frac{T}{c_2 h b^2} \quad \text{with} \quad c_2 = c_1 \left[1 - \frac{8}{\pi^2} \sum_{n=1,3,5,\ldots}^{\infty} \frac{1}{n^2 \cosh\left(\frac{n\pi}{2}\frac{h}{b}\right)} \right]^{-1} . \tag{244}$$

Sometimes it is also necessary to compute the shearing stress in the mid-points of the smaller sides, in order to add it to the stress caused by the shear force. This stress may also be computed from (244), by interchanging the roles of b and h in the computation of c_1 and c_2 and taking into consideration that $h^2 b = \left(\frac{h}{b}\right) h b^2$. A coefficient c_3 may then be defined, which substitutes c_2 in the first of (244). Alternatively, the value of c_3 may be computed by means of the expression [4]

$$c_3 = c_1 \frac{\pi^2}{8} \left[\sum_{n=1,3,5,\ldots}^{\infty} \frac{1}{n^2} (-1)^{\frac{n-1}{2}} \tanh\left(\frac{n\pi}{2}\frac{h}{b}\right) \right]^{-1} .$$

The coefficients c_1, c_2 and c_3 depend only on the relation between the height h and the width b of the cross-section ($h > b$). In the following table the values of these coefficients for some h/b-ratios are given.

h/b	1	1.25	1.5	2	3	5	10	∞
c_1	0.1406	0.1717	0.1958	0.2287	0.2633	0.2913	0.3123	0.3333
c_2	0.2082	0.2212	0.2310	0.2459	0.2672	0.2915	0.3123	0.3333
c_3	0.2082	0.2415	0.2689	0.3093	0.3547	0.3924	0.4207	0.4490

X.4.e Open Thin-Walled Cross-Sections

Considering the membrane analogy, we easily verify that both the slope and the volume delimited by the membrane remain approximately constant when the shape of an open thin-walled cross-section is changed, while the length of the centre line and the wall thickness are unchanged. This means that, neither the shearing stresses, nor the corresponding twisting moment are changed. The torsional stiffness remains constant as well. In fact, the unit rotation θ may be obtained by the principle of energy conservation, which implies that the work done by the torque ($W = \frac{1}{2}T\theta$) is equal to the elastic potential energy ($U = \frac{1}{2}\int_\Omega \tau\gamma\,\mathrm{d}\Omega = \frac{1}{2G}\int_\Omega \tau^2\,\mathrm{d}\Omega$). Thus, if the shearing stresses τ and the torque T remain constant, θ will also remain the same.

For these reasons, a thin-walled cross-section with constant thickness may be analysed as a rectangle with a height h equal to the length of the wall centre line l and a width b equal to the wall thickness e. Since in this case we have $h \gg b$, we may take $c_1 = c_2 = \frac{1}{3}$. The maximum shearing stress and the torsional stiffness are then

$$\tau_{\max} = \frac{3T}{le^2}; \qquad \theta = \frac{3T}{Gle^3} = \frac{T}{GJ} \quad \text{with} \quad J = \frac{le^3}{3}. \tag{245}$$

In many cross-sections not all segments of the walls have the same thickness. In these cases, in accordance with the considerations above based on the membrane analogy, we may assume that the torsional stiffness of each segment with constant thickness is the same that it would have if it were not connected to the remaining cross-section. In fact, the analogous membrane of the disconnected segment only differs from the membrane of the connected segment in the connection zone, as represented in Fig. 152. It is easy to understand that, the smaller is the connection zone, i.e., the smaller is the wall thickness, the smaller will be that difference, i.e., the error introduced by the assumption above, as illustrated by the two examples in Fig. 152.

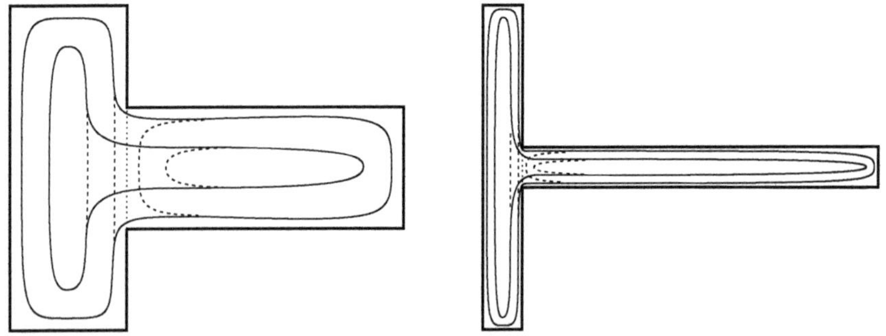

Fig. 152. Level curves of the analogous membrane in cross-sections with two segments: ——— connected segments; ______ disconnected segments

In order to attenuate the stress concentration in concave corners, the usual profile sections have rounded concavities, as in Fig. 153. In accordance with the assumption above, the total torsional stiffness of the cross-section may be obtained by adding the stiffnesses of all segments,[10] and multiplying this sum by a factor which takes the stiffness increase introduced by the connections between the different segments and by the rounding of the concavities into consideration.

$$J_i = \frac{1}{3} l_i e_i^3 \;\Rightarrow\; J = k\frac{1}{3}\sum_{i=1}^{n} l_i e_i^3 \;\Rightarrow\; \theta = \frac{T}{GJ} \,. \tag{246}$$

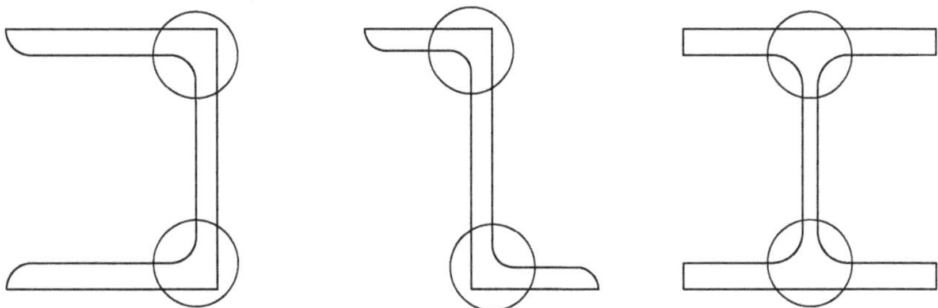

Fig. 153. Rounding the concavities to attenuate the stress concentration in concave corners of profile cross-sections under torsion

The coefficient k takes values between 1.0 and 1.25. Usual values are $k = 1$ for angle sections, $k = 1.1$ for channel- and T-beams, $k = 1.25$ for I- and K-beams, etc. Since the rotation θ is the same in all cross-section segments, the maximum shearing stress in each segment may be obtained from the relation between $\tau_{\max}$ and θ, which may be derived from (245), yielding

$$\tau_i^{\max} = G\theta e_i = \frac{3Te_i}{k\sum_{i=1}^{n} l_i e_i^3} \,. \tag{247}$$

We conclude that the maximum shearing stress in the cross-section occurs in the segment with the greatest thickness.

X.5 Optimal Shape of Cross-Sections Under Torsion

The above considerations relating to Fig. 151 lead to the conclusion that open thin-walled cross-sections have a poor efficiency in terms of resistance

---

[10]The torsional stiffness of a segment is the torque which must be applied to that segment to get a unit value of the rotation per unit length θ. Thus, the total stiffness of the cross-section is the sum of the torsional moments that must be applied to all segments, to introduce that torsional rotation into the bar.

to torsion. In fact, the value of the torque corresponding to a given value of the maximum shearing stress in a thin-walled cross-section is very low, compared with the torque corresponding to a closed cross-section with the same geometry. The torsional stiffness is also much lower in an open cross-section, as we may see, by comparing the values of the geometrical stiffness parameter J given by (242) and (245).[11] Denoting by J_c the value of this parameter in a closed thin-walled cross-section with constant thickness and by J_o its value in an open thin-walled cross-section with the same geometry, we get from those expressions ($\Omega = pe$)

$$\frac{J_c}{J_o} = \frac{\frac{4A^2 e}{p}}{\frac{pe^3}{3}} = \frac{12A^2 e}{p^2 e^3} = 12\frac{A^2}{p^2 e^2} = 12\left(\frac{A}{\Omega}\right)^2 .$$

Since in a thin-walled cross-section the area A of the surface limited by the wall centre line is usually much larger than the cross-section area Ω, we conclude immediately from this expression that the torsional stiffness of a closed cross-section is much higher than that of an open cross-section with the same geometry.

Regarding the solid cross-sections, they may have a good resistance to torsion, but lead generally to an inefficient use of the material, since in the neighbourhood of the point around which the section rotates in the torsion deformation – the torsion centre – the stress is low and the lever arm of the corresponding forces with respect to the same point is also low, as described in the case of circular cross-sections.

From these considerations we conclude that the tubular cross-sections lead to a more efficient use of the material in the resistance to a given torque. As the shear flow is constant, if the wall thickness is constant, the stress will be constant, which corresponds to the optimal stress distribution. In terms of torsional strength, the analysis may be carried out on the basis of (240). Taking into consideration that the area of a constant-thickness thin-walled

---

[11] The torsional stiffness of a bar with an open thin-walled cross-section is considerably increased if the warping of one of its cross-sections is prevented, especially in the case of short bars. In fact, the theory expounded in this chapter is only valid for cross-sections which are far enough from the extremities and points of application of forces to accept the validity of the Saint-Venant principle. Generally, only a small distance is needed, so that this principle may be applied to practically the whole bar. The bars with open thin-walled cross-sections are an exception, since the effects of these singularities are only attenuated slowly, the smaller the wall thickness, compared with the cross-section dimensions, the slower the attenuation (see, e.g., [2], Sect. 10.13).

This effect leads to the appearance of normal stresses in the cross-section, and these vary along the bar, allowing a non-zero shear flow, even in the case of an open cross-section. Besides this, the deformation caused by the twisting moment is much smaller than that given by (246). The analysis of stresses and deformations caused by this effect is, however, beyond the scope of this introductory text, so it is not presented (see, e.g., [12]).

cross-section may be expressed by the product of the thickness e and the perimeter of the wall centre line p, we get, for a given allowable shearing stress τ_{all}, the allowable torque T_{all}

$$T_{\text{all}} = 2\tau_{\text{all}}eA = 2\tau_{\text{all}}\Omega\frac{A}{p} \ .$$

We conclude that, for a constant cross-section's area Ω, the torsional strength depends on the ratio between the area A enclosed by the centre line and the perimeter p. In the case of a constant thickness, p will also be constant. The shape with the highest ratio between area and perimeter is the circle, so that it is the optimal shape. Increasing the diameter and reducing the thickness, so that the cross-section area remains constant, the torsional strength may be increased, since we have

$$T_{\text{all}} = 2\tau_{\text{all}}\Omega\frac{\pi r^2}{2\pi r} = \tau_{\text{all}}\Omega r \ .$$

A limit is imposed on the reduction of the wall thickness by the risk of elastic instability (buckling) caused by the compressive principal stress which acts in the direction defined in Fig. 141.

X.6 Examples and Exercises

X.1. The structure represented in Fig. X.1 is made of a material with a shear modulus G and has a circular cross-section with diameters $2d$ and d respectively in segments AB and BC. In cross-section B the torque M is applied. Determine the distribution of twisting moments using the force and the displacement methods.

Resolution

Force Method

A statically determinate base structure may be obtained by removing the right support. In this case the segment BC does not have internal forces, so that the rotations of cross-sections B and C are equal. Thus, we get

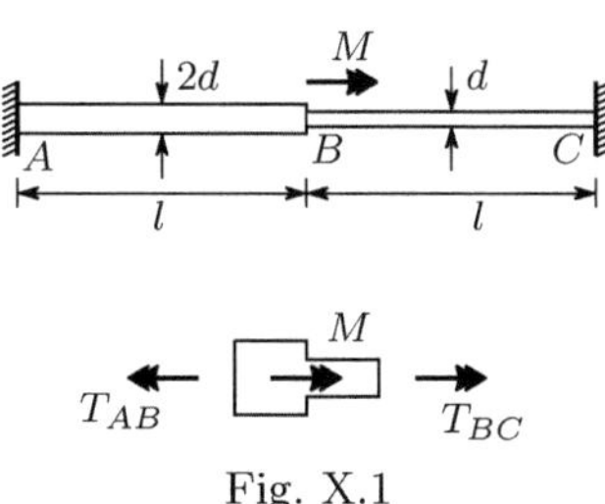

Fig. X.1

$$T_{AB} = M \Rightarrow \varphi_C = \varphi_B = \frac{T_{AB}l}{GI_p^{AB}}$$

$$= \frac{Ml}{G\frac{\pi(2d)^4}{32}} = \frac{2Ml}{G\pi d^4} \ .$$

The hyperstatic unknown is the torque in cross-section C, T_C, which intro-duces a rotation at point C with the value (φ_C' has the opposite direction to φ_C)

$$\varphi_C' = \frac{T_Cl}{GI_p^{AB}} + \frac{T_Cl}{GI_p^{BC}} = \frac{T_Cl}{G}\left(\frac{1}{\frac{\pi(2d)^4}{32}} + \frac{1}{\frac{\pi d^4}{32}}\right) = \frac{34T_{tC}l}{G\pi d^4} \ .$$

The condition of compatibility yields then the value of T_C

$$\varphi_C = \varphi_C' \Rightarrow \frac{2Ml}{G\pi d^4} = \frac{34T_Cl}{G\pi d^4} \Rightarrow T_C = \frac{1}{17}M \ .$$

The twisting moments in the two segments are then

$$T_{AB} = M - T_C = \frac{16}{17}M \ ; \qquad T_{BC} = -T_C = -\frac{1}{17}M \ .$$

Displacement Method

Taking the torsional rotation φ in cross-section B as the kinematic unknown, the twisting moments in segments AB and BC, expressed as functions of φ, are ((237), φ is positive in the direction defined by M)

$$T_{AB} = \frac{GI_p^{AB}}{l}\varphi \qquad T_{BC} = -\frac{GI_p^{BC}}{l}\varphi \ .$$

The condition of equilibrium of moments in node B (Fig. X.1) yields the value of φ ($I_p^{AB} = 16I_p^{BC}$)

$$M = T_{AB} - T_{BC} = \frac{G}{l}\left(I_p^{AB} + I_p^{BC}\right)\varphi \Rightarrow \varphi = \frac{Ml}{G\left(I_p^{AB} + I_p^{BC}\right)} = \frac{Ml}{17GI_p^{BC}} \ .$$

Substituting these values in the expression above, we get

$$T_{AB} = \frac{G16I_p^{BC}}{l}\frac{Ml}{17GI_p^{BC}} = \frac{16}{17}M \qquad T_{BC} = -\frac{GI_p^{BC}}{l}\frac{Ml}{17GI_p^{BC}} = -\frac{1}{17}M \ .$$

X.2. A bar made of a material with a shear modulus G has the closed thin-walled cross-section depicted in Fig. X.2. The cross-section is symmetri-cal with respect to axis BD. The centre line has the shape of a circum-ference. The arcs AB and BC have constant thickness, while in arcs DA and DC the thickness varies as a linear relation of angle α, between a (point D) and $2a$ (points A and C).

Determine the unit rotation θ introduced by a torque T in the bar.

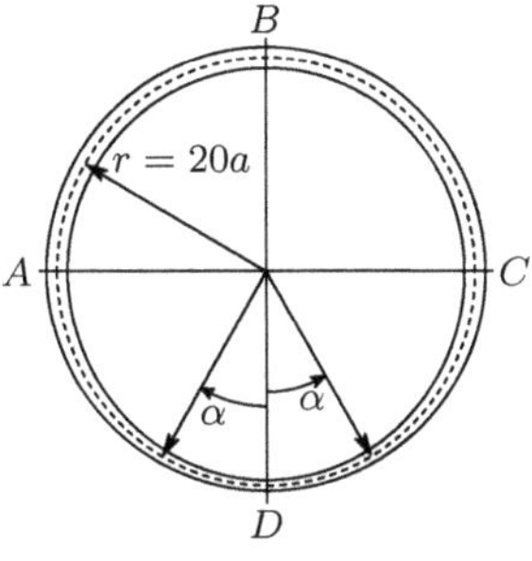

Fig. X.2

Resolution

In arcs DA and DC the thickness may be expressed as a function of angle α by the relation

$$e(\alpha) = a + \frac{2a}{\pi}\alpha \ .$$

The integral $\oint \frac{\mathrm{d}s}{e}$, needed to compute the torsional stiffness, (241) takes the value $(r = 20a)$

$$\oint \frac{\mathrm{d}s}{e} = \frac{\pi r}{2a} + 2\int_0^{\frac{\pi}{2}} \frac{r\,\mathrm{d}\alpha}{a + \frac{2a}{\pi}\alpha}$$

$$= 10\pi + 40\left[\frac{\pi}{2}\ln\left(1 + \frac{2}{\pi}\alpha\right)\right]_0^{\frac{\pi}{2}} = 20\pi\left(\frac{1}{2} + \ln 2\right) \ .$$

Substituting this value and the area limited by the wall's centre line $(A = 400\pi a^2)$ in (241), we get

$$\theta = \frac{T}{4GA^2}\oint\frac{\mathrm{d}s}{e} = \frac{T}{4G\left(400\pi a^2\right)^2}\,20\pi\left(\frac{1}{2} + \ln 2\right) \approx 11.8685 \times 10^{-6}\frac{T}{Ga^4} \ .$$

X.3. The beam illustrated in Fig. X.3-a has the cross-section represented in Fig. X.3-b. The load is in the plane defined by the indicated action axis (a.a.). Determine the rotation of cross-section A around the beam's axis.

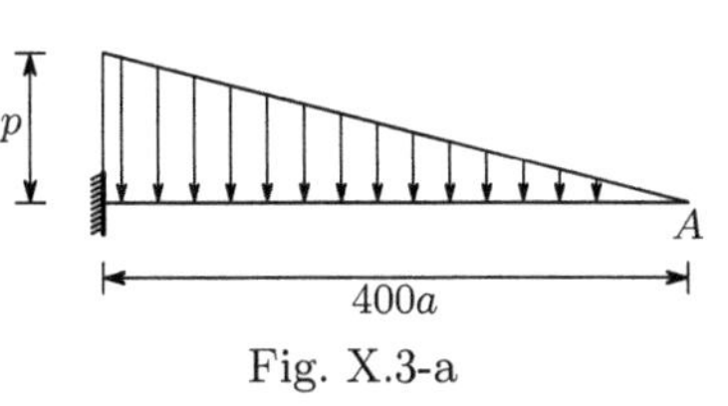

Fig. X.3-a

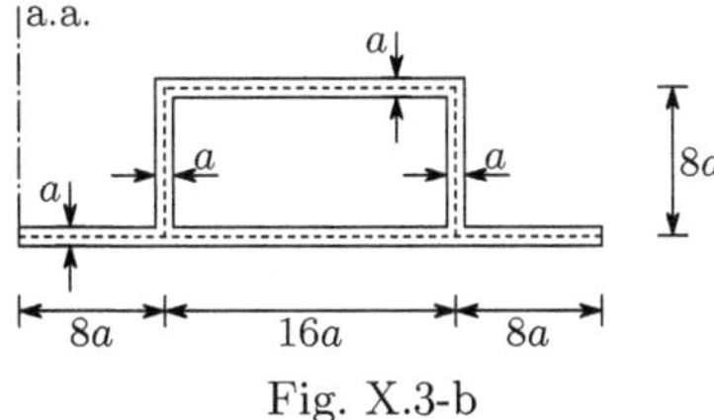

Fig. X.3-b

Resolution

Considering the cross-section at the distance z from point A, the load per unit length there takes the value

$$q(z) = \frac{p}{400a}\, z \; .$$

The shear force, as a function of z, may be determined by integrating this expression, yielding

$$V(z) = \int q\,\mathrm{d}z = \frac{p}{800a}\, z^2 \; .$$

The twisting moment in a given cross-section is the product of the shear force and the distance from its action axis to the shear centre, which is on the cross-section symmetry axis, yielding

$$T(z) = \frac{p}{800a}\, z^2 \times 16a = \frac{p}{50}\, z^2 \; .$$

The rotation of cross-section A may be obtained from (237), substituting I_p with the parameter J defined in (242), yielding

$$\varphi = \frac{1}{GJ}\int_0^l T\,\mathrm{d}z = \frac{1}{G\frac{4(16a\times 8a)^2 a}{48a}}\int_0^{400a} \frac{p}{50} z^2\,\mathrm{d}z = \frac{625}{2}\frac{p}{Ga} \; .$$

X.4. The cantilever beam represented in Fig. X.4 is made of a material with a shear modulus G. Determine the rotation of the right cross-section around the bar axis, caused by the uniformly distributed loading p.

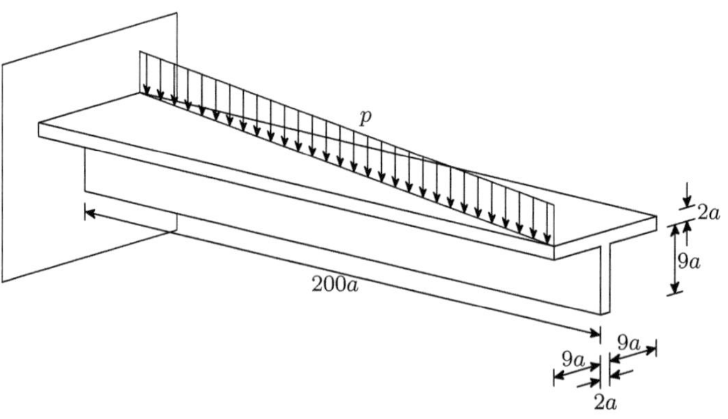

Fig. X.4

Resolution

Denoting by α the angle between the horizontal line where the load is applied and the bar axis, we get

$$\alpha = \arctan \frac{20a}{200a} = \arctan \frac{1}{10} \; .$$

The shear force in a cross-section at a distance z from the right end of the beam is given by the expression

$$V(z) = p\frac{z}{\cos \alpha} \; .$$

This shear force acts at a distance from the shear centre (mid-point of the horizontal segment of the cross-section) which varies with coordinate z and is given by the expression

$$d(z) = 10a - \frac{z}{2}\tan \alpha = 10a - \frac{z}{20} \; .$$

The twisting moment is then given by the expression

$$T(z) = Vd = p\frac{z}{\cos \alpha} \times \left(10a - \frac{z}{20}\right) = \frac{p}{\cos \alpha}\left(10az - \frac{z^2}{20}\right) \; .$$

This expression defines a parabola with the maximum value in the middle of the beam ($z = 100a$) and vanishing values in both extremities. The rotation of the right-end cross-section around the beam's axis may be obtained in the same way as in example X.2, with the parameter J of torsional stiffness defined in (246). In this way, we get

$$J = \frac{1}{3}(20a + 9a) \times (2a)^3 = \frac{232}{3}a^4 \Rightarrow \varphi = \frac{1}{GJ}\int_0^l T\,\mathrm{d}z$$

$$= \frac{1}{G\frac{232}{3}a^4}\int_0^{200a} \frac{p}{\cos \alpha}\left(10az - \frac{z^2}{20}\right)\mathrm{d}z = \frac{25000}{29\cos \alpha}\frac{p}{Ga} \approx 866.369\frac{p}{Ga} \; .$$

X.5. Figure X.5-a represents the thin-walled cross-section of a bar which undergoes a torque $T = 1200\tau_0 a^3$.

    (a) Determine the unit rotation θ, assuming that the bar is made of a material with linear elastic behaviour and a shear modulus G.

    (b) Determine the same rotation, assuming that the material has the constitutive law represented by the diagram depicted in Fig. X.5-b.

Resolution

(a) As the material has linear elastic behaviour, the unit rotation may be computed directly by means of Bredt's second formula (241). In this cross-section we have

$$\begin{cases} A &= (20a)^2 = 400a^2 \\ \oint \dfrac{\mathrm{d}s}{e} &= \dfrac{40a}{2a} + \dfrac{40a}{a} = 60 \end{cases} \Rightarrow J = \frac{4A^2}{\oint \frac{\mathrm{d}s}{e}} = \frac{4 \times \left(400a^2\right)^2}{60} = \frac{32000}{3}a^4 \; .$$

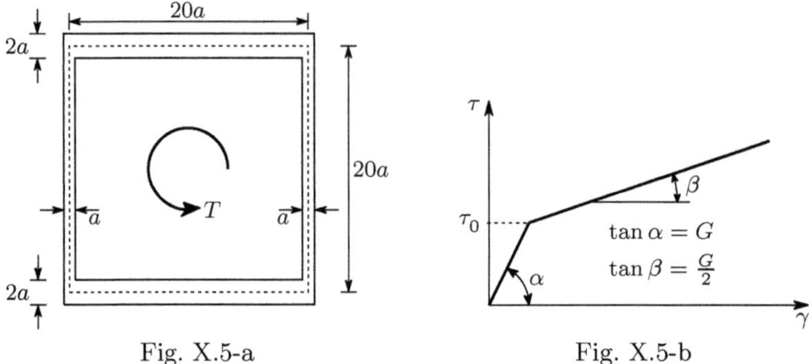

Fig. X.5-a   Fig. X.5-b

The rotation θ is then

$$\theta = \frac{T}{GJ} = \frac{1200\tau_0 a^3}{G\frac{32000}{3}a^4} = 0.1125\frac{\tau_0}{Ga} \ .$$

(b) If the maximum stress exceeds the value τ_0, the constitutive law is not
linear and, as a consequence, the rotation must be obtained by means of
(243). To this end, it is necessary to compute the stress distribution in
the cross-section and the corresponding distortions. Thus, denoting by τ_1
and τ_2 the stresses in the thinnest and thickest walls, respectively, we get
(240)

$$\tau_1 = \frac{T}{2A \times a} = \frac{1200\tau_0 a^3}{2 \times 400a^2 \times a} = 1.5\tau_0 \qquad\qquad \tau_2 = \frac{T}{2A \times 2a} = 0.75\tau_0 \ .$$

The corresponding distortions are

$$\gamma_1 = \frac{\tau_0}{G} + \frac{0.5\tau_0}{\frac{1}{2}G} = 2\frac{\tau_0}{G} \qquad\qquad \gamma_2 = \frac{0.75\tau_0}{G} = \frac{3}{4}\frac{\tau_0}{G} \ .$$

Substituting these values of γ in (243), we get

$$\theta = \frac{\oint \gamma\,ds}{2A} = \frac{1}{2 \times 400a^2}\left(40a \times 2\frac{\tau_0}{G} + 40a \times \frac{3}{4}\frac{\tau_0}{G}\right) = \frac{110}{800}\frac{\tau_0}{Ga} = 0.1375\frac{\tau_0}{Ga} \ .$$

X.6. Generalize Bredt's formulas to composite beams made of two materials
with linear elastic behaviour, in the two following situations:
(a) tubular cross-section with homogeneous wall (Fig. X.6-a);
(b) tubular cross-section with composite wall (Fig. X.6-b).

Resolution

(a) As seen in Sect. X.3, Bredt's first formula (240), expressed in terms of
shear flow, is valid for any thin-walled cross-section with degree of connection two. Since in this case the wall is homogeneous, (240) is also valid
in terms of the shearing stress.

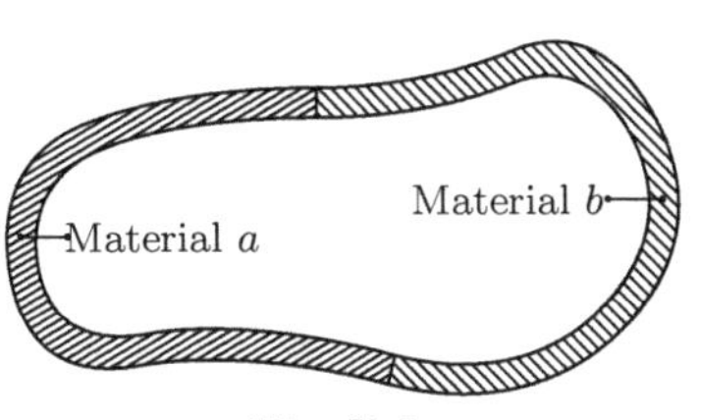
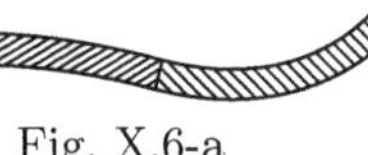

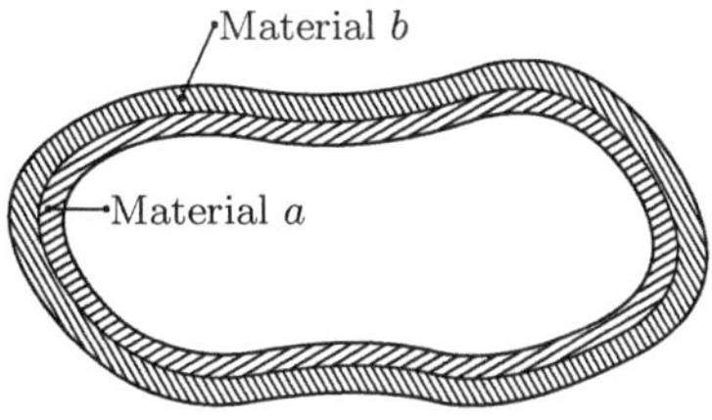

Fig. X.6-a   Fig. X.6-b

In relation to Bredt's second formula, this no longer yields since the shear modulus G is not constant in the whole cross-section. However, the unit rotation θ may also be obtained by the energy conservation principle. The elastic potential energy stored by a unit length of the bar is given by (s is a coordinate along the centre line of the wall and G_a and G_b are the shear moduli of materials a and b, respectively)

$$U = \frac{1}{2}\int_\Omega \tau\gamma \, d\Omega = \frac{1}{2}\int_\Omega \frac{\tau^2}{G}\, d\Omega = \frac{1}{2}\left(\int_a \frac{\tau^2}{G_a}e\,ds + \int_b \frac{\tau^2}{G_b}e\,ds\right)$$

$$= \frac{1}{2}\frac{T^2}{4A^2}\left(\frac{1}{G_a}\int_a \frac{ds}{e} + \frac{1}{G_b}\int_b \frac{ds}{e}\right).$$

The principle of energy conservation requires that this energy is equal to the work done by the torque in the same unit length, $W = \frac{1}{2}T\theta$. From this condition we get

$$W = U \;\Rightarrow\; \theta = \frac{T}{4A^2}\left(\frac{1}{G_a}\int_a \frac{ds}{e} + \frac{1}{G_b}\int_b \frac{ds}{e}\right).$$

By using the homogenization concept, two other forms may be given to this expression. Thus we have

$$\theta = \underbrace{\frac{T}{4A^2 G_a}\left(\int_a \frac{ds}{e} + \int_b \frac{ds}{\frac{G_b}{G_a}e}\right)}_{\text{homogenization in material } a} = \underbrace{\frac{T}{4A^2 G_b}\left(\int_a \frac{ds}{\frac{G_a}{G_b}e} + \int_b \frac{ds}{e}\right)}_{\text{homogenization in material } b}.$$

We conclude that, by multiplying the wall thickness at the part of the cross-section occupied by material b by the homogenizing coefficient of material b in material a, $m_a = \frac{G_b}{G_a}$, we may compute θ, as in the case of a homogeneous cross-section made of material a, and vice-versa for the homogenization in material b.

(b) In the case with the composite wall, Bredt's first formula is no longer valid in terms of stresses. However, since it is valid in terms of stress flow, we may establish the relation (e_a and e_b are the thicknesses of materials a and b, respectively, in a point of the centre line)

$$f = \tau_a e_a + \tau_b e_b = \frac{T}{2A} \, .$$

In this expression A represents the area limited by the centre line of the composite cross-section. This line is defined as the line of action of the resultant of the shearing stress in the wall thickness. As the distortion must be the same in the two materials, we have

$$\gamma = \frac{\tau_a}{G_a} = \frac{\tau_b}{G_b} \Rightarrow \tau_b = \frac{G_b}{G_a}\tau_a \, .$$

Substituting this equation in the previous one, we get τ_a. In the same way, τ_b could be obtained, yielding

$$\tau_a = G_a\frac{T}{2A\left(G_a e_a + G_b e_b\right)} \qquad \tau_b = G_b\frac{T}{2A\left(G_a e_a + G_b e_b\right)} \, . \qquad \text{(X.6-a)}$$

In the same way as in question a), the unit rotation θ may be obtained by means of the energy conservation law. In this case, we have

$$U = W \Rightarrow \frac{1}{2}\oint\left(\frac{\tau_a^2}{G_a}e_a + \frac{\tau_b^2}{G_b}e_b\right)\mathrm{d}s = \frac{1}{2}T\theta \, .$$

Substituting in this expression the values of τ_a and τ_b above, we get

$$\theta = \frac{T}{4A^2}\oint\frac{G_a e_a + G_b e_b}{\left(G_a e_a + G_b e_b\right)^2}\,\mathrm{d}s = \frac{T}{4A^2}\oint\frac{\mathrm{d}s}{G_a e_a + G_b e_b} \, . \qquad \text{(X.6-b)}$$

Here, too, we may give forms corresponding to the use of the homogenization concept to the expression of θ, yielding

$$\theta = \overbrace{\frac{T}{4A^2 G_a}\oint\frac{\mathrm{d}s}{e_a + \frac{G_b}{G_a}e_b}}^{\text{homogenization in material } a} = \underbrace{\frac{T}{4A^2 G_b}\oint\frac{\mathrm{d}s}{\frac{G_a}{G_b}e_a + e_b}}_{\text{homogenization in material } b} \, .$$

Alternatively, the expressions defining the unit rotation θ could be obtained from the relation between θ and the distribution of distortions (243).

X.7. Figure X.7-a represents the cross-section of a composite beam made of two materials with linear elastic behaviour, a and b. The shear moduli of the two materials are, respectively $G_a = 5G$ and $G_b = G$. Determine the maximum shear stress in each of the two materials and the unit rotation θ, caused by a torque T.

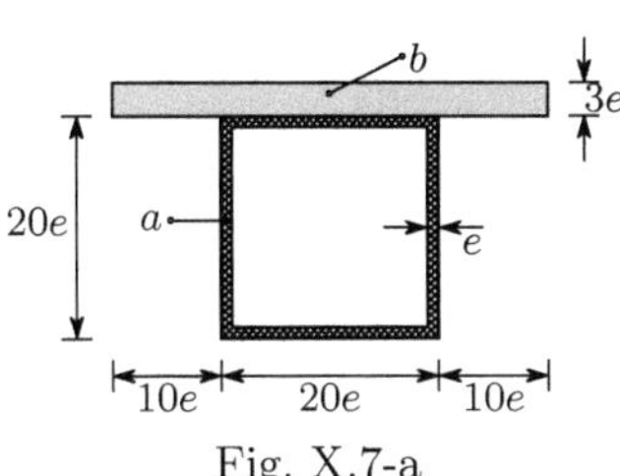

Fig. X.7-a

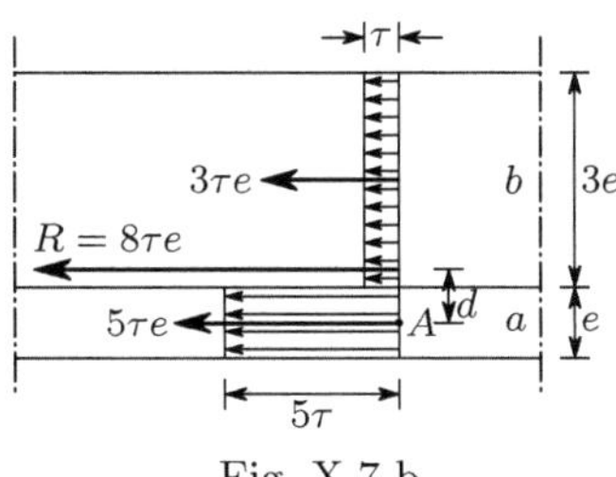

Fig. X.7-b

Resolution

Since, in a thin-walled cross-section with closed and open elements, the latter do not need to be considered, the problem may be solved by direct application of the expressions obtained in example X.6-b.

In order to get the position of the centre line of the composite wall, let us consider a stress τ in material b, which corresponds to a stress 5τ in material a, since $G_a = 5G_b$. In Fig. X.7-b the stress distribution in the wall thickness is represented. The distance d, from the stress resultant to the middle line of the part of the wall occupied by material a, may be obtained by means of the condition of equivalence on the moments with respect to point A (Fig. X.7-b), yielding

$$8\tau e \times d = 3\tau e \times 2e \;\Rightarrow\; d = \frac{3}{4}e \, .$$

Thus, the centre line defines a rectangle with the area $A = 19e \times 19.75e = 375.25e^2$.

The maximum stress in material a occurs in the homogeneous part of the wall, so it can be obtained directly from Bredt's first formula (240), yielding

$$\tau_{a-\text{max}} = \frac{T}{2Ae} = \frac{T}{2 \times 375.25e^2 e} = \frac{T}{2 \times 750.5e^3} \approx 1.33245 \times 10^{-3}\frac{T}{e^3} \, .$$

The stress in material b may be found from the second of X.6-a, yielding

$$\tau_b = G_b\frac{T}{2A\left(G_a e_a + G_b e_b\right)} = \frac{GT}{750.5e^2\left(5Ge + G3e\right)}$$
$$= \frac{T}{6004e^3} \approx 0.166556 \times 10^{-3}\frac{T}{e^3} \, .$$

The unit rotation θ is given by (X.6-b), yielding

$$\theta = \frac{T}{4A^2} \oint \frac{ds}{G_a e_a + G_b e_b}$$
$$= \frac{T}{4\left(375.25e^2\right)^2}\left(\frac{19e + 2 \times 19.75e}{5Ge} + \frac{19e}{5Ge + G3e}\right) \approx 24.9889 \times 10^{-6}\frac{T}{Ge^4} \, .$$

X.8. Generalize Bredt's theory to thin-walled cross-sections with degree of connection three (closed cross-sections with two channels, Fig. X.8).

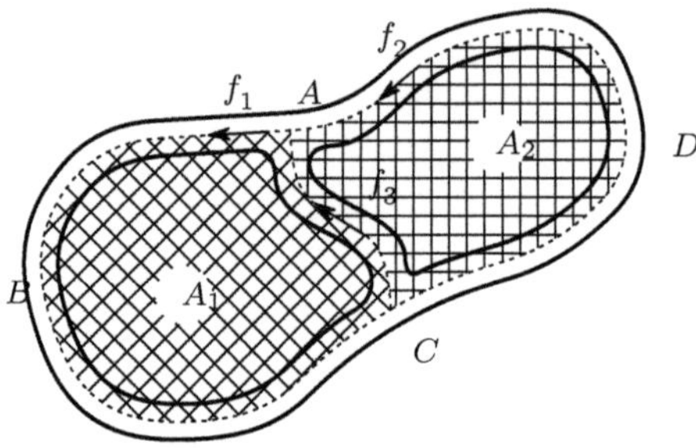

Fig. X.8

Resolution

The problem may be solved by considering three arcs ABC, CDA and CA. The condition of equilibrium of the shear flows explained in Subsect. VIII.3.c, is also valid in the case of shearing stresses caused by the twisting moment. Thus, the flows f_1, f_2 and f_3, with the directions shown in Fig. X.8, are related by the expression

$$f_1 = f_2 + f_3 .$$

As the unit rotations θ of the two channels must be equal, we get from (243) (A_1 and A_2 are the areas limited by the centre lines of the two channels, Fig. X.8)

$$\gamma = \frac{f}{Ge} \Rightarrow \theta = \frac{1}{2A_1 G}\left(f_1 m_1 + f_3 m_3\right) = \frac{1}{2A_2 G}\left(f_2 m_2 - f_3 m_3\right)$$

$$\text{with}\quad m_1 = \int_{ABC} \frac{ds}{e} \; ; \quad m_2 = \int_{CDA} \frac{ds}{e} \quad \text{and} \quad m_3 = \int_{CA} \frac{ds}{e} .$$
(X.8-a)

The minus sign in the last element of the expression of θ is because the shear flow f_3, with the direction indicated, causes a negative rotation in channel 2 of the cross-section. The appearance of this minus sign may also be easily explained by means of the theorem of virtual forces (Chap. XII).

The moment of the shearing stress with respect to any point of the cross-section plane is equal to the applied twisting moment. The simplest way to establish the equation which represents this moment is to decompose the shear flow f_3 into two parts, using the condition of equilibrium of shear flows $f_3 = f_1 - f_2$. In this way, we may consider the component f_1 of arc CA in the computation of the moment corresponding to the shearing stresses acting around channel 1 and the component $-f_2$ in the computation of the stresses acting around channel 2 (with reversed direction, as a consequence of the minus sign). This allows (240) to be applied to each of the two channels, yielding

$$T = 2A_1 f_1 + 2A_2 f_2 .$$
(X.8-b)

The flows f_1 and f_2 may then be computed by solving the system formed by (X.8-a) and (X.8-b),

$$\begin{cases} \dfrac{f_1 m_1}{A_1} + \dfrac{(f_1 - f_2)\, m_3}{A_1} = \dfrac{f_2 m_2}{A_2} - \dfrac{(f_1 - f_2)\, m_3}{A_2} \\[2ex] T = 2A_1 f_1 + 2A_2 f_2 \;, \end{cases}$$

yielding

$$\begin{cases} f_1 = \dfrac{T}{2A_1 + 2A_2 \frac{K_1}{K_2}} \\[2ex] f_2 = \dfrac{T}{2A_1 \frac{K_2}{K_1} + 2A_2} \end{cases} \quad \text{with} \quad \begin{cases} K_1 = \dfrac{m_1}{A_1} + \dfrac{m_3}{A_1} + \dfrac{m_3}{A_2} \\[2ex] K_2 = \dfrac{m_2}{A_2} + \dfrac{m_3}{A_2} + \dfrac{m_3}{A_1} \; . \end{cases} \qquad \text{(X.8-c)}$$

The stresses in each one of the two arcs are then

$$\tau_1 = \frac{f_1}{e_1} \qquad \tau_2 = \frac{f_2}{e_2} \qquad \tau_3 = \frac{f_3}{e_3} = \frac{f_1 - f_2}{e_3} \; . \qquad \text{(X.8-d)}$$

In this expression e_1, e_2 and e_3 represent the (eventually variable) thicknesses of the walls in each arc. Once f_1, f_2 and f_3 are known, the unit rotation may be computed by means of (X.8-a).

X.9. Determine the distribution of shearing stresses and the unit rotation θ introduced by a torque T in the cross-section represented in Fig. X.9-a.

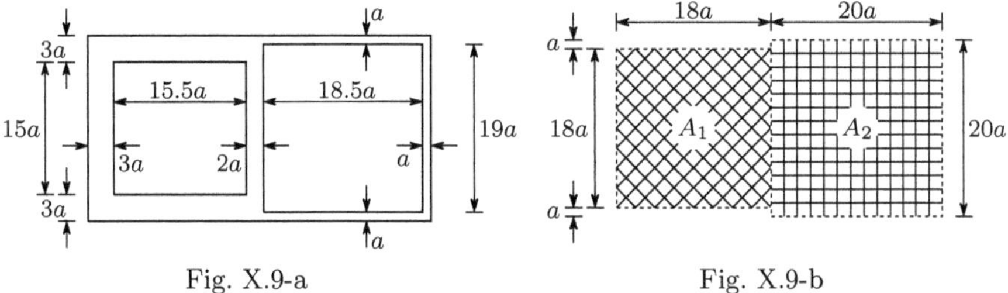

Fig. X.9-a   Fig. X.9-b

Resolution

This problem may be solved by the direct application of the expressions developed in example X.8. The parameters related to the geometry of the cross-section take the values

$$A_1 = (18a)^2 = 324a^2$$

$$A_2 = (20a)^2 = 400a^2$$

$$m_1 = \frac{54a}{3a} = 18$$

$$m_2 = \frac{62a}{a} = 62$$

$$m_3 = \frac{18a}{2a} = 9 \; .$$

Substituting these values in (X.8-c) and (X.8-d), we get the shearing stresses

$$\begin{cases} K_1 = 0.10583\frac{1}{a^2} \\ K_2 = 0.20528\frac{1}{a^2} \end{cases} \Rightarrow \begin{cases} f_1 = 9.4300 \times 10^{-4}\frac{T}{a^2} \\ f_2 = 4.8617 \times 10^{-4}\frac{T}{a^2} \\ f_3 = 4.5682 \times 10^{-4}\frac{T}{a^2} \end{cases} \Rightarrow \begin{cases} \tau_1 = 3.1433 \times 10^{-4}\frac{T}{a^3} \\ \tau_2 = 4.8617 \times 10^{-4}\frac{T}{a^3} \\ \tau_3 = 2.2841 \times 10^{-4}\frac{T}{a^3} \ . \end{cases}$$

From (X.8-a), we get the unit rotation θ

$$\theta = 3.2539 \times 10^{-5}\frac{T}{Ga^4} \ .$$

X.10. Figure X.10 represents the cross-section of a bar made of a material with a shear modulus G, under a positive twisting moment T. Determine the stresses and the unit rotation caused by this internal force.

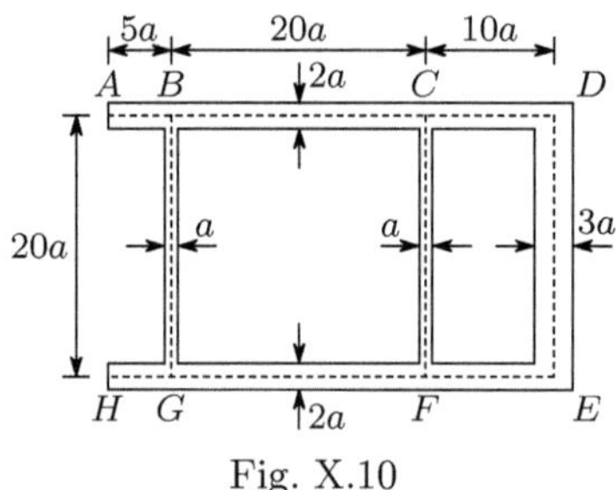

Fig. X.10

Resolution

The simply-connected wall elements AB and GH do not need to be considered. Defining as areas A_1 and A_2 the rectangles $BCFG$ and $CDEF$, respectively, we have $A_1 = (20a)^2 = 400a^2$, $A_2 = 20a \times 10a = 200a^2$ and

$$m_1 = \int_{CBGF} \frac{ds}{e} = \frac{20a}{2a} + \frac{20a}{a} + \frac{20a}{2a} = 40$$

$$m_2 = \int_{FEDC} \frac{ds}{e} = \frac{10a}{2a} + \frac{20a}{3a} + \frac{10a}{2a} = \frac{50}{3} \qquad m_3 = \int_{FC} \frac{ds}{e} = \frac{20a}{a} = 20 \ .$$

Substituting these values in (X.8-c), we get the shear flows

$$\begin{cases} K_1 = \dfrac{1}{4a^2} \\ K_2 = \dfrac{7}{30a^2} \end{cases} \Rightarrow \begin{cases} f_1 = \dfrac{7}{8600}\dfrac{T}{a^2} = 813.953 \times 10^{-6}\dfrac{T}{a^2} \\ f_2 = \dfrac{3}{3440}\dfrac{T}{a^2} = 872.093 \times 10^{-6}\dfrac{T}{a^2} \ . \end{cases}$$

From (X.8-d), the shearing stresses in the different wall segments are obtained

$$\tau_{CB} = \tau_{GF} = \frac{f_1}{2a} = \frac{7}{17200}\frac{T}{a^3} \qquad \tau_{ED} = \frac{f_2}{3a} = \frac{1}{3440}\frac{T}{a^3}$$

$$\tau_{BG} = \frac{f_1}{a} = \frac{7}{8600}\frac{T}{a^3} \qquad\qquad \tau_{CF} = -\tau_{FC} = -\frac{f_3}{a} = -\frac{f_1 - f_2}{a} = \frac{1}{17200}\frac{T}{a^3} \ .$$

$$\tau_{FE} = \tau_{DC} = \frac{f_2}{2a} = \frac{3}{6880}\frac{T}{a^3}$$

The unit rotation may be obtained by the direct substitution in one of (X.8-a)

$$\theta = \frac{1}{2A_1 G}\left(f_1 m_1 + f_3 m_3\right) = \frac{1}{2A_2 G}\left(f_2 m_2 - f_3 m_3\right) = 39.2442 \times 10^{-6}\frac{T}{Ga^4} \ .$$

X.11. The cross-section represented in Fig. X.11 has a constant wall thickness
   a. Determine the maximum shearing stress and the unit rotation θ
   caused by a torque T, considering:
   (a) only the closed part of the cross-section;
   (b) the open and closed parts of the cross-section.

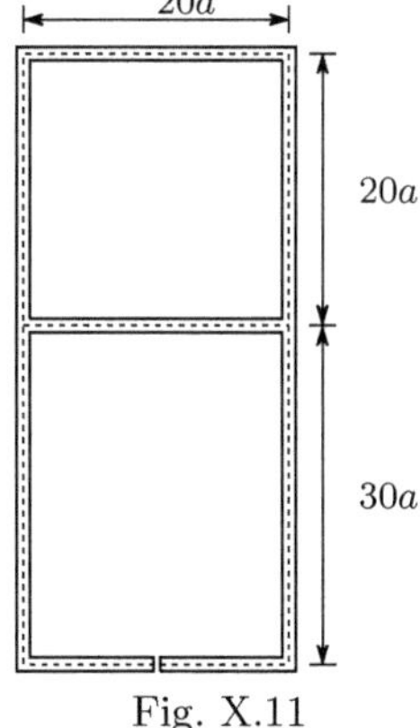

Fig. X.11

Resolution

(a) This question may be answered by the direct application of the Bredt
   formulas. Thus, we get from (240), (241) and (242)

$$\begin{cases} A = 400a^2 \\ e = a \\ p = 80a \end{cases} \Rightarrow \begin{cases} \tau_{\max} = \dfrac{T}{2Ae} = \dfrac{T}{2 \times 400a^2 \times a} = \dfrac{T}{800a^3} = 1.25 \times 10^{-3}\dfrac{T}{a^3} \\[2mm] \theta = \dfrac{Tp}{4GA^2 e} = \dfrac{T \times 80a}{4G(400a^2)^2 a} = \dfrac{T}{8000a^4 G} = 1.25 \times 10^{-4}\dfrac{T}{a^4 G} \ . \end{cases}$$

(b) When the contribution of the whole cross-section to the torsion strength
is considered, we have a statically indeterminate problem, since it is necessary to use the compatibility conditions respecting the deformations of
the open and closed parts of the cross-section. The problem may be solved
by means of the displacement method, taking the unit rotation θ as the
kinematic unknown.

As we have seen in the answer to question a), the torque required to
introduce a unit value of the relative rotation per unit length θ in the
closed part of the cross-section – the torsional stiffness of this part – takes
the value $GJ_c = 8000\, a^4 G$. The geometrical stiffness parameter $J = J_o$ of
the open part of the cross-section may be computed by means of (246),
yielding

$$J_o = \frac{1}{3} 2 \times \left(30a \times a^3 + 10a \times a^3\right) = \frac{80}{3} a^4 \ .$$

The rotation θ may now be computed by dividing the torque T by the
total stiffness of the cross-section (see Footnote 74), yielding

$$\theta = \frac{T}{G\left(J_c + J_o\right)} = \frac{T}{\left(8000 + \frac{80}{3}\right) a^4 G} \approx 1.2458 \times 10^{-4} \frac{T}{a^4 G} \ .$$

This value practically coincides with that obtained by considering only
the closed part of the cross-section.

The shearing stress in the closed part of the cross-section is constant,
since the thickness is constant. This stress may be computed from the
unit rotation θ. To this end, we may combine (240), (241) and (242), so
as to express the stress as a function of θ, yielding

$$\tau_c = \theta \frac{2GA}{p} = \frac{T}{\left(8000 + \frac{80}{3}\right) a^4 G} \frac{2G400a^2}{80a} = \frac{T}{\left(800 + \frac{8}{3}\right) a^3} \approx 1.2458 \times 10^{-3} \frac{T}{a^3} \ .$$

The maximum shearing stress in the open part of the cross-section may
be computed by means of (247), yielding

$$\tau_o^{\max} = G\theta e = \frac{TGa}{\left(8000 + \frac{80}{3}\right) a^4 G} = \frac{1}{10} \tau_c \ .$$

We conclude that the maximum value of the shearing stress occurs in the
closed part of the cross-section and is practically coincident with the value
obtained by disregarding the simply-connected wall elements.

X.12. Compare the torsional strength of two thin-walled cross-sections with
constant thickness e and the same geometry, considering that one is
closed and the other is open.

Resolution

Denoting by τ_{all} the value of the allowable shearing stress, we get the following
values for the torsional strength in the two cases (240) and (245)

$$\tau_f = \frac{T}{2Ae} \;\Rightarrow\; T^f_{\text{all}} = 2Ae\tau_{\text{all}} \qquad \text{(closed cross-section)}$$

$$\tau_a = \frac{3T}{pe^2} \;\Rightarrow\; T^a_{\text{all}} = \frac{1}{3}pe^2\tau_{\text{all}} \qquad \text{(open cross-section)} \, .$$

The relation between these two values may be expressed by the ratio

$$\frac{T^f_{\text{all}}}{T^a_{\text{all}}} = \frac{2Ae}{\frac{pe^2}{3}} = 6\frac{A}{pe} = 6\frac{A}{\Omega} \, .$$

We conclude that the torsional strength of the closed cross-section is much higher than that for the open section, since in this kind of cross-section the area A limited by the wall centre line is usually much higher than the cross-section's area $\Omega = pe$.

X.13. Figure X.13 represents the cross-section of a bar made of two materials. Material a has linear elastic behaviour defined by the shear modulus G, while material b has non-linear behaviour defined by the relation

$$\gamma = 5\frac{\tau}{G}\left[1 + \left(\frac{\tau}{\tau_0}\right)^3\right] \, .$$

Determine the unit rotation θ introduced by a torque $T = 7000\,e^3\tau_0$.

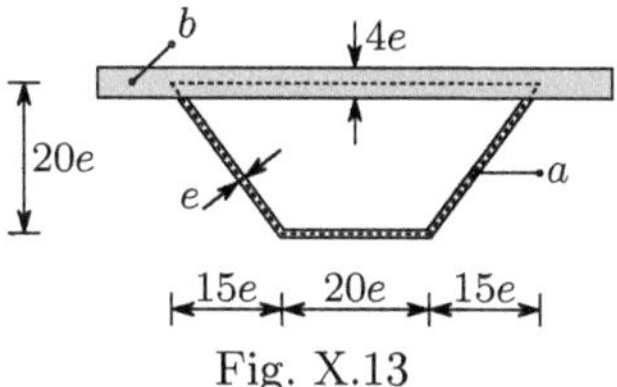

Fig. X.13

Resolution

Considering the wall centre line is represented by the dashed line in Fig. X.13, the area limited by this line and the stress in each of the two materials take the values (the dimensions indicated in the figure refer to this line)

$$A = 20e \times 35e = 700e^2 \;\Rightarrow\; \begin{cases} \tau_a = \dfrac{7000e^3\tau_0}{2\times 700e^2\times e} = 5\tau_0 \\[2mm] \tau_b = \frac{1}{4}\tau_a = \frac{5}{4}\tau_0 \, . \end{cases}$$

Since one material does not have a linear behaviour, the unit rotation must be computed by means of (243). For this it is necessary to compute the distortions, which take the values

$$\gamma_a = 5\frac{\tau_0}{G} \qquad \gamma_b = 5\frac{\frac{5}{4}\tau_0}{G}\left[1 + \left(\frac{\frac{5}{4}\tau_0}{\tau_0}\right)^3\right] \approx 18.4570\frac{\tau_0}{G}\ .$$

From (243) we get then

$$\theta = \frac{\oint \gamma\,ds}{2A} = \frac{1}{1400e^2}\left(50e \times 18.4570\frac{\tau_0}{G} + 70e \times 5\frac{\tau_0}{G}\right) = 0.909180\frac{\tau_0}{Ge}\ .$$

X.14. Prove that a non-circular cross-section of a prismatic bar under uniform torsion cannot remain plane and perpendicular to the bar axis.

Resolution

The proof may be established by means of the following line of reasoning:

(1) since the twisting moment is constant, all the cross-sections deform in the same way; in fact, the Saint Venant principle ensures that the deformation caused by forces that are not close to a given cross-section depend only on the resultant of those forces (in this case the twisting moment), which is the same for all cross-sections, since the torque is constant;

(2) since all cross-sections deform in the same way, if they were to remain plane and perpendicular to the bar's axis, the rotation caused by the twisting moment around the torsion centre would introduce a shearing stress in each point of the cross-section, which would be perpendicular to the line joining the point with the shear centre; as this is not possible in the points belonging to the boundary of the cross-section, unless the boundary is circular, the cross-section cannot remain plane.

X.15. Figure X.15 represents the cross-section of a bar made of two materials with linear elastic behaviour, whose shear moduli are $G_a = 2G$ and $G_b = 3G$. The bar is under uniform torsion. Determine:
(a) the force that must be resisted by the connection between the two materials;
(b) the relative rotation of two cross-sections at a unit distance from each other.

X.16. Figure X.16 represents the cross-section of a bar made of two materials with linear elastic behaviour, whose shear moduli are $G_a = 2G$ and $G_b = 3G$. The bar is under uniform torsion. Determine:
(a) the shearing stresses in each of the two materials;
(b) the relative rotation of two cross-sections at a unit distance from each other.

X.17. A bar has the cross-section represented in Fig. X.17 and is made of a material which has an allowable stress τ_{all} and a shear modulus G.
(a) Determine the increase of torsional strength and torsional stiffness that is obtained by welding the bar in the longitudinal line defined by point A, so that a closed cross-section is obtained.

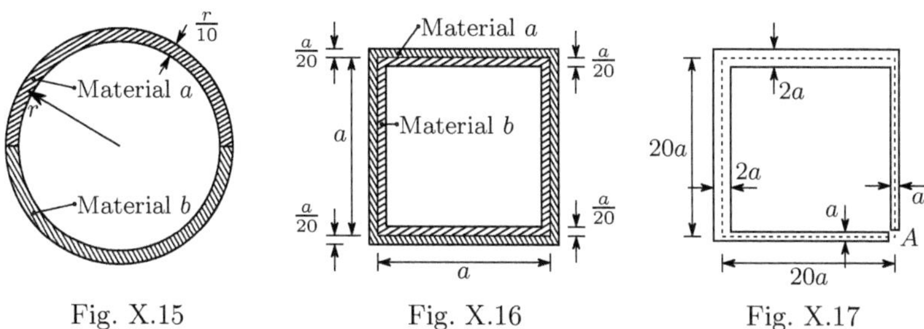

Fig. X.15   Fig. X.16   Fig. X.17

(b) Determine the force that has to be resisted by the welding, when the bar undergoes the maximum allowable torque.

X.18. In the thin-walled cross-section represented in Fig. X.18 determine the error in the computation of the maximum shear force $\tau_{\max}$ and the unit rotation θ caused by a torque T, if the simply-connected wall elements are not considered.

X.19. Figure X.19 represents the cross-section of a bar made of two materials with linear elastic behaviour, whose shear moduli are $G_a = 2G$ and $G_b = 3G$. The bar undergoes a uniform torsion. Determine:
(a) the stress in each of the two materials;
(b) the unit rotation θ.

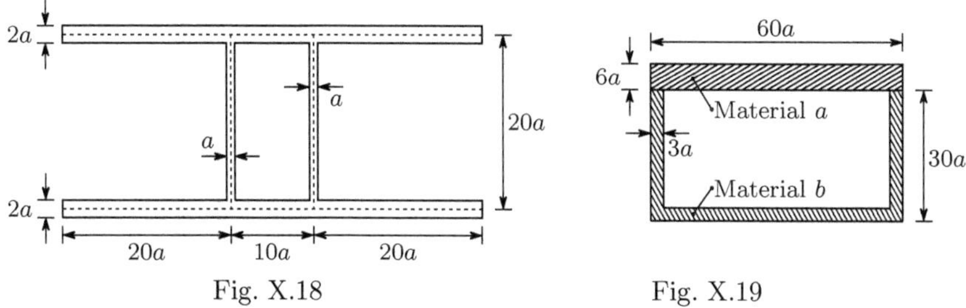

Fig. X.18   Fig. X.19

X.20. The thin-walled cross-section represented in Fig. X.20-a is made of a material with the rheological behaviour represented in Fig. X.20-b, as a function of the rheological parameters τ_0 and γ_0. Determine the unit rotation θ introduced into the bar by a torque $T = 1500\tau_0 a^3$.

X.21. Figure X.21 represents the cross-section of a bar made of a material with non-linear elastic behaviour, whose constitutive law, in terms of shearing stresses and distortions, is given by the relation

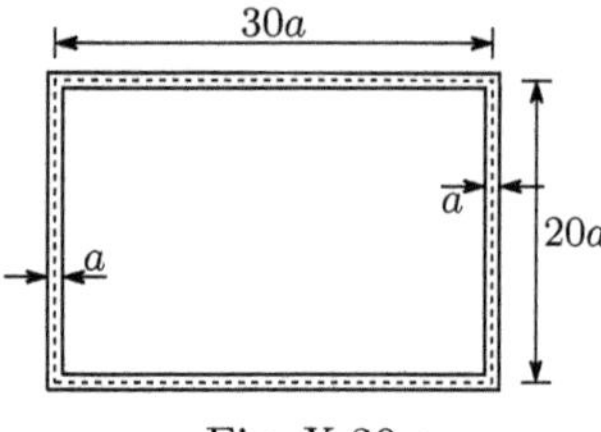

Fig. X.20-a

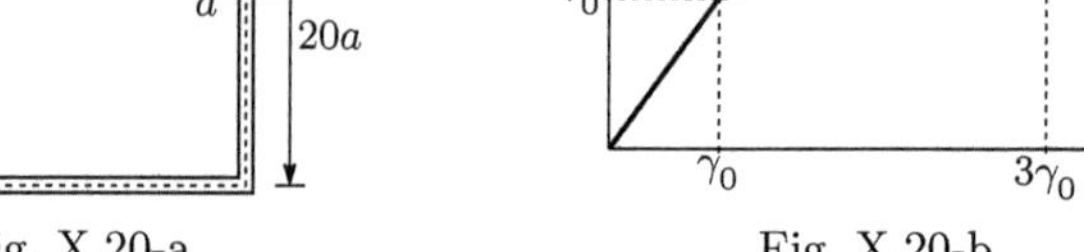

Fig. X.20-b

$$\gamma = -0.01 \ln \left(1 - \frac{\tau}{\tau_0} \right) \,,$$

where τ_0 is a given reference stress. The bar is under uniform torsion.

(a) Knowing that the maximum shearing stress in the cross-section takes the value $\tau_{\max} = 0.8\tau_0$ determine the unit rotation θ.

(b) Determine the corresponding twisting moment.

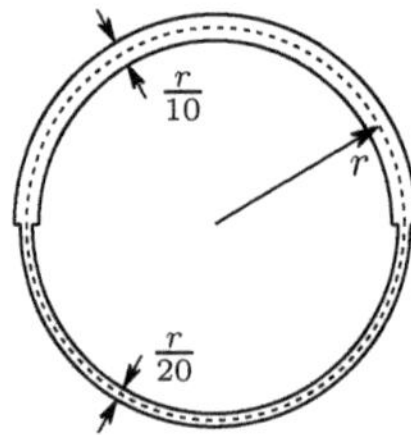

Fig. X.21

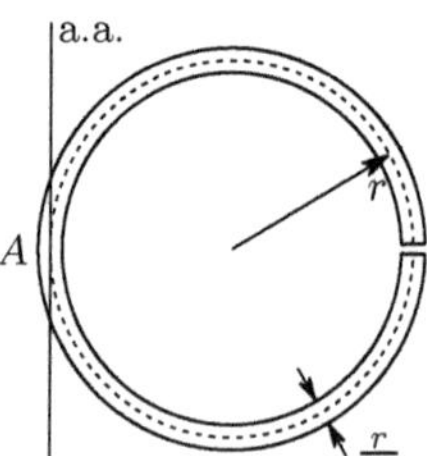

Fig. X.22

X.22. A vertical concentrated load P is applied at the free end of a cantilever beam with a span $20r$. The beam has the cross-section represented in Fig. X.22 and is made of a material with an elasticity modulus E and a Poisson's coefficient $\nu = 0.25$. The line of action of the load P passes through point A of the cross-section. Using the results of examples VIII.3, VIII.11 and IX.8, determine the contributions of the bending moment, the twisting moment and the shear force in the displacement of the point of application of the load P.

XI

Structural Stability

XI.1 Introduction

In the analysis of the behaviour of slender members performed until now, equilibrium and compatibility conditions have been used in order to find the internal forces and deformations. In the simplest cases, a structure's safety is evaluated by confirming that the maximum values computed for the stresses are lower than the allowable stress defined for the material the structure is made of. This is a necessary condition for structural safety, but it may not be sufficient, either because the deformations are limited for some reason, or because there is the risk that the equilibrium configuration of the structure is not stable, i.e., that *buckling* may occur. In this chapter the study of the conditions under which a structure is stable is introduced. Buckling may occur when there are compressive internal forces in the structure. In fact, while tensile forces may only do work if the material deforms or ruptures, for the case of compression there is a third possibility – buckling – which consists of a lateral deflection of the material, in relation to direction of actuation of the compressive forces. The main subject of this chapter is the investigation of the conditions under which buckling occurs in a slender member.

The concept of stability may be easily understood by means of the classical example of a sphere which is in static equilibrium on a horizontal surface, as represented in Fig. 154.

In this example physical evidence makes a deeper analysis unnecessary: in the first case (concavity) the equilibrium is stable, while in the second case (convexity) it is unstable. However, in the study of the stability of deformable bodies, the physical evidence is almost completely lost, since stability or instability are determined by how the internal forces and stiffnesses are distributed inside the body which, in most cases, is not an intuitive matter.

In order to solve this problem, two distinct approaches are usually taken into consideration. Both of them are easily understood by analogy with the example of the sphere.

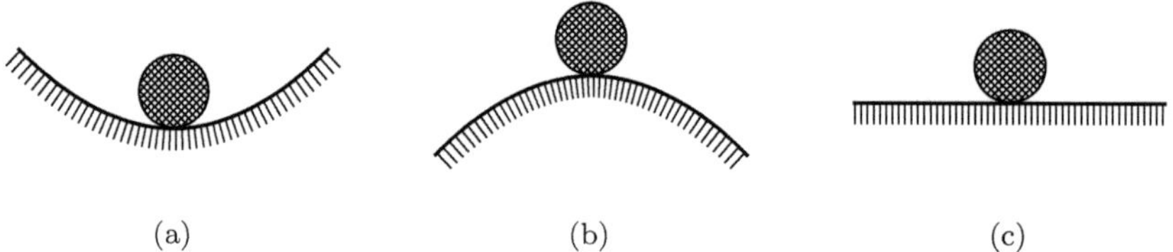

Fig. 154. Equilibrium of a sphere on a horizontal surface:(**a**) Stable equilibrium; (**b**) Unstable equilibrium; (**c**) Neutral equilibrium

- The first is based on the concept of potential energy. In the example of Fig. 154, the height of a point on the supporting surface defines the potential energy. The equilibrium position is an extremum of this energy. If this extremum is a minimum, the equilibrium will be stable (Fig. 154a), if it is a maximum, it will be unstable (Fig. 154b).

  In the equilibrium configuration of a deformed body the potential energy also reaches an extremum (see Sect. XII.5). However, in this case it has two distinct components: the potential energy of the applied forces and the potential energy stored by the elastic deformation of the body. Both vary when the deformation field in the body varies. In the same way as in the example of the sphere, the stability or instability of an equilibrium configuration may be investigated by determining if it corresponds to a minimum or to a maximum of the potential energy.

- The second approach is based on the analysis of the effect of perturbations to the equilibrium configuration. If the perturbation is amplified, the equilibrium state is unstable; if it is damped, the equilibrium will be stable. In the example of the sphere the perturbation may be given by a small displacement from the equilibrium point. It is obvious that in the case of Fig. 154a, when the force causing the displacement disappears, the sphere returns to the equilibrium position, which means that the perturbation is damped, i.e., that the equilibrium is stable. In the case of Fig. 154b any displacement from the equilibrium point, no matter how small it is, will cause the sphere to roll away from the equilibrium configuration, which means that the perturbation is *amplified*. The situation of neutral equilibrium (Fig. 154-c) represents the transition from the the stable to the unstable phase. It corresponds to a situation in which the deformation state of the structure can be changed without disturbing the equilibrium between internal and external forces. This is the so-called *critical phase*.

In the analysis contained in this chapter the second approach will be used, since it allows a better physical understanding of the buckling phenomenon. Furthermore, with this option, the difficulties arising when the energy method is applied in cases where energy dissipation takes place (plastic and viscous deformation), may be circumvented.

The analysis of the effect of perturbations requires the analysis of the equilibrium in a deformed configuration of the structure, as opposite to the analysis of the effects of the internal forces in slender members described in the preceding chapters, where the equilibrium conditions have always been stated in the undeformed configuration. For these reasons, the theories relating to the buckling phenomena are sometimes called *second-order theories*, in which the interaction between the deformations and the internal forces is taken into consideration. It is this interaction (influence of the deformation on the internal forces and vice versa) that may cause instability.

The need to take this interaction into account means that the structural behaviour may not be accepted as geometrically linear, since we cannot consider that the geometry of the structure remains unchanged. As a consequence, the superposition principle cannot generally be used when buckling is analysed.

Since buckling is always associated with compressive stresses, we consider in this Chapter that compressive axial forces and stresses are positive.

XI.2 Fundamental Concepts

Before the main subject of this chapter – the buckling of axially compressed slender members – is tackled, some very simple examples are analysed in detail, in order to introduce fundamental concepts related to the study of structural stability.

XI.2.a Computation of Critical Loads

In accordance with the considerations above, the stability of a structure may be analysed by computing its *critical load*, i.e., the load corresponding to the situation in which a perturbation of the deformation state does not disturb the equilibrium between the external and internal forces. In order to illustrate these considerations, we first analyse the column represented in Fig. 155a.

It is obvious that this structure is stable in the case of a tensile axial force and unstable under compression. In order to give the structure the capacity to withstand compressive forces, a spring with stiffness E is placed in point B, as represented in Fig. 155b. The critical load of this new structure may be found by analysing the force F which is needed to introduce the perturbation represented by the horizontal displacement δ. Considering this displacement as infinitesimal and denoting by θ the rotation angle of the bars, we have $\sin\theta = \frac{\delta}{l} \approx \theta$ and $N \approx P$. From the horizontal equilibrium condition of the forces acting in node B we get ($E\delta$ is the force in the spring)

$$2N\sin\theta + F = E\delta \;\Rightarrow\; F = \left(E - \frac{2P}{l} \right)\delta \,.$$

If force F is positive, i.e., if it takes the same direction as the displacement δ, the structure is stable, since it is necessary to apply a force to disturb the equilibrium configuration. A negative value of force F, however, means that it is necessary to apply a force in order to prevent displacement δ from increasing, i.e., that the perturbation is amplified. Thus, the structure is unstable. The critical load corresponds to the transition situation $(F = 0)$, where the equilibrium is neutral, since the perturbation δ does not affect the equilibrium. The critical load is then

$$F = 0 \;\Rightarrow\; P = P_{cr} = \frac{El}{2} .$$

The quantity $\frac{F}{\delta} = E - \frac{2P}{l}$ represents the horizontal stiffness of the structure in point B. When it becomes negative, the structure becomes unstable.

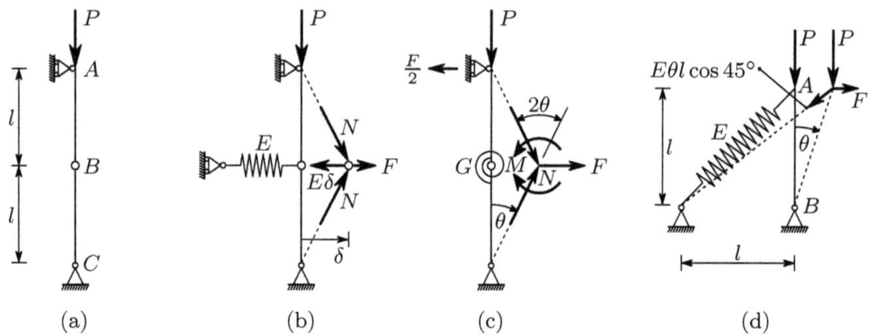

Fig. 155. Simple examples of computation of the critical load

In the structure represented in Fig. 155c, the stability under compressive loading is increased by means of a rotation spring linking the extremities of the two bars converging in point B. The spring has a stiffness G, which means that it is necessary to apply a bending moment $M = 2G\theta$ to introduce the relative rotation 2θ between bars $\overline{AB}$ and $\overline{BC}$. The relation between the load P, the force F and the rotation θ may be found by expressing the bending moment in the rotation spring as a function of the load P. The reaction force in support A is $\frac{F}{2}$, as is easily concluded by means of the balance condition of the moment with respect to point C. Considering the rotation angle θ as infinitesimal, we get

$$\begin{cases} \sin\theta \approx \theta \\ \cos\theta \approx 1 \end{cases} \Rightarrow P \times \theta l + \frac{F}{2} \times l = 2G\theta \;\Rightarrow\; F = \left(\frac{4G}{l} - 2P\right)\theta .$$

In the critical situation the perturbation θ does not disturb the equilibrium, which means that the critical load is that corresponding to $F = 0$, i.e.,

$$F = 0 \;\Rightarrow\; P = P_{cr} = \frac{2G}{l} .$$

From a qualitative point of view, this problem is analogous to the compression of a pin-ended bar. In fact, even without a deeper analysis, we intuitively know that the shorter the bar and the higher its bending stiffness (here represented by G), the higher the load required to induce buckling in a compressed bar.

The critical load of the structure represented in Fig. 155d may be computed in the same way, by considering the infinitesimal perturbation θ caused by the force F. The rotation θ induces an elongation $\theta l \cos 45°$ in the spring. The balance condition of the moment with respect to node B of the forces acting in node A yields

$$F \times l + P \times \theta l + \overbrace{E\theta l \frac{\sqrt{2}}{2}}^{\text{force in the spring}} (\sin 45° \theta l - \cos 45° l) = 0 \ .$$

As in the two other examples, the critical load corresponds to $F = 0$, which leads to (θ may be disregarded in the subtraction $1 - \theta$))

$$F = 0 \ \Rightarrow \ P\theta l = E\theta l \frac{\sqrt{2}}{2} (l - \theta l) \frac{\sqrt{2}}{2} \ \Rightarrow \ P = \frac{El}{2} \ .$$

XI.2.b Post-Critical Behaviour

The three structures analysed in Subsect. XI.2.a exhibit different behaviour after the critical load is reached. In order to analyse the differences it is necessary, not only to consider deformed configurations, but also to abandon the infinitesimal deformation domain and consider arbitrary values for the parameters defining the deformation.

Taking the first of the above analysed structures (Figs. 155a and 156a), now with the deformation defined by angle θ (Fig. 156b), the load P necessary to balance the force acting in the spring for a given value of θ, may be computed by means of the horizontal balance condition of the forces acting in node B. Thus, we have (F_s is the force in the spring)

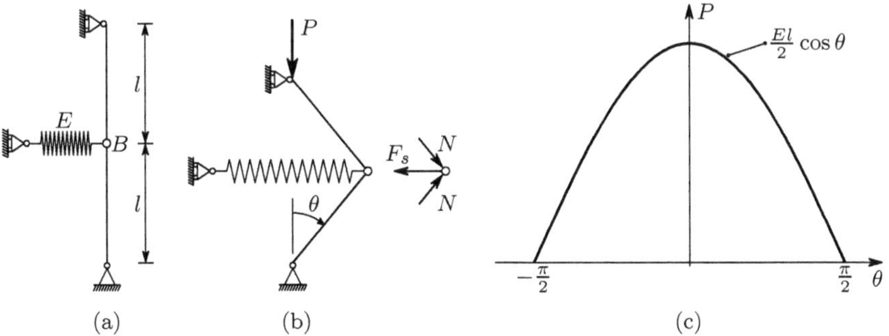

Fig. 156. Unstable post-critical behaviour

$$\begin{cases} N = \dfrac{P}{\cos\theta} \\[2mm] F_s = El\sin\theta \end{cases} \Rightarrow 2N\sin\theta = F_s \Rightarrow 2P\dfrac{\sin\theta}{\cos\theta} = El\sin\theta \;.$$

In the undeformed configuration ($\theta = 0$) equilibrium occurs for any value of the load P, since $\sin\theta = 0$. However, when there is deformation ($\theta \neq 0$), the equilibrium condition requires that P takes the value

$$P = \frac{El}{2}\cos\theta \;.$$

From this expression we conclude that, in a deformed configuration, equilibrium is only possible for values of P that are lower than the critical load $\frac{El}{2}$. Furthermore, we conclude from the relation between P and θ (Fig. 156c) that, in a deformed configuration, the stiffness is negative, since an increase in the deformation θ corresponds to a decrease of the load P. As seen in Subsect. XI.2.a, this means that the deformed situation is unstable. This structure is said to have an *unstable post-critical behaviour*. The critical load obtained for the undeformed configuration is, therefore, the maximum load that can be supported by the structure.

Let us now consider the second of the structures analysed in Subsect. XI.2.a (Figs. 155c and 157a).

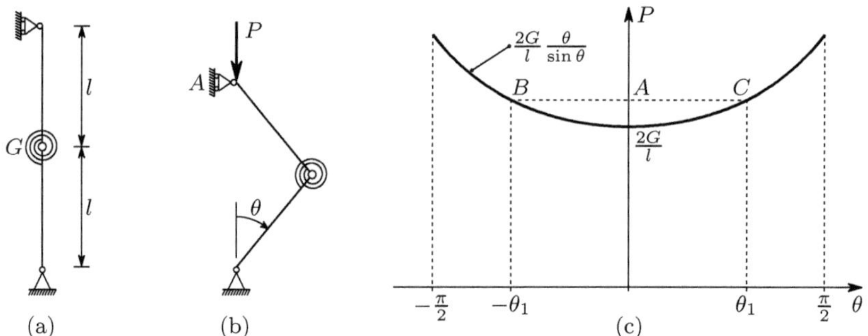

Fig. 157. Stable post-critical behaviour

The equilibrium condition in the deformed configuration defined by a given value of angle θ (Fig. 157b), may be established by computing the bending moment caused by load P in the rotation spring. Since the reaction force in support A vanishes, we have

$$Pl\sin\theta = 2G\theta \Rightarrow P = \frac{2G}{l}\frac{\theta}{\sin\theta} \;. \tag{248}$$

Since θ (measured in radians) is always superior to $\sin\theta$, the only possible equilibrium configuration for values of P that are inferior to $\frac{2G}{l}$ is the undeformed configuration ($\theta = 0$). Furthermore, the first of 248, confirms that the undeformed configuration is always an equilibrium configuration. As seen in Subsect. XI.2.a, this configuration is stable for $P < \frac{2G}{l}$.

When the value of P is higher than $\frac{2G}{l}$, we conclude from (248) and from Fig. 157c that there are three possible equilibrium configurations: the undeformed configuration which is unstable (point A) and the two roots of (248), θ_1 and $-\theta_1$ (points B and C). These two situations are stable, since an increase of the deformation corresponds to an increase of the load P (positive stiffness). We conclude that the loading capacity of this structure is not exhausted when the critical load corresponding to the undeformed configuration is attained, i.e., this structure has a *stable post-critical behaviour*. Nevertheless, for higher values of P than the critical load, a stable equilibrium requires appreciable deformations, as the diagram in Fig. 157c shows.

The post-critical behaviour of these two structures (Figs 156a and 157a) differs, but they do share a common feature: their behaviour is the same for positive and negative values of the deformation. These structures have a *symmetrical post-critical behaviour*. However, this is not always the case. For example, the third of the structures analysed in Subsect. XI.2.a (Figs. 155d and 158a) has a non-symmetrical behaviour. The relation between angle θ and the corresponding value of P may be obtained in the same way as in the previous two cases.

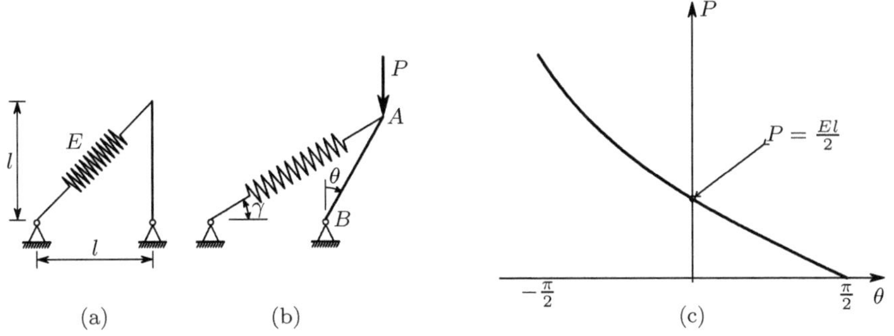

Fig. 158. Non-symmetrical post-critical behaviour

Considering a finite value for the deformation θ (Fig. 158b), relations allowing the computation of P from θ may be obtained

$$\gamma = \arctan \frac{\cos\theta}{1 + \sin\theta} \quad \longrightarrow \quad F_s = El\left(\frac{1 + \sin\theta}{\cos\gamma} - \sqrt{2}\right)$$

$$\longrightarrow \quad P = F_s \frac{\cos\gamma\cos\theta - \sin\gamma\sin\theta}{\sin\theta},$$

where F_s represents the force in the spring. The last expression is obtained by computing the moment of the forces acting in node A, with respect to hinge B, and equating to zero. The relation obtained between P and θ is represented in Fig. 158c. From this diagram we conclude that the structure has a post-critical behaviour which is unstable for $\theta > 0$ and is stable for $\theta < 0$.

XI.2.c Effect of Imperfections

In the two Sub-sections above, ideal structures have been analysed, that is, the three columns represented in Fig. 155 and the loads P are perfectly vertical and the loads act without any eccentricity on the centroids of the upper cross-sections. Under these conditions, no internal forces are necessary in the deformable elements (the springs) to balance the load P, so the analysis has led to the conclusion that these columns do not deform until the critical load is exceeded.

However, the unavoidable imperfections of the structures may influence their stability behaviour considerably, with respect to the value of the critical load, and even in terms of the characteristics of the deformation. In order to find out about this influence, we analyse the three columns considered in the previous Sub-sections, when they are affected by an imperfection represented by a residual deformation.

Let us first consider the column represented in Fig. 159a which corresponds to the column represented in Fig. 156a with an imperfection represented by the residual deformation α, when P vanishes.

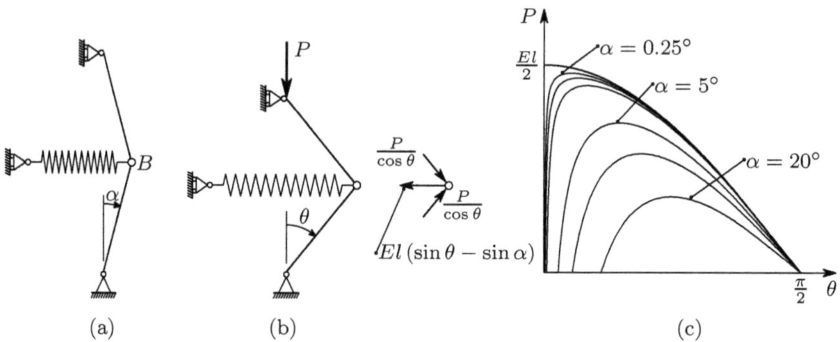

Fig. 159. Effect of an imperfection in a structure with unstable post-critical behaviour

The relation between P and θ may be computed in the same way as in Subsect. XI.2.b. The condition of horizontal equilibrium of the forces acting in node B (Fig. 159b) yields

$$2\frac{P}{\cos\theta}\sin\theta = El(\sin\theta - \sin\alpha) \Rightarrow P = \frac{El}{2}\overbrace{\frac{\cos\theta(\sin\theta - \sin\alpha)}{\sin\theta}}^{\gamma} . \tag{249}$$

In Fig. 159c curves representing this relation for several values of the imperfection parameter α are represented. We find that, contrary to the column without imperfection, the deformation is present for any value of the load, and that, even small values of the imperfection substantially reduce the maximum value of load the P. This value is given in the following table for the same values of α used in Fig. 159c ($\gamma_{\max}$ is the maximum value of parameter γ defined in (249)).

α	$0°$	$0.25°$	$0.5°$	$1°$	$5°$	$10°$	$20°$
$\gamma_{\max}$	1.000	0.960	0.937	0.901	0.720	0.572	0.365

If we consider now the second of the structures analysed in Subsect. XI.2.b with an imperfection represented by angle α (Fig. 160a), the equilibrium condition represented by (248) takes the form (Fig. 160b)

$$Pl \sin \theta = 2G \left(\theta - \alpha \right) \Rightarrow P = \frac{2G}{l} \frac{\theta - \alpha}{\sin \theta} . \tag{250}$$

Figure 160c shows the curves represented this relation, considering the same values of the imperfection parameter α as in Fig. 159c.

Also in this case we verify that the deformation takes place for any value of the load P. But this deformation suffers a significant increase when P approaches the value corresponding to the critical load of the perfect column. When compared with the column in Fig. 159, the principal difference is that in this case a maximum value of P is not found, even for large values of the imperfection. This difference is a consequence of the fact that this column has a stable post-critical behaviour, while the column in Fig. 159 has unstable behaviour. However, in actual structures the large deformations caused by the imperfection, when the load gets close to the critical value, do limit the loading capacity, even in the case of stable post-critical behaviour, as we will see in Sub-Sects. XI.2.d and XI.4.a.

The analysis of the third column with imperfections may be performed in a similar way. However, since in this case the post-critical behaviour is not symmetric, we must consider imperfections for both sides of the equilibrium configuration. Defining the imperfection by means of the angle α represented in Fig. 161a and using a similar line of reasoning as in the perfect structure (Fig. 158), we arrive at the expressions which allow the computation of the load P corresponding to the deformed configuration defined by angle θ (Fig. 161b)

$$\gamma = \arctan \frac{\cos \theta}{1 + \sin \theta} \quad \longrightarrow \quad F_s = El \left(\frac{1 + \sin \theta}{\cos \gamma} - \frac{1 + \sin \alpha}{\cos \gamma_0} \right)$$

$$\longrightarrow \quad P = F_s \frac{\cos \gamma \cos \theta - \sin \gamma \sin \theta}{\sin \theta} \quad \text{with} \quad \gamma_0 = \arctan \frac{\cos \alpha}{1 + \sin \alpha} .$$

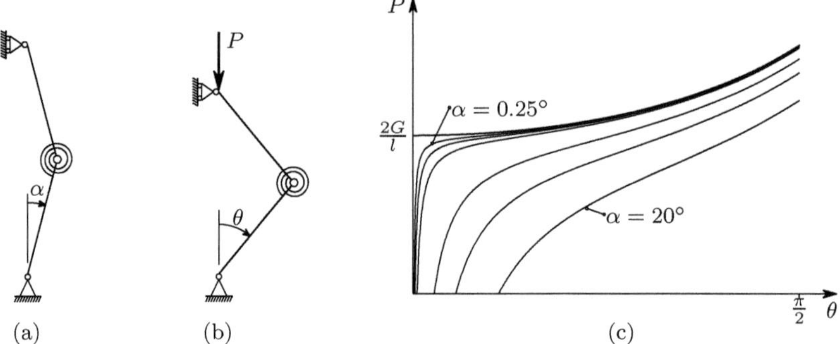

Fig. 160. Effect of an imperfection in a structure with stable post-critical behaviour

From these expressions, using the same values of the imperfection parameter α as in the two previous examples, we get the curves shown in Fig. 161c. We conclude that the effect of the imperfection is similar to that found in the first example (Fig. 159), in the case of positive values of α (unstable post-critical behaviour), while for negative values of the imperfection parameter α the behaviour is similar to that of the second example (Fig. 160 – stable post-critical behaviour).

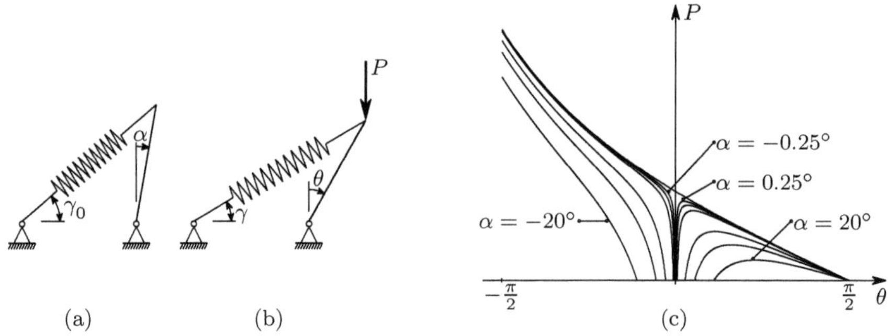

Fig. 161. Effect of an imperfection in a structure with non-symmetrical post-critical behaviour

The analysis expounded in this and the previous Sub-sections illustrates two kinds of structural instability:

– in the perfect structures considered forces are not necessary in the deformed elements, in order to guarantee the equilibrium, so that, before the critical load is reached, no deformation takes place and there is only one equilibrium configuration; when the critical value is attained more than one equilibrium configuration is possible for the same external load; this kind of instability is called *instability by equilibrium bifurcation;*

– in the imperfect structures the deformation appears for any value of the load, suffering a large increase when the load gets close to the critical value of the corresponding perfect structure; in this case, we have the so-called *divergence instability*;[1]

XI.2.d Effect of Plastification of Deformable Elements

When the deformable elements of a compressed structure enter the elasto-plastic regime, the corresponding loss of stiffness usually causes a considerable reduction in the maximum load of the structure. In order to get an idea about the importance of this reduction and the conditions under which it occurs, let us consider the example depicted in Figs. 155c and 157a, with the elasto-plastic behaviour of the deformable element – the rotation spring – given by the moment-rotation diagram depicted in Fig. 162.

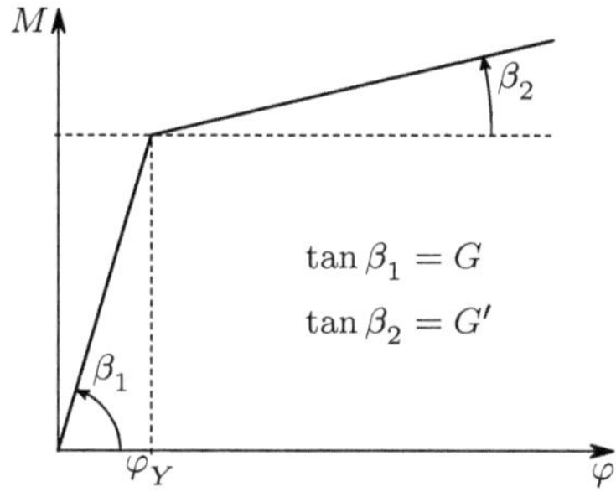

Fig. 162. Elasto-plastic behaviour of the rotation spring in the column represented in Fig. 157a.

If θ is smaller than $\frac{\varphi_Y}{2}$, the relation between θ and the load P is given by (248). When the rotation exceeds the value corresponding to yielding, φ_Y, which happens for higher values of θ than $\frac{\varphi_Y}{2}$ (Fig. 155c), this relation takes the form

$$\theta > \frac{\varphi_Y}{2} \quad \Rightarrow \quad Pl\sin\theta = G\varphi_Y + G'\left(2\theta - \varphi_Y\right) \quad \Rightarrow \quad P = \frac{2G\,\frac{\varphi_Y}{2}\left(1-\gamma\right)+\theta\gamma}{l}\frac{}{\sin\theta} ,$$

---

[1] This definition is inspired by the small deformation theory where a post-critical analysis is not carried out. For this reason, some authors do not consider this divergence as instability, considering instead another kind of instability called "snap-through", which corresponds to a vanishing stiffness, as in the imperfect cases depicted in Fig. 159. According to this definition, instability never occurs in the imperfect cases represented in Fig. 160. The definition of divergence instability is retained here, since it is useful in the definition of the point where the interaction between deformation and internal forces becomes important.

with $\gamma = \frac{G'}{G}$. In Fig. 163 the relations between P and θ, described by this equation for $\theta > \frac{\varphi_Y}{2}$ and by (248), for $\theta \le \frac{\varphi_Y}{2}$, are depicted taking several values of γ and φ_Y.

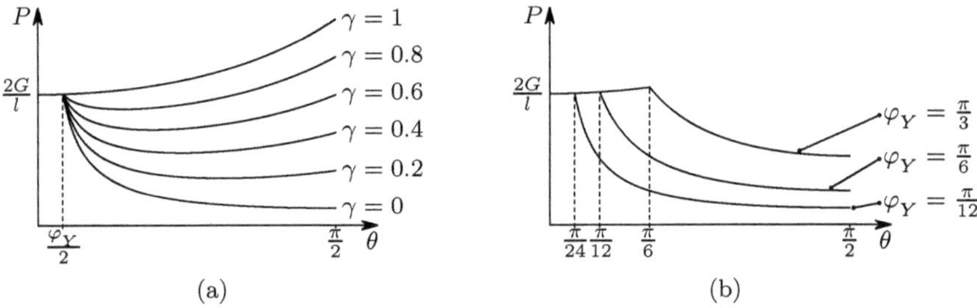

Fig. 163. P-θ relations of the column represented in Fig. 157a, with the constitutivelaw of the rotation spring defined in Fig. 162, for several values of the (**a**) elasto-plastic stiffness of the spring, $G' = \gamma G$; (**b**) yielding rotation of the spring, φ_Y, for $\gamma = 0$.

We conclude that yielding transforms the stable post-critical behaviour into unstable, since, after yielding, an increase in the deformation causes a decrease of the corresponding load P. Furthermore, especially for small values of the elasto-plastic stiffness $G' = \gamma G$ and of the yielding rotation of the spring, φ_Y, the failure of the column has the characteristics of a brittle failure, although the behaviour of the spring is ductile. In fact, the yielding of the spring substantially reduces the loading capacity of the column, which practically vanishes when γ and φ_Y take small values.

Since the post-critical behaviour may become unstable when elasto-plastic deformations take place, it is very important to investigate the influence of imperfections on the loading capacity of the column. To this end, let us consider the same column with the imperfection represented by the residual deformation α depicted in Fig. 160a. For smaller values of θ (Fig. 160b) than $\alpha + \frac{\varphi_Y}{2}$ the relation between the load P and the rotation θ is given by (250). For larger deformations the same relation takes the form

$$\theta > \alpha + \frac{\varphi_Y}{2} \Rightarrow Pl\sin\theta = G\varphi_Y + G'\left[2\theta - (2\alpha + \varphi_Y)\right]$$

$$\Rightarrow P = \frac{2G}{l}\frac{\frac{\varphi_Y}{2} + \gamma\left(\theta - \alpha - \frac{\varphi_Y}{2}\right)}{\sin\theta} \quad \text{with} \quad \gamma = \frac{G'}{G}.$$

In Fig. 164 this relation is represented for the particular case of elastic perfectly plastic behaviour ($\gamma = 0$), considering two values for the yielding rotation of the spring, φ_Y, and the same values for the imperfection parameter, α, which have been used in Figs. 159 to 161.

We conclude that even small values of the imperfection cause generally a considerable reduction in the loading capacity of the column. However, in the particular case of a high yielding rotation φ_Y and of a very small value of the

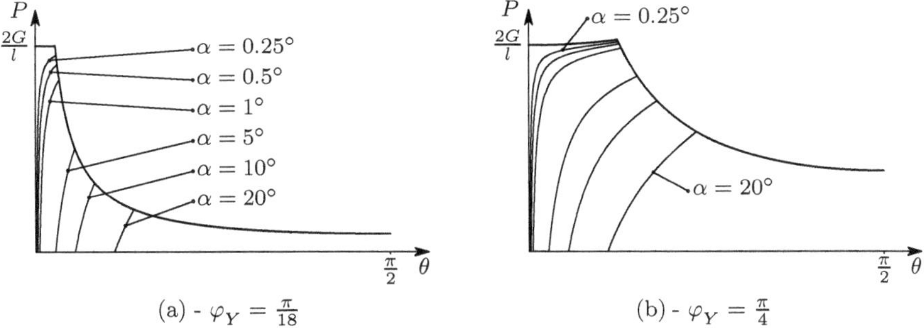

Fig. 164. Effect of an imperfection α, in the case of elastic perfectly plastic behaviour

imperfection parameter α, it is possible to have a maximum load superior to the critical load of the perfect structure ($\frac{2G}{l}$), as shown in Fig. 164-b.

The conclusions reached in relation to the behaviour of the column represented in Fig. 157-a are also valid qualitatively for the case of a compressed prismatic bar, which is of great practical interest and is analysed in the next sections.

XI.3 Instability in the Axial Compression of a Prismatic Bar

XI.3.a Introduction

The main difference between the buckling of an axially compressed prismatic bar and the instability behaviour of the simple examples analysed in Sect. XI.2 is that in the prismatic bar the deformation is distributed throughout the whole bar, while in the simple examples it is concentrated in a single deformable element. Furthermore, those examples have a degree of kinematic indeterminacy of one, which means that only one quantity is needed to define the deformed configuration, while in the prismatic bar the degree of kinematic indeterminacy is infinite, as we will see. As a consequence the analysis of the buckling phenomenon is substantially more complex than the simple examples analysed in Sect. XI.2.

For these reasons, in the analysis presented in the remaining of this Chapter only small deformations are considered, which means that the post-critical behaviour cannot be analysed. However, as mentioned in Subsect. XI.2.a, the buckling of a prismatic bar is qualitatively analogous to the example presented in Figs. 155-c and 157, displaying also a stable post-critical behaviour [11].

XI.3.b Euler's Problem

Euler has found the critical load of an axially compressed bar by directly analysing the critical state. As we saw in Sect. XI.1 and confirmed in Subsect. XI.2.a, in the transition from a stable to an unstable state – the critical state – we have a state of neutral equilibrium, which means that no forces are needed to introduce a small change in the deformation state of the structure. Thus, the critical axial load of a prismatic bar is a force which balances the internal forces (the bending moment, in this case) in a slightly deflected configuration of the bar, without the need of transversal forces, as represented in Fig. 165-a.

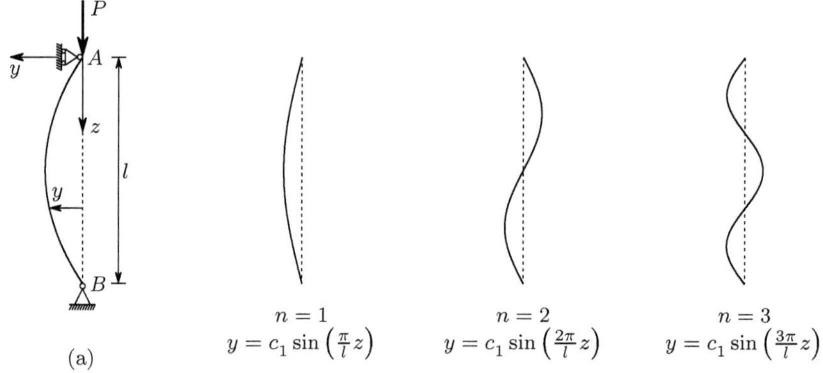

Fig. 165. Euler's problem: problem definition and buckling shapes

In the deformed configuration depicted in Fig. 165-a the bending moment is given by the expression $M = Py$. Since we are considering only infinitesimal displacements, the rotations will be small. Furthermore, taking only linear elastic material behaviour into consideration, the relation between the bending moment and the deflection y is given by the second of (210). Thus, the deformed configuration of the bar under the conditions represented in Fig. 165-a is given by the solution of the differential equation

$$\begin{cases} M = Py \\ \dfrac{|M|}{EI} = \left|\dfrac{\mathrm{d}^2 y}{\mathrm{d}z^2}\right| \end{cases} \Rightarrow Py = -EI\dfrac{\mathrm{d}^2 y}{\mathrm{d}z^2} \Rightarrow \dfrac{\mathrm{d}^2 y}{\mathrm{d}z^2} + \dfrac{P}{EI}y = 0 . \tag{251}$$

The minus sign affecting the bending stiffness results from the fact that a positive value of y leads to a negative value of $\frac{\mathrm{d}^2 y}{\mathrm{d}z^2}$ and vice versa. Equation (251) is a linear homogeneous equation with constant coefficients, whose solution takes the form

$$y = C_1 \sin(kz) + C_2 \cos(kz) \quad \text{with} \quad k = \sqrt{\dfrac{P}{EI}} . \tag{252}$$

The compatibility conditions at the supports yield the relations[2]

$$z = 0 \;\Rightarrow\; y = 0 \;\Rightarrow\; C_2 = 0$$
$$z = l \;\Rightarrow\; y = 0 \;\Rightarrow\; C_1 \sin(kl) = 0 \;\Rightarrow\; C_1 = 0 \quad \vee \quad kl = n\pi \;.$$

(253)

The condition of zero deflection in support B shows that the integration constant C_1 is either zero, or indeterminate ($\sin(kl) = 0$). The first possibility means that the straight configuration is always an equilibrium configuration (although it may be unstable). In the second possibility, the fact that C_1 is indeterminate reflects the fact that the equilibrium is neutral, which means that the amplitude of the deflections does not influence the equilibrium. The last of (253) yields a relation between the load P, the bending stiffness EI and the bar's length l

$$kl = n\pi \;\Rightarrow\; k^2 = \frac{n^2\pi^2}{l^2} = \frac{P}{EI} \;\Rightarrow\; P = \frac{n^2\pi^2 EI}{l^2} \;.$$

This expression furnishes an infinite number of values of the axial load P ($n = 1, 2, 3, \ldots, \infty$), for which P balances the bending moments in a slightly deflected configuration. The deformed bar has a sinusoidal shape as indicated by (252) ($kl = n\pi$ and $C_2 = 0$). In Fig. 165 these shapes are shown for the three smallest values of n. The critical load is that corresponding to $n = 1$, since it is the minimum value of P which satisfies the necessary (but not sufficient) condition of neutral equilibrium. It may be shown that the configurations corresponding to values of n superior to one are unstable, unless the deflection of the points with $y = 0$ is prevented. The critical load and the corresponding critical stress are then

$$P_{cr} = \frac{\pi^2 EI}{l^2} \;\Rightarrow\; \sigma_{cr} = \frac{P_{cr}}{\Omega} = \frac{\pi^2 E}{\lambda^2} \quad \text{with} \quad \lambda = \frac{l}{\sqrt{\dfrac{I}{\Omega}}} = \frac{l}{i} \;. \tag{254}$$

The first expression is known as *Euler's formula* and the value P_{cr} as the *Euler buckling load*. The non-dimensional quantity λ is called the *slenderness ratio*. It depends only on the geometry of the bar, represented by the length l and the radius of gyration i of its cross-section. Associated with the modulus of elasticity of the material it completely defines the critical stress.

XI.3.c Prismatic Bars with Other Support Conditions

Equation (254) is only valid for bars with the support conditions shown in Fig. 165-a, i.e., for pin-ended bars. However, in some cases with different support conditions, symmetry considerations allow the generalization of Euler's

---

[2]In this analysis the axial deformation does not need to be considered, since the instability is caused by the interaction between the bending deformation and the bending moment caused by the axial force P in a deflected configuration. The axial deformation does not play a role in this interaction.

solution (254), so that the corresponding differential equations do not need to be solved.

In the case of the cantilever represented in Fig. 166-a the critical configuration has the same shape as a pin-ended bar with the length $2l$, which means that the relation between the load P and the bending moment is the same. Thus, the corresponding critical load takes the value

$$P_{cr} = \frac{\pi^2 EI}{(2l)^2} = \frac{\pi^2 EI}{l_e^2} \quad \text{with} \quad l_e = 2l \ . \tag{255}$$

This value will be confirmed below in the study of the case of composed bending (Subsect. XI.4.a). The quantity l_e is called the *effective length* of the column. It may be defined as the length that a pin-ended bar with the same cross-section would need, in order to have the same critical load as the bar under consideration.

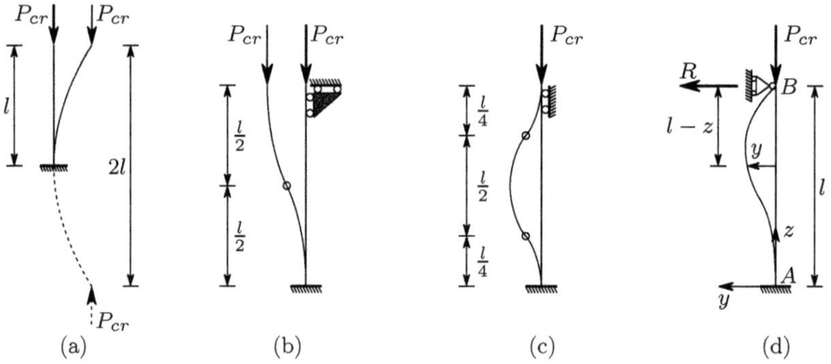

Fig. 166. Determination of effective lengths

Symmetry and antisymmetry considerations easily lead to the conclusion that the effective lengths of the bars represented in Figs. 166-b and 166-c are, respectively $l_e = 2\frac{l}{2} = l$ and $l_e = \frac{l}{2}$.

The critical load of the column represented in Fig. 166-d cannot be obtained from Euler's solution (254), since there are no symmetry or antisymmetry conditions. This load must, therefore, be computed by solving the differential equation of equilibrium in the critical phase, as was done for Euler's problem. To this end, let us consider the reference axes shown in Fig. 166-d. Considering as positive the bending moments corresponding to a positive curvature in this reference system and the reaction force R with the shown direction, we get

$$\begin{cases} M(z) = -Py + R(l - z) \\[2mm] \dfrac{M}{EI} = \dfrac{\mathrm{d}^2 y}{\mathrm{d}z^2} \end{cases} \Rightarrow \frac{\mathrm{d}^2 y}{\mathrm{d}z^2} + \frac{P}{EI}y = \frac{R}{EI}(l - z) \ .$$

This equation has a homogeneous part which coincides with (251) and admits the particular integral $y = \frac{R}{P}(l - z)$, so that its general solution takes the form

$$y = C_1 \cos(kz) + C_2 \sin(kz) + \frac{R}{P}(l - z) \quad \text{with} \quad k^2 = \frac{P}{EI} \,. \tag{256}$$

The support conditions in the built-in end A define conditions expressing the integration constants as functions of the reaction force R, yielding

$$z = 0 \Rightarrow \begin{cases} y = 0 \Rightarrow C_1 = -\dfrac{Rl}{P} \\[2mm] \dfrac{\mathrm{d}y}{\mathrm{d}z} = 0 \Rightarrow C_2 k - \dfrac{R}{P} = 0 \Rightarrow C_2 = \dfrac{R}{Pk} \,. \end{cases}$$

By substituting these values in the general solution (256) and using the support condition in end B, we get

$$z = l \Rightarrow y = 0 \Rightarrow \frac{R}{P}\left[\frac{\sin(kl)}{k} - l\cos(kl)\right] = 0$$
$$\Rightarrow \quad R = 0 \quad \vee \quad \frac{\sin(kl)}{k} - l\cos(kl) = 0 \Leftrightarrow kl = \tan(kl) \,. \tag{257}$$

The first possibility, $R = 0$, corresponds to the equilibrium without buckling, since it leads to $y(z) = 0$. The second possibility, $kl = \tan(kl)$, may be used to find the critical load, since it corresponds to a deflected configuration, as R may be different from zero. In this case the value of R is indeterminate, which reflects the fact that the equilibrium is neutral in the critical phase. The value of the critical load may be computed by solving the transcendental equation $kl = \tan(kl)$. The minimum value of kl which satisfies this condition is $kl = 4.493409$. The corresponding critical load may be expressed as a function of an effective length l_e, yielding

$$P_{cr} = EI k^2 = \frac{EI}{l^2} 4.493409^2 = \frac{\pi^2 EI}{\left(\dfrac{\pi}{4.493409}l\right)^2} = \frac{\pi^2 EI}{l_e^2}$$

$$\text{with} \quad l_e = \frac{\pi}{4.493409}l = 0.6992l \approx 0.7l \,.$$

Euler's formula may thus be used for this column, provided that an effective length $l_e = 0.7l$ is considered.

The critical load of columns with other support or loading conditions, such as bars with intermediate supports, distributed axial loads, variable cross-section, etc., may be computed by solving the corresponding differential equations (see examples XI.11 to XI.13).

XI.3.d Safety Evaluation of Axially Compressed Members

The theory expounded so far in this Section is valid for a perfectly prismatic bar with a perfectly axial force, i.e., a force whose line of action coincides with the line defined by the centroids of the cross-sections. Under these conditions, the theory indicates that the bar axis remains straight until the critical load is attained. However, actual members always have imperfections, both in the way the load is applied (eccentricity with respect to the centroid of the cross-sections or inclination with respect to the bar axis) and with respect to the geometry of the bar (residual curvature, non perfectly constant cross-section, etc.). As a consequence of these unavoidable perturbations, the axial force causes bending even when it takes a value which is smaller that the critical load, as shown in the introductory examples with imperfections (Subsect. XI.2.c), and this will be confirmed in the study of the case of bending caused by an eccentric axial force (Sect. XI.4). The bending deformation introduces additional stresses, which become larger when the load gets close to the critical value. As a consequence, the critical load predicted by Euler's formula is usually not reached, since plastic deformations or material failure take place before this point. Residual stresses introduced into the bar by the manufacturing process are also an imperfection, since fibres with a residual compressive stress may reach the limit of proportionality before the computed critical stress (254) is attained, which would influence the value of the critical load (see below).

In addition, as seen in Subsect. XI.2.d, the buckling failure of a bar has the characteristics of a brittle failure, since its axial strength after buckling is practically reduced to zero, even in the case of ductile material behaviour.

For these two reasons, in the safety evaluation of long members (bars in which Euler's critical stress is smaller than the proportionality limit stress) by means of Euler's formula, a supplementary safety coefficient, ψ, is used. Thus, in the design of long axially compressed bars, the following condition must be satisfied:

$$P \leq \frac{\pi^2 EI}{\psi l_e^2} \Rightarrow \sigma \leq \frac{\pi^2 E}{\psi \lambda^2} \, .$$

Coefficient ψ is usually considerably greater than one. A current value is $\psi = 1.8$.

In the case of columns with intermediate slenderness, i.e., bars which are stable for stresses higher than the proportionality limit, Euler's theory may still be used, provided that the tangent modulus of elasticity corresponding to the critical stress is used. Since this stress itself depends on the elasticity modulus, Euler's critical load must be computed by means of an iterative approach (see example XI.16). In order to avoid this process, and also because it is not easy to define the exact value of the tangent modulus of elasticity, the safety of columns with intermediate slenderness is commonly evaluated by means of experimentally obtained curves which give the critical buckling stress as a function of the slenderness ratio.

In the case of steel structures, the so-called Tetmeyer's line is often recommended, especially in the codes of European countries. Tetmeyer has approximated by straight lines the experimental results for the relation between the slenderness ratio and critical stress in columns with intermediate slenderness.

These lines pass through the point where Euler's theory ceases to be valid, i.e., the point whose ordinate is the proportionality limit stress. For very low values of the slenderness ratio (short columns) this line may give results which are higher than the yielding stress of the material, so that Tetmeyer's line is not applicable to such columns. In these cases there is no buckling failure, which means that the limit stress is the maximum allowable stress for the material under consideration.

As an alternative to the Tetmeyer's line, an approximation by means of a parabola, known as Johnson's parabola, is also used. This curve has a vanishing tangent at the point corresponding to zero slenderness ($\lambda = 0$, $\sigma_{cr} = \sigma_Y$), is tangent to the hyperbola defined by Euler's theory $\sigma_{cr} = \frac{\pi^2 E}{\lambda^2}$ and is valid for columns with a lower slenderness than that corresponding to the contact point between these two curves. It may easily be verified that the coordinates of this point are $\lambda = \pi\sqrt{\frac{2E}{\sigma_Y}}$ and $\sigma_{cr} = \frac{\sigma_{all}}{2}$ and that the parabola has the equation

$$\sigma_{cr} = \sigma_Y - \frac{\sigma_Y^2}{4\pi^2 E}\lambda^2 \ .$$

Figure 167 gives the two approximations. When used for safety evaluation, the values defined by these curves must be affected by safety coefficients which, as mentioned above, should have higher values in the case of long members.

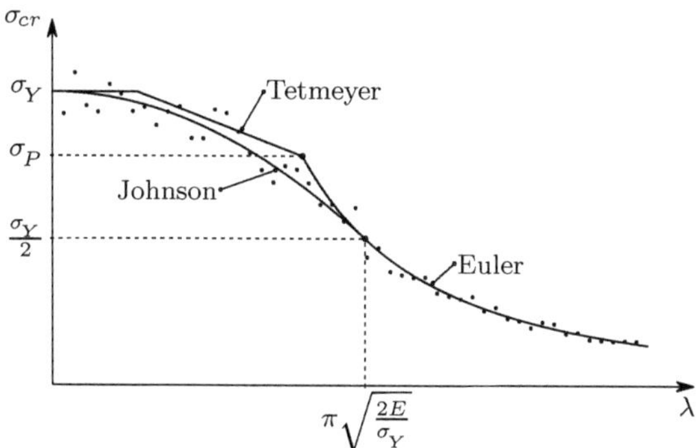

Fig. 167. Approximation of experimental values in axially compressed members · · · · · · experimental —— results

In the case of Tetmeyer's approximation, it is usually considered that for slenderness ratios below 20 there is no buckling failure. The Tetmeyer's straight line is thus used for slenderness ratios between 20 and the value corresponding to the limit of proportionality, σ_p, which depends on the steel type under consideration. A value of $\sigma_p \approx 0.8\sigma_Y$ is considered in some codes, which leads to the following values of the slenderness ratio

$$\frac{\pi^2 E}{\lambda^2} = 0.8\,\sigma_Y \;\Rightarrow\; \lambda = \pi\sqrt{\frac{E}{0.8\,\sigma_Y}} \;\Rightarrow\; \begin{cases} \lambda &=& 105\,(S\,235) \\ \lambda &=& 96\;\;(S\,275) \\ \lambda &=& 85\;\;(S\,355)\,, \end{cases}$$

where $S\,235$, $S\,275$ and $S\,355$ are current steel grades, which have the nominal yielding stresses of 235, 275 and 355 MPa, respectively. These considerations are summarized in the design curves represented in Fig. 168.

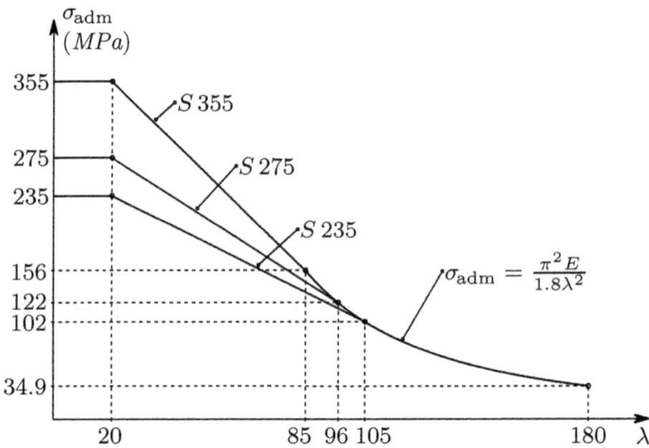

Fig. 168. Design curves for axial compression for current steel grades ($\psi = 1.8$)

From these curves we conclude that the use of high-strength steel in bars with a larger than 105 slenderness ratio does not increase the loading capacity. This is because the only material parameter entering Euler's formula is the modulus of elasticity E which takes the same value for all steel grades.

If the bar has the same effective length in both principal bending planes, buckling takes place by bending around the principal axis of inertia with the lowest value of the moment of inertia, since the maximum slenderness corresponds to the minimum value of the radius of gyration (254). However, very often the effective length is not the same in the two principal bending planes, and the smallest effective length corresponds to the smallest moment of inertia. In these cases buckling takes place in the bending plane with the highest slenderness ratio (see example XI.8).

If the cross-section of the column does not have two symmetry axes, the centroid and the shear centre will have distinct positions. In these cases, buckling may be accompanied by torsion, especially in bars with a small torsional stiffness, such as open thin-walled cross-sections. This torsional effect is due to small inclinations of the axial force with respect to the bar axis causing a small shear force which induces torsion because it does not pass through the shear centre.

There is also the possibility of torsional buckling, even in the case of doubly symmetric cross-sections. This kind of instability may occur in columns with very thin cross-section walls and higher bending stiffness. The general study of these phenomena is beyond the scope of this text, however, because it is a relatively complex analysis which has no practical relevance, since, in usual cross-sections, bending instability occurs before torsional instability. However, an example of the computation of the torsional buckling load in a cruciform cross-section is included (see example XI.15).

XI.3.e Optimal Shape of Axially Compressed Cross-Sections

From what has been explained above, we conclude that in most cases buckling takes place by bending around the principal axis of inertia which corresponds to the highest value of the slenderness ratio. If the cases with very different effective lengths in the two principal bending planes are excluded, we conclude that, in order to get high values of buckling strength, high flexural stiffness in both principal bending planes is required. Since the material around the centroid of the cross-section makes a lower contribution to the bending stiffness, the cross-section of compressed members should have the most part of the area far from the centroid and should have similar values in the two principal moments of inertia. Furthermore, torsional effects in buckling may be avoided by giving the cross-section a high torsional stiffness. A closed thin-walled cross-section, such as a square or circular tube, satisfies all of these conditions.

XI.4 Instability Under Composed Bending

XI.4.a Introduction and General Considerations

Instability phenomena under composed bending with a compressive axial force may be analysed using the same tools as in the case of purely axial compression. There is, however, a much larger variety of differential equations to be solved, since a distinct equation is obtained for each kind of transversal displacement. For this reason, alternative, generally semi-empirical, expressions are frequently used in the safety evaluation of compressed and bent members. However, before some of those alternative expressions are presented, we first analyse in detail a particular case of composed bending with compressive axial force, in order to facilitate the physical understanding of the phenomenon of interaction between deformation and internal forces.

Let us consider the cantilever column represented in Fig. 169 which is eccentrically compressed by a force P. The differential equation expressing the interaction between the deformation and the bending moment may be obtained from the relations between bending moment and curvature and between bending moment and deflection. Considering only small rotations, we have

$$\begin{cases} M = P\left(\delta + e - y\right) \\ M = EI\dfrac{\mathrm{d}^2 y}{\mathrm{d}z^2} \end{cases} \Rightarrow \frac{\mathrm{d}^2 y}{\mathrm{d}z^2} + k^2 y = k^2\left(\delta + e\right) \quad \text{with} \quad k^2 = \frac{P}{EI}\,. \tag{258}$$

This equation admits the particular integral $y = \delta + e$ and has the same homogeneous solution as (251). The general solution is then

$$y = C_1 \sin(kz) + C_2 \cos(kz) + \delta + e\,.$$

The integration constants are computed from the support conditions in cross-section A, yielding

$$z = 0 \Rightarrow \begin{cases} y = 0 \Rightarrow C_2 = -\delta - e \\ \dfrac{\mathrm{d}y}{\mathrm{d}z} = 0 \Rightarrow C_1 = 0 \end{cases} \Rightarrow y = (\delta + e)\left[1 - \cos(kz)\right]\,.$$

The relation between the maximum deflection δ and load P is then

$$z = l \Rightarrow y = \delta = (\delta + e)\left[1 - \cos(kl)\right] \Rightarrow \delta = e\frac{1 - \cos(kl)}{\cos(kl)}\,. \tag{259}$$

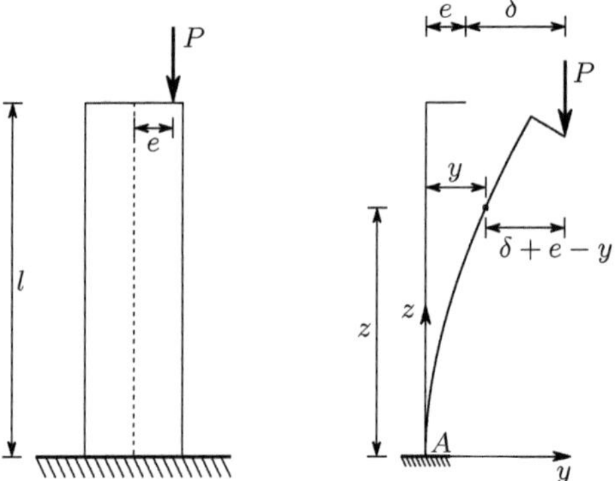

Fig. 169. Eccentric compression of a column

We conclude that when kl reaches the value $\frac{\pi}{2}$ the displacement δ becomes infinite. This means that the critical load has been attained, since the column has lost stiffness to oppose the bending moment caused by load P. In the case of Euler's problem this bending moment vanishes, and so we have neutral equilibrium in the critical phase. Remembering the analogy with the example of the sphere (Fig. 154), this situation corresponds to having a small horizontal force applied to the sphere: in the stable situation this force causes a small displacement in relation to the point with zero slope; in the critical situation (Fig. 154-c) this force would cause an infinite displacement. This kind of instability, where the deformation gradually increases, going to infinite values when the load attains the critical load, corresponds to the divergence instability which has been defined at the end of Subsect. XI.2.c. The fact that an infinite deformation is obtained for $P = P_{cr}$ is a consequence of the assumption of small rotations, which actually makes a post-critical analysis impossible. As seen above, in the perfect bars under perfectly axial loading there is no deformation until the critical load is attained. When this happens, there is equilibrium in both the undeformed and slightly deformed configurations. In this case we have the bifurcation instability defined in Subsect. XI.2.c. As already mentioned (Subsect. XI.2.c), actual columns always exhibit divergence instability, since imperfections are unavoidable.[3]

The critical load is then

$$kl = \frac{\pi}{2} \ \Rightarrow \ P_{cr} = \frac{\pi^2 EI}{(2l)^2} \ .$$

This value coincides with that given by (255), which means that the bending moment introduced by the eccentricity of the load does not influence the critical load. This conclusion may be generalized to other loading cases, since, with the exception of the cases where the applied loads depend on the deformation (non-conservative loads), the bending moment only influences the particular solution of the differential equation, while the critical load is determined by $\sin(kz)$ or $\cos(kz)$, which appear in the solution of the homogeneous equation. A physical interpretation of the same phenomenon is that there is no interaction between this bending moment and the deformation, that is, the deformation does not increase the additional bending moment introduced by the eccentricity of the axial load.

In composed bending with a compressive axial force the maximum applicable load is always smaller than the critical load, even in the case of long members, since, as a consequence of the additional stresses introduced by bending, the maximum allowable stress is reached before the load attains the critical value. The maximum stress introduced by eccentric loading may be computed by means of the so-called *secant formula*, which is obtained from (259), yielding

$$\begin{cases} M_{\max} = P(\delta + e) = \frac{Pe}{\cos(kl)} \\ \sigma_{\max} = \frac{P}{\Omega} + \frac{M_{\max}}{\frac{I}{v}} \end{cases} \Rightarrow \sigma_{\max} = \frac{P}{\Omega} + \frac{Pe}{\frac{I}{v}\cos(kl)} \ . \tag{260}$$

We conclude that the interaction between the bending moment and the deformation causes an increase in the stress caused by bending which is represented by the factor $\frac{1}{\cos(kl)} = \sec(kl)$.[4]

---

[3]If we take $e = 0$ in (259) (purely axial loading), we also get an indeterminate value for the deflection δ, if $kl = n\frac{\pi}{2}$.

[4]The stresses caused by the axial force and by the bending moment may be computed separately and added together, although the superposition principle, in its

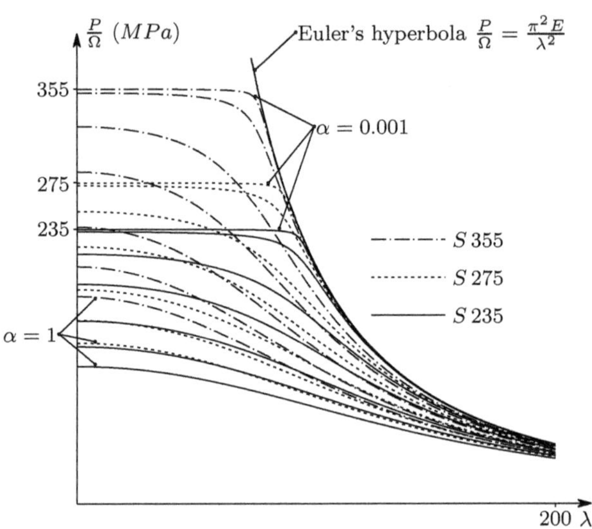

Fig. 170. Eccentric axial loads which may be applied to steel columns for several values of the eccentricity ratio α. α=0.001 α=0.01 α=0.1 α=0.25 α=0.5 α=0.75 α=1

The maximum value of the eccentric load P, for a given value of the allowable stress, σ_{all}, may be related to the slenderness ratio of the bar. This relation may be obtained from (260), yielding

$$
\begin{cases}
kl = \sqrt{\dfrac{Pl^2}{EI}} = \sqrt{\dfrac{Pl_e^2}{4E\Omega i^2}} = \sqrt{\dfrac{\lambda^2}{4E}\dfrac{P}{\Omega}} \\[2mm]
\sigma_{\max} = \sigma_{\text{all}}
\end{cases}
\Rightarrow
\quad
\left|
\begin{aligned}
&\frac{P}{\Omega} + \alpha\,\frac{\frac{P}{\Omega}}{\cos\sqrt{\dfrac{\lambda^2}{4E}\dfrac{P}{\Omega}}} = \sigma_{\text{all}} \\[2mm]
&\text{with}\quad \alpha = \frac{ve}{i^2}\,.
\end{aligned}
\right.
$$

$$(261)$$

The relation between $\frac{P}{\Omega}$, λ and σ_{all} depends on the eccentricity of the load which is expressed by the non-dimensional parameter α, called the *eccentricity ratio*. By ascribing fixed values to α and σ_{all}, a relation between λ and $\frac{P}{\Omega}$ is obtained. This is actually a transcendental equation, so an explicit expression for $\frac{P}{\Omega}$ cannot be found. The value of $\frac{P}{\Omega}$ corresponding to a given value of λ may, however, be obtained by numerical means using, for example, the Newton-Raphson algorithm. This has been done for the three steel grades considered above (Fig. 168), yielding the results represented by the curves depicted in Fig. 170.

---

most general form, cannot be applied when the interaction between internal forces and deformations is taken into account. In fact, the non-linearity of the problem appears in the relation between external and internal forces (P and $M_{\max}$ in this case) and not in the relation between internal forces and stresses.

From these curves we conclude that, when the eccentricity goes to zero, the load capacity of the column tends to the value defined by Euler's formula, if the corresponding critical stress is smaller than the maximum allowable stress σ_{all} (i.e., if $\frac{\pi^2 E}{\lambda^2} \leq \sigma_{\text{all}}$), while in less slender columns it tends to the maximum allowable stress of the steel grade under consideration. Obviously, these curves do not take into consideration the fact that the proportionality limit stress may be smaller than the maximum allowable stress.

The maximum deflection δ may be expressed as a function of the ratio between the applied and the critical loads, $\frac{P}{P_{cr}}$, and of the eccentricity e. From (259), we get

$$kl = \sqrt{\frac{Pl^2}{EI}} = \sqrt{P\left(\frac{\pi}{2}\right)^2 \frac{(2l)^2}{\pi^2 EI}} = \frac{\pi}{2}\sqrt{\frac{P}{P_{cr}}} \Rightarrow \delta = e\frac{1 - \cos\left(\frac{\pi}{2}\sqrt{\frac{P}{P_{cr}}}\right)}{\cos\left(\frac{\pi}{2}\sqrt{\frac{P}{P_{cr}}}\right)} .$$

Dividing both members of this expression by the length of the column, l, the non-dimensional relations depicted in Fig. 171-a are obtained.

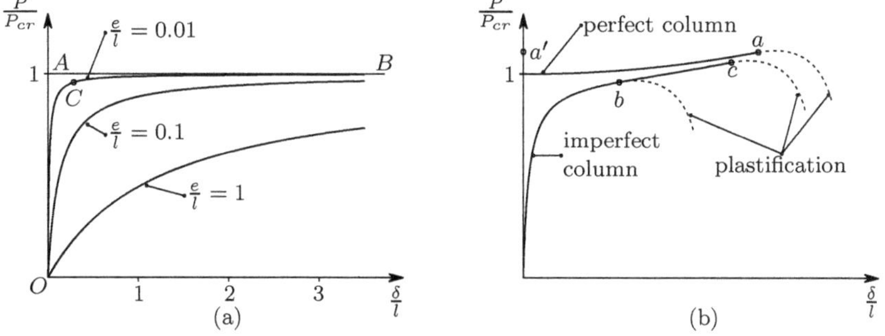

Fig. 171. Deformations caused by an eccentric compression: (a) infinitesimal rotations (b) finite rotations

The maximum value of the stress is attained when the deflection δ, reaches a certain value, since $M = P(\delta + e)$. The diagrams in Fig. 171-a show that, for a small eccentricity, the load-displacement curve is very close to Euler's solution (line OAB) and that the maximum allowable load gets very close to Euler's critical load (point C in the diagram).

Furthermore, we must remember that the differential (258) has been established under the condition of infinitesimal rotations. The exact equation[5]

---

[5]The equation obtained without the restriction to infinitesimal rotations contains, in the place of the second derivative of y with respect to z, the curvature expression defined in (208) or (215). The solution of this equation may be found in reference [11]. This solution has been obtained by means of a Lagrangian formulation of the problem (see Footnote 61 in Chap. IX).

yields the solution represented in a qualitative approach in Fig. 171-b. From this curve we conclude that in a perfect column (that is, without eccentricity) the maximum allowable load is always higher than Euler's critical load (point a on the diagram). This is due to the fact that an elastic column exhibits stable post-critical behaviour (note the similarity between the curves represented in Fig. 171-b and the curves depicted in Fig. 160-c).

The imperfect column generally reaches the maximum allowable stress before the load attains the critical value, as mentioned above (point b on the diagram). However, in the same way as in the simple example analysed in Subsect. XI.2.d, if the column is flexible enough and if the imperfections are small, it may withstand a higher axial force than Euler's critical load, even with a small eccentricity (point c on the diagram).

In the same way as in the second of the simple examples analysed in Subsect. XI.2.b, the bifurcation of the equilibrium state which occurs in the perfect bar, when the load exceeds the critical value, becomes more clear in the case of finite deformations. From Fig. 171-b we conclude that for higher loads than the critical value, there are two equilibrium configurations: one, corresponding to the undeformed configuration of the column (point a') is unstable; the other is stable and corresponds to the deformed configuration represented by the abscissa of point a.

XI.4.b Safety Evaluation

Expressions of the type of (260) might be found for other loading cases (see example XI.14). Some of these solutions are given in reference [3]. In the case of inclined bending, the expression of the stress will have another term, corresponding to the other principal bending moment. In the case of eccentric compression with respect to the two principal axes of inertia of the cross-section, (260) is replaced by the following expression (e_x and e_y are the eccentricities with respect to the principal axes)[6]

$$\sigma_{\max} = \frac{P}{\Omega} + \frac{Pe_x}{\left(\frac{I}{v}\right)_x \cos\left(k_x l\right)} + \frac{Pe_y}{\left(\frac{I}{v}\right)_y \cos\left(k_y l\right)} \quad \text{with} \quad \begin{cases} k_x^2 = \frac{P}{EI_x} \\ k_y^2 = \frac{P}{EI_y} \end{cases}.$$

---

[6]A cross-section with a rectangular convex contour and a symmetry axis is assumed when the section moduli with respect to the principal axes x and y, are used.

It should be noted that both this expression and (260) are deduced without using the superposition principle, since this principle is not valid when the interaction between deformation and internal forces is taken into account, although the linear analysis of composed bending (155) has been carried out on the basis of that principle. In axial compression that is eccentric with respect to the two principal axes of the cross-section, two independent equations of the type of (258) are established. Thus, they may be solved separately, yielding the results $\delta_x = e_x \frac{1-\cos(k_x l)}{\cos(k_x l)}$ and $\delta_y = e_y \frac{1-\cos(k_y l)}{\cos(k_y l)}$ for the displacements of the upper end of the cantilever column in directions y and x, respectively.

When expressions like this are used in the safety evaluation, the following three points must be taken into consideration:

- A semi-probabilistic approach (Subsect. V.9.d) should be used, that is, the load is multiplied by a factor, instead of a safety coefficient being used for the stresses, since they are not proportional to the load.
- These expressions are valid only in the linear elastic phase, i.e., only if the stresses do not exceed the proportionality limit, and lead to an overestimation of the safety degree (that is, the error is disadvantageous to safety), when they are used above this limit.
- When the eccentricity vanishes, only the term $\sigma = \frac{P}{\Omega}$ remains, which obviously does not take the buckling risk into account. For this reason a residual eccentricity in both principal bending planes shall be considered, which corresponds to considering imperfections in how the load is applied or in the geometry of the column. As an alternative, the methods described in Subsect. XI.3.d for axially compressed members may of course be used.

As mentioned at the beginning of this Section, semi-empirical expressions are often used in the safety evaluation of compressed and bent members. The more straightforward ones are based on the generalization of the linear expression for composed bending, (155), using the so-called *buckling coefficient.* This coefficient is defined as the ratio between the maximum allowable stress in axial compression, σ_{adm} (Fig. 168) and the nominal value of the allowable stress for the material the bar is made of, σ_{all}, that is

$$\varphi = \frac{\sigma_{\mathrm{adm}}}{\sigma_{\mathrm{all}}} \leq 1 \,.$$

The simplest equation limits the value of the maximum stress to the value obtained by multiplying the nominal value of the allowable stress defined for the material by the buckling coefficient, which leads to the expression

$$\frac{P}{\Omega\,\varphi\,\sigma_{\mathrm{all}}} + \frac{M_x}{\left(\frac{I}{v}\right)_x \varphi\,\sigma_{\mathrm{all}}} + \frac{M_y}{\left(\frac{I}{v}\right)_y \varphi\,\sigma_{\mathrm{all}}} \leq 1 \,.$$

However, in bars with a large bending moment and a small axial force, this expression leads to exaggerated dimensions, since instability is caused only by the interaction between the moment caused by the axial force and the bending deformation. In these cases, the so-called *interaction formula* may be used, in which the buckling coefficient affects only the term corresponding to the axial force, yielding

$$\frac{P}{\Omega\,\varphi\,\sigma_{\mathrm{all}}} + \frac{M_x}{\left(\frac{I}{v}\right)_x \sigma_{\mathrm{all}}} + \frac{M_y}{\left(\frac{I}{v}\right)_y \sigma_{\mathrm{all}}} \leq 1 \,.$$

It must be noted that this approximate expression may overestimate the loading capacity of the bar (see example XI.17).

In the design codes for metallic structures more elaborate expressions are usually given, which take into account the way the bar is connected to the rest of the structure, the risk of lateral buckling (buckling of the compressed part of a bent beam), etc., as well as design rules for avoiding local instability of elements of profile cross-sections, such as web and flanges.

XI.4.c Composed Bending with a Tensile Axial Force

In the case of a tensile axial force, the interaction between internal forces and deformations (second-order effects) may be analysed with the same tools described above for the case of compression. The results obtained are similar, with the difference that hyperbolic functions appear in place of trigonometric functions ($\sinh(kz)$ and $\cosh(kz)$ instead of $\sin(kz)$ and $\cos(kz)$). Obviously, there are no critical loads, since these hyperbolic functions never vanish when the argument is other than zero.

From a physical point of view, it is obvious that the interaction between the bending moment caused by the axial force and the bending deformation reduces this deformation and, as a consequence, the stresses caused by bending. Thus, the error introduced when the second-order effects are not taken into account is advantageous for safety, since larger stresses than the actual ones are computed. For this reason, the second-order effects are usually not considered, when the safety of members under composed bending with a tensile axial force is analysed.

When the global critical load of framed structures, where tensile axial forces appear, is to be computed, the beneficial effects of the tensile internal forces on global stability shall be considered, in order to get an accurate estimate for that load. In Sect. XI.6 a global stability analysis of framed structures using the displacement method is introduced, which takes the influence of the tensile forces in the bending stiffness into account.

XI.5 Examples and Exercises

XI.1. Determine the critical loads of the plane structures represented in Figs. XI.1-a to XI.1-d.

Resolution

(a) The critical load may be obtained from the equilibrium condition between the applied load and the forces in the springs in a slightly deformed configuration, as depicted in Fig. XI.1-e.

Considering that the deformation represented by angle θ is infinitesimal, and disregarding higher order infinitesimal quantities, the moment balance condition with respect to point A (Fig. XI.1-e), yields

$$P_{cr} \times \theta l = E\theta l \times l \Rightarrow P_{cr} = El \ .$$

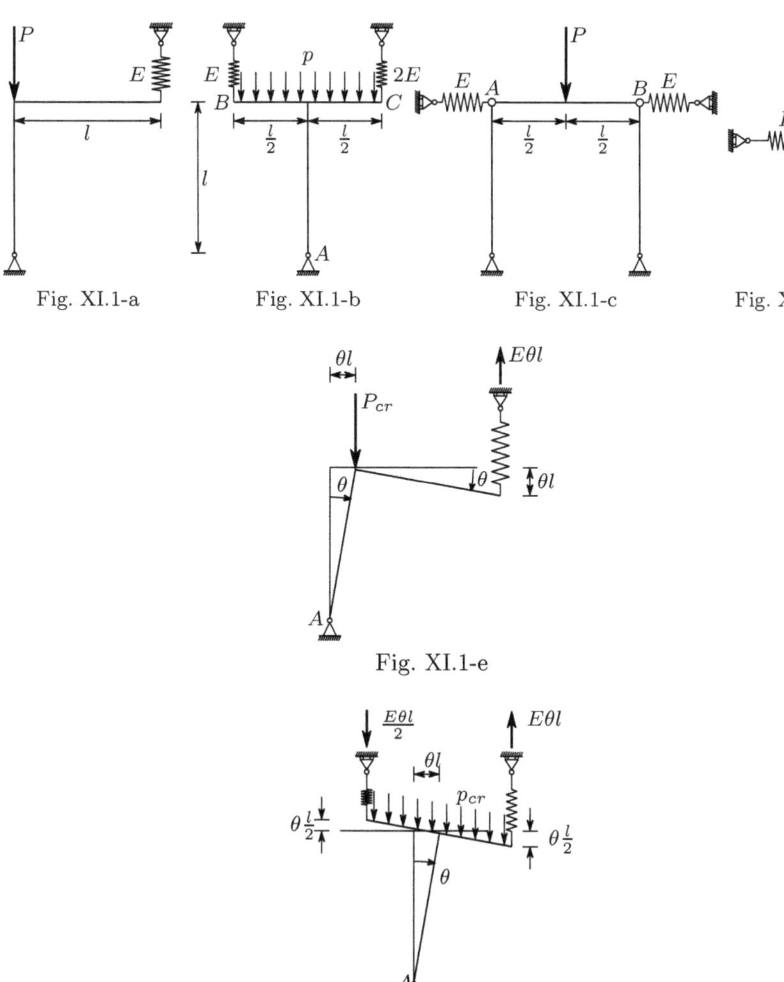

Fig. XI.1-a   Fig. XI.1-b   Fig. XI.1-c   Fig. XI.1-d

Fig. XI.1-e

Fig. XI.1-f

(b) The critical load of this structure may be computed in the same way as
the previous one. The condition of moment balance with respect to point
A leads to (Fig. XI.1-f)

$$p_{cr}l \times \theta l = \frac{E\theta l}{2} \times \frac{l}{2} + E\theta l \times \frac{l}{2}$$

$$\Rightarrow \quad p_{cr} = \frac{3}{4}E \quad .$$

(c) In this structure the critical load may be computed by means of the horizontal balance condition of the forces acting on bar AB. From Fig. XI.1-g

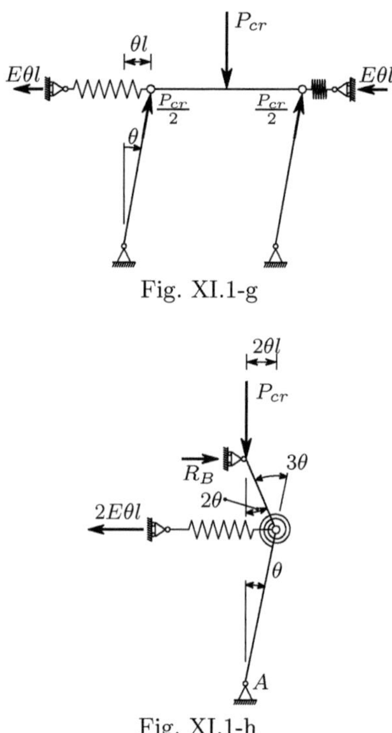

Fig. XI.1-g

Fig. XI.1-h

we conclude that, for an infinitesimal value of θ, this condition may be
expressed by

$$2E\theta l = 2\frac{P_{cr}}{2}\theta \;\Rightarrow\; P_{cr} = 2El \;.$$

(d) The critical load of this structure may also be computed by means of the
balance conditions of the forces acting on the structure in the deformed
configuration defined by angle θ (Fig. XI.1-h). The moment balance condition with respect to point A yields the value of the reaction force in
support B (the elongation of the spring is $2\theta l$)

$$R_B \times 3l = 2E\theta l \times 2l \;\Rightarrow\; R_B = \frac{4}{3}E\theta l \;.$$

The critical load may then be computed by means of the bending moment
needed to introduce a rotation 3θ into the rotation spring. Expressing this
moment as a function of the forces acting on node B, we get

$$P_{cr} \times 2\theta l - \frac{4}{3}E\theta l \times l = 3\theta G \;\Rightarrow\; P_{cr} = \frac{3}{2}\frac{G}{l} + \frac{2}{3}El \;.$$

XI.2 Consider the mechanism represented in Fig. XI.2-a under the action of
forces p and F. The stability of this structure depends on the value of

force F. Considering the pin-ended bars as axially rigid, determine the minimum value of force F, in order to have a stable structure.

Resolution

This problem may be solved with the same line of reasoning that was used in problem XI.1. Thus, since in the critical situation an infinitesimal perturbation does not disturb the equilibrium, we may find the minimum value of force F by means of equilibrium considerations in the forces acting on the mechanism, when it suffers the perturbation defined by the infinitesimal angle θ (Fig. XI.2-b). The horizontal balance condition of the forces acting in bars AB and BC yields

$$4\frac{pl}{2} \times \theta = F_{\min} \times \frac{2}{3}\theta \Rightarrow F_{\min} = 3pl \ .$$

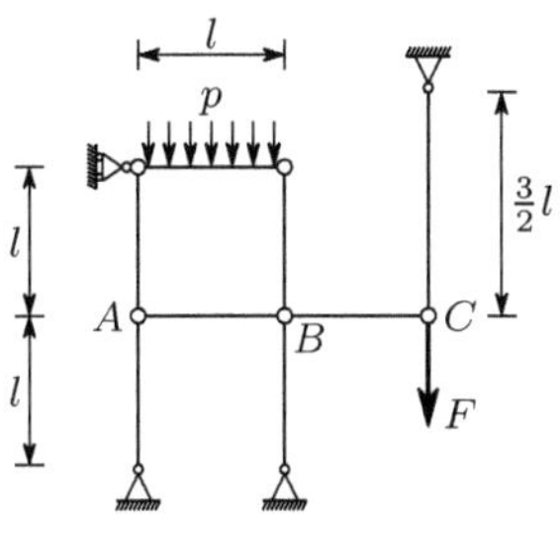

Fig. XI.2-a

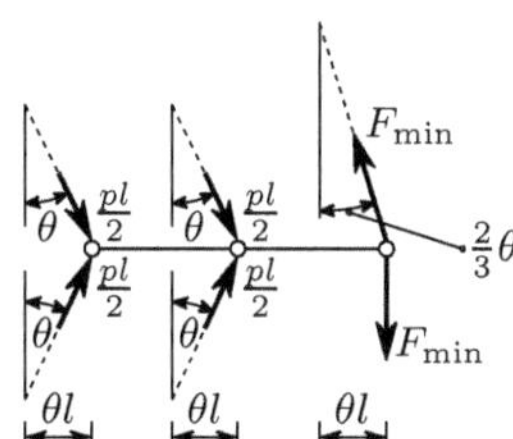

Fig. XI.2-b

XI.3 Find the relation between the lengths l and l', so that the mechanism represented in Fig. XI.3-a is stable.

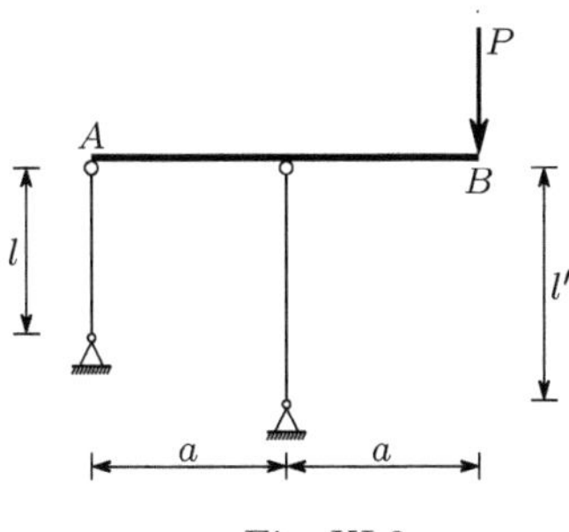

Fig. XI.3-a

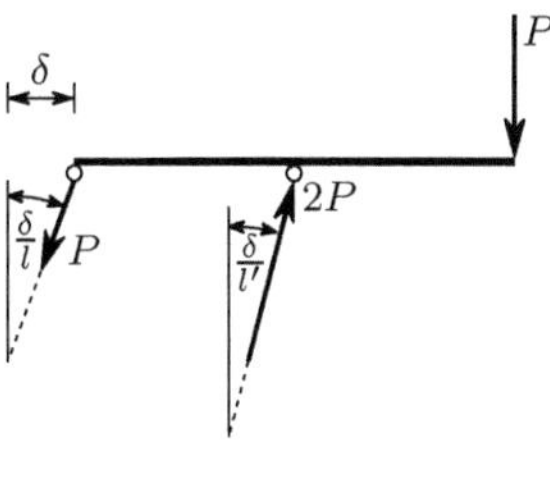

Fig. XI.3-b

Resolution

Also in this case the stability analysis may be performed by means of the horizontal balance condition of the forces acting on bar AB in a slightly disturbed configuration. Thus, considering the perturbation defined by the infinitesimal displacement δ, we conclude that the tensile axial force in the left pin-ended bar, P, opposes the perturbation, while the compressive force $2P$ in the right vertical bar has a component which tends to increase the perturbation. Therefore, the structure will be stable if the horizontal component of the tensile force is greater than the horizontal component of the compressive force, that is

$$P\frac{\delta}{l} \geq 2P\frac{\delta}{l'} \;\Rightarrow\; l' \geq 2l \;.$$

We conclude that the stability of this structure does not depend on the value of P, which is a consequence of the fact that the structure does not have deformable elements opposing the deformation.

XI.4 In the structure represented in Fig. XI.4-a, load P may be displaced along the beam. Without considering any deformation in the bars, determine the region of the beam where the load may be to have a stable situation.

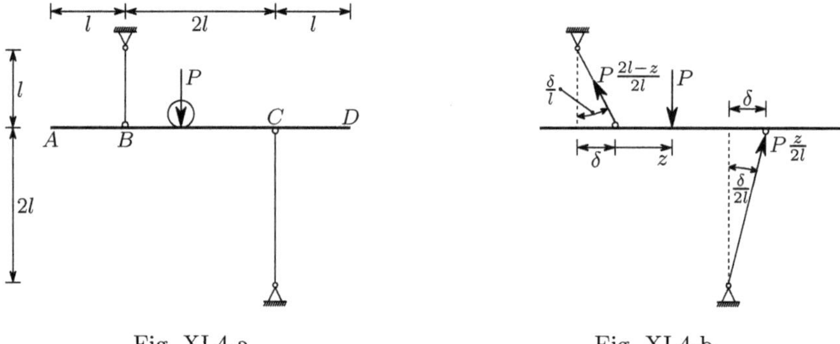

Fig. XI.4-a   Fig. XI.4-b

Resolution

When the load P is on beam segment AB, the structure is stable, since the two vertical bars have tensile axial forces. But if load P is on beam segment CD, the equilibrium is unstable, since both vertical bars have compressive axial forces. If the load is in segment BC, the structure will be stable if the stabilizing force introduced by the tensile axial force in the left vertical bar is larger than the destabilizing force corresponding to the compressive axial force in the right vertical bar. Thus, considering the infinitesimal perturbation δ represented in in Fig. XI.4-b, we conclude that stability requires the following condition to be satisfied

$$P\frac{2l - z}{2l}\frac{\delta}{l} > P\frac{z}{2l}\frac{\delta}{2l} \Rightarrow 2l - z > \frac{z}{2} \Rightarrow z < \frac{4}{3}l \approx 1.333l \ .$$

Thus, the structure will be stable if the load is on beam segment AB, or in beam segment BC at a smaller distance than $1.333l$ of point B.

XI.5 Determine the critical load of the structure represented in Fig. XI.1-b, supposing that the load p remains perpendicular to bar BC.

Resolution

Under these conditions, the structure is stable for any value of p, since the equilibrium between load p and the reaction force in support A is not affected by a perturbation like that represented in Fig. XI.1-f.

XI.6 Determine the critical load of the structure represented in Fig. XI.6.

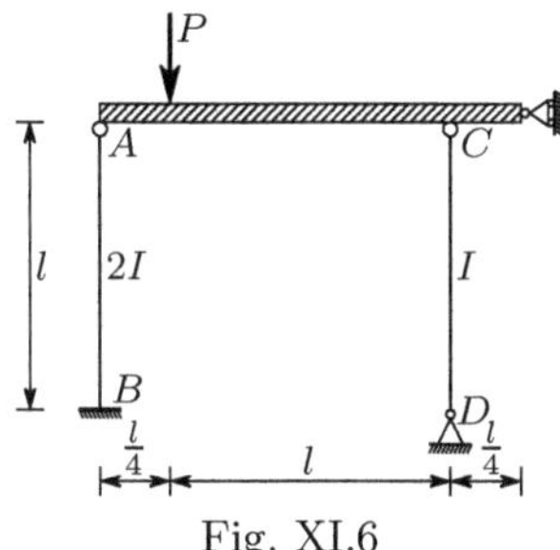

Fig. XI.6

Resolution

Buckling failure of this structure takes place when one of the vertical bars buckles. Since the horizontal displacements in points A and C are prevented, there is no interaction between the two vertical bars, that is, they buckle independently of each other. Therefore, the critical load is the smallest of the two values of P which correspond to the critical axial force in each column. Taking the different effective lengths, support conditions and moments of inertia into consideration, the critical axial forces of the two columns are

$$N_{cr}^{AB} = \frac{\pi^2 E 2I}{(0.7l)^2} = \frac{2}{0.7^2}\frac{\pi^2 EI}{l^2}; \qquad N_{cr}^{CD} = \frac{\pi^2 EI}{l^2} \ .$$

Since the reaction forces in the supports B are $D \frac{4}{5}P$ and $\frac{1}{5}P$, respectively, the values of P which correspond to these two axial forces are, respectively

$$\frac{4}{5}P = \frac{2}{0.7^2}\frac{\pi^2 EI}{l^2} \Rightarrow P = 5.102\frac{\pi^2 EI}{l^2}; \qquad \frac{1}{5}P = \frac{\pi^2 EI}{l^2} \Rightarrow P = 5\frac{\pi^2 EI}{l^2} \ .$$

The critical load of this structure is the smallest of these two values, i.e., $P_{cr} = 5\frac{\pi^2 EI}{l^2}$.

XI.7 Determine the increase of buckling strength that is obtained in the structure represented in Fig. XI.7 when a support is added which prevents the horizontal displacement of point A.

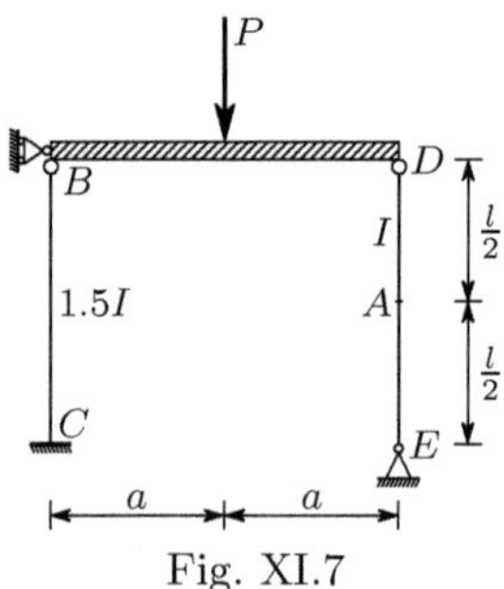

Fig. XI.7

Resolution

In the original situation (without the support in A) the right vertical buckles at first. The corresponding value of force P is

$$N_{DE} = \frac{P}{2} = \frac{\pi^2 EI}{l^2} \Rightarrow P = 2\frac{\pi^2 EI}{l^2}.$$

When the horizontal displacement of point A is prevented, the first buckling mode (Fig. 165, $n = 1$) cannot occur. Thus, the critical axial force of the bar is that corresponding to the second mode (see Fig. 165 with $n = 2$ and example XI.13), that is, $N_{cr} = \frac{4\pi^2 EI}{l^2}$. Under these conditions, the values of P corresponding to the critical axial forces in each column are

$$N_{BC} = N_{cr}^{BC} \Rightarrow \frac{P}{2} = \frac{\pi^2 E \, 1.5I}{(0.7l)^2} \Rightarrow P \approx 6.122\frac{\pi^2 EI}{l^2}$$

$$N_{DE} = N_{cr}^{DE} \Rightarrow \frac{P}{2} = \frac{4\pi^2 EI}{l^2} \Rightarrow P = \frac{8\pi^2 EI}{l^2}.$$

Since the critical load of the structure is the smallest of these two values, we conclude that the horizontal support in point A raises the critical load from $2\frac{\pi^2 EI}{l^2}$ to $6.122\frac{\pi^2 EI}{l^2}$.

XI.8 The column of the structure represented in Fig. XI.8 has a rectangular cross-section, with the dimension a in the plane of the structure and $2a$ in the perpendicular direction. The displacement of point A in the direction perpendicular to the structure's plane is not prevented. Determine the critical load of this structure.

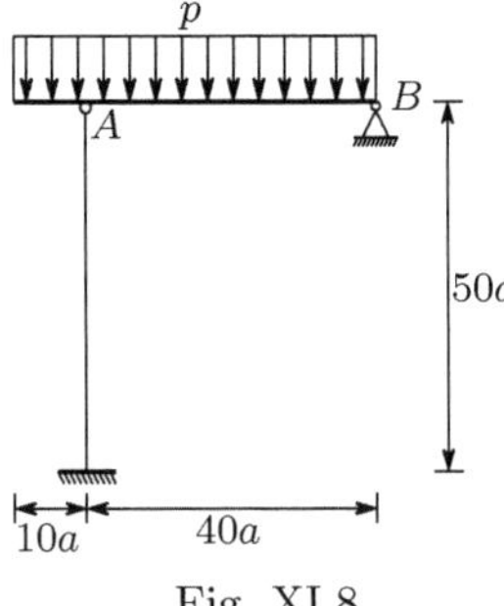

Fig. XI.8

Resolution

The column may buckle in the plane of the structure, with the buckling shape depicted in Fig. 166-d, since the beam prevents the displacement of point A in direction AB. Another possibility is buckling in the perpendicular plane with the buckling shape depicted in Fig. 166-a. Of these two possibilities, the actual buckling mode is that corresponding to the smallest value of the slenderness ratio. In the first case, we have, since the cross-section has a height $h = a$ and a width $b = 2a$

$$\begin{cases} i^2 = \dfrac{I}{\Omega} = \dfrac{2a \times a^3}{12} \times \dfrac{1}{2a^2} = \dfrac{a^2}{12} \\ l_e = 0.7l = 0.7 \times 50a = 35a \end{cases} \Rightarrow \lambda = \dfrac{l_e}{i} = \dfrac{35a}{\frac{a}{\sqrt{12}}} \approx 121.24 \ .$$

In the second case, the cross-section has a width $b = a$ and a height $h = 2a$, yielding

$$\begin{cases} i^2 = \dfrac{I}{\Omega} = \dfrac{a \times (2a)^3}{12} \times \dfrac{1}{2a^2} = \dfrac{a^2}{3} \\ l_e = 2l = 2 \times 50a = 100a \end{cases} \Rightarrow \lambda = \dfrac{l_e}{i} = \dfrac{100a}{\frac{a}{\sqrt{3}}} \approx 173.21 \ .$$

We conclude that the second possibility corresponds to the actual buckling mode. Since the axial force in the column is $N = 31.25pa$, the critical load takes the value

$$\lambda = \dfrac{100a}{\frac{a}{\sqrt{3}}} \Rightarrow \sigma_{cr} = \dfrac{\pi^2 E}{\lambda^2} = \dfrac{\pi^2 E}{30000} \ ;$$

$$\sigma = \sigma_{cr} \Rightarrow \dfrac{N}{\Omega} = \dfrac{31.25 p_{cr} a}{2a^2} = \dfrac{\pi^2 E}{30000} \Rightarrow p_{cr} = \dfrac{\pi^2 E a}{468750} \ .$$

XI.9 Consider a bar with two built-in ends, with length l and a cross-section with area Ω and moment of inertia I, made of a material with an elasticity modulus E and a thermal expansion coefficient α. Determine the value of a uniform temperature increase Δt which causes a transversal deflection of the bar.

Resolution

The uniform temperature increase Δt introduces into the bar a compressive axial force with the value (see example VI.3)

$$N = E\Omega\alpha\Delta t \, .$$

The transversal deflection will take place when this value reaches the critical load of the bar. Since the effective length of a bar with two built-in ends is $\frac{l}{2}$, we have

$$N = N_{cr} \;\Rightarrow\; E\Omega\alpha\Delta t = \frac{\pi^2 EI}{\left(\frac{l}{2}\right)^2} \;\Rightarrow\; \Delta t = \frac{4\pi^2 I}{l^2 \Omega\alpha} = \frac{4\pi^2}{\left(\frac{l}{i}\right)^2 \alpha} \, .$$

Considering, for example, a steel bar ($\alpha = 1.2\times10^{-5}/^\circ C$) with a ratio $\frac{l}{i} = 300$ (slenderness ratio $\lambda = \frac{l}{2i} = 150$), a temperature increase $\Delta t = 36.55^\circ C$, is enough to cause the transversal deflection.

XI.10 The plane structure represented in Fig. XI.10-a is stabilized by means of the cables with cross-section area Ω represented in the Figure. The columns and the cables are made of the same material which has an elasticity modulus E. The cables are not connected to each other in their intersection. The cross-section of the columns has the moment of inertia $I = \frac{\Omega l^2}{125}$.

Determine the critical load of this structure and indicate if it may be able to withstand a higher load than the critical one.

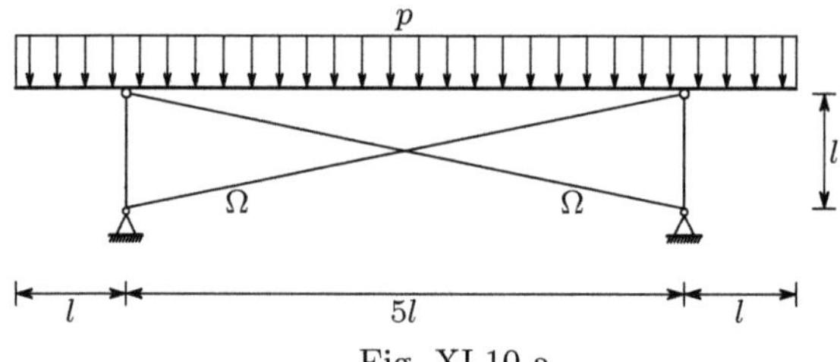

Fig. XI.10-a

Resolution

This structure may become unstable, either by buckling of the columns, or by a horizontal translation of the beam with elongation of one of the cables. The first situation will occur if load p reaches the value corresponding to the buckling axial force of the columns. In this case, we have

$$N_p = \frac{7pl}{2} = \frac{\pi^2 EI}{l^2} \;\Rightarrow\; \frac{7pl}{2} = \frac{\pi^2 E\frac{\Omega l^2}{125}}{l^2} \;\Rightarrow\; p = p_1 = \frac{2\pi^2}{875}\frac{E\Omega}{l} \approx 0.0225591\frac{E\Omega}{l} \, .$$

In the buckling mode corresponding to the horizontal translation of the beam, the motion represented in Fig. XI.10-b takes place.

The compressive axial forces in the columns, N_p, have a destabilizing horizontal component, since it acts in the direction of the perturbation θl, while the axial force, N_c, in the elongated cable has a stabilizing effect, since it opposes the perturbation. The elongation of this cable and the corresponding axial force take the values

$$\Delta l = \theta l \cos\left(\arctan\frac{l}{5l}\right) \ \Rightarrow\ N_c = \Delta l\frac{E\Omega}{\sqrt{l^2 + (5l)^2}} = \frac{\cos\left(\arctan\frac{1}{5}\right)}{\sqrt{26}}E\Omega\theta \ .$$

In the critical situation the horizontal components of the compressive axial forces in the columns equilibrate the horizontal component of the tensile axial force in the cable. This condition may be used to get the critical load corresponding to this buckling mode, yielding

$$2N_p\theta = N_c \cos\left(\arctan\frac{1}{5}\right) \ \Rightarrow\ 2\frac{7pl}{2}\theta = \frac{\cos^2\left(\arctan\frac{1}{5}\right)}{\sqrt{26}}E\Omega\theta$$

$$\Rightarrow\ p = p_2 = \frac{\cos^2\left(\arctan\frac{1}{5}\right)}{7\sqrt{26}}\frac{E\Omega}{l} \approx 0.0269390\frac{E\Omega}{l} \ .$$

Since $p_1 < p_2$, we conclude that the critical load of the structure takes the value $p = \frac{2\pi^2}{875}\frac{E\Omega}{l}$. Furthermore, we may conclude that this structure may be able to withstand a value of p greater than this (but smaller than p_2), since a compressed bar has a stable post-critical behaviour. If the second mode of instability were to take place ($P_{cr} = p_2$), the post-critical behaviour would be of the same type as the examples in Figs. 155-d and 158-b, i.e., it would be unstable.

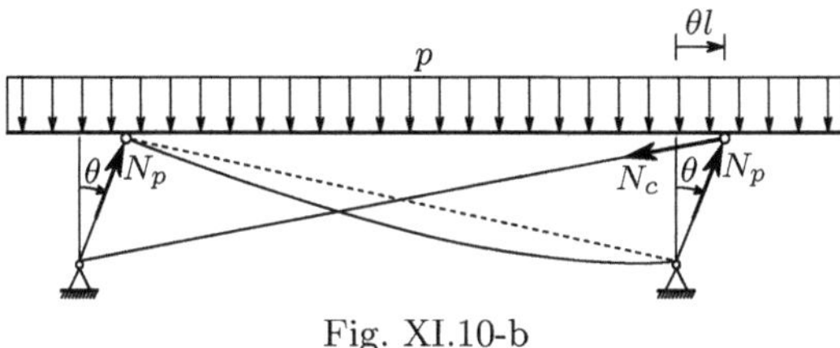

Fig. XI.10-b

XI.11 Considering as rigid the segment AB of the column represented in Fig. XI.11-a, determine its effective length.

Resolution

The problem may be solved by means of a procedure which is similar to that used to compute the effective length of the column represented in Fig. 166-d.

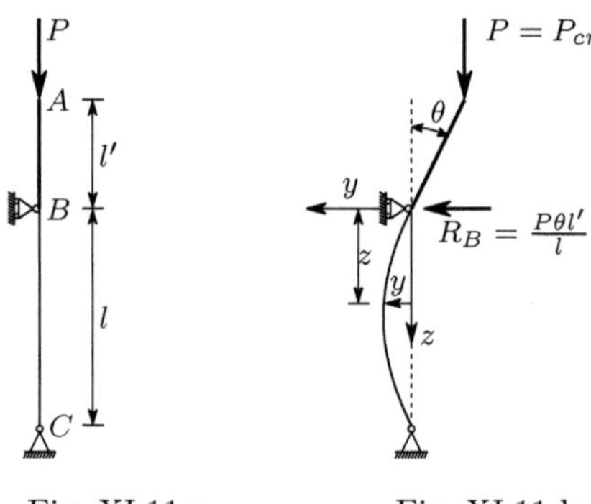

Fig. XI.11-a   Fig. XI.11-b

Considering the deformed configuration defined by the infinitesimal angle θ (Fig. XI.11-b), the reaction force in support B may be computed by means of the condition of a vanishing bending moment in hinge C, yielding

$$R_B \times l = P \times \theta l' \;\Rightarrow\; R_B = \frac{P\theta l'}{l} \; .$$

The differential equation defining the interaction between bending moments and bending deformations may be found by expressing the bending moment in the cross-section at the distance z from support B (M is positive when it induces a positive curvature in reference system yz) as a function of the curvature, yielding

$$\begin{cases} M(z) = -P \times (y + \theta l') + \frac{P\theta l'}{l} \times z \\[2mm] \dfrac{M}{EI} = \dfrac{\mathrm{d}^2 y}{\mathrm{d}z^2} \end{cases} \Rightarrow \quad \frac{\mathrm{d}^2 y}{\mathrm{d}z^2} + \frac{P}{EI}y = \frac{P}{EI}\theta l'\left(\frac{z}{l}-1\right) \; .$$

This equation admits the particular integral $y = \theta l'\left(\frac{z}{l}-1\right)$. Thus, the general solution takes the form

$$y = C_1 \cos(kz) + C_2 \sin(kz) + \theta l'\left(\frac{z}{l}-1\right) \quad \text{with} \quad k^2 = \frac{P}{EI} \; .$$

Differentiating with respect to z, we get the slope of the deflection line which is given by the expression

$$\frac{\mathrm{d}y}{\mathrm{d}z} = -kC_1 \sin(kz) + kC_2 \cos(kz) + \theta \frac{l'}{l} \; .$$

By means of the compatibility condition in support B, C_1 and C_2 may be expressed as functions of θ, yielding

$$z = 0 \;\Rightarrow\; \begin{cases} y = 0 \;\Rightarrow\; C_1 = \theta l' \\[2mm] \dfrac{\mathrm{d}y}{\mathrm{d}z} = 0 \;\Rightarrow\; kC_2 + \theta\frac{l'}{l} = 0 \;\Rightarrow\; C_2 = \theta\frac{l-l'}{kl} \; . \end{cases}$$

Substituting these values in the general solution, we get

$$y = \theta \left[l' \cos(kz) + \frac{l - l'}{kl} \sin(kz) + l' \left(\frac{z}{l} - 1 \right) \right] .$$

The condition of a vanishing deflection in point C leads to the conclusion

$$z = l \Rightarrow y = 0 \Rightarrow \theta = 0 \quad \vee \quad l' \cos(kl) + \frac{l - l'}{kl} \sin(kl) = 0 .$$

The first alternative ($\theta = 0$) corresponds to the undeformed configuration. The second possibility corresponds to the equilibrium in a slightly deformed configuration, allowing the computation of the critical load. This equation takes especially simple forms in two cases. In the first, $l' = 0$, we get Euler's solution for a pin-ended bar, $\sin(kl) = 0$, as expected. In the second, $l' = l$, we have

$$l' \cos(kl) = 0 \;\Rightarrow\; kl = \frac{\pi}{2} \;\Rightarrow\; \frac{P}{EI} l^2 = \left(\frac{\pi}{2} \right)^2 \;\Rightarrow\; P = \frac{\pi^2 EI}{l_e^2} \quad \text{with} \quad l_e = 2l .$$

In the general case the value of kl must be found by numerical means, since it is a transcendental equation. Determining the least value of kl which satisfies the condition

$$\frac{l'}{l} \cos(kl) + \left(1 - \frac{l'}{l} \right) \frac{\sin(kl)}{kl} = 0 ,$$

the effective length is given by the expression

$$(kl)^2 = \frac{Pl^2}{EI} \;\Rightarrow\; P = \frac{\pi^2 EI}{l_e^2} \quad \text{with} \quad l_e = \frac{\pi}{kl} l .$$

The following Table gives the results obtained for some given values of the ratio $\frac{l'}{l}$.

$\frac{l'}{l}$	0	0.1	0.5	1	2	4
kl	π	2.836	2.029	1.571	1.166	0.845
l_e	l	$1.108l$	$1.549l$	$2l$	$2.695l$	$3.719l$

XI.12 Determine the critical load of the column represented in Fig. XI.12-a (do not consider the possibility of the buckling of bar AB alone).

Resolution

In the deformed configuration defined by the infinitesimal displacement Δ in hinge B (Fig. XI.12-b) the forces acting on the upper end of bar BC are those represented in Fig. XI.12-c.

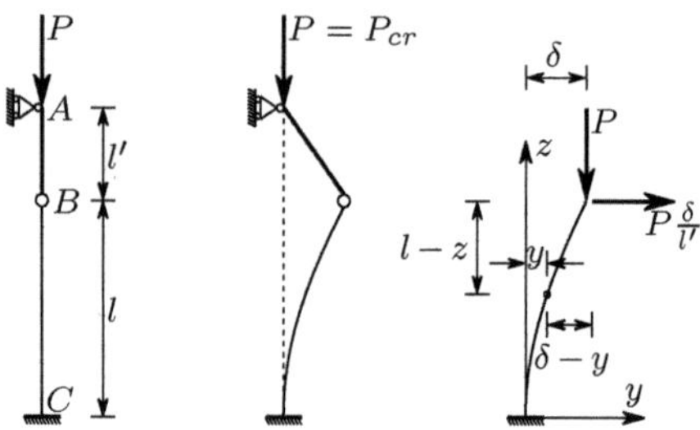

Fig. XI.12-a   Fig. XI.12-b   Fig. XI.12-c

The differential equation of this problem is

$$\begin{cases} M(z) = P\left(\delta - y\right) + P\frac{\delta}{l'}\left(l - z\right) \\ \dfrac{M}{EI} = \dfrac{\mathrm{d}^2 y}{\mathrm{d}z^2} \end{cases} \Rightarrow \frac{\mathrm{d}^2 y}{\mathrm{d}z^2} + \frac{P}{EI}y = \frac{P}{EI}\delta\left(1 + \frac{l-z}{l'}\right).$$

This equation admits the particular integral $y = \delta\left(1 + \frac{l-z}{l'}\right)$. Thus, the general solution is

$$y = C_1\cos(kz) + C_2\sin(kz) + \delta\left(1 + \frac{l-z}{l'}\right) \quad \text{with} \quad k^2 = \frac{P}{EI}.$$

Derivation with respect to z yields the slope of the deflection line

$$\frac{\mathrm{d}y}{\mathrm{d}z} = -kC_1\sin(kz) + kC_2\cos(kz) - \frac{\delta}{l'}.$$

The compatibility conditions in support C enable the integration constants C_1 and C_2 to be expressed as functions of δ, yielding

$$z = 0 \Rightarrow \begin{cases} y = 0 \Rightarrow C_1 + \delta\left(1 + \frac{l}{l'}\right) = 0 \Rightarrow C_1 = -\delta\frac{l+l'}{l'} \\ \frac{\mathrm{d}y}{\mathrm{d}z} = 0 \Rightarrow kC_2 - \frac{\delta}{l'} = 0 \Rightarrow C_2 = \frac{\delta}{kl'}. \end{cases}$$

Substituting these values in the general solution, we get

$$y = \delta\left[\frac{\sin(kz)}{kl'} - \frac{l+l'}{l'}\cos(kz) + 1 + \frac{l-z}{l'}\right].$$

In point B this equation must yield the value δ. Thus, we have

$$z = l \Rightarrow y = \delta \Rightarrow \delta\left[\frac{\sin(kl)}{kl'} - \frac{l+l'}{l'}\cos(kl) + 1\right] = \delta$$

$$\Rightarrow \quad \delta\left[\frac{\sin(kl)}{kl'} - \frac{l+l'}{l'}\cos(kl)\right] = 0 \Rightarrow \delta = 0 \vee \frac{\sin(kl)}{kl'} - \frac{l+l'}{l'}\cos(kl) = 0.$$

Like example XI.11 the first possibility $(\Delta = 0)$ corresponds to an undeformed configuration, so that it cannot be used to compute the critical load.

The second alternative corresponds to the equilibrium in a slightly deformed configuration. Thus, the critical load may be obtained from the condition

$$\frac{\sin(kl)}{kl'} = \frac{l+l'}{l'}\cos(kl) \;\Rightarrow\; \left(1+\frac{l'}{l}\right)kl = \tan(kl) \,.$$

Note that for $l' = 0$ we get to the same problem as that represented in Fig. 166-d, $kl = \tan(kl)$, $l_e = 0.7l$. With $l' = \infty$ the example in Fig. 166-a is obtained $(kl = \frac{\pi}{2})$, $l_e = 2l$, since the pin-ended bar AB remains vertical for any value of Δ. For other values of l', we conclude that the higher the value of l', the higher is the critical load, as indicated in the following table.

$\frac{l'}{l}$	0	0.1	0.2	0.5	1	10	∞
kl	4.493	0.5175	0.6955	0.9674	1.166	1.511	$\frac{\pi}{2}$
l_e	0.6992l	6.071l	4.517l	3.247l	2.695l	2.080l	2l

XI.13 Find the critical load of the column depicted in Fig. XI.13-a.

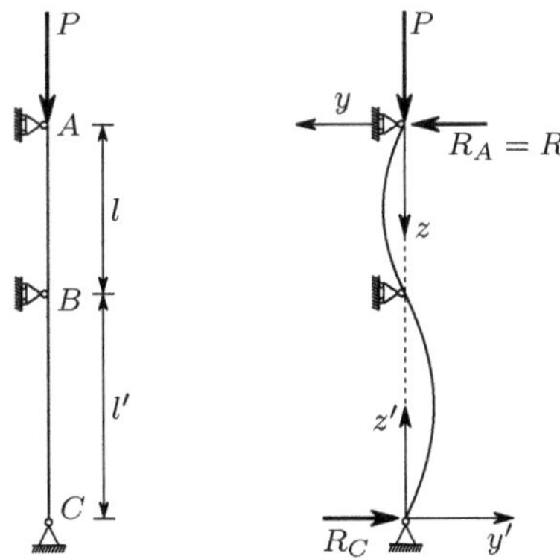

Fig. XI.13-a   Fig. XI.13-b

Resolution

Considering the deformed configuration and the two reference systems represented in Fig. XI.13-b, the differential equation corresponding to segment AB takes the form $(M(z) = -Py + R_A z)$

$$\frac{d^2y}{dz^2} + \frac{P}{EI}y = \frac{R_A}{EI}z \,.$$

This equation admits the particular integral $y = \frac{R_A}{P}z$. Thus, its general solution is given by the expression

$$y = C_1 \cos(kz) + C_2 \sin(kz) + \frac{R_A}{P} z \quad \text{with} \quad k^2 = \frac{P}{EI} \ .$$

Using the conditions of zero deflection in supports A and B, the integration constants may be related to the reaction force R_A, yielding

$$z = 0 \ \Rightarrow \ y = 0 \ \Rightarrow \ C_1 = 0$$

$$z = l \ \Rightarrow \ y = 0 \ \Rightarrow \ C_2 \sin(kl) + \frac{R_A}{P} l = 0 \ \Rightarrow \ C_2 = -\frac{R_A}{P} \frac{l}{\sin(kl)} \ .$$

Substituting these values in the general solution and derivating, we get the rotation as a function of coordinate z

$$y = \frac{R_A}{P} \left[z - \frac{l}{\sin(kl)} \sin(kz) \right] \ \Rightarrow \ \frac{dy}{dz} = \frac{R_A}{P} \left[1 - \frac{kl}{\sin(kl)} \cos(kz) \right] \ .$$

In cross-section B this rotation takes the value

$$\left(\frac{dy}{dz} \right)_{z=l} = \frac{R_A}{P} \left[1 - \frac{kl}{\tan(kl)} \right] = \frac{R}{P} \left[1 - \frac{kl}{\tan(kl)} \right] \ .$$

Note that by equating this rotation to zero, we get the problem represented in Fig. 166-d.

This rotation may also be obtained from the deformation of the column segment BC. The procedure is exactly the same as in segment AB, so we may simply substitute y by y', z by z', l by l' and R_A by R_C. Using the moment balance condition with respect to point B, this reaction force may be related to R, yielding $R_C = -R\frac{l}{l'}$. The minus sign is a consequence of the positive directions adopted for the reaction forces, whose objective was that of having analogous equations in segments AB and BC. The rotation of cross-section B is then also given by the expression

$$\left(\frac{dy'}{dz'} \right)_{z'=l'} = \frac{R_C}{P} \left[1 - \frac{kl'}{\tan(kl')} \right] = -\frac{R}{P} \frac{l}{l'} \left[1 - \frac{kl'}{\tan(kl')} \right] \ .$$

The condition of continuity in point B leads to the condition

$$\left(\frac{dy}{dz} \right)_{z=l} = \left(\frac{dy'}{dz'} \right)_{z'=l'} \ \Rightarrow \ \frac{R}{P} \left[1 - \frac{kl}{\tan(kl)} \right] = -\frac{R}{P} \frac{l}{l'} \left[1 - \frac{kl'}{\tan(kl')} \right] \ .$$

If the reaction force R is not zero, this equation is equivalent to

$$1 - \frac{kl}{\tan(kl)} = -\frac{l}{l'} + \frac{kl}{\tan(kl\frac{l'}{l})} \ \Rightarrow \ \frac{kl}{\tan(kl)} + \frac{kl}{\tan\left(\frac{kl}{\alpha}\right)} = 1 + \alpha \quad \text{with} \quad \alpha = \frac{l}{l'} \ .$$

The smallest of the roots of this transcendental equation is the value of kl which corresponds to the critical load. However, it must be noted that this expression does not completely define the problem. In fact, if we have $l = l'$, we will have $R = 0$, since the deformed configuration is antisymmetric, which means that the bending moment vanishes in cross-section B. This corresponds to the second buckling mode in Fig. 165 ($n = 2$). In the case of $l = l'$ the smallest of the roots of this equation is $kl = 4.4934$. However, the equilibrium in a slightly deformed configuration requires that integration constant C_2 does not vanish ($C_2 \neq 0$), since $R_A = 0$. This implies that $\sin(kl) = 0$, i.e., $kl = n\pi$. When $l \neq l'$, R cannot vanish, since this would mean that two pin-ended columns with different lengths would have the same critical load. In the following Table the smallest roots are given for some values of α.

α	0.01	0.05	0.1	0.5	0.99	1.01	2	5	10	∞
kl	0.04479	0.2211	0.4352	1.928	3.126	3.157	3.857	4.223	4.352	4.493

XI.14 Determine the maximum bending moment in the column represented in Fig. XI.14.

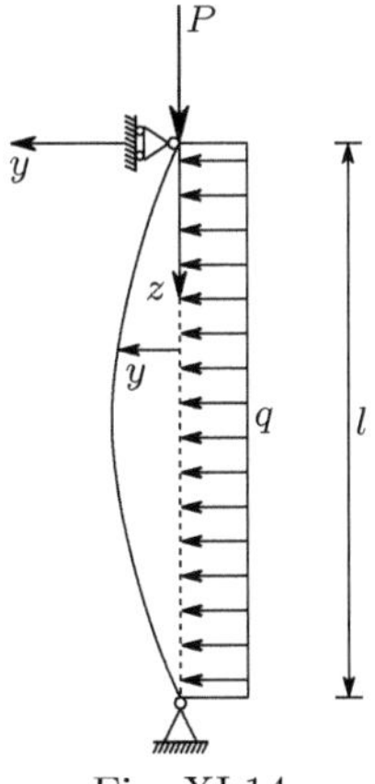

Fig. XI.14

Resolution

Considering as positive the bending moments which correspond to a positive curvature in reference system yz, the bending moment in the cross-section defined by coordinate z is given by the expression

$$M = -Py - \frac{ql}{2}z + \frac{q}{2}z^2 \,.$$

Since the moment-curvature relation in the same reference system may also be expressed by $M = EI\frac{\mathrm{d}^2 y}{\mathrm{d}z^2}$, the differential equation which defines the deformation of the column takes the form

$$\frac{\mathrm{d}^2 y}{\mathrm{d}z^2} + \frac{P}{EI}y = \frac{q}{EI}y\left(-\frac{l}{2}z + \frac{z^2}{2}\right) .$$

Since this equation admits the particular integral

$$y = \frac{q}{P}\left(\frac{z^2}{2} - \frac{l}{2}z\right) - \frac{EI}{P^2}q ,$$

the general solution is

$$y = C_1 \sin(kz) + C_2 \cos(kz) + \frac{q}{P}\left(\frac{z^2}{2} - \frac{l}{2}z\right) - \frac{q}{Pk^2} \quad \text{with} \quad k^2 = \frac{P}{EI} .$$

The integration constants may be found by means of the support conditions, yielding

$$z = 0 \Rightarrow y = 0 \Rightarrow C_2 = \frac{q}{Pk^2}$$

$$z = l \Rightarrow y = 0 \Rightarrow C_1 \sin(kl) + \frac{q}{Pk^2}\cos(kl) - \frac{q}{Pk^2} = 0$$

$$\Rightarrow \quad C_1 = \frac{q}{Pk^2}\frac{1 - \cos(kl)}{\sin(kl)} .$$

Substituting these quantities in the general solution, we get the displacement y in the cross-section defined by coordinate z

$$y = \frac{q}{Pk^2}\left[\frac{1 - \cos(kl)}{\sin(kl)}\sin(kz) + \cos(kz) - 1\right] + \frac{q}{P}\left(\frac{z^2}{2} - \frac{l}{2}z\right) .$$

This function reaches a maximum for $z = \frac{l}{2}$, as may be concluded by symmetry considerations, and is confirmed by differentiation

$$\frac{\mathrm{d}y}{\mathrm{d}z} = \frac{q}{Pk^2}\left[\frac{1 - \cos(kl)}{\sin(kl)}k\cos(kz) - k\sin(kz)\right] + \frac{q}{p}\left(z - \frac{l}{2}\right) ;$$

$$z = \frac{l}{2} \Rightarrow \frac{\mathrm{d}y}{\mathrm{d}z} = \frac{q}{Pk}\left[\frac{1 - \cos(kl)}{\sin(kl)}\cos\left(k\frac{l}{2}\right) - \sin\left(k\frac{l}{2}\right)\right]$$

$$= \frac{q}{Pk}\left[\frac{2\sin^2\left(\frac{kl}{2}\right)}{2\sin\left(\frac{kl}{2}\right)\cos\left(\frac{kl}{2}\right)}\cos\left(k\frac{l}{2}\right) - \sin\left(k\frac{l}{2}\right)\right] = 0 .$$

Thus, the maximum displacement occurs for $z = \frac{l}{2}$ and takes the value

$$z = \frac{l}{2} \Rightarrow y = y_{\max} = \frac{q}{Pk^2}\left[\frac{\sin^2\left(\frac{kl}{2}\right)}{\cos\left(\frac{kl}{2}\right)} + \cos\left(\frac{kl}{2}\right) - 1\right] - \frac{q}{P}\frac{l^2}{8} ,$$

since $\frac{1 - \cos(kl)}{\sin(kl)} = \frac{\sin\left(\frac{kl}{2}\right)}{\cos\left(\frac{kl}{2}\right)}$. The maximum value of the bending moment is then

$$M_{\max} = Py_{\max} + \frac{ql^2}{8} = \frac{q}{k^2}\left[\frac{\sin^2\left(\frac{kl}{2}\right)}{\cos\left(\frac{kl}{2}\right)} + \cos\left(\frac{kl}{2}\right) - 1\right] = \frac{q}{k^2}\frac{1 - \cos\left(\frac{kl}{2}\right)}{\cos\left(\frac{kl}{2}\right)} .$$

Instability by divergence takes place for $kl = \pi$ ($M_{\max} = \infty$), as expected, since the transversal load q does not influence the value of the critical load. If we consider $k = 0$, the expression of $M_{\max}$ becomes indeterminate. This problem may be solved by means of L'Hôpital's rule, yielding, as expected, the expression of the maximum bending moment in a simply supported beam with span l under the action of a uniformly distributed load q,

$$\lim_{k\to 0} M_{\max} = \lim_{k\to 0} q\frac{1 - \cos\left(\frac{kl}{2}\right)}{k^2\cos\left(\frac{kl}{2}\right)} = \frac{ql^2}{8} .$$

XI.15 Figure XI.15 represents the cross-section of a bar which supports an axial compressive force P. Determine:
  (a) the value of P which causes torsional buckling of the bar;
  (b) the maximum slenderness ratio, so that torsional buckling occurs before bending buckling, considering $d = 20e$ and $\nu = 0.3$.

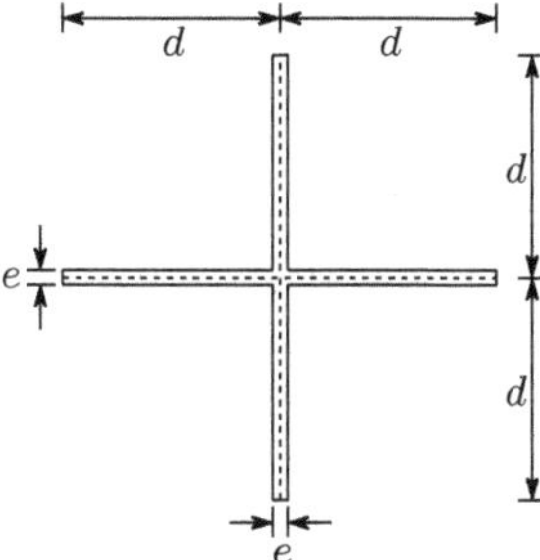

Fig. XI.15

Resolution

(a) The critical load for torsional buckling may be found by means of the equilibrium conditions of the forces applied to a bar segment with unit length in the deformed configuration defined by the infinitesimal angle θ (Fig. XI.15-a). The vector representing the force per surface unit acting on each point of the upper cross-section may be decomposed into two components: one parallel to the fibres (Fig. XI.15-b) and other in the plane defined by the wall's centre lines of the cross-section (Fig. XI.15-c). The first one does not cause torsion, since it is balanced by the normal stresses acting on facets that are perpendicular to the fibres of the deformed bar (it may be accepted that the shearing stresses perpendicular to the wall centre line vanish in all these facets, since this cross-section has thin walls).

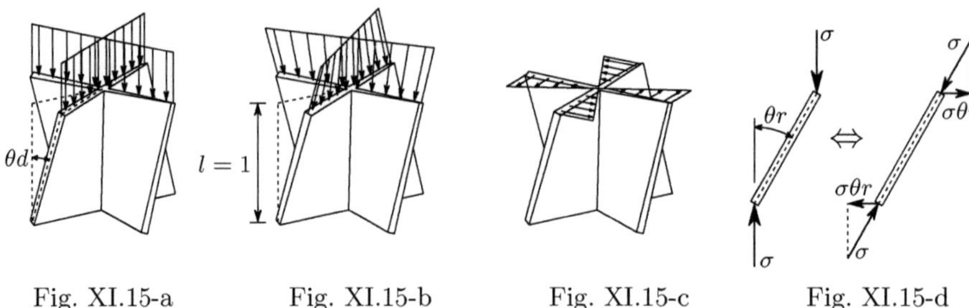

Fig. XI.15-a   Fig. XI.15-b   Fig. XI.15-c   Fig. XI.15-d

The second component, on the other side, introduces a twisting moment in the same direction as the rotation θ.

The rotation of the fibres caused by the torsional deformation θ is θr, where r is the distance of the fibre to the bar's axis ($0 \leq r \leq d$). Thus, the second component takes the value $\sigma\theta r$ (Fig. XI.15-d). The total twisting moment needed to keep the torsional deformation defined by θ is then (GJ is the torsional stiffness, as defined by (246))

$$T = GJ\theta - 4\int_0^d r\sigma\theta r e\,\mathrm{d}r = G\theta 4\frac{1}{3}de^3 - 4\theta\sigma e\frac{d^3}{3} = \frac{4}{3}\theta de\left(Ge^2 - \sigma d^2\right) \ .$$

In the critical phase this twisting moment vanishes. Thus, the critical load for torsional buckling takes the value

$$T = 0 \Rightarrow \sigma = \sigma_{cr}^t = G\left(\frac{e}{d}\right)^2 \Rightarrow P_{cr}^t = \Omega\sigma_{cr}^t = 4de\sigma_{cr}^t = 4G\frac{e^3}{d} \ .$$

(b) For $d = 20e$ the critical stress for torsional buckling takes the value

$$d = 20e \Rightarrow \sigma_{cr}^t = \frac{G}{400} \ .$$

In the case of bending buckling the critical stress takes the value

$$I = \frac{2de^3}{12} + \frac{e\left(2d\right)^3}{12} = \frac{de}{6}\left(e^2 + 4d^2\right)$$

$$\Rightarrow i^2 = \frac{I}{\Omega} = \frac{\frac{de}{6}\left(e^2 + 4d^2\right)}{4de} = \frac{1}{24}\left(e^2 + 4d^2\right) = \frac{1601}{24}e^2$$

$$\Rightarrow \sigma_{cr}^f = \frac{\pi^2 E i^2}{l^2} = \frac{\pi^2 E}{l^2}\frac{1601}{24}e^2 \ .$$

Torsional buckling will take place if the corresponding critical stress is smaller than the value corresponding to bending buckling, that is, if

$$\sigma_{cr}^f > \sigma_{cr}^t \Rightarrow \frac{\pi^2 E}{l^2}\frac{1601}{24}e^2 > \frac{E}{2\left(1+\nu\right)}\frac{1}{400}$$

$$\Rightarrow l^2 < \pi^2 e^2\frac{1601}{24} \times 400 \times 2\left(1+\nu\right) \ .$$

The given value for the Poisson coefficient ($\nu = 0.3$) yields

$$\nu = 0.3 \;\Rightarrow\; l < 827.48e = 41.37d \;.$$

The slenderness ratio corresponding to this length is

$$\lambda = \frac{l}{i} = \frac{827.48}{\sqrt{\frac{1601}{24}}} = 101.31 \;.$$

XI.16 The stress-strain relation for aluminium may be defined by means of the so-called Ramberg-Osgood equation which takes the form (Fig. XI.16)

$$\varepsilon = \frac{\sigma}{E} + 0.002 \left(\frac{\sigma}{\sigma_{0.2}} \right)^n \quad \text{with} \quad \begin{cases} E & = 70\,000 \; MPa \\ \sigma_{0.2} & = 277 \; MPa \\ n & = 18.55 \;. \end{cases}$$

(a) Verify the stability of a pin-ended bar with length $l = 3m$ and hollow square cross-section with outside side-length of $15cm$ and wall-thickness of $2.5mm$, under a compressive axial force $N = 250\,kN$.

(b) Determine the critical load.

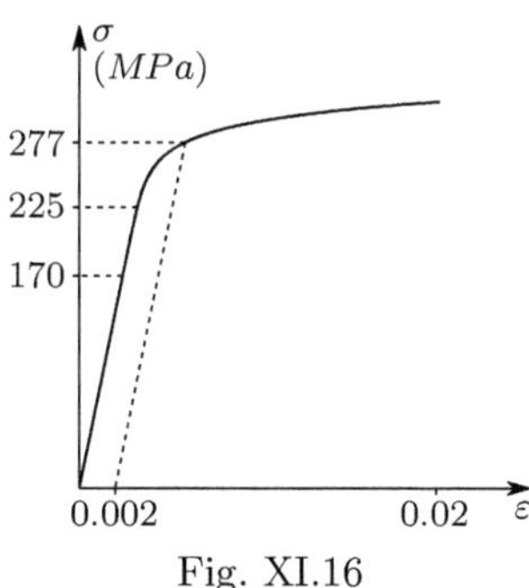

Fig. XI.16

Resolution

(a) The area and the moment of inertia of the bar's cross-section are, respectively

$$\Omega = 0.15^2 - 0.145^2 = 1.475 \times 10^{-3} m^2$$

$$I = \frac{0.15^4 - 0.145^4}{12} = 5.350 \times 10^{-6} m^4 \;.$$

Since the constitutive law is not linear, it is necessary to compute the tangent modulus of elasticity corresponding to the stress acting on the bar. Since the strain is defined as a function of the stress, the computation may be performed by the expressions

$$\sigma = \frac{N}{\Omega} \longrightarrow E_t = \frac{d\sigma}{d\varepsilon} = \left(\frac{d\varepsilon}{d\sigma}\right)^{-1} = \left[\frac{1}{E} + 0.002\, n \left(\frac{\sigma}{\sigma_{0.2}}\right)^{n-1} \frac{1}{\sigma_{0.2}}\right]^{-1} .$$

Substituting the parameters contained in these expressions by the given values, we get

$$\sigma = 169.492\, MPa \longrightarrow E_t = 69.882\, GPa .$$

If this value were to remain constant, the critical load of the bar would be

$$P_{cr} = \frac{\pi^2 E_t I}{l^2} = 409988\, N .$$

This value exceeds the applied axial load, $N = 250\, kN$. This is not the actual value of the critical load, since the tangent elasticity modulus corresponding to this axial force ($409988\, N$) is smaller than the value used which corresponds to $250\, kN$. However, the fact that this value is larger than the applied axial force indicates that the actual value of the critical load lies between these two values, which means that the bar is stable.

(b) The critical load may be obtained by means of successive approximations, or by more sophisticated numerical methods for solving non-linear equations. Convergence is attained, when the critical load computed by means of the procedure used in answer to question a) is equal to the value of the axial force used to compute the tangent modulus of elasticity.
In the present case the solution has been computed by increasing the value of N by 5% of the difference between N and the value of P_{cr} corresponding to $E_t(N)$. Convergence has been reached after 60 iterations, yielding the results

$$P_{cr} = 331483\, N, \qquad \sigma_{cr} = 224.734\, MPa \qquad \text{and} \qquad E_t = 56.501\, GPa .$$

It must be noted that the simple successive substitution (substitution of N by the value of P_{cr} obtained from it) does not converge in this case.

Remark: Although in the diagram presented in Fig. XI.16 (which exactly represents the Ramberg-Osgood equation given) the difference between the elasticity moduli corresponding to the given ($\sigma \approx 170\, MPa$) and critical ($\sigma \approx 225\, MPa$) stresses is too small to be observed, the actual critical load is substantially smaller than that corresponding to the first value. This example illustrates the error introduced when Euler's formula is used above the proportionally limit without considering the reduction of the elasticity modulus.

XI.17 Compare the values obtained for the maximum load under eccentric compression using the exact solution (261) and the interaction formula. Consider a grade $S\,355$ steel and the values 0.001, 0.25 and 1 (Fig. 170) for the eccentricity ratio α.

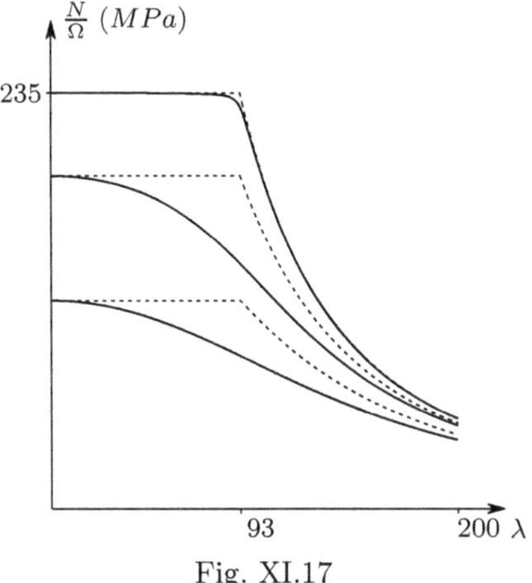

Fig. XI.17

Resolution

In the case of plane composed bending caused by an eccentric compression
the interaction formula may be given the form

$$\left\{\begin{array}{l} M = Pe \\ I = \Omega i^2 \\ \dfrac{P}{\Omega\varphi} + \dfrac{M}{\frac{I}{v}} \le \sigma_{\text{all}} \end{array}\right. \Rightarrow \frac{P}{\Omega}\left(\frac{1}{\varphi} + \frac{ve}{i^2}\right) \le \sigma_{\text{all}} \Rightarrow \frac{P}{\Omega}\left(\frac{1}{\varphi} + \alpha\right) \le \sigma_{\text{all}}\,.$$

In order to facilitate the comparison, no safety coefficients are considered and
the allowable stress σ_{all} is assumed to be under the proportionality limit.
Under these conditions φ is given by Euler's formula for values of λ higher
than

$$\frac{\pi^2 E}{\lambda^2} = \sigma_{\text{all}} \Rightarrow \lambda = \pi\sqrt{\frac{E}{\sigma_{\text{all}}}} = \pi\sqrt{\frac{206000}{235}} = 93.0\,.$$

Thus, we get the following expressions for parameter φ

$$\lambda \le 93 \Rightarrow \varphi = 1\,; \qquad\qquad \lambda > 93 \Rightarrow \varphi = \frac{\pi^2 E}{\lambda^2 \sigma_{\text{all}}}\,.$$

The maximum load that may be applied, so that the stress does not exceed
the value σ_{all}, may thus be defined by the expression

$$\frac{P}{\Omega} = \frac{\sigma_{\text{all}}}{\frac{1}{\varphi} + \alpha}\,.$$

The dashed lines in Fig. XI.17 represent the curves which define $\frac{P}{\Omega}$ as a function of λ for the given values of α. The solid lines represent the exact curves

which are those depicted in Fig. 170. We find that for a small eccentricity
($\alpha = 0.001$) the curves practically coincide, since in this case the axial force is
virtually centred. For the other two values of α, considerable differences are
observed. Furthermore, the error has an adverse effect on structural safety,
since the loading capacity is overestimated. However, this error is covered by
the larger additional safety coefficients used in the safety evaluation of long
members. Furthermore, the error attains a maximum in the region where
Euler's curve is usually substituted by smoother approximation curves (Fig.
167), which reduces the peak corresponding to $\lambda = 93$.

XI.18 Determine the eccentricities corresponding to the values of parameter
α considered in Fig. 170 in a rectangular cross-section with width b.

XI.19 In the structure represented in Fig. XI.19 the spring has a stiffness E
and bars AB and AC may be considered as rigid. Support C prevents
displacement in the direction of bar AC. Determine the critical load of
the structure.

XI.20 In the plane structure represented in Fig. XI.20 the column AB is
stabilized by means of cables BC and BD, which have a cross-section
area Ω and are made of a material with an elasticity modulus E. The
column has a bending stiffness EI. Determine the minimum value of
Ω, so that the structure has a stable post-critical behaviour.

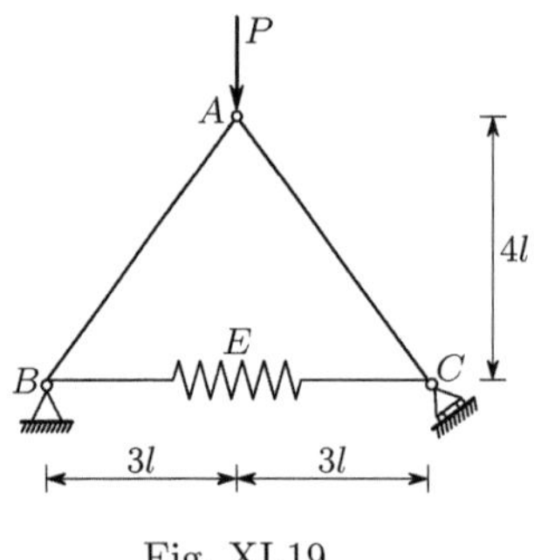

Fig. XI.19

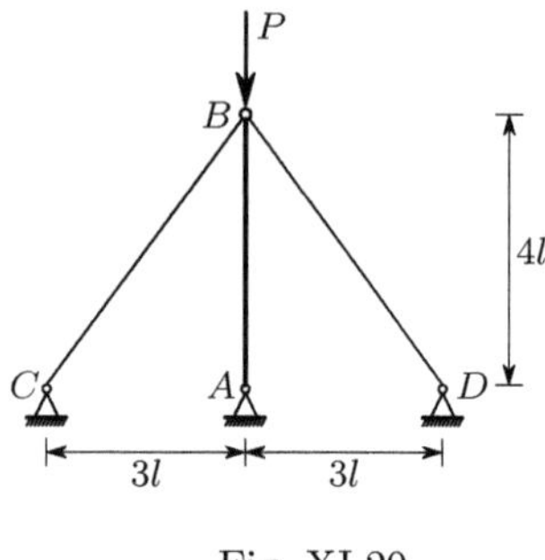

Fig. XI.20

XI.21 Determine whether the spring is necessary to stabilize the plane structure represented in Fig. XI.21. If it is, determine the critical load of the
structure.

XI.22 Using the results of example XI.13, determine the critical load of the
structure represented in Fig. XI.22.

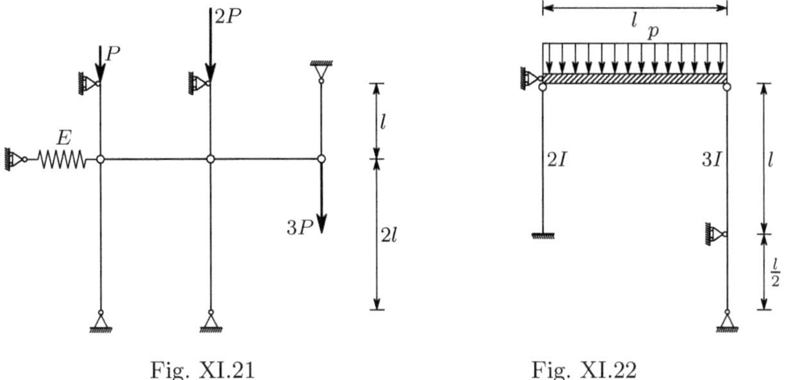

Fig. XI.21   Fig. XI.22

XI.23 The beam represented in Fig. XI.23 has a bending stiffness EI. Taking the interaction between deformation and internal forces into consideration determine the maximum bending moment.

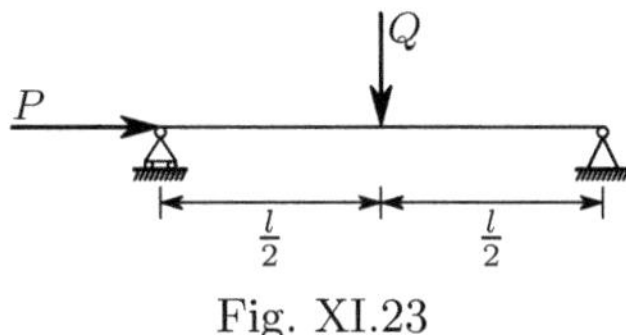

Fig. XI.23

XI.6 Stability Analysis by the Displacement Method

XI.6.a Introduction

In Sects. XI.2 to XI.5 problems of stability have been analysed whose degrees of kinematic indeterminacy are one (introductory examples in Sect. XI.2) or infinite (Euler's theory furnishes an infinite number of buckling shapes). However, in the modern computational analysis of structures, a finite number of degrees of freedom is always considered, either in naturally discrete structures, such as framed structures, or in finite-element discretizations of two- or three-dimensional structures.

In this section we will introduce the global stability analysis of framed structures by means of the displacement method. However, this analysis does not belong to the traditional field of Strength of Materials, since it is based on the matrix formulation used in the systematization of the displacement method, which is usually taught after the Strength of Materials has been studied, in the disciplines of Structural Analysis. This is why this analysis is explained as an appendix to Chap. XI.

XI.6.b Simple Examples

As a first example of a naturally discrete structure with a degree of kinematic indeterminacy superior to one, we shall analyse the stability behaviour of the column represented in Fig. 172-a.

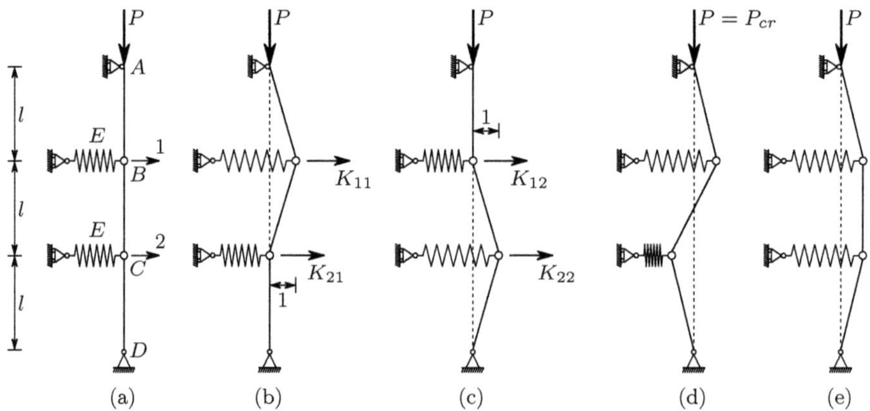

Fig. 172. Stability analysis of a structure with degree two of kinematic indeterminacy

If the vertical bars are assumed to be axially non-deformable, this structure has a degree two of kinematic indeterminacy, since, once the horizontal displacements of points B and C are known (coordinates 1 and 2 in Fig. 172-a), the deformed configuration of the structure is completely defined. Assuming that there are no imperfections, no deformations will take place as long as the structure is stable, since no forces are needed in the deformable elements to balance the external force P.

Let us consider now that two forces, F_1 and F_2, are applied in coordinates 1 and 2, respectively. These forces will cause displacements D_1 and D_2. Conversely, we may state that, in order to get the deformed configuration defined by displacements D_1 and D_2, corresponding forces F_1 and F_2 mus be applied in coordinates 1 and 2. The force-displacement relations may be represented by the expression

$$\begin{cases} F_1 = F_1\,(D_1, D_2) \\ F_2 = F_2\,(D_1, D_2)\ . \end{cases}$$

If the displacements D_1 and D_2 are infinitesimal, the corresponding forces are given by the expressions

$$\begin{cases} \mathrm{d}F_1 = \dfrac{\partial F_1}{\partial D_1}\,\mathrm{d}D_1 + \dfrac{\partial F_1}{\partial D_2}\,\mathrm{d}D_2 \\[2ex] \mathrm{d}F_2 = \dfrac{\partial F_2}{\partial D_1}\,\mathrm{d}D_1 + \dfrac{\partial F_2}{\partial D_2}\,\mathrm{d}D_2\ . \end{cases} \tag{262}$$

Using a matrix formulation, the expression may be written as

$$\left\{ \begin{array}{c} \mathrm{d}F_1 \\ \mathrm{d}F_2 \end{array} \right\} = \underbrace{\left[\begin{array}{cc} \dfrac{\partial F_1}{\partial D_1} = K_{11} & \dfrac{\partial F_1}{\partial D_2} = K_{12} \\ \dfrac{\partial F_2}{\partial D_1} = K_{21} & \dfrac{\partial F_2}{\partial D_2} = K_{22} \end{array} \right]}_{[K]} \left\{ \begin{array}{c} \mathrm{d}D_1 \\ \mathrm{d}D_2 \end{array} \right\} .$$

Matrix $[K]$ is the so-called *stiffness matrix* of the structure. It may be shown that this matrix is symmetrical (see Sect. XII.4).

In the critical situation the equilibrium state is neutral, that is, it is not disturbed when the equilibrium configuration suffers an infinitesimal perturbation, which may be defined by the displacements $\mathrm{d}D_1$ and $\mathrm{d}D_2$. This means that the infinitesimal forces corresponding to these displacements, $\mathrm{d}F_1$ and $\mathrm{d}F_2$, vanish when force P attains the critical value. This condition may be expressed by the relation

$$\left[\begin{array}{cc} K_{11} & K_{12} \\ K_{21} & K_{22} \end{array} \right] \left\{ \begin{array}{c} \mathrm{d}D_1 \\ \mathrm{d}D_2 \end{array} \right\} = \left\{ \begin{array}{c} 0 \\ 0 \end{array} \right\} . \tag{263}$$

This system of equations is homogeneous, which means that it only has non-simultaneously vanishing solutions if matrix $[K]$ is singular, i.e., if its determinant is zero. Since this is what happens in the critical phase – a deformed configuration is possible, without forces being applied in the nodal points – the condition $|K| = 0$ may be used to find the critical load of the structure.

As an alternative to the derivation of the force-displacement relations (262), which would require those functions to be found, the elements of the stiffness matrix may be determined directly by means of equilibrium considerations, assuming given values for the displacements $\mathrm{d}D_1$ and $\mathrm{d}D_2$. Thus, we get from (262)

$$\left\{ \begin{array}{l} \mathrm{d}D_1 = 1 \\ \mathrm{d}D_2 = 0 \end{array} \right. \Rightarrow \left\{ \begin{array}{l} \mathrm{d}F_1 = \dfrac{\partial F_1}{\partial D_1} = K_{11} \\ \mathrm{d}F_2 = \dfrac{\partial F_2}{\partial D_1} = K_{21} . \end{array} \right. \tag{264}$$

From these expressions we conclude that the stiffness coefficients K_{11} and K_{21} are the forces that must be applied in coordinates 1 and 2, to get the deformed configuration depicted in Fig. 172-b. Since the unit displacement $\mathrm{d}D_1$ is infinitesimal, the axial force in the bars is P and the rotations of bars AB and BC are $\frac{1}{l}$. Therefore, the stiffness coefficient K_{11} is the force needed to induce a unit elongation in the spring and to balance the horizontal components of the axial forces in bars AB and BC. K_{21} is the force required to balance the horizontal component of the axial force in bar BC. Thus, these coefficients take the values

$$K_{11} = E - 2\frac{P}{l} \qquad K_{21} = \frac{P}{l} .$$

The two remaining coefficients may be found in a similar way, by considering the deformed configuration represented in Fig. 172-c ($\mathrm{d}D_1 = 0$, $\mathrm{d}D_2 = 1$). The complete stiffness matrix is then

$$[K] = \begin{bmatrix} E - 2\frac{P}{l} & \frac{P}{l} \\ \frac{P}{l} & E - 2\frac{P}{l} \end{bmatrix} = \underbrace{E \begin{bmatrix} 1 & 0 \\ 0 & 1 \end{bmatrix}}_{\substack{\text{material} \\ \text{stiffness}}} + \underbrace{P \begin{bmatrix} -2\frac{1}{l} & \frac{1}{l} \\ \frac{1}{l} & -2\frac{1}{l} \end{bmatrix}}_{\substack{\text{geometrical} \\ \text{stiffness}}} . \tag{265}$$

The *material stiffness* is the component of the stiffness matrix that depends on the stiffness of the deformable elements of the structure: the stiffness of the springs in this case, the modulus of elasticity and the Poisson's coefficient in the case of materials with linear elastic behavior. The *geometrical stiffness* represents the influence of the geometry change of the structure, when it deforms, that is, the change in the nodal forces which equilibrate the internal forces, caused by the change in the directions of the internal forces as a consequence of the rotations. Thus, the geometrical stiffness depends on the internal forces. Tensile internal forces cause a positive geometrical stiffness, while compressive internal forces lead to a negative geometrical stiffness. Buckling occurs when the negative influence of the geometrical stiffness caused by compressive internal forces compensates the positive influence of the material stiffness and of the geometrical stiffness corresponding to tensile internal forces.

Since the stiffness matrix depends on the forces applied to the structure, when it includes the geometrical stiffness, the critical load may be computed by means of the condition of a vanishing determinant, $|K| = 0$, yielding

$$\begin{vmatrix} E - 2\frac{P}{l} & \frac{P}{l} \\ \frac{P}{l} & E - 2\frac{P}{l} \end{vmatrix} = E^2 - \frac{4E}{l}P + \frac{3}{l^2}P^2 = 0 \Rightarrow P = \frac{El}{3} \quad \lor \quad P = El .$$

$$\tag{266}$$

Thus, there are two values of the load P which satisfy the condition of neutral equilibrium which must be fulfilled in the transition from stable to unstable equilibrium: the same internal forces balance the external loads both in the undeformed and in a slightly deformed configuration. The critical load corresponds to the smallest of these values, that is, the critical load takes the value $P_{cr} = \frac{El}{3}$.

However, the equilibrium in the slightly deformed configuration does not occur for an arbitrary deformation shape, but only if some relations between the nodal displacements are satisfied. These relations define the so-called *buckling modes*. In Euler's problem the buckling modes have sinusoidal shapes, as seen in Subsect. XI.3.b. The shape corresponding to $P = P_{cr}$ defines the deformation of the structure when buckling takes place. This shape, i.e., the relation between $\mathrm{d}D_1$ and $\mathrm{d}D_2$ may be found by substituting P by P_{cr} in the stiffness matrix and assuming a given value for one of the nodal displacements.[7] Thus, if we take $dD_1 = 1$, we get from (265) and (263)

$$\left\{ \begin{array}{l} P = P_{cr} = \dfrac{El}{3} \\[2mm] dD_1 = 1 \end{array} \right\} \Rightarrow \begin{bmatrix} \dfrac{E}{3} & \dfrac{E}{3} \\[2mm] \dfrac{E}{3} & \dfrac{E}{3} \end{bmatrix} \left\{ \begin{array}{l} 1 \\ dD_2 \end{array} \right\} = \left\{ \begin{array}{l} 0 \\ 0 \end{array} \right\}.$$

Any of the two equations of this system yields the value -1 for dD_2. Therefore, when the load attains the critical value, the equilibrium takes place both in the undeformed configuration and in the slightly deformed configuration represented in Fig. 172-d. This configuration defines the *first buckling mode*. The second mode, corresponding to the second root of (266), may be found in the same way as the first one, i.e., by substituting P by El in the stiffness matrix, yielding the shape represented in Fig. 172-e. Obviously, this configuration is unstable, since it corresponds to a larger load than the critical one. This is demonstrated by the negative values of the diagonal elements of the stiffness matrix, $P = El \Rightarrow K_{11} = K_{22} = -E$ (see Subsect. XII.5.c).

As a second example of stability analysis in discrete problems with a degree of kinematic indeterminacy superior to one, we shall analyse the column represented in Fig. 173-a, where the stability is guaranteed by the bending stiffness of the two rotational springs, B and C, which have stiffness G.

The line of reasoning used in the first example is obviously also valid in this case, which means that the elements of the stiffness matrix corresponding to the two represented degrees of freedom, 1 and 2, are the forces that must be applied in these coordinates in order to keep the column in the deformed configurations represented in Figs. 173-b ($dD_1 = 1$, $dD_2 = 0$) and 173-c ($dD_1 = 0$, $dD_2 = 1$).

The computation of the stiffness coefficients is a little lengthier than in the first example, since in this case the bars are not under purely axial loading, when dD_1 or dD_2 are different from zero. For the computation of K_{11} and K_{21} by means of equilibrium considerations we may first compute the reaction force R_A (Fig. 173-b), using the expression of the bending moment in spring B as a function of R_A, $M_B = P \times 1 - R_A l = G\frac{2}{l}$. Then K_{11} may be found using the expression of the bending moment in spring C. Once R_A and K_{11} are known, K_{21} may be obtained from the condition of zero bending moment in hinge D. The same procedure, applied to the configuration represented in Fig. 173-c, yields the values of K_{12} K_{22}. The stiffness matrix is then given by

$$[K] = \begin{bmatrix} \dfrac{5G}{l^2} - \dfrac{2P}{l} & -\dfrac{7G}{2l^2} + \dfrac{P}{l} \\[3mm] -\dfrac{7G}{2l^2} + \dfrac{P}{l} & \dfrac{13G}{4l^2} - \dfrac{3P}{2l} \end{bmatrix} = \underbrace{\dfrac{G}{l^2} \begin{bmatrix} 5 & -\dfrac{7}{2} \\[2mm] -\dfrac{7}{2} & \dfrac{13}{4} \end{bmatrix}}_{[K_m]} + \underbrace{\dfrac{P}{l} \begin{bmatrix} -2 & 1 \\[2mm] 1 & -\dfrac{3}{2} \end{bmatrix}}_{[K_g]}.$$

---

[7] It is not possible to compute dD_1 and dD_2 simultaneously because the two equations of the system defined by (263) become linearly dependent for $P = P_{cr}$. This means that it is not possible to find the amplitude of the deformation, which reflects the fact that equilibrium is neutral in the critical phase.

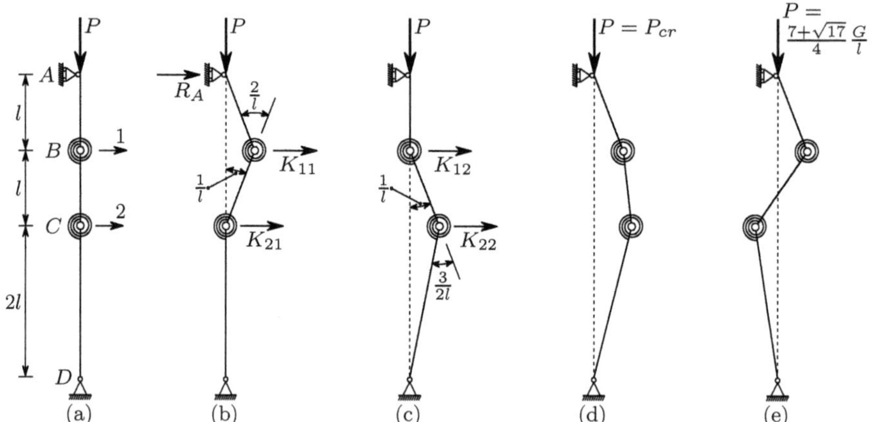

Fig. 173. Example of stability analysis in a column with bending stiffness

$[K_m]$ and $[K_g]$ represent the material and geometrical components of the stiffness matrix. The condition of a vanishing determinant yields the two values of P, for which equilibrium between the internal forces and P exists in a slightly deformed configuration

$$\begin{vmatrix} \dfrac{5G}{l^2} - \dfrac{2P}{l} & -\dfrac{7G}{2l^2} + \dfrac{P}{l} \\[2mm] -\dfrac{7G}{2l^2} + \dfrac{P}{l} & \dfrac{13G}{4l^2} - \dfrac{3P}{2l} \end{vmatrix} = 0 \;\Rightarrow\; P = \frac{7 - \sqrt{17}}{4}\frac{G}{l} \quad \vee \quad P = \frac{7 + \sqrt{17}}{4}\frac{G}{l}\,.$$

The critical load is thus $P_{cr} = \frac{7-\sqrt{17}}{4}\frac{G}{l}$. The equilibrium configurations which correspond to these two roots may be found in the same way as in the first example, yielding the vectors

$$P = P_{cr} = \frac{7 - \sqrt{17}}{4}\frac{G}{l} \;\Rightarrow\; \begin{cases} \mathrm{d}D_1 = 1 \\ \mathrm{d}D_2 = 1.281 \end{cases}$$

and

$$P = \frac{7 + \sqrt{17}}{4}\frac{G}{l} \;\Rightarrow\; \begin{cases} \mathrm{d}D_1 = 1 \\ \mathrm{d}D_2 = -0.781 \end{cases}.$$

The configurations corresponding to these two displacement-vectors are represented in Figs. 173-d and 173-e. Thus, when buckling occurs, the column deforms with the shape given in Fig. 173-d.

In structures with a high degree of kinematic indeterminacy the critical load and the buckling shapes are not computed in the same way as in the two simple examples analysed. Generally, a factor λ is required, by which the applied load must be multiplied in order to find the critical load. It may be shown that, if the material has linear elastic behaviour, if the displacements are small enough to be considered as infinitesimal, and if the displacement of a point of the structure may be expressed as a linear function of the displacements of the nodal points, as happens in the two simple examples above and also in the most used finite-element discretizations, *the material stiffness is constant and the geometrical stiffness is proportional to the applied loads.* Under these conditions, the problem may be formulated as shown below ($\{\mathrm{d}D\}$ is the vector containing the n kinematic unknowns and $\{0\}$ is a vector with n zeros)

$$[K]\{\mathrm{d}D\} = \{0\} \quad \Rightarrow \quad ([K_m] + \lambda\,[K_g])\,\{\mathrm{d}D\} = \{0\}$$
$$\Rightarrow \quad [K_m]\,\{\mathrm{d}D\} = -\lambda\,[K_g]\,\{\mathrm{d}D\} \ . \tag{267}$$

The last equality in (267) may be reduced to the form of the algebraic *generalized symmetric eigenvalue problem*

$$[A]\,\{z\} = \lambda'\,[B]\,\{z\} \ , \tag{268}$$

where the matrices $[A]$ and $[B]$ are symmetrical and $[B]$ is positive definite. The matrices $[K_m]$ and $[K_g]$ are symmetrical. Furthermore, $[K_m]$ is positive definite. Thus, rearranging the last of 267, we may put it in the form corresponding to the generalized eigenvalue problem (268), that is

$$\begin{cases} [A] &=& [K_g] \\ [B] &=& [K_m] \\ \{z\} &=& \{\mathrm{d}D\} \\ \lambda' &=& -\frac{1}{\lambda} \end{cases} \Rightarrow \left| \begin{array}{l} [K_g]\,\{\mathrm{d}D\} = -\frac{1}{\lambda}[K_m]\,\{\mathrm{d}D\} \\ \Leftrightarrow \quad [A]\,\{z\} = \lambda'\,[B]\,\{z\} \ . \end{array} \right.$$

The eigenvectors represent the buckling modes. The resolution of the two simple problems (Figs. 172 and 173) with this method is left as an exercise for the reader.

XI.6.c Framed Structures Under Bending

When the stability of a framed structure is guaranteed by the bending stiffness of its members, the computation of the critical load requires the stiffness of the bars to be expressed as functions of the corresponding axial forces. In this Sub-section we develop expressions to compute the stiffness matrix of a bar with four and three degrees of freedom, disregarding axial deformations (see Footnote 77), and assuming that the rotations are small (see the last part of Subsect. IX.1.b for an analysis of the error introduced by this assumption). The last part includes examples of both the determination of critical loads and of the approximate computation of displacements in framed structures, taking the interaction between internal forces and deformations into consideration.

XI.6.c.i Stiffness Matrix of a Compressed Bar

The stiffness matrix of a bar with the four degrees of freedom represented in Fig. 174-a may be found by integration of the differential equation which defines the relations between forces applied in the coordinates and the bending moments, when the bending deformation is taken into account. To this end, let us consider the column represented in Fig. 174-b, under the action of the compressive axial force N, the transversal load V and the moment M. The differential equation of this problem may be established considering the curvature corresponding to the bending moment in the cross-section defined by coordinate z (Fig. 174-b).

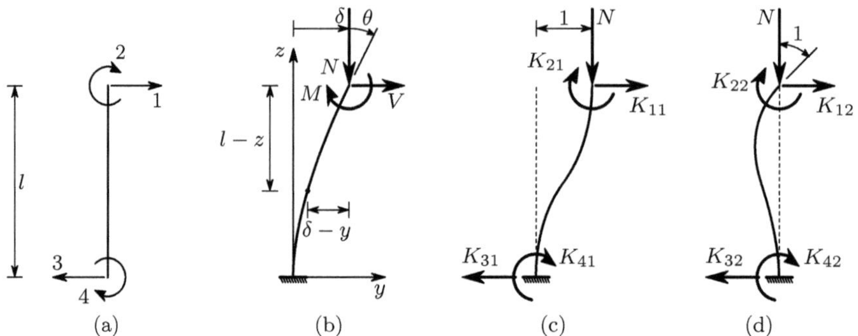

Fig. 174. Determination of the stiffness matrix of a compressed bar

Denoting by Δ the displacement of the upper cross-section (Fig. 174-b), we get

$$\frac{1}{\rho} = \frac{\mathrm{d}^2 y}{\mathrm{d}z^2} = \frac{1}{EI} \overbrace{[M + N\,(\delta - y) + V\,(l - z)]}^{\text{bending moment}} \tag{269}$$

$$\Rightarrow \quad \frac{\mathrm{d}^2 y}{\mathrm{d}z^2} + \frac{N}{EI} y = \frac{1}{EI} \left[M + N\Delta + V\,(l - z) \right] \; .$$

We may easily verify that this equation admits the particular integral

$$y = \frac{M}{N} + \frac{V}{N}\,(l - z) + \delta \; .$$

Thus its general solution is

$$y = C_1 \cos(kz) + C_2 \sin(kz) + \frac{M}{N} + \frac{V}{N}\,(l - z) + \delta \quad \text{with} \quad k^2 = \frac{N}{EI} \; .$$

Derivating this expression with respect to z, we get the equation of the rotations

$$\frac{\mathrm{d}y}{\mathrm{d}z} = -kC_1 \sin(kz) + kC_2 \cos(kz) - \frac{V}{N} \; .$$

The support conditions in the built-in end allow the determination of the integration constants. Thus, we have

$$z = 0 \Rightarrow \begin{cases} y \ = 0 \Rightarrow C_1 = -\dfrac{M}{N} - \dfrac{Vl}{N} - \delta \\[3mm] \dfrac{\mathrm{d}y}{\mathrm{d}z} = 0 \Rightarrow kC_2 - \dfrac{V}{N} = 0 \Rightarrow C_2 = \dfrac{V}{kN} \ . \end{cases}$$

Substituting these values in the previous expressions, we get

$$\begin{aligned} y \ &= \frac{V}{kN}\sin(kz) - \left(\delta + \frac{M+Vl}{N}\right)\cos(kz) + \frac{M}{N} + \frac{V}{N}(l-z) + \delta \\[3mm] \frac{\mathrm{d}y}{\mathrm{d}z} &= \frac{V}{N}\cos(kz) + \left(\delta + \frac{M+Vl}{N}\right)k\sin(kz) - \frac{V}{N} \ . \end{aligned}$$

$$(270)$$

The displacement Δ and the rotation θ of the upper end of the column may be related to the applied forces, yielding

$$z = l \ \Rightarrow \begin{cases} y \ = \delta \ \Rightarrow \ \dfrac{V}{kN}\sin(kl) - \left(\delta + \dfrac{M+Vl}{N}\right)\cos(kl) + \dfrac{M}{N} + \delta = \delta \\[4mm] \dfrac{\mathrm{d}y}{\mathrm{d}z} = \theta \ \Rightarrow \ \dfrac{V}{N}\cos(kl) + \left(\delta + \dfrac{M+Vl}{N}\right)k\sin(kl) - \dfrac{V}{N} = \theta \ . \end{cases}$$

Rearranging this system of equations, the following form may be given to it

$$\begin{cases} \dfrac{k\sin(kl)}{N}M + \dfrac{kl\sin(kl)+\cos(kl)-1}{N}V = \theta - \delta k\sin(kl) \\[4mm] \dfrac{1-\cos(kl)}{N\cos(kl)}M + \dfrac{\tan(kl)-kl}{kN}V \quad = \delta \ . \end{cases}$$

By solving this system of equations, we get the force V and the moment M required to introduce the displacement δ and the rotation θ in the upper cross-section, for a given value of the axial force N. This solution takes the form

$$M = \frac{c_1 b_2 - b_1 c_2}{a_1 b_2 - a_2 b_1} \qquad\qquad V = \frac{a_1 c_2 - c_1 a_2}{a_1 b_2 - a_2 b_1} \qquad\qquad (271)$$

with

$$a_1 = \frac{k\sin(kl)}{N} \qquad b_1 = \frac{kl\sin(kl)+\cos(kl)-1}{N} \qquad c_1 = \theta - \delta k\sin(kl)$$

$$a_2 = \frac{1-\cos(kl)}{N\cos(kl)} \qquad b_2 = \frac{\tan(kl)-kl}{kN} \qquad c_2 = \delta \ .$$

From these expressions we get, after some manipulation,

$$a_1 b_2 - a_2 b_1 = \frac{2\left[1 - \cos(kl)\right] - kl\sin(kl)}{N^2\cos(kl)}$$

$$c_1 b_2 - b_1 c_2 = \frac{\theta\left[\tan(kl) - kl\right] + \Delta\left[k - k\cos(kl) - k\sin(kl)\tan(kl)\right]}{kN} \tag{272}$$

$$a_1 c_2 - c_1 a_2 = \frac{\theta\left[\cos(kl) - 1\right] + \Delta k\sin(kl)}{N\cos(kl)} \; .$$

The same line of reasoning as used in the first example in Subsect. XI.6.b (264), leads to the conclusion that the elements K_{11} and K_{21} of the stiffness matrix are the transversal force V and the moment M required to induce the deformation represented in Fig. 174-c, i.e., $\delta = 1$ and $\theta = 0$. Substituting these values in (272) and the result of this substitution in (271), we get

$$V = K_{11} = \frac{N}{2 - 2\cos(kl) - kl\sin(kl)} k\sin(kl)$$

$$M = K_{21} = \frac{N}{2 - 2\cos(kl) - kl\sin(kl)} \left[\cos(kl) - 1\right] \; .$$

Elements K_{31} and K_{41} may then be found by means of equilibrium considerations, yielding

$$K_{31} = K_{11} \quad \text{and} \quad K_{41} = -K_{11}l - K_{21} - N = K_{21} \; .$$

The last equality is obtained by a moment equation with respect to the built-in end. The antisymmetry of the deformation also leads directly to this equality.

The elements K_{12} and K_{22} are the transversal force V and the moment M needed to induce the deformation represented in Fig. 174-d, i.e., $\delta = 0$ and $\theta = 1$. The same procedure as above leads to

$$V = K_{12} = \frac{N}{2 - 2\cos(kl) - kl\sin(kl)} \left[\cos(kl) - 1\right] = K_{21}$$

$$M = K_{22} = \frac{N}{2 - 2\cos(kl) - kl\sin(kl)} \frac{\sin(kl) - kl\cos(kl)}{k} \; .$$

Elements K_{32} are K_{42} also obtained by means of equilibrium considerations, which yield

$$K_{32} = K_{12} \quad \text{and} \quad K_{42} = -K_{22} - K_{12}l = \frac{N}{2 - 2\cos(kl) - kl\sin(kl)} \frac{kl - \sin(kl)}{k} \; .$$

The two remaining columns of the stiffness matrix could be obtained by considering unit displacements in coordinates 3 and 4. It is, however, obvious that the results would the same as those obtained by interchanging the roles of indices 1 and 2 with indices 3 and 4, respectively. Thus, the complete stiffness matrix takes the form

$$[K] = \frac{N}{2 - 2\cos(kl) - kl\sin(kl)}$$
$$\times \begin{bmatrix} k\sin(kl) & \cos(kl) - 1 & k\sin(kl) & \cos(kl) - 1 \\ \cos(kl) - 1 & \frac{\sin(kl) - kl\cos(kl)}{k} & \cos(kl) - 1 & \frac{kl - \sin(kl)}{k} \\ k\sin(kl) & \cos(kl) - 1 & k\sin(kl) & \cos(kl) - 1 \\ \cos(kl) - 1 & \frac{kl - \sin(kl)}{k} & \cos(kl) - 1 & \frac{\sin(kl) - kl\cos(kl)}{k} \end{bmatrix} . \quad (273)$$

Sometimes it is also useful to know the stiffness matrix of a compressed bar with the three coordinates represented in Fig. 175-a.

The corresponding stiffness coefficients may easily be obtained by solving the problems represented in Figs. 175-b and 175-c. Since these problems do not have any additional difficulty compared with those in Fig. 174, their detailed analysis is left as an exercise to the reader. The resulting stiffness matrix takes the form

$$\frac{N}{\sin(kl) - kl\cos(kl)} \begin{bmatrix} k\cos(kl) & -\sin(kl) & k\cos(kl) \\ -\sin(kl) & l\sin(kl) & -\sin(kl) \\ k\cos(kl) & -\sin(kl) & k\cos(kl) \end{bmatrix} . \quad (274)$$

All examples of determination of the critical load presented in Sect. XI.3, and also examples XI.11, XI.12 and XI.13, may easily be confirmed by analysing particular elements of the two stiffness matrices above (273) and (274).

Thus, the critical load of a pin-ended bar (Euler's problem) is attained when coefficient K_{22} of the matrix in (274) vanishes, that is, when the bar ceases to have sufficient stiffness to resist the rotation of the upper cross-section. The smallest non-zero value of kl for which this coefficient vanishes is π, which coincides with Euler's solution.

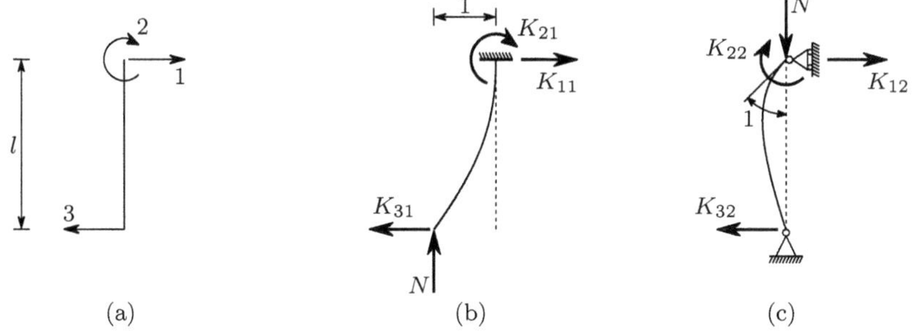

Fig. 175. Determination of the stiffness matrix of a bar with three coordinates

The critical load of the example presented in Fig. 166-a is attained when coefficient K_{33} of the matrix in (274) vanishes, which takes place for $\cos(kl) = 0$, i.e., when the condition $kl = (2n-1)\frac{\pi}{2}$ is satisfied, which leads to the effective length $2l$.

The critical load of the example in Fig. 166-b corresponds to a vanishing value of coefficient K_{11} of (273), which takes place for $\sin(kl) = 0$.

In the example of Fig. 166-c the critical load is reached when the bar loses the stiffness needed to resist the transversal displacement of the mid-height cross-section. Considering the column to be divided in two bars with equal length $\frac{l}{2}$, the horizontal stiffness of the column in the connection point between the two bars is given by twice the coefficient K_{11} of (273), with the length of the bar reduced to $\frac{l}{2}$. Thus, buckling takes place when $\sin(k\frac{l}{2}) = 0$, that is, when $k\frac{l}{2} = \pi$.

The critical load of the column depicted in Fig. 166-d is attained when the coefficient K_{22} in (273) vanishes, which happens for $\sin(kl) = kl\cos(kl)$. This is exactly the expression used in Subsect. XI.3.c to find the effective length of this column.

In example XI.11 the rotational stiffness of cross-section B has two components, one corresponding to the deformation of column segment BC and other corresponding to the rotation of the rigid segment AB. The first component is given by coefficient K_{22} of the matrix in (274). The second takes the value $-Pl'$, since, for a unit rotation θ, this is the moment needed in cross-section B, to balance the moment caused by the load P. Thus, the total rotational stiffness of cross-section B vanishes when the following condition is satisfied

$$K = \frac{Pl\sin(kl)}{\sin(kl) - kl\cos(kl)} - Pl' = 0 \Rightarrow l'\cos(kl) + \frac{l - l'}{kl}\sin(kl) = 0 \,,$$

which coincides with the expression obtained in example XI.11.

Also in example XI.12, the horizontal stiffness of the column in point B has two components, one corresponds to the deformation of segment BC and the other is the force needed to balance the horizontal component of the pin-ended bar AB, when point B suffers a unit displacement. The first component coincides with coefficient K_{33} of the matrix in (274). The second component takes the value $-\frac{P}{l'}$. Buckling occurs when the total stiffness vanishes, that is, when the following condition is satisfied

$$K = \frac{Pk\cos(kl)}{\sin(kl) - kl\cos(kl)} - \frac{P}{l'} = 0 \Rightarrow \left(1 + \frac{l'}{l}\right)kl = \tan(kl) \,.$$

Finally, in example XI.13 the rotational stiffness of cross-section B has the components which correspond to the deformation of segments AB and BC. These components are given by coefficient K_{22} of the matrix in (274), with the lengths l and l', respectively. The critical load may thus be obtained by solving the equation

$$K = \frac{Pl\sin(kl)}{\sin(kl) - kl\cos(kl)} + \frac{Pl'\sin(kl')}{\sin(kl') - kl'\cos(kl')} = 0 \,.$$

This expression defines the problem completely, as opposed to the equation obtained in example XI.13. In fact, if we have $l = l'$, we can see immediately that $\sin(kl) = 0 \Rightarrow kl = n\pi$ is a solution of the equation. The expression found in XI.13 may be obtained from this one by dividing both terms (numerator and denominator) of the fraction representing the first component by $\sin(kl)$ and both terms of the second component by $\sin(kl')$. When we have $l = l'$, this operation eliminates the roots of the equation that are defined by $kl = n\pi$.

XI.6.c.ii Stiffness Matrix of a Tensioned Bar

The stiffness of a tensioned bar with the degrees of freedom represented in Fig. 174-a may be obtained in the same way as in the case of the compressed bar. The differential equation defining the relations between the forces applied in the coordinates and the bending moments, when a tensile force is considered, takes a similar form to (269), but with a reversed sign in the elements containing N (Fig. 174-b with N pointing upwards)

$$\frac{1}{\rho} = \frac{\mathrm{d}^2 y}{\mathrm{d}z^2} = \frac{1}{EI}\left[M - N\left(\delta - y\right) + V\left(l - z\right)\right]$$

$$\Rightarrow \quad \frac{\mathrm{d}^2 y}{\mathrm{d}z^2} - \frac{N}{EI}y = \frac{1}{EI}\left[M - N\delta + V\left(l - z\right)\right].$$

This equation admits the particular integral

$$y = \delta - \frac{M}{N} - \frac{V}{N}\left(l - z\right),$$

Since in this case the homogeneous equation does not have imaginary roots, its general solution takes the form

$$y = C_1 \mathrm{e}^{kz} + C_2 \mathrm{e}^{-kz} + \delta - \frac{M}{N} - \frac{V}{N}\left(l - z\right) \quad \text{with} \quad k^2 = \frac{N}{EI}. \tag{275}$$

In the same way as in the case of axial compression, the integration constants may be eliminated by means of the support conditions in the built-in end (Fig. 174-b). In this case we have

$$z = 0 \Rightarrow \begin{cases} y = 0 \Rightarrow C_1 + C_2 + \delta - \dfrac{M}{N} - \dfrac{Vl}{N} = 0 \\[2ex] \dfrac{\mathrm{d}y}{\mathrm{d}z} = 0 \Rightarrow kC_1 - kC_2 + \dfrac{V}{N} = 0. \end{cases}$$

Solving this system of equations, we get

$$C_1 = \frac{M}{2N} - \frac{V}{2kN} + \frac{Vl}{2N} - \frac{\delta}{2} \quad \text{and} \quad C_2 = \frac{M}{2N} + \frac{V}{2kN} + \frac{Vl}{2N} - \frac{\delta}{2}.$$

Substituting these expressions in the general solution (275) and using the definitions of hyperbolic sine and cosine ($\sinh x = \frac{e^x - e^{-x}}{2}$, $\cosh x = \frac{e^x + e^{-x}}{2}$), we may give (275) the following form

$$y = \left(\frac{M + Vl}{N} - \delta \right) \cosh(kz) - \frac{V}{kN} \sinh(kz) + \delta - \frac{M}{N} - \frac{V}{N}(l - z) \ .$$

Derivating with respect to z, we get

$$\frac{dy}{dz} = \left(\frac{M + Vl}{N} - \delta \right) k \sinh(kz) - \frac{V}{N} \cosh(kz) + \frac{V}{N} \ .$$

These two expressions are very similar to those obtained in the compressive case (270), showing, in addition to the substitution of sine by hyperbolic sine and of cosine by hyperbolic cosine, only sign differences in some elements. The treatment of these expressions until the stiffness coefficients are obtained is the same as in the case of axial compression, so it is not repeated here. The resulting stiffness matrix takes the form

$$[K] = \frac{N}{2 - 2\cosh(kl) + kl\sinh(kl)}$$

$$\times \begin{bmatrix} k\sinh(kl) & 1 - \cosh(kl) & k\sinh(kl) & 1 - \cosh(kl) \\ 1 - \cosh(kl) & \frac{kl\cosh(kl) - \sinh(kl)}{k} & 1 - \cosh(kl) & \frac{\sinh(kl) - kl}{k} \\ k\sinh(kl) & 1 - \cosh(kl) & k\sinh(kl) & 1 - \cosh(kl) \\ 1 - \cosh(kl) & \frac{\sinh(kl) - kl}{k} & 1 - \cosh(kl) & \frac{kl\cosh(kl) - \sinh(kl)}{k} \end{bmatrix} \ .$$

$$(276)$$

In the case of the coordinates defined in Fig. 175-a the stiffness matrix for a tensile axial force takes the form

$$[K] = \frac{N}{kl\cosh(kl) - \sinh(kl)} \begin{bmatrix} k\cosh(kl) & -\sinh(kl) & k\cosh(kl) \\ -\sinh(kl) & l\sinh(kl) & -\sinh(kl) \\ k\cosh(kl) & -\sinh(kl) & k\cosh(kl) \end{bmatrix} \ . \quad (277)$$

XI.6.c.iii Linearization of the Stiffness Coefficients

The stiffness matrices defined by (273), (274), (276) and (277) include the influences of the bending stiffness (by means of the term EI contained in k) and of the axial force. However, the stiffness coefficients are not used in this form, as a rule, since the fact that it is not possible to decompose these matrices in a component which is independent of the axial force (the material stiffness), and in a component which is independent of the elasticity modulus of the material the bar is made of (the geometrical stiffness), prevents the use of the algorithm based on the reduction of the problem to eigenvalue form (267).[8]

This problem may be circumvented by linearizing the stiffness coefficients. This may be achieved by computing the value of these coefficients and their derivatives with respect to the axial force, for a vanishing axial force. These operations are rather lengthy, since the elements of the stiffness matrices and their derivatives become indeterminate for $k = 0$. However, by using a computational program for symbolic manipulation, this virtually ceases to be a problem. Performing these operations in the expressions relating to compression (273) and (274) and tension (276) and (277), we conclude that the functions expressing the stiffness coefficients and their derivatives are continuous in the point defined by a vanishing axial force. For example, in the case of coefficient K_{11} in (273) and (276) we get

$$\lim_{N \to 0} \left[\frac{N}{2 - 2\cos(kl) - kl\sin(kl)} k\sin(kl) \right] = \frac{12EI}{l^3}$$

$$\lim_{N \to 0} \left[\frac{N}{2 - 2\cosh(kl) + kl\sinh(kl)} k\sinh(kl) \right] = \frac{12EI}{l^3}$$

$$\lim_{N \to 0} \left\{ \frac{\mathrm{d}}{\mathrm{d}N} \left[\frac{N}{2 - 2\cos(kl) - kl\sin(kl)} k\sin(kl) \right] \right\} = -\frac{6}{5l}$$

$$\lim_{N \to 0} \left\{ \frac{\mathrm{d}}{\mathrm{d}N} \left[\frac{N}{2 - 2\cosh(kl) + kl\sinh(kl)} k\sinh(kl) \right] \right\} = \frac{6}{5l} \ .$$

The difference in the sign between the two last expressions is a consequence of the fact that in the first one a compressive axial force is considered to be positive, while in the second one the tensile force is positive. The linearized form of this coefficient is then (a tensile axial force is considered as positive)

$$K_{11} = \frac{12EI}{l^3} + N\frac{6}{5l} \ .$$

Carrying out the same operations for all the stiffness coefficients, the following matrices are obtained

$$[K] = EI \begin{bmatrix} \frac{12}{l^3} & -\frac{6}{l^2} & \frac{12}{l^3} & -\frac{6}{l^2} \\ -\frac{6}{l^2} & \frac{4}{l} & -\frac{6}{l^2} & \frac{2}{l} \\ \frac{12}{l^3} & -\frac{6}{l^2} & \frac{12}{l^3} & -\frac{6}{l^2} \\ -\frac{6}{l^2} & \frac{2}{l} & -\frac{6}{l^2} & \frac{4}{l} \end{bmatrix} + N \begin{bmatrix} \frac{6}{5l} & -\frac{1}{10} & \frac{6}{5l} & -\frac{1}{10} \\ -\frac{1}{10} & \frac{2l}{15} & -\frac{1}{10} & -\frac{l}{30} \\ \frac{6}{5l} & -\frac{1}{10} & \frac{6}{5l} & -\frac{1}{10} \\ -\frac{1}{10} & -\frac{l}{30} & -\frac{1}{10} & \frac{2l}{15} \end{bmatrix} \tag{278}$$

and

$$[K] = EI \begin{bmatrix} \frac{3}{l^3} & -\frac{3}{l^2} & \frac{3}{l^3} \\ -\frac{3}{l^2} & \frac{3}{l} & -\frac{3}{l^2} \\ \frac{3}{l^3} & -\frac{3}{l^2} & \frac{3}{l^3} \end{bmatrix} + N \begin{bmatrix} \frac{6}{5l} & -\frac{1}{5} & \frac{6}{5l} \\ -\frac{1}{5} & \frac{l}{5} & -\frac{1}{5} \\ \frac{6}{5l} & -\frac{1}{5} & \frac{6}{5l} \end{bmatrix} \ , \tag{279}$$

---

[8] In this problem the displacement of a point of the bar is not a linear function of the displacements of the coordinates, as required for (267) to be valid.

respectively for the coordinates indicated in Figs. 174-a and 175-a. The first elements in each equation represent the material stiffness (proportional to the elasticity modulus E) and the second ones are the geometrical stiffness (proportional to the axial force N).

In Fig. 176 the exact and linearized forms of coefficient K_{11} of the first matrix, as functions of N, are represented. The infinite peaks appearing in the compressive part of the first diagram correspond to the values of kl for which the denominator $2 - 2\cos(kl) - kl\sin(kl)$ (273) vanishes. However, the very high values of the stiffness coefficients do not have a physical correspondence. In fact, the above theory is only valid for small rotations, since only under these conditions may we accept that $\frac{1}{\rho} = \frac{d^2 y}{dz^2}$ (see the last part of Subsect. IX.1.b). When the quantity $2 - 2\cos(kl) - kl\sin(kl)$ gets close to zero, the stiffness coefficients take high values, which leads to high rotation values, as may be concluded from the second of (270) (M and V take the values of the stiffness coefficients).

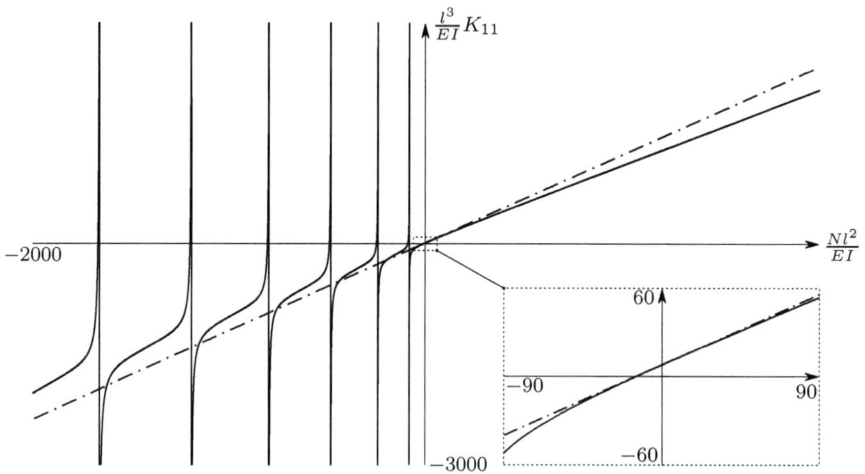

Fig. 176. "Exact" (*solid lines*) and linearized (*dashed-dotted line*)forms of stiffness coefficient K_{11} (Fig. 174-a)

Obviously, the linearization of the stiffness coefficient introduces errors into the computation of buckling forces and shapes. However, these errors are not large if only the critical load and the first buckling mode are required, and if the kinematic coordinates are well-chosen. In the case of higher modes, considerable errors may be introduced.

When the stiffness coefficients are used in their exact form (273), (274), (276) and (277), we get an infinite number of buckling modes. As a consequence of the linearization, the number of buckling modes obtained becomes equal to the degree of kinematic indeterminacy, as may be concluded from the algebraic form of the problem (267).

When the linearized form of the stiffness coefficients is used, the correct choice of the kinematic coordinates is very important, since local buckling modes may only be captured with sufficient accuracy if the chosen coordinates allow it. For this reason, in a systematic computational analysis it is advisable to consider additional nodes in the middle or even in each third of the bars, although these nodes are not needed to compute the internal forces.

In Subsect. XI.6.c.*iv* examples are presented to illustrate these considerations.

XI.6.c.iv Examples of Application

As a first example of stability analysis of a framed structure using the displacement method, let us consider the plane frame represented in Fig. 177-a.

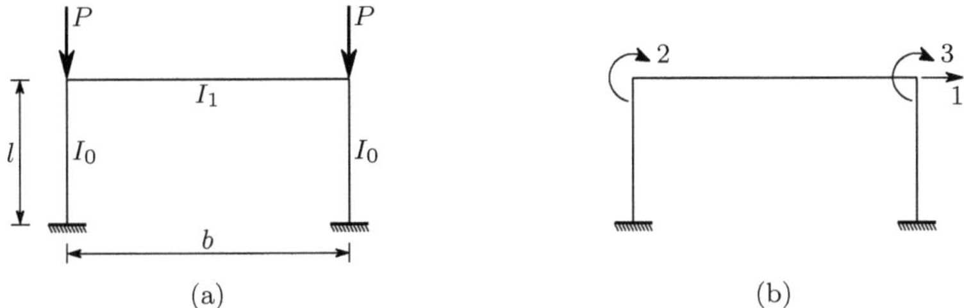

(a)   (b)

Fig. 177. Stability analysis of a plane frame

This problem has an analytical solution which may be compared with the solution yielded by the displacement method, both in the exact and in the linearized formulation. This solution is defined by the expression (see, e.g., [11], Sect. 2.4)

$$\frac{kl}{\tan(kl)} = -\frac{6lI_1}{bI_0} \quad \text{with} \quad k = \sqrt{\frac{P}{EI_0}}.$$

Giving numerical values to the geometrical parameters, we get the critical load ($E = 206\,GPa$)

$$
\begin{aligned}
l &= 5m \\
b &= 10m \\
I_0 &= 20 \times 10^{-6} m^4 \\
I_1 &= 50 \times 10^{-6} m^4
\end{aligned}
\quad \Rightarrow \quad
\begin{aligned}
&\frac{kl}{\tan(kl)} = -7.5 \Rightarrow kl = 2.785931 \\
&\Rightarrow P_{cr} = (kl)^2 \frac{EI_0}{l^2} = 1279081N = 1279.081kN .
\end{aligned}
$$

To resolve this problem by means of the displacement method, the coordinates represented in Fig. 177-b may be used. Denoting by K^0 and K^1 the stiffness coefficients of the vertical and horizontal bars, respectively (Fig. 174-a), we conclude immediately that the forces required to introduce the deformation depicted in Fig. 178-a ($D_1 = 1$, $D_2 = D_3 = 0$) take the values

$$K_{11} = 2K_{11}^0 \qquad K_{21} = K_{12}^0 \qquad K_{31} = K_{12}^0 \ .$$

Proceeding in the same way in the configurations represented in Figs. 178-b ($D_2 = 1$, $D_1 = D_3 = 0$) and 178-c ($D_1 = D_2 = 0$, $D_3 = 1$), we get the other elements of the stiffness matrix of this structure, which takes the form

$$[K] = \begin{bmatrix} K_{11} & K_{12} & K_{13} \\ K_{21} & K_{22} & K_{23} \\ K_{31} & K_{32} & K_{33} \end{bmatrix} = \begin{bmatrix} 2K_{11}^0 & K_{12}^0 & K_{12}^0 \\ K_{12}^0 & K_{22}^0 + K_{22}^1 & K_{24}^1 \\ K_{12}^0 & K_{24}^1 & K_{22}^0 + K_{22}^1 \end{bmatrix} \ .$$

The critical load is the smallest value of P which makes this matrix singular ($|K| = 0$).

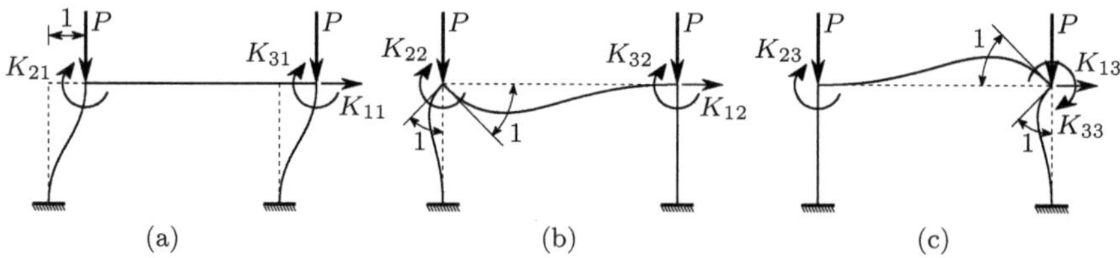

Fig. 178. Elements of the stiffness matrix

Writing a small computer program with the expression of the determinant of matrix $[K]$, taking the stiffness coefficients K^0 from (273) and the coefficients K^1 from the first matrix of (278) (the elements of (273) become indeterminate for $N = 0$), we conclude that the smallest value of P that leads to a vanishing determinant coincides with the above solution. When the linearized form of the stiffness coefficients is used (278), we get the slightly higher value P_{cr}^l

$$P_{cr}^l = 1290528N = 1290.528kN \ \Rightarrow \ \frac{P_{cr}^l}{P_{cr}} = \frac{1290.528}{1279.081} = 1.0089 \ .$$

The first buckling mode is defined by the displacement vector that satisfies the condition

$$[K]\{D\} = \{0\} \quad \text{with} \quad P = P_{cr} \Rightarrow |K| = 0 \ .$$

Since we have only three coordinates, the components of vector $\{D\}$ may be computed by means of the same algorithm that has been used in the resolution of exercise II.1. Thus, using the same line of reasoning as described there, we easily conclude that a vector with the components

$$D_1 = \begin{vmatrix} K_{12} & K_{13} \\ K_{22} & K_{23} \end{vmatrix} \qquad D_2 = - \begin{vmatrix} K_{11} & K_{13} \\ K_{21} & K_{23} \end{vmatrix} \qquad D_3 = \begin{vmatrix} K_{11} & K_{12} \\ K_{21} & K_{22} \end{vmatrix}$$

satisfies the equation above. The results obtained with these expressions are practically the same, either taking the exact stiffness coefficients ((273), or the linearized form (278). Multiplying the elements of this vector by a constant, so that the displacement D_1 takes the value of $10\,\mathrm{cm}$, we get for the rotations $D_2 = D_3 = 0.010$ radians. This buckling mode corresponds to the deformed configuration represented in Fig. 179-a.

When the load corresponding to the second buckling mode is computed, very different values are obtained from the exact and linearized forms of the stiffness coefficients. Thus, using (273) we get $P = 4304kN$, while (278) gives $P = 8034kN$. However, the corresponding buckling modes are similar, defining the configuration represented in Fig. 179-b.

To illustrate the correct choice of the kinematic coordinates, when the linearized form of the stiffness coefficients is used, let us consider the plane structure represented in Fig. 180-a, where the horizontal and vertical bars have cross-sections with moments of inertia $5I$ and I, respectively.

Since, when buckling occurs, a rotation of node B takes place, the corresponding stiffness may be used to compute the critical value of load P. The rotational stiffness of node B is the moment K which is needed to cause a unit rotation of the ends of the bars converging in node B, that is, it is the sum of

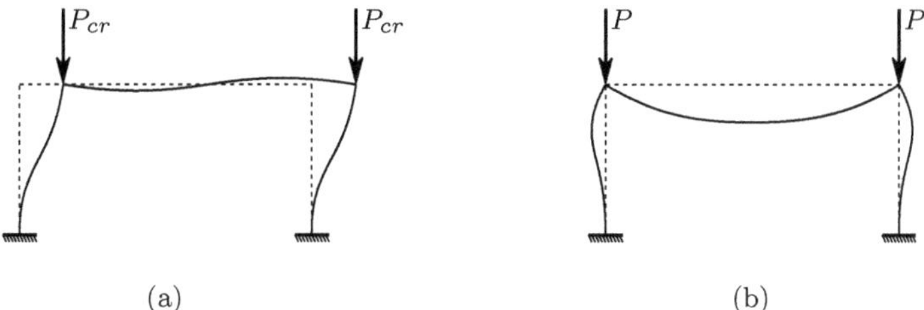

(a)   (b)

Fig. 179. First and second buckling modes of the frame represented in Fig. 177-a

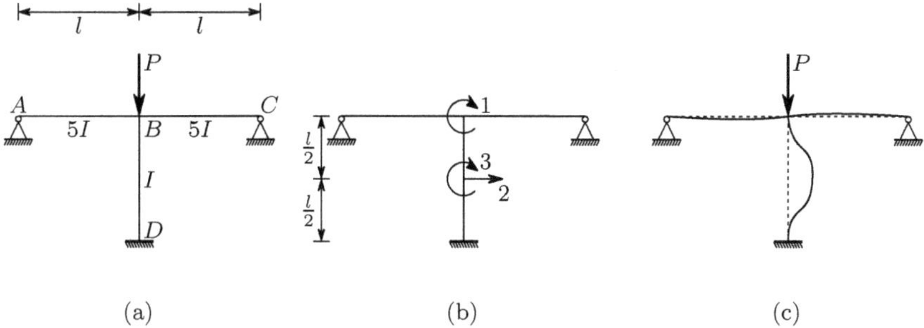

(a)   (b)   (c)

Fig. 180. Example illustrating the importance of the correct choice of the kinematic coordinates, when the linearized form of the stiffness coefficients is used

the stiffness coefficients K_{22} both in the case represented in Fig. 174-a for the vertical bar, and in the case represented in Fig. 175-a for the horizontal bars. Thus, we get from the first matrix in (279) (horizontal bars, $N = 0$) and from (273) (vertical bar)

$$K = 2 \times \frac{3E5I}{l} + \frac{P}{k} \frac{\sin(kl) - kl\cos(kl)}{2 - 2\cos(kl) - kl\sin(kl)} \, .$$

Since we have $P = k^2 EI$, we may give this expression the following form

$$K = \frac{EI}{l} \left[30 + kl\frac{\sin(kl) - kl\cos(kl)}{2 - 2\cos(kl) - kl\sin(kl)} \right] \, .$$

A simple numerical investigation shows that the smallest value of load P that leads to a vanishing value of this stiffness is

$$K = 0 \Rightarrow kl = 6.083065 \Rightarrow P = P_{cr} = 37.00368\frac{EI}{l^2} \, . \qquad (280)$$

Repeating this analysis with the linearized form of the stiffness coefficients of the bars, we get from (278) and (279)

$$K = 2 \times \frac{3E5I}{l} + \frac{4EI}{l} - P\frac{2l}{15} = 0 \Rightarrow P = 255\frac{EI}{l^2} \, .$$

We find that the solution yielded by the linearized form of the stiffness coefficients leads to an enormous error.

However, if instead of considering only the rotation coordinate in node B, we consider the stiffness matrix corresponding to the three coordinates represented in Fig. 180-b, the results obtained using the linearized forms of the stiffness coefficients come much closer to the exact solution. The procedure leading to the computation of the critical load and the first buckling mode is the same as in the case of the frame represented in Fig. 177, so it is not described here. The critical load computed using the exact form of the stiffness coefficients confirms the value given in (280). The linearized stiffness yields the value $P_{cr} = 37.453265\frac{EI}{l^2}$. The error of this solution is approximately 1%. The displacement vector corresponding to this load shows that instability is caused almost exclusively by the buckling of the vertical bar: If we multiply the displacement vector corresponding to the first buckling mode by a constant factor, so that we get $D_2 = \frac{l}{10}$, we get for the rotations D_1 and D_3 the values -0.06035 radians $(-3.4578°)$ and 0.025357 radians $(1.4528°)$, respectively. Figure 180-c shows the deformed configuration represented by this displacement vector.

The fact that the rotation of node B plays a small role in the buckling of the structure is confirmed by an approximate computation of the critical load in which a vanishing rotation of node B is assumed. Under this condition, the buckling load of the vertical bar may be computed with Euler's formula, considering an effective length $l_e = \frac{l}{2}$, which yields the value

$$P_{cr} = \frac{\pi^2 EI}{\left(\frac{l}{2}\right)^2} = 39.47842 \frac{EI}{l^2} \ .$$

It is the fact that this buckling mode, which has a very local character, cannot be accurately described by a rotation coordinate in node B that leads to the large error of the linearized solution. However, this fact does not affect the accuracy of the solution yielded by the exact form of the stiffness coefficients.

In the two examples above the geometrical component of the stiffness matrix has been used to compute the critical load. The next example illustrates the influence of the geometrical stiffness on the correct computation of the displacements of frames with heavily compressed members. An example of considerable practical interest is the computation of the displacements caused by horizontal forces in frames with large compressive axial forces in their vertical members. To this end let us consider the same frame that was considered in the first example (Fig. 177), now under the action of the forces represented in Fig. 181-a.

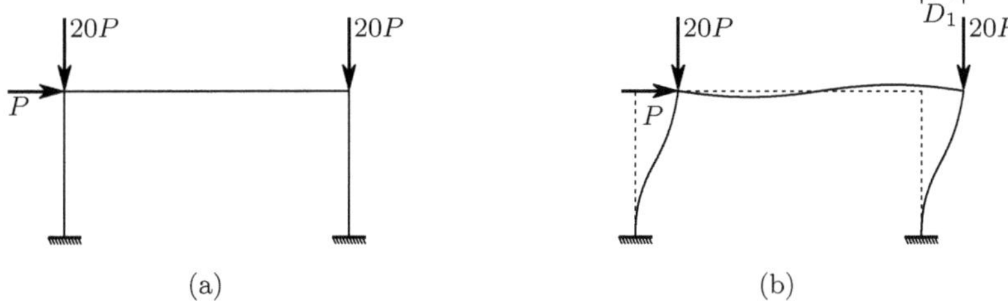

Fig. 181. Computation of the deformations caused by a horizontal load in a frame

As mentioned at the beginning of Subsect. XI.6.b, to introduce a deformation into a structure, which is represented by a given set of displacements in chosen kinematic coordinates, a set of forces must be applied at the same coordinate set. These forces are, therefore, functions of the given displacements. In the case of a system with three degrees of freedom, like the structure represented in Fig. 181, these functions may be expressed by

$$F_1 = F_1\left(D_1, D_2, D_3\right)$$
$$F_2 = F_2\left(D_1, D_2, D_3\right)$$
$$F_3 = F_3\left(D_1, D_2, D_3\right) \ .$$

Conversely, a set of forces P_1, P_2, P_3 applied at the same coordinates introduces a set of displacements, D_1, D_2, D_3, which must satisfy the conditions

$$F_1\left(D_1, D_2, D_3\right) = P_1$$
$$F_2\left(D_1, D_2, D_3\right) = P_2$$
$$F_3\left(D_1, D_2, D_3\right) = P_3 \ ,$$

which represent the equilibrium conditions between internal and external forces. In order to compute the values of D_1, D_2, D_3 it is therefore necessary to solve this system of equations, which is not linear, when the geometrical stiffness is considered. One of the most effective methods for solving non-linear systems of equations is the Newton-Raphson method. According to this method, given a set of estimated solutions, $\{D_1, D_2, D_3\}_i$, another set with improved approximation, $\{D_1, D_2, D_3\}_{i+1}$, may be computed by means of the algorithm

$$
\underbrace{\left\{\begin{array}{c} D_1 \\ D_2 \\ D_3 \end{array}\right\}_{i+1}}_{\{D\}_{i+1}} = \underbrace{\left\{\begin{array}{c} D_1 \\ D_2 \\ D_3 \end{array}\right\}_i}_{\{D\}_i} + \underbrace{\begin{bmatrix} \frac{\partial F_1}{\partial D_1} & \frac{\partial F_1}{\partial D_2} & \frac{\partial F_1}{\partial D_3} \\ \frac{\partial F_2}{\partial D_1} & \frac{\partial F_2}{\partial D_2} & \frac{\partial F_2}{\partial D_3} \\ \frac{\partial F_3}{\partial D_1} & \frac{\partial F_3}{\partial D_2} & \frac{\partial F_3}{\partial D_3} \end{bmatrix}^{-1}}_{[K]_i^{-1}} \underbrace{\left\{\begin{array}{c} P_1 - F_{1i} \\ P_2 - F_{2i} \\ P_3 - F_{3i} \end{array}\right\}}_{\{P-F\}_i},
$$

where F_{1i}, F_{2i}, F_{3i} are the forces that must be applied at the coordinates in order to introduce the displacements $\{D_1, D_2, D_3\}_i$. Since, for small rotations and constant axial forces, the nodal forces are linear functions of the displacements, as shown by (271) and (272,) those forces may be computed by means of the matrix operation

$$
\underbrace{\left\{\begin{array}{c} F_1 \\ F_2 \\ F_3 \end{array}\right\}_i}_{\{F\}_i} = \underbrace{\begin{bmatrix} \frac{\partial F_1}{\partial D_1} = K_{11} & \frac{\partial F_1}{\partial D_2} = K_{12} & \frac{\partial F_1}{\partial D_3} = K_{13} \\ \frac{\partial F_2}{\partial D_1} = K_{21} & \frac{\partial F_2}{\partial D_2} = K_{22} & \frac{\partial F_2}{\partial D_3} = K_{23} \\ \frac{\partial F_3}{\partial D_1} = K_{31} & \frac{\partial F_3}{\partial D_2} = K_{32} & \frac{\partial F_3}{\partial D_3} = K_{33} \end{bmatrix}_i}_{[K]_i} \underbrace{\left\{\begin{array}{c} D_1 \\ D_2 \\ D_3 \end{array}\right\}_i}_{\{D\}_i},
$$

provided that the axial forces corresponding to the displacements $\{D\}_i$ are used to compute the derivatives. Matrix $[K]_i$ coincides with the definition of the stiffness matrix and depends on the axial forces installed in the bars. Vector $\{P\}$ has the components $P_1 = P$ and $P_2 = P_3 = 0$, since the vertical forces are directly balanced by the axial forces in the vertical bars, and so they do not need forces in the coordinates to be balanced.

The axial forces in the bars also depend on the displacements in the kinematic coordinates (Fig. 177-b). Thus, the axial force in the left vertical bar depends on the load $20P$ and on the shear force at the left end of the horizontal bar. This force depends on displacements D_2 and D_3. Applying the same line of reasoning to the other two bars, and assuming that the displacements may be considered as infinitesimal, we conclude that the axial forces in the

P (N)	D_1 (cm) (exact stiffness)	D_1 (cm) (linearized stiffness)	D_1 (cm) (linear analysis)
1 000	0.17373	0.17373	0.17103
2 000	0.35303	0.35301	0.34207
5 000	0.92709	0.92705	0.85507
10 000	2.0247	2.0243	1.7103
20 000	4.9645	4.9588	3.4207
30 000	9.6289	9.5960	5.1310
40 000	18.182	18.024	6.8413
50 000	39.009	38.091	8.5517
55 000	66.855	63.925	9.4068
60 000	161.26	144.31	10.262
65 000	–	–	11.117

left (N_l), horizontal (N_h) and right (N_r) bars take the values given by the expressions

$$\begin{cases} N_l = -20P - K_{12}^1 D_2 - K_{14}^1 D_3 \\[2mm] N_h = -P + K_{11}^0 D_1 + K_{12}^0 D_2 \\[2mm] N_r = -20P + K_{32}^1 D_2 + K_{34}^1 D_3 \, , \end{cases}$$

where K^0 and K^1 represent the same quantities as in the first example (Figs. 177 and 178) and are obtained from the previous iteration $(i - 1)$. Once the axial forces are defined, the new stiffness coefficients may be computed.

The iterative procedure may be summarized by the following scheme

$$\{D\}_i \; \to \; \begin{cases} N_l, N_h, N_r \to [K]_i \\[3mm] \{F\}_i \end{cases} \to \{D\}_{i+1} = \{D\}_i + [K]_i^{-1} \{P - F\}_i \, .$$

The process converges in a very small number of iterations, which depends on the importance of the geometrical stiffness. In the following Table the results obtained for the horizontal displacement D_1 (Fig. 181-b) are presented, considering several values of P and the following alternatives for the stiffness coefficients: exact form (2nd column), linearized form (3rd column) and constant values, i.e., no geometrical stiffness (4th column)

When the geometrical stiffness is considered it is not possible to find the displacement corresponding to the last value of P, since it is larger than the critical load of the structure yielded by the procedure presented here (see final comment). This load is slightly smaller than that computed in the example without the horizontal load, which is because this load introduces a compression in the horizontal bar and increases the axial force in the right vertical bar. The determinant of the stiffness matrix of this structure vanishes during the iterative process when the load exceeds the value

$$P_{cr} = 63132N \Rightarrow 20P_{cr} = 1262640N \ .$$

This value has been computed using the exact form of the stiffness coefficients. If the linearized form is used instead, the value $P_{cr} = 63804N$ is obtained.

In most framed structures the displacements are relatively small, so the influence of the deformation in the distribution of axial forces is also small. For this reason, an insignificant error is generally introduced if the iterative procedure described above is replaced by a direct computation, assuming that the axial forces remain constant and equal to the values yielded by a linear computation. This procedure reduces the non-linear problem to a sequence of two linear problems: the usual linear computation of the axial forces disregarding the effect of the geometrical stiffness, followed by another linear computation, in which the displacements are computed by means of a corrected stiffness matrix, where the influence of the axial forces (computed in the first step) in the stiffness is taken into consideration. An even simpler alternative is to take the axial forces obtained by locking the kinematic coordinates, when the bars are considered as axially non-deformable.

Applying the latter alternative to the present problem, we have $N_l = -20P$, $N_r = -20P$ and $N_h = -P$. Determining the stiffness matrix which corresponds to these axial forces, we get the following values of displacement D_1

$P \ (N)$	$D_1 \ (cm)$ (exact stiffness)	$D_1 \ (cm)$ (linearized stiffness)
1 000	0.17373	0.17373
2 000	0.35303	0.35303
5 000	0.92719	0.92714
10 000	2.0252	2.0247
20 000	4.9672	4.9614
30 000	9.6389	9.6058
40 000	18.219	18.059
50 000	39.192	38.265
55 000	67.508	64.516
60 000	170.04	150.65

We find that, with exception of the last value, the differences are small in relation to the results obtained by means of the iterative method. The difference in the last value is a consequence of the size of the displacement. This value, however, is no longer accurate, since the entire theory is based on the assumption of small rotations (269), which is no longer acceptable for $P = 60\,000 \ N$. An accurate computation, carried out by means of an algorithm which may be applied irrespective of rotation size [14], yields the results shown in the last column of the following table

P (N)	D_1 (cm) (small rotations)	D_1 (cm) (no rotation limit)
1 000	0.17373	0.17373
2 000	0.35303	0.35303
5 000	0.92709	0.92714
10 000	2.0247	2.0249
20 000	4.9645	4.9652
30 000	9.6289	9.6293
40 000	18.182	18.165
50 000	39.009	38.636
55 000	66.855	64.229
60 000	161.26	123.31
65 000	–	215.35

The second column of this table contains the results yielded by the above described iterative method, considering the exact stiffness for small rotations. In the computation without limiting the size of the rotations, the bars have been considered as axially deformable with a cross-section area of $2\,000\,cm^2$. Figure 182 shows the relations between the load P and the displacement

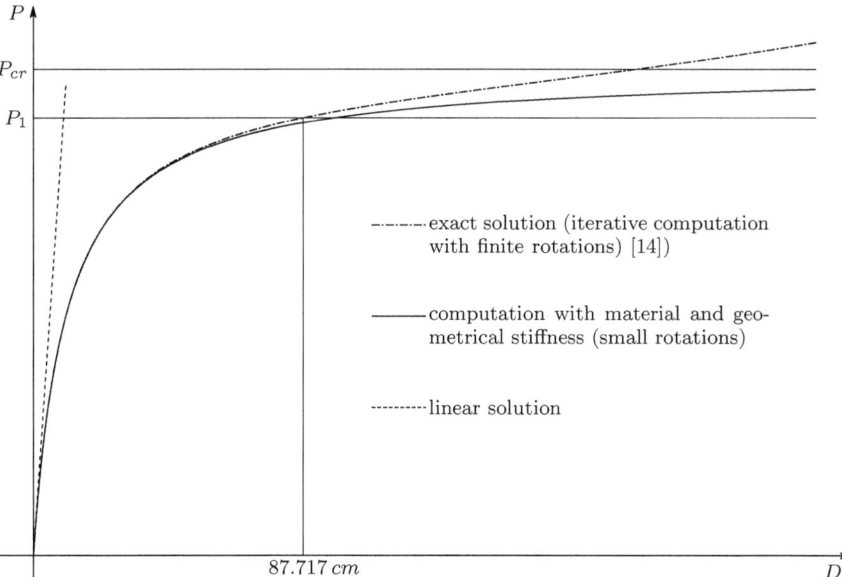

Fig. 182. Relation between P and D_1 in the structure represented in Fig. 181 $(P_1 = 57\,500\,N \approx 0.9\,P_{cr})$

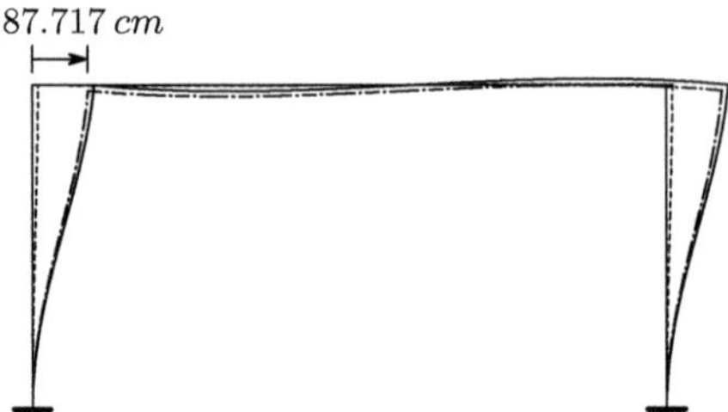

Fig. 183. Deformed configurations of the structure represented in Fig. 181, corresponding to the solution represented in Fig. 182, for $P = 57\,500\,N$

D_1 that have been obtained by the three main types of analysis: linear (only material stiffness); with material and geometrical stiffness and small rotations (the analysis described here), and the analysis without limitation of the size of displacements and rotations [14]. Figure 183 gives the corresponding deformed configurations.

Comparing the results yielded by the methods based on both the exact and the linearized stiffness for small rotations, with the results presented in the third column of the last table, we conclude that all of them give good results until about 90% of the critical load of the structure, $(P_{cr}, \text{Fig.182})$. This is the load that leads to a vanishing determinant of the stiffness matrix, which is obtained assuming small rotations. Actually, this frame supports higher loads without instability, since the tangent stiffness matrix obtained without restriction on the size of the rotations does not become singular, even for substantially higher values of P. This is the reason why it was possible to include the displacement corresponding to $P = 65\,000\,N$ in the last table $(215.35\,cm)$.

XII

Energy Theorems

XII.1 General Considerations

When forces are applied to a body it deforms, as a consequence of the internal forces. As the body deforms, the points of application of the external forces suffer a displacement, doing work. The physical law of energy conservation implies that the potential energy of the external forces, which is consumed when that work is done, is either dissipated, i.e., transformed into heat, or stored as other energy form.

From a general point of view, we may summarize the problem, by stating that the work done by the external forces that deform a body, which is initially in a resting state and free of internal forces, is mainly transformed into the following four parts:

- *Kinetic energy* – Energy stored in a body when its mass suffers a motion. A supported body does not have rigid body motions, but vibrations may occur.
- *Elastic potential energy* – Energy stored by the recoverable part of the deformation (see Sect. V.4). Note that in a vibrating body kinetic energy is being continuously transformed into elastic potential energy and vice versa: when the amplitude of the oscillatory motion reaches a maximum, the velocity vanishes, which means that the kinetic energy also vanishes, while the deformation attains a maximum value, that is, the elastic potential energy reaches a maximum. Conversely, when the deformation vanishes, the elastic potential energy also vanishes, but the velocity and, therefore, the kinetic energy, attain a maximum.
- *Energy dissipation by external friction* – This kind of friction occurs in supports that allow some motion, since it is not possible to manufacture a hinge without any rotation friction or a roller support without rolling friction. The friction between a vibrating structure, and the air around it, is also an external friction which helps to damp the vibration.

– *Energy dissipation by internal friction* – In a solid body this kind of energy dissipation occurs in plastic and viscous deformations, either through friction between the particles which constitute the body, or via other transformations of its internal structure in which heat is set free.

In addition to these, there are other forms of energy dissipation, such as sound, infrared radiation caused by a rise in temperature, etc. However, in the usual problems of Solid Mechanics and Structural Analysis these components do not play a significant role.

In the main part of this chapter we shall study the cases where the external forces are applied sufficiently slowly for it to be possible to disregard the dynamic effects (inertial forces) and where internal and external friction dissipates only very small quantities of energy, so that it may be disregarded. This is equivalent to considering only static loads applied to bodies made of elastic materials (that is, without internal friction) and without external friction. Furthermore, the analysis is limited to linear structural behaviour, i.e., linear elastic material behaviour and geometrical linearity (small deformations). But Sects. XII.4 and XII.5 are not accommodated by all these limiting conditions, since in the first (Sect. XII.4: theorems of virtual forces and virtual displacements) internal friction may occur, while in the second (Sect. XII.5: elementary analysis of impact loads) the load is applied suddenly, so that kinetic energy must also be taken into consideration.

XII.2 Elastic Potential Energy in Slender Members

The elastic potential energy stored in a body made of materials with linear elastic behaviour may be expressed by (see (95))[1]

$$U = \frac{1}{2} \int_V \left(\sigma_x \varepsilon_x + \sigma_y \varepsilon_y + \sigma_z \varepsilon_z + \tau_{xy} \gamma_{xy} + \tau_{zx} \gamma_{zx} + \tau_{yz} \gamma_{yz} \right) \mathrm{d}V \ . \tag{281}$$

In the analysis developed in the chapters on slender members we have concluded that the axial force and the bending moment introduce normal stresses in the cross-sections (σ_z), while the shear force and the twisting moment introduce shearing stresses (τ_{zx} and τ_{zy}). The other components of the stress tensor are either zero, or sufficiently small to be disregarded. Thus, in the special case of a slender member (281) takes the form

$$U = \frac{1}{2} \int_l \int_\Omega \left[\left(\sigma_z^N + \sigma_z^M \right) \left(\varepsilon_z^N + \varepsilon_z^M \right) + \left(\tau_{zx}^V + \tau_{zx}^T \right) \left(\gamma_{zx}^V + \gamma_{zx}^T \right) \right.$$

$$\left. + \left(\tau_{zy}^V + \tau_{zy}^T \right) \left(\gamma_{zy}^V + \gamma_{zy}^T \right) \right] \mathrm{d}\Omega \, \mathrm{d}l \ . \tag{282}$$

---

[1]In Sects. IV.6, IV.7 and V.4, U and W represent, respectively, the energy and the work per volume unit. In this Chapter these quantities refer to the whole volume of the body.

In this expression the upper indices denote the internal force that causes the stress or strain under consideration. A first inspection of this expression indicates that, in principle, the energy cannot be split into components caused only by the axial force, or only by the bending moment, since there are the mixed products $\sigma_z^N \varepsilon_z^M$ and $\sigma_z^M \varepsilon_z^N$. However, by substituting, in these products, the stresses and strains by their expressions in terms of the corresponding internal forces ((115) and (150), for example), we conclude immediately that these mixed products vanish (see example XII.1)

When it comes to the shearing stresses and strains caused by the shear force and the twisting moment, that verification is more complex, since there are no general formulas for the computation of shearing stresses in slender members. However, since the shear force does not cause rotation of the cross-sections around the bar's axis, and the twisting moment does not cause shear displacement (relative transversal displacement of the cross-sections), the terms $\int_\Omega \left(\tau_{zx}^V \gamma_{zx}^T + \tau_{zy}^V \gamma_{zy}^T \right) \mathrm{d}\Omega$ and $\int_\Omega \left(\tau_{zx}^T \gamma_{zx}^V + \tau_{zy}^T \gamma_{zy}^V \right) \mathrm{d}\Omega$ vanish, since they represent, respectively, the work done by the shear force in the torsional deformation and the work done by the twisting moment in the deformation caused by shear (see example XII.1). Obviously, this line of reasoning may also be applied to the components of the elastic potential energy which correspond to the axial force and to the bending moment.

According to these considerations, (282) is equivalent to

$$U = \frac{1}{2} \int_l \int_\Omega \left(\sigma_z^N \varepsilon_z^N + \sigma_z^M \varepsilon_z^M + \tau_{zx}^V \gamma_{zx}^V + \tau_{zy}^V \gamma_{zy}^V + \tau_{zx}^T \gamma_{zx}^T + \tau_{zy}^T \gamma_{zy}^T \right) \mathrm{d}\Omega \, \mathrm{d}l \ .$$

$$(283)$$

The components corresponding to the axial force and to the bending moment are easily deduced. In the case of homogeneous members we get (x and y are the central principal axes of inertia of the cross-section)

$$\int_\Omega \sigma^N \varepsilon^N \, \mathrm{d}\Omega = \int_\Omega \frac{N}{\Omega} \frac{N}{E\Omega} \, \mathrm{d}\Omega = \frac{N^2}{E\Omega}$$

$$\int_\Omega \sigma^{M_x} \varepsilon^{M_x} \, \mathrm{d}\Omega = \int_\Omega \frac{M_x}{I_x} y \frac{M_x}{EI_x} y \, \mathrm{d}\Omega = \frac{M_x^2}{EI_x^2} \int_\Omega y^2 \, \mathrm{d}\Omega = \frac{M_x^2}{EI_x}$$

$$\int_\Omega \sigma^{M_y} \varepsilon^{M_y} \, \mathrm{d}\Omega = \frac{M_y^2}{EI_y} \ .$$

The elastic potential energy corresponding to the deformation caused by the shear force and by the twisting moment had already been analysed when these deformations were studied. Generalizing (227) to the two principal axes of inertia, we get

$$\int_\Omega \left(\tau_{zx}^{V_x} \gamma_{zx}^{V_x} + \tau_{zy}^{V_x} \gamma_{zy}^{V_x} \right) \mathrm{d}\Omega = \frac{V_x^2}{G\Omega_{rx}}$$

$$\int_\Omega \left(\tau_{zx}^{V_y} \gamma_{zx}^{V_y} + \tau_{zy}^{V_y} \gamma_{zy}^{V_y} \right) \mathrm{d}\Omega = \frac{V_y^2}{G\Omega_{ry}} \ .$$

In these expressions Ω_{rx} and Ω_{ry} represent the reduced areas of the cross-section which correspond to shear forces that are parallel to axes x and y, respectively. In the same way, we may find the elastic potential energy corresponding to the twisting moment by generalizing the procedure leading to the computation of the torsional stiffness of closed thin-walled cross-sections (Bredt's second formula) to other types of cross-section, yielding (J represents the geometrical parameter of the torsional stiffness, GJ)

$$\int_{\Omega} \left(\tau_{zx}^{T} \gamma_{zx}^{T} + \tau_{zy}^{T} \gamma_{zy}^{T} \right) \, \mathrm{d}\Omega = \frac{T^2}{GJ} \; .$$

Thus, (283) may be rewritten in the form

$$U = \frac{1}{2} \int_{l} \left(\frac{N^2}{E\Omega} + \frac{M_x^2}{EI_x} + \frac{M_y^2}{EI_y} + \frac{V_x^2}{G\Omega_{rx}} + \frac{V_y^2}{G\Omega_{ry}} + \frac{T^2}{GJ} \right) \, \mathrm{d}l \; . \tag{284}$$

XII.3 Theorems for Structures with Linear Elastic Behaviour

In the case of structures with linear elastic behaviour the superposition principle is valid and may be used to demonstrate some useful theorems about the energy stored in a deformed elastic body.

XII.3.a Clapeyron's Theorem

Let us consider a body made of a material with linear elastic behaviour, free from residual stresses, under the action of n concentrated loads, $P_1, P_2, \ldots, P_n$. These forces cause small displacements, $\delta_1, \delta_2, \ldots, \delta_n$, along their lines of action. Clapeyron's theorem states that the elastic potential energy stored by the body may be computed by the expression

$$U = \frac{1}{2} \left(P_1 \delta_1 + P_2 \delta_2 + \cdots + P_n \delta_n \right) = \frac{1}{2} \sum_{i=1}^{n} P_i \delta_i \; . \tag{285}$$

This theorem has a very simple demonstration. In fact, the elastic potential energy stored by a body depends only on the stresses and strains installed inside it. Since linear elastic behaviour is assumed, the superposition principle leads to the conclusion that the final stresses and strains do not depend on the order of application of the loads. Therefore, we may assume that the n loads are applied simultaneously and that they vary proportionally to a parameter α ($0 \leq \alpha \leq 1$), from zero until they reach their final values. Under these conditions (285) represents the work done by the external forces P_i during the deformation process. Finally, the principle of energy conservation leads to the conclusion that this work is transformed into elastic potential energy, which demonstrates the theorem.

The loads P_i may be forces, moments, pairs of forces or pairs of moments, and are called *generalized forces*. The corresponding displacements are called *generalized displacements*. In fact, in the same way as a force does work, when its point of application moves along its line of action, a moment does work when the region where it is applied rotates around the vector which represents the moment, a pair of equal forces with the same line of action and opposite directions does work when the distance between their points of application varies, and two equal moments, represented by parallel vectors with opposite directions, do work when the regions where they are applied undergo a relative rotation around the direction of the two vectors. Generalized displacements therefore include displacements, rotations, relative displacements and relative rotations.

Clapeyron's theorem is also valid in the case of distributed loads on the boundary or in the mass of the body. In this case, the mathematical expression of Clapeyron's theorem takes the form

$$U = \frac{1}{2} \int_V (Xu + Yv + Zw) \, \mathrm{d}V + \frac{1}{2} \int_S \left(\overline{X}u + \overline{Y}v + \overline{Z}w \right) \mathrm{d}S \ .$$

In this expression S represents the limiting surface of the body, V represents its volume, $\overline{X}$, $\overline{Y}$ and $\overline{Z}$ and X, Y and Z are the components in the reference axes, of the loads distributed on the surface and in the volume of the body, respectively, and u, v and w are the components of the displacement of a point in the directions of the reference axes, x, y and z, respectively.

XII.3.b Castigliano's Theorem

There are versions of Castigliano's theorems which are valid for bodies with both linear and non-linear elastic behaviour. However, in its original version, it is only valid for linear behaviour [3]. This version is studied here. It is also known as the *second Castigliano theorem*.

Let us consider a body or a structure which is deformed by a set of n independent generalized forces, P_i, which are applied at the generalized coordinates $i = 1, 2, \ldots, n$. The elastic potential energy stored by the body may be considered as a function of the generalized forces applied, $U(P_1, P_2, \ldots, P_n)$. In fact, expressing the generalized displacements $\delta_1, \delta_2, \ldots, \delta_n$ in (285) as functions of the forces $P_1, P_2, \ldots, P_n$, we get an expression for the energy U, where the variables are the forces P_i. Castigliano's theorem states that the generalized displacement in coordinate i is equal to the derivative of the elastic potential energy with respect to the generalized force P_i, that is

$$\delta_i = \frac{\partial U}{\partial P_i} \ . \tag{286}$$

In order to demonstrate this theorem, let us first consider an infinitesimal variation dP_i of the generalized force P_i. This variation causes an infinitesimal variation in the elastic potential energy which may be represented by the expression

$$dU_1 = \frac{\partial U}{\partial P_i}\, dP_i \ .$$

Let us consider now that the infinitesimal generalized force dP_i is first applied to the structure, i.e., that it is already acting when forces $P_1, P_2, \ldots, P_n$ are applied. The increase in elastic potential energy caused by the application of the infinitesimal load dP_i is, in this case

$$dU_2 = \frac{1}{2}\, dP_i\, d\delta_i + dP_i\, \delta_i \ .$$

The first term of dU_2, $\frac{1}{2}\, dP_i\, d\delta_i$, corresponds to the work done by dP_i when it is first applied to the structure. The second term, $dP_i\, \delta_i$, is the work done by dP_i when the structure deforms, as a consequence of the application of the generalized forces $P_1, P_2, \ldots, P_n$. The first term is a second-order infinitesimal quantity which may be disregarded when added to the first-order infinitesimal quantity $dP_i\, \delta_i$.

The superposition principle guarantees that the order of application of the forces does not influence the final value of the stored elastic potential energy. This means that dP_i must cause the same increase in the elastic potential energy both when it is applied at the same time as load P_i, and when it is first applied to the structure, that is, dU_1 and dU_2 must be equal. This condition leads directly to (286).

Note that the condition of independence of the loads in set $P_1, P_2, \ldots, P_n$ leads to the conclusion that the application of Castigliano's theorem to a pair of forces yields the relative displacement of their points of application, as represented in Fig. 184. In the same way, the relative rotation of two infinitesimal regions may be computed by applying this theorem to a pair of moments.

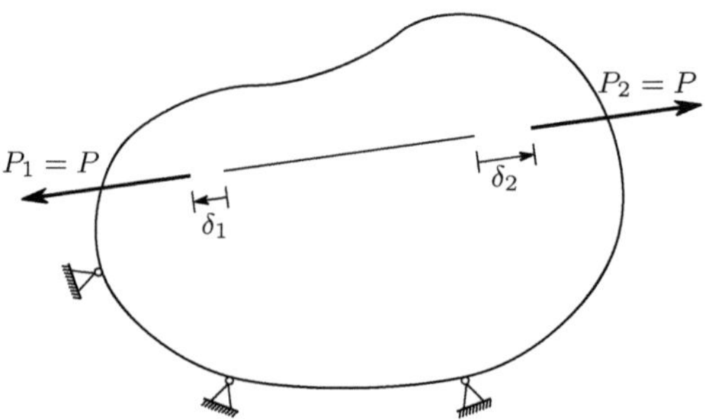

Fig. 184. Computation of the relative displacement of two points using Castigliano's theorem: $\frac{\partial U}{\partial P} = \frac{\partial U}{\partial P_1} + \frac{\partial U}{\partial P_2} = \delta_1 + \delta_2 = \delta_r$

Slender members

In slender members the elastic potential energy is given by (284). Therefore, in the case of framed structures, Castigliano's theorem may be applied using the following expression (m is the number of bars of the structure and z_j is a coordinate taking the direction of the axis of bar j)

$$\delta_i = \frac{\partial U}{\partial P_i} = \sum_{j=1}^{m} \int_0^{l_j} \left(\frac{N}{E\Omega} \frac{\partial N}{\partial P_i} + \frac{M_x}{EI_x} \frac{\partial M_x}{\partial P_i} + \frac{M_y}{EI_y} \frac{\partial M_y}{\partial P_i} \right.$$
$$\left. + \frac{V_x}{G\Omega_{rx}} \frac{\partial V_x}{\partial P_i} + \frac{V_y}{G\Omega_{ry}} \frac{\partial V_y}{\partial P_i} + \frac{T}{GJ} \frac{\partial T}{\partial P_i} \right) \mathrm{d}z_j \ . \tag{287}$$

Since the structure has linear elastic behaviour, the internal forces in the bars, N, M_x, M_y, V_x, V_y and T are linear functions of the applied loads. Thus, the derivatives of the internal forces with respect to force P_i are the internal forces N_u, M_{xu}, M_{yu}, V_{xu}, V_{yu} and T_u which are introduced into the structure by a dimensionless unit force applied in coordinate i. In the case of a rotation, the unit force is substituted by a dimensionless unit moment. If the displacement is to be computed in a coordinate where an external force is not applied, a fictitious infinitesimal force may be considered to be acting at that coordinate. Since it is infinitesimal, it does not influence the internal forces, but the elastic potential energy function may be derived with respect to it, which is equivalent to applying a dimensionless unit force there, as seen above. Summarizing, the displacement at coordinate i may be computed by integrating into the whole structure the product of the internal forces due to the actual loading and the internal forces generated by a dimensionless unit force applied at coordinate i, divided by the corresponding stiffness (287).

In order to illustrate these considerations, we use Castigliano's theorem to compute the vertical displacement of the free end of the cantilever beam represented in Fig. 185, considering the components corresponding to the bending moment and to the shear force (see example IX.5).

The shear force and the bending moment caused by the distributed force and by a vertical downwards dimensionless unit force applied in point B are, respectively

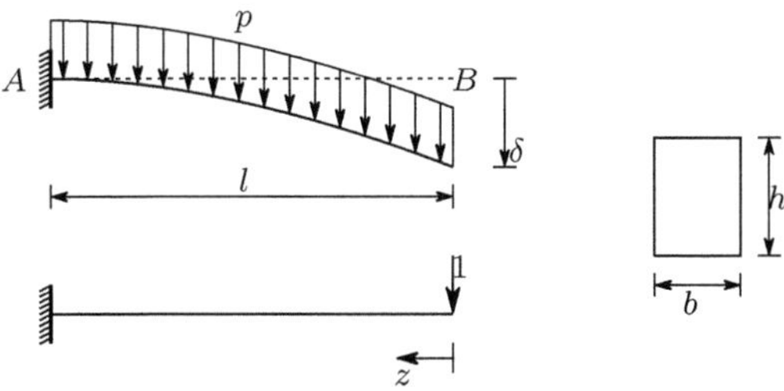

Fig. 185. Computation of a displacement in a cantilever beam using Castigliano's theorem

$$\begin{cases} V = pz \\ M = p\dfrac{z^2}{2} \end{cases} \quad \text{and} \quad \begin{cases} V_u = 1 \\ M_u = z \,. \end{cases}$$

The vertical displacement of point B, computed by means of (287), is then

$$\delta = \int_0^l \left(\frac{M}{EI} M_u + \frac{V}{G\Omega_r} V_u \right) \mathrm{d}z = \int_0^l \frac{pz^3}{2EI}\,\mathrm{d}z + \int_0^l \frac{pz}{G\Omega_r}\,\mathrm{d}z = \underbrace{\frac{pl^4}{8EI}}_{\delta_M} + \underbrace{\frac{pl^2}{2G\Omega_r}}_{\delta_V} \,.$$

δ_M and δ_V represent the contributions of the bending moment and the shear force, respectively, to the displacement δ. Particularizing these expressions for the given rectangular cross-section, we get

$$\begin{cases} I = \dfrac{bh^3}{12} \\ \Omega_r = \dfrac{bh}{1.2} \end{cases} \Rightarrow \begin{cases} \delta_M = \dfrac{3}{2}\dfrac{pl^4}{bh^3 E} \\ \delta_V = \dfrac{pl^2}{2\frac{E}{2(1+\nu)}\frac{bh}{1.2}} = 1.2\,(1+\nu)\dfrac{pl^2}{bhE} \end{cases} \Rightarrow \frac{\delta_V}{\delta_M} = 0.8\,(1+\nu)\left(\frac{h}{l}\right)^2 \,.$$

As mentioned in Chap. IX, in slender members we have $l \gg h$, which is why the contribution of the shear force to the deflection is very small, so that it is usually disregarded. For example, for $\nu = 0.25$ and $l/h = 10$, in this cantilever we get $\delta_V/\delta_M = 1\%$. However, in bending-optimized cross-sections, as I-beams and channel beams, this contribution may be considerably higher, especially in the case of short beams.

When Castigliano's theorem is used to compute displacements in statically indeterminate structures, the internal forces caused by the unit forces may be computed in any statically determinate structure that is obtained from the actual structure by removing connections, instead of being computed in the actual statically indeterminate structure, which simplifies the computation. This procedure is easily justified, since it corresponds to the computation of the displacement under consideration by superposing the displacements caused in a statically determinate base structure by the actual load and by the forces which are taken as hyperstatic unknowns (forces acting in the released connections), when the force method is used (see examples XII.6 and XII.7).

The considerations about the perturbation introduced in the deformation of a beam by the fact that free warping is prevented at the supports, which have been established in Subsect. IX.2.d (Fig. 130), may now be demonstrated. In fact, if the forces needed to keep the built-in cross-section plane have a vanishing resultant, they have a local influence on the stress distribution, as a consequence of Saint-Venant's principle. Thus, there are no additional internal forces resulting from the forces needed to avoid the warping of that cross-section. This is reflected in the expression of Castigliano's theorem for framed structures (287) which is not affected by such forces. This fact demonstrates that the perturbation only affects the region of the beam where the theory of slender members and, as a consequence, (287), are not valid, that is, the region which is close to the support. More precisely, this perturbation may be represented by a constant term to be added to (287), whose relative importance decays as the length of the bar increases. This means that the perturbation also has a local character in relation to the computation of displacements.

XII.3.c Menabrea's Theorem or Minimum Energy Theorem

Menabrea's theorem is a particular case of Castigliano's theorem. If we consider the forces applied to a structure as constants and the hyperstatic unknowns X_i as variables (force method), the elastic potential energy may be expressed as a function of the hyperstatic unknowns, $U = U(X_1, X_2, \ldots, X_n)$. The compatibility conditions of the deformations in the released connections may be established by means of Castigliano's theorem, yielding

$$\begin{cases} \delta_1 = 0 \\ \\ \delta_2 = 0 \\ \\ \quad\vdots \\ \\ \delta_n = 0 \end{cases} \Rightarrow \begin{cases} \dfrac{\partial U}{\partial X_1} = 0 \\ \\ \dfrac{\partial U}{\partial X_2} = 0 \\ \\ \quad\vdots \\ \\ \dfrac{\partial U}{\partial X_n} = 0 \,. \end{cases}$$

The hyperstatic unknowns may then be computed by solving this system of equations. The same system of equations may also be used to find the point where the function $U(X_1, X_2, \ldots, X_n)$ has an extremum. Since, in the elastic domain, the elastic potential energy U can always be increased by increasing the forces X_i, this extremum must be a minimum. Thus, Menabrea's theorem or minimum energy theorem may be stated as follows:

in a statically indeterminate structure the hyperstatic unknowns take values which minimize the elastic potential energy.

It should be noted that Menabrea's theorem is only valid for the hyperstatic unknowns, since equilibrium must be guaranteed for arbitrary values of the forces X_i, that is, the structure must be supported. If equilibrium is not guaranteed, the work done by the external forces will not be totally transformed into elastic potential energy, since those forces will cause accelerations, so that that work is at least partly transformed into kinetic energy.

XII.3.d Betti's Theorem

Let us consider a supported body, to which two distinct systems of forces are applied, $P_1, P_2, \ldots, P_n$ and $Q_1, Q_2, \ldots, Q_m$. The body has linear elastic behaviour, so that the superposition principle is valid.

Now, let us first consider a loading sequence in which forces P are first applied, and that they remain constant during the subsequent application of forces Q. The evolution of the elastic potential energy stored by the body in this load sequence is as follows:

– application of forces P:

$$U_1 = \frac{1}{2} \sum_{i=1}^{n} P_i \, \delta_{iP} \; ;$$

– application of forces Q: when the body deforms, as a consequence of the application of forces Q, forces P do work, so that when the two systems are acting, the elastic potential energy takes the value

$$U = U_1 + \frac{1}{2} \sum_{i=1}^{m} Q_i \, \delta_{iQ} + \sum_{i=1}^{n} P_i \, \delta_{iQ}$$
$$= \frac{1}{2} \left(\sum_{i=1}^{n} P_i \, \delta_{iP} + \sum_{i=1}^{m} Q_i \, \delta_{iQ} \right) + \sum_{i=1}^{n} P_i \, \delta_{iQ} \; . \tag{288}$$

In these expressions the second subscript of δ denotes the system of forces which causes the displacement under consideration.

Let us consider now that the order in which the two systems are applied is reversed, i.e., that forces Q are already acting when forces P are applied. In this case, the evolution of the elastic potential energy is

– application of forces Q:

$$U_1' = \frac{1}{2} \sum_{i=1}^{m} Q_i \, \delta_{iQ} \; ;$$

– application of forces P:

$$U = U_1' + \frac{1}{2} \sum_{i=1}^{n} P_i \, \delta_{iP} + \sum_{i=1}^{m} Q_i \, \delta_{iP}$$
$$= \frac{1}{2} \left(\sum_{i=1}^{n} P_i \, \delta_{iP} + \sum_{i=1}^{m} Q_i \, \delta_{iQ} \right) + \sum_{i=1}^{m} Q_i \, \delta_{iP} \; . \tag{289}$$

The superposition principle ensures that the stresses and thus the elastic potential energy do not depend on the order of application of the loads. Therefore, (288) and (289) must yield the same result which implies

$$\sum_{i=1}^{n} P_i \, \delta_{iQ} = \sum_{i=1}^{m} Q_i \, \delta_{iP} \; . \tag{290}$$

This expression defines Betti's theorem which states that

in a structure under the action of two distinct force systems, P and Q, the work done by forces P in the deformation caused by forces Q is equal to the work done by forces Q in the deformation caused by forces P.

Influence lines

Betti's theorem has many applications in linear structural analysis. One of the most important is the analytical or experimental determination of *influence lines*. These lines are diagrams expressing the value taken by a given quantity (a support reaction, a bending moment or other internal force in a given cross-section, a stress in a point, the displacement or the rotation of a cross-section, etc.), when a load is applied to a structure, as a function of the position of that load. As those quantities are proportional to the load, the ordinate of the influence line in a point is usually defined as the value taken by the quantity for which the influence line is drawn, when a unit load is applied in that point. Thus, the value taken by a quantity is given by the ordinate of its influence line, in the point where the load is applied, multiplied by the value of the load (see example XII.9-a).

As an applied example let us consider the beam represented in Fig. 186. Suppose, at first, that we want to find the influence line of the reaction force in support B.

To this end, let us consider the two force systems represented in Figs. 186-a and 186-b which act in a beam which is equal to the given one, with the exception that support B has been removed. The first system (Fig. 186-a) does not cause displacement in point B, since force X takes the value corresponding

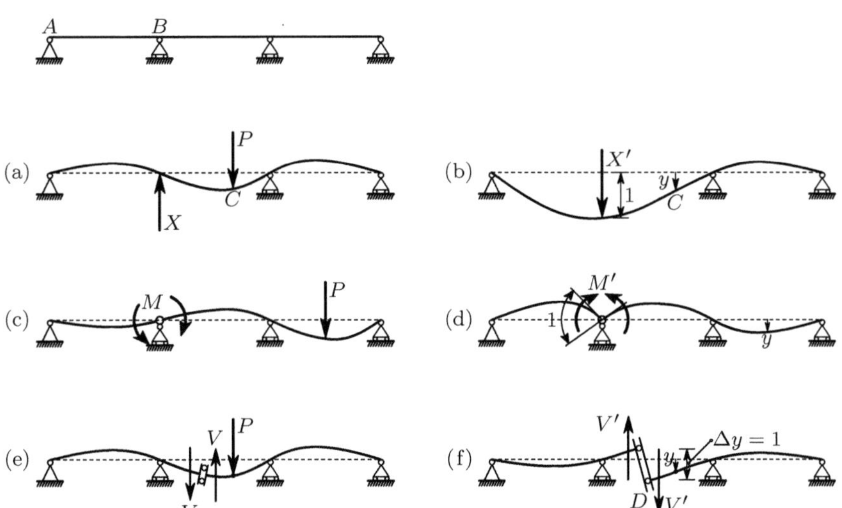

Fig. 186. Examples of determination of influence lines in a continuous beam

to the reaction force, when force P acts in the original beam. The second system has only one force, X', which causes a vertical unit displacement in the direction opposite to that defined as positive for the reaction force X. Applying Betti's theorem to these two systems, we get

$$\begin{cases} P_1 = X & \delta_{1Q} = -1 \\ P_2 = P & \delta_{2Q} = y \\ Q_1 = X' & \delta_{1P} = 0 \end{cases} \Rightarrow \quad \left| \begin{array}{l} P_1\,\delta_{1Q} + P_2\,\delta_{2Q} = Q_1\,\delta_{1P} \\ \Rightarrow -X + Py = 0 \Rightarrow X = Py \,. \end{array} \right.$$

This expression shows that the reaction force in support B, when the load P is applied at the generic point C, may be computed by multiplying the value of load P by the displacement y that takes place in point C, when point B suffers a unit displacement in the direction opposite to that defined as positive for the reaction force. Thus, the deflection line in Fig. 186-b represents the influence line of the reaction force in support B.[2]

Betti's theorem is also valid for generalized forces and displacements. Therefore, the influence line of the pair of moments which defines the bending moment in a cross-section may be found by imposing a unit value to the corresponding generalized displacement, that is, by introducing a hinge in the cross-section whose bending moment's influence line is required (cross-section B in this case) and deforming the beam with a pair of moments applied to the left and to the right of the hinge, so that a negative unit relative rotation is obtained. This procedure is represented in Figs. 186-c and 186-d. In this case we have

$$\begin{cases} P_1 = M & \delta_{1Q} = -1 \\ P_2 = P & \delta_{2Q} = y \\ Q_1 = M' & \delta_{1P} = 0 \end{cases} \Rightarrow -M + Py = 0 \Rightarrow M = Py \,.$$

In Figs. 186-e and 186-f a similar procedure leading to the determination of the influence line of the shear force in cross-section D is justified. In this case, the corresponding generalized displacement is the relative deflection $\Delta y = 1$, with the same rotation to the left and to the right of cross-section D, as represented in Fig. 186-f. Such a displacement would be obtained by means of the mechanism represented in Figs. 186-e and 186-f.

In the case of a statically determinate structure, the removal of the connection corresponding to the quantity whose influence line is required, transforms the structure into a mechanism. However, the equilibrium of the forces acting in that mechanism is not affected, since the removal of the connection is accompanied by the application of the force that, if applied at the coordinate corresponding to the released connection, balances the applied force P. For this reason, Betti's theorem may also be applied to this mechanism, although it has been demonstrated for supported bodies.

---

[2] This application of Betti's theorem to the definition of influence lines is often referred as the *Müller-Breslau principle*.

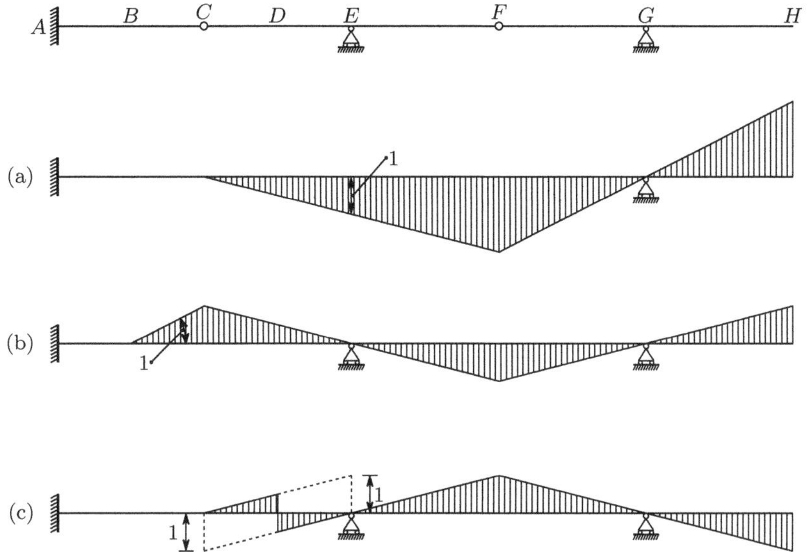

Fig. 187. Examples of determination of influence lines in a statically determinate beam: (**a**) influence line of the reaction force in support E; (**b**) influence line of the bending moment in cross-section B; (**c**) influence line of the shear force in cross-section D

The force needed to induce the unit displacement Q_1 (X', M' or V' in the examples in Fig. 186) vanishes, since it is not necessary to deform the bars of the mechanism to introduce that displacement. For this reason, the influence lines of statically determinate framed structures with straight bars consist of straight lines, as in the statically determinate beam represented in Fig. 187.

XII.3.e Maxwell's Theorem

Maxwell's theorem is a particular case of Betti's theorem. If each of the systems P and Q has only one force, with force P applied at coordinate 1 and force Q at coordinate 2, (290) reduces to

$$P\,\delta_{1Q} = Q\,\delta_{2P}\,. \tag{291}$$

If the forces P and Q take unit values, we have $\delta_{1Q} = \delta_{2P}$. This equality represents Maxwell's theorem which states:

In a structure with linear elastic behaviour, a generalized unit force applied at a coordinate $\underline{a}$ causes a generalized displacement in other coordinate $\underline{b}$, which is numerically equal to the displacement caused in coordinate $\underline{a}$ by a generalized unit force applied at coordinate $\underline{b}$.

According to Betti's theorem the two coordinates do not need to be of the same kind. For this reason, the equality is referred to the numerical value of the generalized displacement, since a rotation cannot be physically equal to a displacement.

Maxwell's theorem has several applications. For example, it makes it possible to demonstrate that the matrix which relates the stress and strain tensors is symmetrical, when the material has linear elastic behaviour, even if it is not isotropic. To this end, let us consider an infinitesimal rectangular parallelepiped with sides dx, dy and dz, in whose faces the normal and shearing stresses act that define the stress tensor. If we consider $\sigma_x = 1$ this stress will introduce a strain ε_y which, as a consequence of Maxwell's theorem, must be equal to the strain ε_x which is introduced by a stress $\sigma_y = 1$.

Using Betti's or Maxwell's theorems it is possible to demonstrate that in a prismatic bar made of a material with linear elastic behaviour, isotropic or not, the shear and torsion centres coincide. To this end, let us consider the cantilever beam represented in Fig. 188 where two coordinates are considered. The first one is a translation coordinate, has an arbitrary direction contained in the cross-section's plane and is applied at the torsion centre. Coordinate 2 defines a rotation around a fibre of the bar and is applied in an arbitrary point of the cross-section.

Applying a generalized force in coordinate 2, that is, applying a twisting moment to the bar, we conclude that the displacement in coordinate 1 must vanish, since this coordinate is defined on the point around which the cross-section rotates in the torsional deformation of the bar (the torsion centre). Equation (291) leads immediately to the conclusion that the application of a force in coordinate 1 will not cause a displacement in coordinate 2, i.e., the rotation of the cross-section around the bar's axis vanishes. This means that the twisting moment is zero, which implies that the shear force (force in coordinate 1) acts on the shear centre. Since, by definition, coordinate 1 is in the torsion centre, the shear and torsion centres coincide.

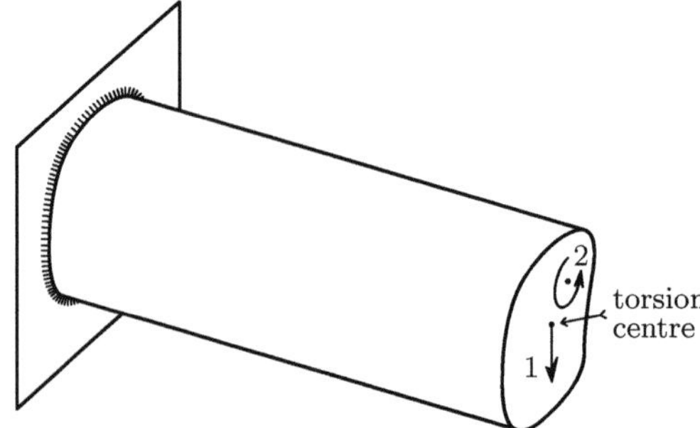

Fig. 188. Demonstration of coincidence of the shear and torsion centres

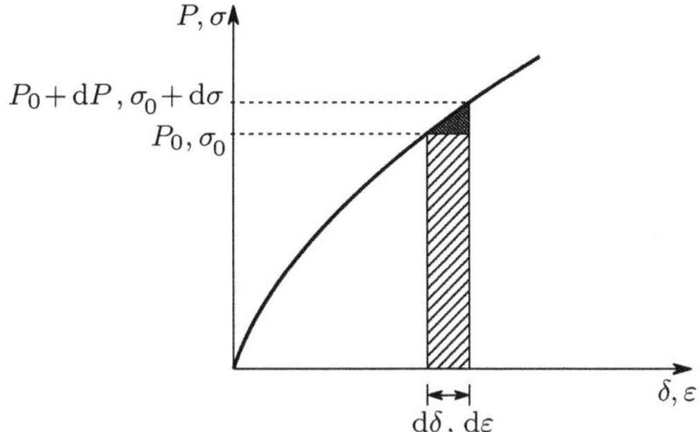

Fig. 189. Increments of external work and elastic potential energy plus dissipated energy, caused by infinitesimal displacements $\mathrm{d}\delta$

XII.4 Theorems of Virtual Displacements and Virtual Forces

The theorems of virtual displacements and virtual forces together define the theorem of virtual work. They are the only theorems about the deformation energy of bodies made of materials with non-linear behaviour that are studied here. These theorems have a relatively complex mathematical demonstration, especially in the cases where displacements and/or strains are not small enough to be treated as infinitesimal quantities. However, their demonstration on the basis of the principle of energy conservation is relatively simple.

XII.4.a Theorem of Virtual Displacements

In order to demonstrate the theorem of virtual displacements, let us consider a body which is deformed under the action of external forces, P, distributed in its boundary surface and in its volume.[3] Let us suppose now that additional infinitesimal forces $\mathrm{d}P$ are applied to this body. These forces will introduce additional stresses $\mathrm{d}\sigma$, strains $\mathrm{d}\varepsilon$ and displacements $\mathrm{d}\delta$. All these quantities are infinitesimal. Furthermore, the displacements $\mathrm{d}\delta$ are obviously compatible with the support conditions. The law of energy conservation implies that the increase of the work done by the external forces is either stored as elastic potential energy, or dissipated in the viscous and plastic deformations, since no external friction is taken into account and the external forces are assumed to be statically applied. Note that in an infinitesimal deformation the material behaviour may always be accepted as linear (Fig. 189).

The work done by the increments $\mathrm{d}P$ of the external forces, $\mathrm{dd}W = \frac{1}{2}\mathrm{d}P\,\mathrm{d}\delta$, is an infinitesimal quantity of higher order, when compared with

---

[3]Actually, there are no concentrated forces. Treating the external forces as concentrated presupposes that the validity of Saint-Venant's principle is accepted, which allows systems of forces, such as distributed loads, to be substituted by their resultants.

the work $dW = P\,d\delta$, done by forces P in the deformation $d\delta$, so that it may be disregarded. The same happens with the sum of elastic potential and dissipated energies that corresponds to the infinitesimal increments of the stresses $d\sigma$, $ddU = \frac{1}{2}d\sigma\,d\varepsilon$, which may be disregarded when added to the sum of elastic potential and dissipated energies that correspond to the finite stresses σ, $dU = \sigma\,d\varepsilon$, as is concluded from Fig. 189. In this Figure, the area of the hatched triangle represents the work done by forces dP (or the sum of elastic potential and dissipated energies corresponding to the stress increments $d\sigma$). The area of the hatched rectangle represents the work done by the finite external forces P in the deformation $d\delta$ (or the sum of elastic potential and dissipated energies, $\sigma\,d\varepsilon$). It is evident that for infinitesimal values of $d\delta$ and $d\varepsilon$, the area of the triangle may be disregarded in the sum with the area of the rectangle.

Representing by du, dv and dw the components of the displacement field $d\delta$ in the reference axes and by $\overline{X}, \overline{Y}, \overline{Z}$ and X, Y, Z the components of the external surface loads and body forces, respectively, the increment of work done by the external forces may be expressed by

$$dW = \int_S \left(\overline{X}\,du + \overline{Y}\,dv + \overline{Z}\,dw\right) dS + \int_V \left(X\,du + Y\,dv + Z\,dw\right) dV \ .$$

In this expression the first integral represents the increment of the work done by the external forces applied to the boundary surface S of the body. The second integral represents the increment of the work done by the external forces distributed in the volume V of the body.

In the same way, the increment of elastic potential energy plus dissipated energy may be expressed by

$$dU = \int_V \left(\sigma_x\,d\varepsilon_x + \sigma_y\,d\varepsilon_y + \sigma_z\,d\varepsilon_z + \tau_{xy}\,d\gamma_{xy} + \tau_{zx}\,d\gamma_{zx} + \tau_{yz}\,d\gamma_{yz}\right) dV \ .$$

The equality $dW = dU$ represents the theorem of virtual displacements. Since the infinitesimal forces dP are arbitrary, the infinitesimal displacements $d\delta$ are also arbitrary, provided that they are compatible with the support conditions. Thus, the theorem of virtual displacements may be stated as follows:

> *if, in a structure under the action of external forces, a field of infinitesimal displacements $d\delta$ – the virtual displacements – which is compatible with the support conditions, is applied, the increment of the sum of elastic potential energy plus dissipated energy is equal to the work done by the external forces applied to the structure – the virtual work – when their points of application are displaced by $d\delta$.*

The theorem of virtual displacements has many applications in structural analysis, especially in the methods based on the displacement method. It is mainly used to find relations between internal forces or stresses and the corresponding external forces.

As an applied example, the collapse loads corresponding to the internal forces in the collapse configurations represented in Fig. 138 (kinematic method) are computed. As external forces P are to be related to the internal forces M_p, the problem may be solved by means of the theorem of virtual displacements. Of all possible virtual displacements, a field in which bending deformation occurs only in the plastic hinges leads to the simplest numerical treatment. Considering an infinitesimal relative rotation $\mathrm{d}\theta$ of the cross-sections that are immediately to the left and to the right of the plastic hinge, the integral $\int_V \sigma_z \, \mathrm{d}\varepsilon_z \, \mathrm{d}V$ in this zone is equal to the product $M_p \, \mathrm{d}\theta$, where M_p represents the plastic moment of the cross-section. We straightaway arrive at this conclusion by means of energy conservation considerations on the piece of bar defined by the limits of the plastic hinge.

Applying the theorem of virtual displacements to the three possible collapse configurations represented in Fig. 138-b, we get, considering as virtual displacements the collapse configurations defined by the infinitesimal angle θ

$$i) - \begin{cases} \mathrm{d}W = P\dfrac{\theta l}{2} + 2P\dfrac{\theta l}{4} \\[2mm] \mathrm{d}U = M_p\theta + M_p 2\theta \end{cases} ; \qquad \mathrm{d}U = \mathrm{d}W \Rightarrow P = 3\dfrac{M_p}{l}$$

$$ii) - \begin{cases} \mathrm{d}W = P\dfrac{\theta l}{2} + 2P\dfrac{3\theta l}{4} \\[2mm] \mathrm{d}U = M_p\theta + M_p 4\theta \end{cases} ; \qquad \mathrm{d}U = \mathrm{d}W \Rightarrow P = 2.5\dfrac{M_p}{l}$$

$$iii) - \begin{cases} \mathrm{d}W = 2P\dfrac{\theta l}{4} \\[2mm] \mathrm{d}U = M_p\theta + M_p 2\theta; \end{cases} ; \qquad \mathrm{d}U = \mathrm{d}W \Rightarrow P = 6\dfrac{M_p}{l} .$$

These values coincide with those found by means of equilibrium considerations.

The theorem of virtual displacements can be used to demonstrate that the tangential stiffness matrix of a structure (see Subsect. XI.6.b) is symmetric, even in the case of non-linear structural behaviour.

To this end let us assume that the structure is under the action of a system of n concentrated forces, $P_1, P_2, \ldots, P_n$, and that we introduce a field of virtual displacements in which the displacements in the points of application of the loads vanish, with exception of one, i, which takes the value $\mathrm{d}\delta_i$. Under these conditions, the virtual work of the external forces reduces to $\mathrm{d}W = P_i \, \mathrm{d}\delta_i$. The theorem of virtual displacements leads to the conclusion

$$\mathrm{d}W = \mathrm{d}U \Rightarrow \mathrm{d}U = P_i \, \mathrm{d}\delta_i .$$

Since the sum of elastic potential and dissipated energies, U, may be expressed as a function of the displacements $\delta_1, \delta_2, \ldots, \delta_n$ of the points of application of the external loads in the directions of the loads, and since only the load P_i suffers a displacement $\mathrm{d}\delta_i$, we conclude that

$$dU = \sum_{j=1}^{n} \frac{\partial U}{\partial \delta_j}\, d\delta_j = \frac{\partial U}{\partial \delta_i}\, d\delta_i = P_i\, d\delta_i \;\Rightarrow\; P_i = \frac{\partial U}{\partial \delta_i} \,. \qquad (292)$$

As explained in Subsect. XI.6.b, the element K_{ij} of the stiffness matrix is the derivative of force P_i with respect to displacement δ_j, that is

$$K_{ij} = \frac{\partial P_i}{\partial \delta_j} = \frac{\partial^2 U}{\partial \delta_i\, \partial \delta_j} \,.$$

If this line of reasoning is repeated in order to get the element K_{ji} of the stiffness matrix (derivative of P_j with respect to δ_i), it is obvious that the same expression is obtained, which leads to the conclusion that $K_{ij} = K_{ji}$, i.e., that the stiffness matrix is symmetrical, since i and j are any two coordinates.

XII.4.b Theorem of Virtual Forces

As mentioned above, in addition to the theorem of virtual displacements, there is also the *theorem of virtual forces*. This theorem may be stated as follows:

If, in a deformed structure, infinitesimal variations of the external forces and the stresses $(d\sigma_x,\ldots,d\tau_{yz})$ *are considered, such that the equilibrium conditions between the variations of the external forces and of the stresses are satisfied, the following relation is valid*

$$\int_S \left(d\overline{X}\, u + d\overline{Y}\, v + d\overline{Z}\, w \right) dS + \int_V \left(dX\, u + dY\, v + dZ\, w \right) dV$$

$$= \int_V \left(d\sigma_x\, \varepsilon_x + d\sigma_y\, \varepsilon_y + d\sigma_z\, \varepsilon_z + d\tau_{xy}\, \gamma_{xy} + d\tau_{zx}\, \gamma_{zx} + d\tau_{yz}\, \gamma_{yz} \right) dV \,,$$

where $d\overline{X}, d\overline{Y}, d\overline{Z}$ *and* dX, dY, dZ *represent the components in the reference axes of the infinitesimal variations of the external boundary and mass forces, respectively.*

The infinitesimal variations of the external forces are the *virtual forces* and the infinitesimal stresses in equilibrium with them are the *virtual stresses*. This theorem is usually demonstrated using the notions of *complementary work* and *complementary deformation energy*. In fact, the first term of the equality defines the increment of complementary work of the external forces and the second represents the increment of complementary deformation energy. Since these are not usual physical quantities, for which general conservations laws would be available, it is not possible to use the law of conservation of complementary energy directly to demonstrate the theorem, although it can be demonstrated that, in an elastic body, the complementary work of the external forces is equal to the complementary deformation energy. However, it is also possible to demonstrate the theorem of virtual forces by means of the usual law of energy conservation.

To this end, let us assume that, before the external loads P, distributed in the boundary surface of the body (components $\overline{X}, \overline{Y}$ and $\overline{Z}$) and in its volume (components X, Y and Z), were applied, forces ΔP (volume components ΔX, ΔY and ΔZ and surface components $\Delta \overline{X}, \Delta \overline{Y}$ and $\Delta \overline{Z}$) which introduce stresses and strains $\Delta \sigma$ and $\Delta \varepsilon$, respectively, were already acting. The forces P, which introduce stresses σ and strains ε, are subsequently applied. These two loading phases are represented in Fig. 190 (lines AB and BC).

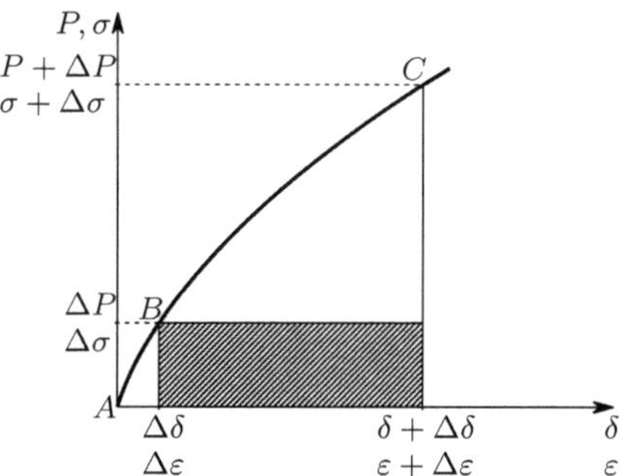

Fig. 190. Increments of external work and elastic potential energy plus dissipated energy, caused by initial forces ΔP

The work done by the external forces in the second loading phase may be symbolically represented by

$$W = \Delta P\, \delta + \int_{\Delta \delta}^{\delta + \Delta \delta} P\, \mathrm{d}\delta \ .$$

The first term $(\Delta P\, \delta)$ represents the work done by the external initial forces ΔP, when the structure deforms under the action of forces P. The second term represents the work done by forces P in the same deformation.

In the same way, the sum of potential elastic and dissipated energies may be symbolically represented by

$$U = \Delta \sigma\, \varepsilon + \int_{\Delta \varepsilon}^{\varepsilon + \Delta \varepsilon} \sigma\, \mathrm{d}\varepsilon \ .$$

The law of energy conservation requires that $U = W$. Furthermore, the following condition must be satisfied

$$\int_{\Delta \delta}^{\delta + \Delta \delta} P\, \mathrm{d}\delta = \int_{\Delta \varepsilon}^{\varepsilon + \Delta \varepsilon} \sigma\, \mathrm{d}\varepsilon \ ,$$

since this equality represents the law of energy conservation in a structure with the same geometry as the one considered, but without initial loads, and made of a material whose constitutive law in each point is given by the line BC of the stress-strain diagram (Fig. 190). We conclude, therefore, that the following relation is valid

$$\Delta P\, \delta = \Delta \sigma\, \varepsilon\,,$$

or, in a more exact formulation, using the components of P, δ, σ and ε

$$\int_S \left(\Delta \overline{X}\, u + \Delta \overline{Y}\, v + \Delta \overline{Z}\, w\right)\, \mathrm{d}S + \int_V \left(\Delta X\, u + \Delta Y\, v + \Delta Z\, w\right)\, \mathrm{d}V$$

$$= \int_V \left(\Delta \sigma_x\, \varepsilon_x + \Delta \sigma_y\, \varepsilon_y + \Delta \sigma_z\, \varepsilon_z + \Delta \tau_{xy}\, \gamma_{xy} + \Delta \tau_{zx}\, \gamma_{zx} + \Delta \tau_{yz}\, \gamma_{yz}\right)\, \mathrm{d}V\,.$$

$$(293)$$

If, instead of the finite initial loads ΔP, infinitesimal loads $\mathrm{d}P$ are considered, the final displacements will be only infinitesimally different from those which would be obtained without the initial loads, that is, the displacements in the actual structure. Substituting in (293) the symbols Δ by d, the expression of the theorem of virtual forces is obtained.

As seen in this demonstration, the theorem of virtual forces is also valid in the case of non-linear structural behaviour (in Fig. 190 a non-linear stress-strain relation is considered). This theorem is mainly used to relate internal deformations with the corresponding displacements. For this reason, it is very useful in structural analysis when the force method is used, since in this method the displacements in the coordinates where the hyperstatic unknowns are applied (the released connections) must be computed.

The application of this theorem to the computation of generalized displacements follows the same pattern as the application of Castigliano's theorem. In fact, the displacement of a point in a structure may be computed by considering a virtual unit force in that point, which plays exactly the same role in the computation as the unit load used in Subsect. XII.3.b.

It should be noted that the rheological behaviour of the material does not play a role in the demonstration when the virtual forces are applied (line AB in Fig. 190). Thus, the theorem of virtual forces is valid irrespective of the material behaviour, with the only restriction that the equilibrium conditions are satisfied between the virtual external forces and the virtual stresses.

This conclusion is in accordance with the considerations drawn in Sect. X.3 (see Footnote 69). In fact, since the relation between strains and displacements is purely kinematic, it cannot depend on the rheological behaviour of the materials the structure is made of. For this reason, the results yielded by the theorem of virtual forces must be the same, either when the virtual stresses $\mathrm{d}\sigma$ are those that actually appear in the structure when the external virtual forces are applied (these stresses depend generally on the material behaviour), or any other stress field $\mathrm{d}\sigma'$ that is in equilibrium with the virtual external forces. For example, in a statically indeterminate structure, we may take as a virtual stress distribution the stresses caused by the external virtual forces in any statically determinate structure that is obtained from the actual structure by releasing the necessary number of connections.

Note that also in this case there is a similarity with the considerations drawn about the use of Castigliano's theorem in statically indeterminate structures.

XII.5 Considerations About the Total Potential Energy

XII.5.a Definition of Total Potential Energy

When a force moves along its line of action in the direction of the force, it does work and simultaneously it looses *potential energy*. This physical quantity represents the capacity of a force to produce work and may be defined as the product of the force and the maximum displacement that it may suffer. However, in the case of a force applied to a deformable structure, it is not easy to define the value of that displacement, and so the potential energy of a force is not easily computed. However, in the applications of this notion to Structural Mechanics the absolute value of that energy is not important, but the way as is varies as the structure deforms is. For this reason, a relative definition of the potential energy of a force may be used. Usually we consider that a force applied to a structure has a vanishing potential energy in the undeformed configuration of the structure. This option has the advantage that both the potential energy of the external forces and the elastic potential energy stored by the structure are zero in the undeformed configuration, provided that there are no residual stresses. Thus, we may define the total potential energy of a structure under the action of n concentrated loads $P_1, P_2, \ldots, P_n$ with the expression

$$V = -\sum_{i=1}^{n} P_i\, \delta_i + U \ . \tag{294}$$

The minus sign in the first term of the right side indicates that the potential energy of a force increases when it suffers a displacement in the opposite direction, i.e., when it moves "backwards" (a force and a displacement are positive when they have the same direction as the corresponding coordinate). If, instead of concentrated forces, distributed surface and volume loads are considered, the total potential energy is given by the expression

$$V = -\left[\int_V (Xu + Yv + Zw)\, \mathrm{d}V + \int_S \left(\overline{X}u + \overline{Y}v + \overline{Z}w\right) \mathrm{d}S\right] + U \ .$$

Remark: in the two last expressions, as also in the whole Sect. XII.5, U represents only the elastic potential energy, while in Sect. XII.4 it has represented the sum of elastic potential energy and dissipated energy. In fact the considerations developed in this Section, in general are only valid if there is no energy dissipation, i.e., when the rheological behaviour of the materials is elastic, although it may be non-linear.

XII.5.b Principle of Stationarity of the Potential Energy

In an elastic body in equilibrium the total potential energy remains constant, when the equilibrium configuration suffers a perturbation represented by a field of infinitesimal displacements $d\delta$. This statement represents the principle of stationarity of the potential energy and is easily demonstrated by means of the theorem of virtual displacements. In fact, as seen in Sect. XII.4, if we consider a virtual displacement field $d\delta$, we have (294)

$$dW = dU \;\Rightarrow\; -dW + dU = dV = 0 \quad \text{with} \quad \begin{cases} dW = \displaystyle\sum_{i=1}^{n} P_i \, d\delta_i \\ dU = \displaystyle\sum_{i=1}^{n} \frac{\partial U}{\partial \delta_i} \, d\delta_i \; . \end{cases}$$

The first term $(-dW)$ represents the increase in potential energy of the external forces in the perturbation $d\delta$, while the second represents the variation of elastic potential energy. Since the sum of these two terms vanishes, and as it simultaneously represents the variation in the total potential energy, we conclude that the perturbation $d\delta$ does not alter this energy. Since the theorem of virtual displacements is valid for any structure under static equilibrium, the statement above is demonstrated. The infinitesimal displacements $d\delta$ are arbitrary, so we have $\frac{\partial V}{\partial \delta_i} = 0$.

XII.5.c Stability of the Equilibrium

In Subsect. XII.5.b we have shown that the equilibrium configuration of a structure under the action of forces corresponds to values of the displacement field for which the variation of the total potential energy vanishes. This means that, in an equilibrium configuration, the function describing the total potential energy either reaches an extremum (a maximum or a minimum), or it obeys the condition $\frac{\partial V}{\partial \delta_i} = 0$, for every i, without an extremum (the potential energy function has a saddle point). In this Sub-section we show that, in the case of stable equilibrium, the potential energy function has a minimum.

In a mechanical system where only conservative forces act and where no energy dissipation takes place, the sum of potential energy, V, and kinetic energy, U_k, is constant, that is,

$$V + U_k = C \; .$$

Let us consider that the system is in equilibrium and that the kinetic energy vanishes. In this system a perturbation is introduced which has the form of a small amount of kinetic energy dU_k. At the instant the perturbation is introduced, we have

$$V + dU_k = C \; .$$

The perturbation introduces motion into the system, so that a deviation from the equilibrium configuration takes place. However, if this configuration corresponds to a minimum of the potential energy, the deviation increases this energy, consuming the kinetic energy dU_k. The motion stops when the perturbation dU_k is totally transformed into potential energy dV, i.e., when

$$V + dV = C \quad \text{with} \quad dV = dU_k \ .$$

But if the equilibrium configuration does not correspond to a minimum of the potential energy, the deviation from that configuration may release some potential energy, which is transformed into kinetic energy, increasing the velocity of deviation with respect to the equilibrium configuration. In fact, denoting by ΔV the decrease in potential energy with respect to the equilibrium configuration, we have

$$V - \Delta V + \underbrace{\Delta U_k + dU_k}_{\text{kinetic energy}} = C \quad \text{with} \quad \Delta V = \Delta U_k \ .$$

In the first case we have a stable state of equilibrium, since the perturbation is absorbed, while in the second case the equilibrium state is unstable, since the perturbation is amplified.

In a stable structure the stiffness matrix is positive definite. In order to demonstrate this statement, let us consider a structure, where n kinematic coordinates are defined, in which given external loads $P_1, P_2, \ldots, P_n$ are acting. In the equilibrium state, these forces are equal to the forces $F_1, F_2, \ldots, F_n$ which are defined as the forces applied at the kinematic coordinates that always balance the internal forces. The difference between these two sets of forces is that the first one $(P_1, P_2, \ldots, P_n)$ remains constant when the equilibrium is disturbed, while forces F are affected by the perturbation, since it changes the displacement field, and, as a consequence, the strains, stresses and their resultants (the forces F). Thus, forces F are functions of the displacements in the kinematic coordinates $\delta_1, \delta_2, \ldots, \delta_n$. The elastic potential energy stored by the structure may also be expressed as a function of these displacements. As seen in Sect. XII.4, forces F_i may be obtained by derivating the elastic potential energy with respect to the displacements δ_i (292) (it was not necessary in that Section to distinguish between forces P and F, since only the equilibrium state was being analysed, where $P = F$).

According to these considerations, the variation occurring in the total potential energy when the equilibrium configuration is disturbed by infinitesimal displacements in the kinematic coordinates, $d\delta_1, d\delta_2, \ldots, d\delta_n$, may be expressed by

$$dV = -\sum_{i=1}^{n} P_i \, d\delta_i + \sum_{i=1}^{n} \frac{\partial U}{\partial \delta_i} \, d\delta_i = -\sum_{i=1}^{n} P_i \, d\delta_i + \sum_{i=1}^{n} F_i \, d\delta_i \ .$$

As seen in Subsect. XII.5.b, in the equilibrium configuration this variation vanishes, since we have $P_i = F_i$. However, in order to analyse the conditions under which the equilibrium configuration corresponds to a minimum of the total potential energy, it is necessary to see how quantity dV varies when the equilibrium state is disturbed, i.e., it is necessary to analyse the second variation of V, ddV. This quantity may be determined by finding the variation introduced by the perturbation $d\delta_i$ in the first variation dV. Since this perturbation affects only forces F, we get

$$ddV = -\sum_{i=1}^{n} P_i\, d\delta_i + \sum_{i=1}^{n} (F_i + dF_i)\, d\delta_i$$
$$-\left(-\sum_{i=1}^{n} P_i\, d\delta_i + \sum_{i=1}^{n} F_i\, d\delta_i \right) = \sum_{i=1}^{n} dF_i\, d\delta_i\ .$$

Expressing the quantities dF_i as functions of the displacements $d\delta_i$, we get

$$dF_i = \sum_{j=1}^{n} \frac{\partial F_i}{\partial \delta_j}\, d\delta_j\ \Rightarrow\ ddV = \sum_{i=1}^{n}\sum_{j=1}^{n} \frac{\partial F_i}{\partial \delta_j}\, d\delta_j\, d\delta_i = \sum_{i=1}^{n}\sum_{j=1}^{n} K_{ij}\, d\delta_j\, d\delta_i\ .$$

The quantities $K_{ij} = \frac{\partial F_i}{\partial \delta_j}$ are the elements of the stiffness matrix corresponding to the n kinematic coordinates. This expression may be given the matrix form

$$ddV = \underbrace{\begin{bmatrix} d\delta_1 & d\delta_2 & \cdots & d\delta_n \end{bmatrix}}_{\{d\delta\}^t} \underbrace{\begin{bmatrix} K_{11} & K_{12} & \cdots & K_{1n} \\ K_{21} & K_{22} & \cdots & K_{2n} \\ \vdots & \vdots & & \vdots \\ K_{n1} & K_{n2} & \cdots & K_{nn} \end{bmatrix}}_{[K]} \underbrace{\begin{Bmatrix} d\delta_1 \\ d\delta_2 \\ \vdots \\ d\delta_n \end{Bmatrix}}_{\{d\delta\}}\ .$$

If ddV takes positive values for any vector $\{d\delta\}$, the equilibrium configuration corresponds to a minimum of the total potential energy, which means that it is stable. From linear algebra we know that this happens if matrix $[K]$ is positive definite. One of the methods to check the positive definiteness of a matrix consists of computing the n determinants

$$\begin{vmatrix} K_{11} & \cdots & K_{1j} \\ \vdots & & \vdots \\ K_{j1} & \cdots & K_{jj} \end{vmatrix} \quad \text{with}\quad j = 1,\dots,n$$

and confirm if they are all positive. Since the first determinant corresponds to K_{11} and taking into account that any diagonal element may be put in this position by changing the numeration of the kinematic coordinates, we conclude that a negative diagonal element in the stiffness matrix is enough to have an unstable situation. However, a unstable situation may also occur with positive values in all diagonal elements (see, for example, (266) with $P = 0.4El$).

When energy dissipation takes place in the system, as it happens in the case of plastic or viscous deformations, the considerations set forth in this Section (XII.5) are not valid, if seen in their generality. In fact, in the presence of non-elastic deformations, the stiffness coefficients cannot be obtained by derivating the elastic potential energy. Besides, they may be different in loading and unloading.

However, the theorem of virtual displacements shows that the stationarity principle is also verified for the sum of total potential energy and dissipated energy. Furthermore, as seen in Sect. XI.6, the stability of a structure depends only on its stiffness matrix. If it is obtained taking non-elastic deformations into consideration, the conclusions drawn here about the characteristics of the stiffness matrix in a stable structure remain valid.

XII.6 Elementary Analysis of Impact Loads

In this section we use simple energy considerations to analyse the internal forces and deformations caused by the impact of a mass with weight P on a structure with linear elastic behaviour. This analysis is based on an important approximation which consists of assuming that *during the deformation caused by the impact, the inertial forces corresponding to the mass of the structure are small enough to be disregarded*. Nevertheless, in most impact problems the deformations occur rapidly, which implies large accelerations and, therefore, important inertial forces, as required by the fundamental law of dynamics (force = mass × acceleration). For this reason, the analysis presented here may often be only an inaccurate approximation, but it does have the advantages of being very simple and of giving the student a first idea about the problems that arise when the load is not statically applied. Still, we should bear in mind that the larger the mass of the impact body, compared with the mass of the structure, and the slower the impact, the better the approximation of this strongly simplified analysis.

As a consequence of the simplifying hypothesis established above, the shape of the deformation caused by impact is the same as would occur in the case of a statically applied load, with the only difference being that the amplitude of the deformation is larger. Under these conditions, an equivalent load, P', may be introduced which is defined as the statically applied load that would cause the same deformation as that caused by impact.

When a moving mass with weight P collides with a supported structure, the structure suffers a deformation which increases until the velocities of the mass and the structure vanish. When this happens, the deformation reaches a maximum and the whole kinetic energy of the mass has been transferred to the structure and stored as elastic potential energy. Denoting by δ the displacement of the impact point in the impact's direction, this displacement will be proportional to the equivalent load P', which means that the elastic potential energy may be expressed by $U_e = \frac{1}{2}P'\delta$, as stated by Clapeyron's theorem. The proportionality constant between between P' and δ, K, is the stiffness of the structure in the coordinate where P' acts. The elastic potential energy may be expressed as a function of K, yielding ($\delta = \frac{P'}{K}$)

$$U_e = \frac{1}{2}P'\delta = \frac{P'^2}{2K}\,.\tag{295}$$

In the case of descending vertical impact, the mass produces positive work during the deformation of the structure. In this case, the elastic potential energy stored by structure, U_e, results not only from the kinetic energy of the mass at the moment of impact, but also from that work, that is, $U_e = U_k + P\delta = U_k + P\frac{P'}{K}$. From this expression and from (295) we get

$$\frac{P'^2}{2K} = U_k + P\frac{P'}{K} \quad \Rightarrow \quad P' = P + \sqrt{P^2 + 2K\,U_k} \quad \Rightarrow \quad \frac{P'}{P} = 1 + \sqrt{1 + \frac{2K}{P^2}U_k}\,.\tag{296}$$

Taking into account that a statically applied load P would introduce the elastic potential energy $U_s = \frac{P^2}{2K}$ (295), (296) may be rewritten in the form

$$\psi = \frac{P'}{P} = 1 + \sqrt{1 + \frac{U_k}{U_s}}\,.\tag{297}$$

The quantity ψ is called the *dynamic coefficient*. This coefficient represents the increase in the stresses and deformations caused by the fact that load P is not statically applied.

From (296) or (297) we can see that the dynamic coefficient takes the value 2, if the kinetic energy at the moment of impact vanishes. Expressed in another way, this means that if a vertical load is applied suddenly (constant load during the deformation), the stresses and deformation caused by it take twice the value that they would take if the load were to be applied statically, i.e., with the load proportional to the deformation.

Furthermore, we conclude from (296) that, the higher the stiffness K of the structure, the higher the dynamic coefficient. This is because a more deformable structure stores more potential energy for the same stress, which means that it can absorb more kinetic energy from the impact mass.

In the case of a mass with horizontal motion, the above expressions are simplified, since the mass does not produce work during the deformation of the structure. From (296) we conclude that the impact of a mass with a kinetic energy U_k causes the same stresses and deformations as the equivalent load P' given by the expression

$$U_e = \frac{P'^2}{2K} = U_k \quad \Rightarrow \quad P' = \sqrt{2K\,U_k}\,.\tag{298}$$

It must be noted that in the deformed configuration analysed here there is no equilibrium between the load P and the resultant P' of the internal forces in the structure. For this reason, the structure commences an oscillatory motion which begins with the acceleration $a = \frac{P'-P}{m}$, where m represents the mass of the impact body.

XII.7 Examples and Exercises

XII.1. Demonstrate the equivalence of (282) and (283).

Resolution

Considering separately the bending moments around the principal axes of inertia of the cross-section, M_x and M_y, the first term of the integral in (282) takes the form

$$\int_\Omega \left(\sigma_z^N + \sigma_z^{M_x} + \sigma_z^{M_y}\right)\left(\varepsilon_z^N + \varepsilon_z^{M_x} + \varepsilon_z^{M_y}\right) d\Omega = \int_\Omega \left(\sigma_z^N \varepsilon_z^N + \sigma_z^N \varepsilon_z^{M_x} + \sigma_z^N \varepsilon_z^{M_y}\right.$$
$$\left. + \sigma_z^{M_x}\varepsilon_z^N + \sigma_z^{M_x}\varepsilon_z^{M_x} + \sigma_z^{M_x}\varepsilon_z^{M_y} + \sigma_z^{M_y}\varepsilon_z^N + \sigma_z^{M_y}\varepsilon_z^{M_x} + \sigma_z^{M_y}\varepsilon_z^{M_y}\right) d\Omega .$$

Substituting the stresses and strains in the second term, $\sigma_z^N \varepsilon_z^{M_x}$, by their expressions in terms of the corresponding internal forces, we get

$$\int_\Omega \sigma_z^N \varepsilon_z^{M_x} d\Omega = \int_\Omega \frac{N}{\Omega}\frac{M_x y}{EI} d\Omega = \frac{N}{\Omega}\frac{M_x}{EI}\int_\Omega y\, d\Omega = 0 ,$$

since the last integral represents the first area moment of the cross-section with respect to an axis which passes through the centroid. In the same way, it could be proved that the following integrals vanish: $\int_\Omega \sigma_z^{M_x}\varepsilon_z^N d\Omega$, $\int_\Omega \sigma_z^N\varepsilon_z^{M_y} d\Omega$ and $\int_\Omega \sigma_z^{M_y}\varepsilon_z^N d\Omega$. In the case of the sixth term, $\sigma_z^{M_x}\varepsilon_z^{M_y}$, we get

$$\int_\Omega \sigma_z^{M_x}\varepsilon_z^{M_y} d\Omega = \int_\Omega \frac{M_x y}{I_x}\frac{M_y x}{EI_y} d\Omega = \frac{M_x M_y}{EI_x I_y}\int_\Omega xy\, d\Omega = 0 ,$$

since the last integral represents the product of inertia with respect to the principal axes of inertia. In the same way, we could show that the eighth term, $\int_\Omega \sigma_z^{M_y}\varepsilon_z^{M_x} d\Omega$ also vanishes.

In order to prove the equivalence of the expressions in relation to the shearing stresses and distortions, let us consider that the transversal forces are applied in the line defined by the shear centres of the cross-sections, so that they do not introduce a twisting moment. Under these conditions, the elastic potential energy stored by the shearing deformations in a piece of bar with infinitesimal length dl may be expressed by (see Subsect. IX.2.a)

$$dU_1 = \frac{1}{2}\int_\Omega \left(\tau_{zx}^V \gamma_{zx}^V + \tau_{zy}^V \gamma_{zy}^V\right) d\Omega\, dl = \frac{1}{2}V\gamma_m\, dl .$$

In a second step, we assume that the transversal loads are moved to their actual positions, so that the twisting moment is introduced without altering the shear force. The twisting moment introduces additional stresses and distortions. The final elastic potential energy is given by

$$dU_2 = dU_1 + \underbrace{\int_\Omega \left(\tau_{zx}^V \gamma_{zx}^T + \tau_{zy}^V \gamma_{zy}^T\right) d\Omega\, dl}_{=0} + \underbrace{\frac{1}{2}\int_\Omega \left(\tau_{zx}^T \gamma_{zx}^T + \tau_{zy}^T \gamma_{zy}^T\right) d\Omega\, dl}_{\frac{1}{2}T\theta} .$$

As seen in Chap. X, the last term represents the work done by the twisting moment in the torsional deformation. Since the first term (dU_1) represents the work done by the shear force in the shear deformation, the second term – work done by the shearing stresses caused by the shear force in the deformation caused by the twisting moment – represents the work done by the shear force in the twisting deformation. This work must vanish, since the twisting moment does not cause the transversal displacement of the cross-sections (see example XII.8-b). Interchanging the order of application of the loading, we likewise conclude that the integral $\int_\Omega \left(\tau_{zx}^T \gamma_{zx}^V + \tau_{zy}^T \gamma_{zy}^V \right) d\Omega$ represents the work done by the twisting moment in the deformation caused by the shear force, which also vanishes, since the shear force does not cause rotation of the cross-sections (see example XII.8-a).

XII.2. Determine the elastic potential energy stored in the deformation caused by the bending moment in the beam depicted in Fig. XII.2, using:
(a) Equation (284);
(b) Clapeyron's theorem.

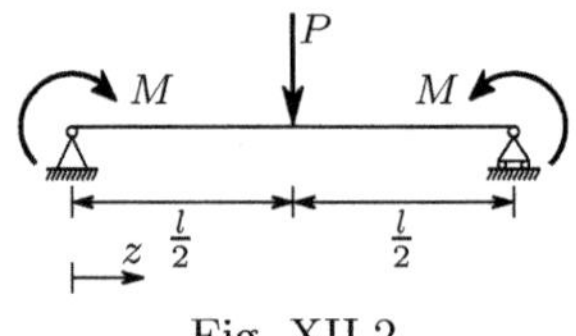

Fig. XII.2

Resolution

(a) Defining the bending moment as a function of coordinate z (Fig. XII.2), we get from (284), since the beam is symmetric,

$$U = 2 \times \frac{1}{2} \int_0^{\frac{l}{2}} \left(M + \frac{P}{2} z \right)^2 \frac{1}{EI} \, dz = \frac{1}{EI} \left[\frac{M^2 l}{2} + PM \frac{l^2}{8} + \frac{P^2 l^3}{96} \right] .$$

(b) In order to apply Clapeyron's theorem, it is necessary to compute the displacement of the point of application of load P and the rotation of the end cross-sections. These quantities are easily obtained by means of the second moment-area theorem, yielding, respectively for the rotation and for the displacement

$$\theta = \frac{1}{EI} \left(\frac{Pl^2}{16} + \frac{Ml}{2} \right) ; \qquad \delta = \frac{1}{EI} \left(\frac{Pl^3}{48} + \frac{Ml^2}{8} \right) .$$

Substituting these values in (285), we get the same result as in answer a), i.e.

$$U = \frac{1}{2} \left(2M\theta + P\delta \right) = \frac{1}{EI} \left[\frac{M^2 l}{2} + PM \frac{l^2}{8} + \frac{P^2 l^3}{96} \right] .$$

XII.3. Solve the same problem (example XII.2) in the beam represented in
Fig. XII.3.

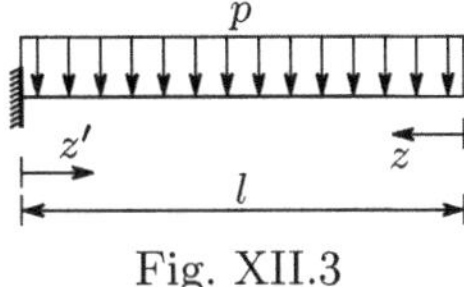

Fig. XII.3

Resolution

(a) Defining the bending moment as a function of coordinate z, we get from
(284)

$$M(z) = -\frac{pz^2}{2} \;\Rightarrow\; U = \frac{1}{2}\int_l \frac{M^2}{EI}\,\mathrm{d}z = \frac{1}{2EI}\int_0^l \frac{p^2 z^4}{4}\,\mathrm{d}z = \frac{p^2 l^5}{40EI}\,.$$

(b) Solving this problem by means of Clapeyron's theorem is a little more
complex than example XII.2, since we have a distributed load with variable displacement. Thus, (285) must be applied to an infinitesimal length
and integrated into the whole length of the beam. Defining the bending
moment as a function of coordinate z' (Fig. XII.3), the method of integration of the curvature equation yields the equation of the deflection
line

$$y = \frac{p}{EI}\left(\frac{l^2}{4}z'^2 - \frac{l}{6}z'^3 + \frac{1}{24}z'^4\right)$$

Clapeyron's theorem then yields

$$U = \frac{1}{2}\int_0^l yp\,\mathrm{d}z' = \frac{p^2}{2EI}\int_0^l \left(\frac{l^2}{4}z'^2 - \frac{l}{6}z'^3 + \frac{1}{24}z'^4\right)\mathrm{d}z' = \frac{p^2 l^5}{40EI}\,.$$

XII.4. Determine the elastic potential energy stored by a cantilever beam
with length $l = 10h$, with the cross-section represented in Fig. 118-a,
for $b = h$ and $e = \frac{h}{10}$, under the action of a vertical load P which
is applied at the free end, with a line of action passing through the
middle of the flange. Consider a Poisson's coefficient $\nu = 0.25$. Solve
the problem using:
(a) Equation (284);
(b) Clapeyron's theorem.

Resolution

To resolve the problem it is necessary to compute the bending and torsional
stiffnesses, the reduced area of the cross-section and the position of the shear
centre. These quantities have already been found in a more general form in

Chaps. VIII and IX. Thus, the distance from the shear centre to the centre line of the web may be obtained from (205), yielding

$$b = h \;\Rightarrow\; d = \frac{3}{7}h \;.$$

The reduced area is given by (231)

$$e = \frac{h}{10} \;\Rightarrow\; \Omega_r = \frac{eh}{1.494} = \frac{h^2}{14.94}\;.$$

The moment of inertia is indicated in the text that precedes (231), that is

$$e = \frac{h}{10} \;\Rightarrow\; I = \frac{7}{12}eh^3 = \frac{7}{120}h^4\;.$$

The geometrical component of the torsional stiffness is defined by parameter J given in (245) which takes the value

$$\begin{cases} l = 3h \\ e = \frac{h}{10} \end{cases} \;\Rightarrow\; J = \frac{le^3}{3} = \frac{h^4}{1000}\;.$$

Defining the internal forces as functions of a coordinate z with origin in the free end of the cantilever beam, we get the expressions

$$V = P\;; \qquad M = -Pz\;; \qquad T = -V\left(d + \frac{h}{2}\right) = -P\left(\frac{3}{7}h + \frac{h}{2}\right) = -\frac{13}{14}Ph\;.$$

(a) Denoting by U_V, U_M and U_T the components of the elastic potential energy which correspond to the three acting internal forces, we get from (284)

$$U_V = \frac{1}{2}\int_0^{10h} \frac{V^2}{G\Omega_r}\,dz = \frac{1}{2}\frac{2(1+\nu)}{E}\int_0^{10h} \frac{14.94 P^2}{h^2}\,dz = 149.4\frac{1+\nu}{E}\frac{P^2}{h}\;,$$

$$\nu = 0.25 \;\Rightarrow\; U_V = 186.75\frac{P^2}{Eh}\;;$$

$$U_M = \frac{1}{2}\int_0^{10h} \frac{M^2}{EI}\,dz = \frac{1}{2}\int_0^{10h} \frac{P^2 z^2}{E\frac{7}{120}h^4}\,dz = 2857.1\frac{P^2}{Eh}\;;$$

$$U_T = \frac{1}{2}\int_0^{10h} \frac{T^2}{GJ}\,dz = \frac{1}{2}\int_0^{10h} \frac{\left(\frac{13}{14}Ph\right)^2}{\frac{E}{2(1+\nu)}\frac{h^4}{1000}}\,dz = 8622.4\,(1+\nu)\frac{P^2}{Eh}\;,$$

$$\nu = 0.25 \;\Rightarrow\; U_T = 10778\frac{P^2}{Eh}\;;$$

(b) In order to apply Clapeyron's theorem, it is necessary to find the displacement of the application point of load P. This displacement may be decomposed in the three components which correspond to the three internal forces, δ_V, δ_M and δ_T. These components take the values (in the computation of δ_M the second moment-area theorem is used)

$$\delta_V = l\gamma_m = 10h\frac{V}{G\Omega_r} = \frac{10hP}{\frac{E}{2(1+\nu)}\frac{h^2}{14.94}} = 373.5\frac{P}{Eh} \; ;$$

$$\delta_M = \frac{1}{EI}\frac{1}{2}10Ph \times 10h \times \frac{2}{3}10h = \frac{1}{E\frac{7}{120}h^4}\frac{1000}{3}Ph^3 = 5714.3\frac{P}{Eh} \; ;$$

$$\delta_T = \theta l\left(d + \frac{h}{2}\right) = \frac{T}{GJ}l\left(d + \frac{h}{2}\right) = \frac{\frac{13}{14}Ph}{\frac{E}{2(1+\nu)}\frac{h^4}{1000}}10h\frac{13}{14}h = 21556\frac{P}{Eh} \; .$$

The three components of the elastic potential energy take then the values

$$U_V = \frac{1}{2}P\left(\delta_V + \delta_M + \delta_T\right) = (186.75 + 2857.1 + 10778)\frac{P^2}{Eh} \; .$$

XII.5. The beam represented in Fig. XII.5 has a cross-section with a moment of inertia I and a reduced area Ω_r. Using Castigliano's theorem in the form given in (286), determine the displacement of point A.

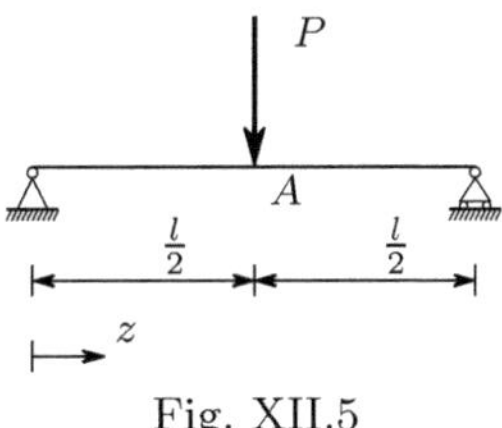

Fig. XII.5

Resolution

Defining the bending moment as a function of coordinate z (Fig. XII.5), we have $M = \frac{P}{2}z$ and $V = \frac{P}{2}$. Substituting these expressions in (284) we get

$$U = 2\frac{1}{2}\int_0^{\frac{l}{2}}\left(\frac{M^2}{EI} + \frac{V^2}{G\Omega_r}\right)\mathrm{d}z = \int_0^{\frac{l}{2}}\left[\frac{\left(\frac{P}{2}z\right)^2}{EI} + \frac{\left(\frac{P}{2}\right)^2}{G\Omega_r}\right]\mathrm{d}z = \frac{P^2l^3}{96EI} + \frac{P^2l}{8G\Omega_r} \; .$$

Equation (286) yields then

$$\delta = \frac{\partial U}{\partial P} = \frac{Pl^3}{48EI} + \frac{Pl}{4G\Omega_r} \; .$$

The first term represents the deformation caused by the bending moment, since it depends on the bending stiffness EI. The second represents the influence of the shear force.

XII.6. In the beam represented in Fig. XII.6-a the deflection of the mid cross-section is to be computed by means of the unit-load method provided by Castigliano's theorem. Show that this computation may be carried out using the internal forces introduced by the unit load in the cantilever beam represented in the same figure.

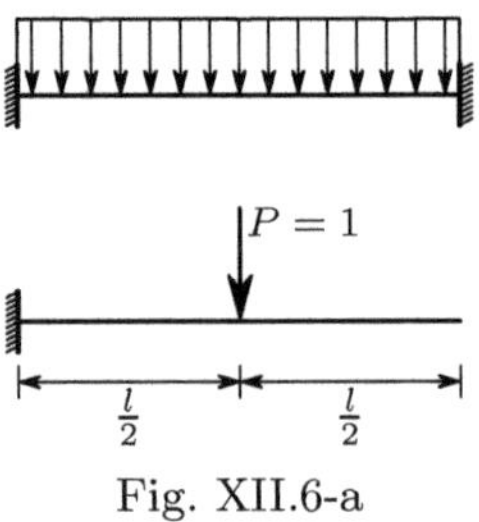

Fig. XII.6-a

Resolution

Considering the beam without the right support as statically determinate base structure, the computation of the displacement of the mid cross-section may be carried out by superposing the displacements δ_1, δ_2 and δ_3 represented in Fig. XII.6-b, caused, respectively, by the applied load, by the vertical reaction R in the right support, and by the reaction moment M in the same support.

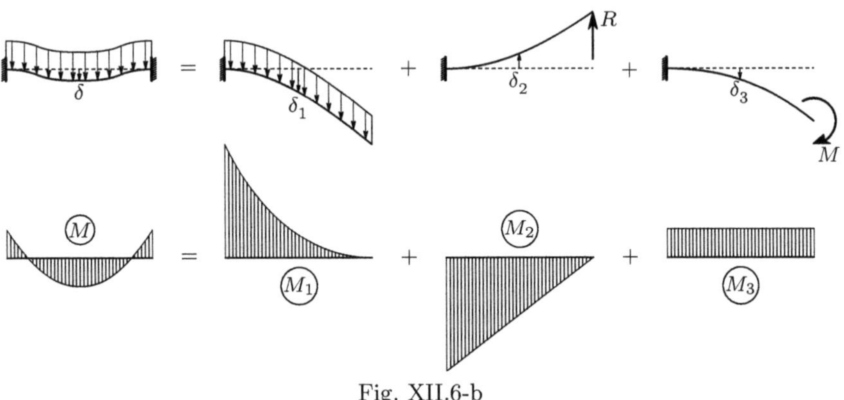

Fig. XII.6-b

Denoting by M_u the bending moment caused by the unit load in the cantilever beam (Fig. XII.6-a) and disregarding the deformation caused by the shear force, we get, using Castigliano's theorem for the computation of δ_1, δ_2 and δ_3, and taking into consideration that $M = M_1 + M_2 + M_3$

$$\delta = \delta_1 + \delta_2 + \delta_3 = \int_l \frac{M_1 M_u}{EI}\,\mathrm{d}l + \int_l \frac{M_2 M_u}{EI}\,\mathrm{d}l + \int_l \frac{M_3 M_u}{EI}\,\mathrm{d}l$$

$$= \int_l \frac{(M_1 + M_2 + M_3)\,M_u}{EI}\,\mathrm{d}l = \int_l \frac{M M_u}{EI}\,\mathrm{d}l \ .$$

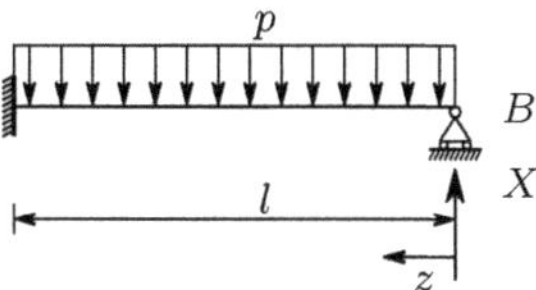

Fig. XII.7

XII.7. In the statically indeterminate beam depicted in Fig. XII.7 the cross-section has a bending stiffness EI and the deformations caused by the shear force are small enough to be disregarded.

(a) Using Menabrea's theorem, find the reaction force in the right support.

(b) Using Castigliano's theorem, determine the rotation of cross-section B.

Resolution

(a) Taking the required reaction force as the hyperstatic unknown, and denoting it by X, the expression of the bending moment as a function of a coordinate z with origin in support B (Fig. XII.7) takes the form

$$M = Xz - p\frac{z^2}{2} \ .$$

Since only the bending deformation needs to be considered, the elastic potential energy is given by the expression

$$U = \frac{1}{2EI} \int_0^l \left(Xz - p\frac{z^2}{2} \right)^2 \mathrm{d}z = \frac{1}{2EI}\left(\frac{X^2 l^3}{3} - \frac{pX l^4}{4} + \frac{p^2 l^5}{20} \right) \ .$$

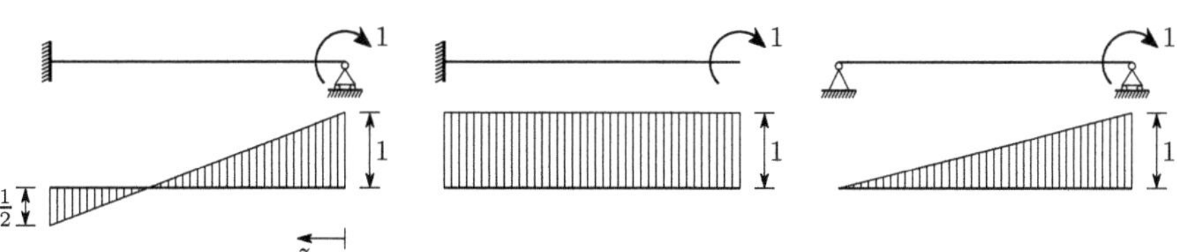

Fig. XII.7-a   Fig. XII.7-b   Fig. XII.7-c

Menabrea's theorem then yields the value of the hyperstatic unknown

$$\frac{\partial U}{\partial X} = 0 \ \Rightarrow\ \frac{2}{3}l^3 X = \frac{pl^4}{4} \ \Rightarrow\ X = \frac{3}{8}pl \ .$$

(b) Since the beam is statically indeterminate, we may compute the internal forces caused by the unit load, either in the actual beam, or in any statically determinate beam obtained from the actual beam by releasing connections. In order to illustrate this possibility, we compute the required rotation using the actual beam (Fig. XII.7-a) and two statically determinate beams (Figs. XII.7-b and XII.7-c).

When the unit moment acts in the cross-section B of the actual beam, it is necessary to solve a statically indeterminate problem to find the corresponding internal forces. This problem may be solved, for example, by means of the force method, in conjunction with the second moment-area theorem, which yields the diagram of bending moments represented in Fig. XII.7-a.

The bending moments introduced by the distributed load and by the unit moment are then, respectively (see answer to question a)

$$M = \frac{3}{8}plz - \frac{1}{2}pz^2 \ ; \qquad M_u = -1 + \frac{3}{2}\frac{z}{l} \ .$$

Castigliano's theorem, in the form given in (287), yields

$$\theta = \frac{1}{EI}\int_0^l MM_u \,\mathrm{d}z = \frac{1}{EI}\int_0^l \left(\frac{3}{8}plz - \frac{1}{2}pz^2\right)\left(-1 + \frac{3}{2}\frac{z}{l}\right)\mathrm{d}z = -\frac{pl^3}{48EI} \ .$$

If instead of the actual beam, the statically determinate beam represented in Fig. XII.7-b is used to compute the bending moments introduced by the unit moment, we get the same value for the rotation of cross-section B, as is easily confirmed

$$M_u = -1 \ \Rightarrow\ \theta = \frac{1}{EI}\int_0^l \left(\frac{3}{8}plz - \frac{1}{2}pz^2\right)(-1)\,\mathrm{d}z = -\frac{pl^3}{48EI} \ .$$

The same happens if the statically determinate structure represented in Fig. XII.7-c is used. In this case, we have

$$M_u = \frac{z}{l} - 1 \ \Rightarrow\ \theta = \frac{1}{EI}\int_0^l \left(\frac{3}{8}plz - \frac{1}{2}pz^2\right)\left(\frac{z}{l} - 1\right)\mathrm{d}z = -\frac{pl^3}{48EI} \ .$$

XII.8. (a) Using Castigliano's theorem, show that the shear force does not cause rotation of the cross-sections around the bar's axis.

(b) Using the result of question a) and Maxwell's theorem, show that the twisting moment does not cause transversal displacement of the cross-sections.

Resolution

(a) Let us consider a prismatic bar with a vanishing twisting moment. In this case, (287) takes the form

$$\frac{\partial U}{\partial P_i} = \int_0^l \left(\frac{N}{E\Omega}\frac{\partial N}{\partial P_i} + \frac{M_x}{EI_x}\frac{\partial M_x}{\partial P_i} + \frac{M_y}{EI_y}\frac{\partial M_y}{\partial P_i} + \frac{V_x}{G\Omega_{rx}}\frac{\partial V_x}{\partial P_i} + \frac{V_y}{G\Omega_{ry}}\frac{\partial V_y}{\partial P_i} \right) dz \ .$$

Note that the validity of this expression (as opposed to (287)) is not conditioned by the second part of the proof presented in exercise XII.1, since there are no shearing stresses caused by the twisting moment.

To find the rotation θ around the bar axis of any infinitesimal area element of any cross-section, we may consider a unit moment M_u applied in that area element. The only internal force introduced by this moment is a twisting moment. This means that the quantities $\frac{\partial N}{\partial M}$, $\frac{\partial M_x}{\partial M}$, $\frac{\partial M_y}{\partial M}$, $\frac{\partial V_x}{\partial M}$ and $\frac{\partial V_y}{\partial M}$ vanish, that is, that the rotation $\theta = \frac{\partial U}{\partial M}$ is zero. This leads to the immediate conclusion that the fibres remain parallel to each other when the twisting moment vanishes.

(b) If the shear force does not cause displacement in the coordinate where the twisting moment acts (rotation coordinate around the bar's axis), Maxwell's theorem leads straight to the conclusion that the twisting moment does not cause a displacement in the coordinate corresponding to the shear force (translation coordinate in the cross-section plane passing through the shear centre), i.e., it does cause a transversal displacement of the cross-sections.

XII.9. Draw the influence lines of the reaction force in support A and of the bending moment in cross-section B of the beam represented in Fig. XII.9-a, using:
(a) the definition of influence line;
(b) the method which results from the application of Betti's theorem.

Resolution

(a) The equation of the influence line of the reaction force in support A is immediately found by means of the balance condition of the moments with respect to the right support of the beam represented in XII.9-b, yielding

$$Pz = Rl \ \Rightarrow \ R = \frac{P}{l}z \ .$$

Fig. XII.9-c shows the influence represented by this equation, for a unit value of load P.

The influence line of the bending moment in cross-section B is described by two expressions: one is valid for $z \leq \frac{l}{2}$ (Fig. XII.9-d) and the other for $z \geq \frac{l}{2}$ (Fig. XII.9-e). These expressions are

$$z \leq \frac{l}{2} \ \Rightarrow \ M_B = \frac{P}{2}z \qquad\qquad z \geq \frac{l}{2} \ \Rightarrow \ M_B = \frac{P}{2}(l-z) \ .$$

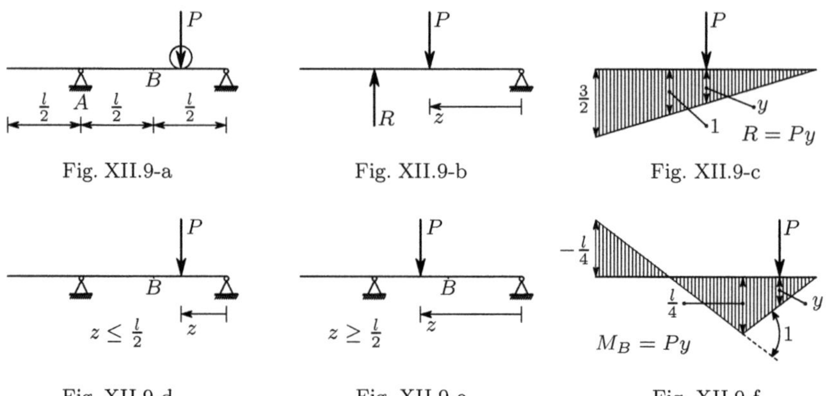

Fig. XII.9-a   Fig. XII.9-b   Fig. XII.9-c

Fig. XII.9-d   Fig. XII.9-e   Fig. XII.9-f

The diagram represented in Fig. XII.9-f is the influence line described by these two equations. The bending moment in cross-section B, for $P = 1$, when the load is applied at a given cross-section is equal to the ordinate of the influence line in that cross-section.

(b) The method based on Betti's theorem indicates that the influence line of the reaction force in support A may be obtained by removing this support and applying a unit displacement to point A in the opposite direction to that assumed as positive for the reaction force. This operation at once gives the diagram represented in Fig. XII.9-c.

In the case of the bending moment in cross-section B, the same method indicates that a hinge should be introduced in this cross-section. The influence line corresponds to the deflection line of the beam when a relative unit rotation is introduced in the left and right cross-sections of the hinge. This operation yields the deflection line represented in Fig. XII.9-f.

XII.10. The beam represented in Fig. XII.10 has a cross-section with a moment of inertia I. The contribution of the shear force to the deformation is sufficiently small to be disregarded.

(a) Draw the influence line of the bending moment in cross-section C, identify the positions of load P, so that this moment takes maximum positive and negative values, and determine the corresponding values of the bending moment.

(b) Determine the positions where load P must be placed in order to get the maximum values of the vertical reaction forces in supports B and C, and find the corresponding values of these forces.

(c) Determine the positions where load P must be placed, in order to get the maximum positive and negative values of the shear force in the mid cross-section of beam segment BC, and find the values of these shear forces.

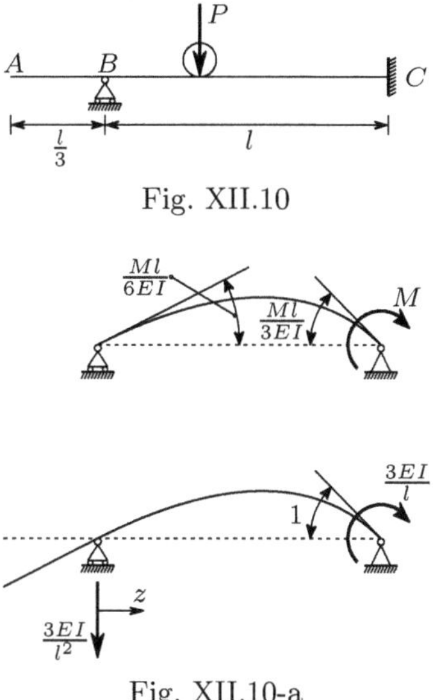

Fig. XII.10

Fig. XII.10-a

Resolution

(a) The influence line of the bending moment in cross-section C corresponds
to the deflection line of the beam which is obtained by introducing a hinge
in that cross-section, and applying a pair of moments, which introduces a
unit relative rotation in the cross-sections immediately to the left and to
the right of the hinge. Since, in this case, the right cross-section cannot
rotate, it is enough to apply a moment which introduces a unit rotation in
the left cross-section, as represented in Fig. XII.10-a. Since a moment M
applied to an extremity of a simply-supported beam introduces a rotation
$\frac{Ml}{3EI}$ (Fig. XII.10-a), the moment needed to introduce a unit rotation takes
the value

$$\frac{Ml}{3EI} = 1 \Rightarrow M = \frac{3EI}{l} \ .$$

The equation of the influence line may be obtained by means of the
method of integration of the curvature equation. Defining the bending
moment as a function of coordinate z (Fig. XII.10-a), we get

$$M = -\frac{3EI}{l^2}z \Rightarrow \theta = \frac{3}{l^2}\frac{z^2}{2} + C_1 \Rightarrow y = \frac{1}{2}\frac{z^3}{l^2} + C_1 z + C_2 \ .$$

The support conditions yield the values of the integration constants, i.e.,

$$z = 0 \Rightarrow y = 0 \Rightarrow C_2 = 0 \ ; \qquad z = l \Rightarrow y = 0 \Rightarrow C_1 = -\frac{1}{2} \ .$$

The rotation and the displacement of segment BC are then

$$\theta = \frac{3}{l^2}\frac{z^2}{2} - \frac{1}{2} \;; \qquad y = \frac{1}{2}\frac{z^3}{l^2} - \frac{1}{2}z \;.$$

In segment AB the influence line is straight, since this segment is statically determinate. The equation of this straight line is

$$y = C_1 z = -\frac{1}{2}z \;.$$

The influence line (Fig. XII.10-a) shows that the maximum negative bending moment in cross-section C is obtained when the load P acts in the point corresponding to a vanishing slope, while the maximum positive bending moment occurs when the load acts in point A. This bending moment and the maximum value of the ordinate of the influence line take the values $(z = 0 \;\Rightarrow\; \theta = -\frac{1}{2})$

$$z = -\frac{l}{3} \;\Rightarrow\; y = \frac{1}{2}\frac{l}{3} = \frac{l}{6} \;\Rightarrow\; M^{+}_{\text{max}} = \frac{Pl}{6} \approx 0.1667Pl \;;$$

$$\theta = 0 \;\Rightarrow\; z = \frac{l}{\sqrt{3}} \;\Rightarrow\; y = y_{\text{max}} = \frac{l}{2\sqrt{27}} - \frac{l}{2\sqrt{3}} \approx -0.1925l \;.$$

Thus, the maximum negative bending moment occurs when the load P acts at a distance $\frac{l}{\sqrt{3}} \approx 0.5774l$ from support B, and takes the value $M^{-}_{\text{max}} \approx -0.1925Pl.$

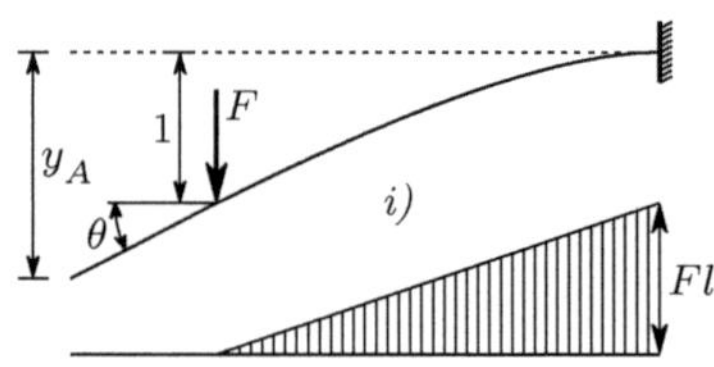

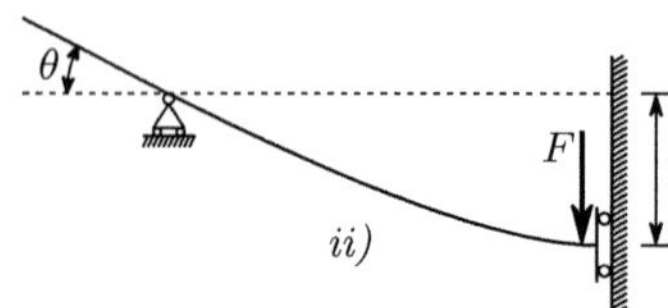

Fig. XII.10-b

(b) The influence line of the reaction force in support B takes the form represented in Fig. XII.10-b-i. Clearly, the maximum value of the ordinate occurs at point A. In order to find this ordinate, the rotation θ is needed. Using the second moment-area theorem, we conclude that force F takes the value

$$\frac{Fl^2}{2EI} \times \frac{2}{3}l = 1 \Rightarrow F = \frac{3EI}{l^3} \ .$$

The rotation θ may then be computed using of the first moment-area theorem, yielding

$$\theta = \frac{Fl^2}{2EI} = \frac{3EI}{l^3}\frac{l^2}{2EI} = \frac{3}{2l} \ .$$

Thus, the maximum ordinate of the influence line and the corresponding value of the reaction force in support B take the values

$$y_A = y_{\max} = 1 + \theta\frac{l}{3} = 1 + \frac{3}{2l}\frac{l}{3} = \frac{3}{2} \Rightarrow R_{B-\max} = \frac{3}{2}P \ .$$

The influence line of the vertical reaction force in support C takes the form shown in Fig. XII.10-b-ii. This curve has the same shape as the influence line of the reaction force of support B. The maximum negative ordinate is clearly smaller than the positive one ($y_A = \theta\frac{l}{3} = 0.5$), so the maximum value of the reaction force is obtained with the load immediately to the left of support C, i.e., $R_{C-\max} = P$.

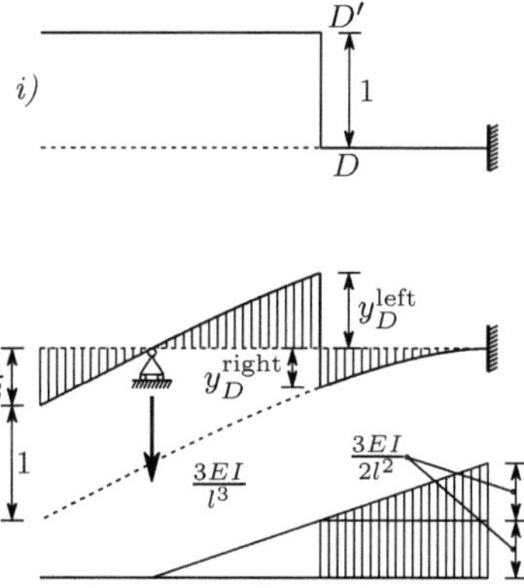

Fig. XII.10-c

(c) The influence line of the shear force in the mid cross-section of beam segment BC (section D) could be obtained by releasing the shear connection using a device like that represented in Figs. 186-e and 186-f and finding a deflection line with a unit relative deflection of the left and right cross-sections of the connection. However, in this case it is simpler to use another method. If we remove support B, a statically determinate beam is obtained, and so the deformation represented in Fig. XII.10-c-i may be introduced without the need of any forces. Subsequently, a vertical downwards force is applied at point B to bring it back to the support. In this way, a deflection line is obtained which has a unit difference between the displacements of the left and right cross-sections of point D. Since the unit deformation DD' is infinitesimal, the deflection line has the same shape as that represented in Fig. XII.10-b-i, with the exception of the discontinuity in cross-section B. For this reason, force F takes the value $\frac{3EI}{l^3}$ and the diagram of bending moments takes the form shown in Fig. XII.10-c.

The ordinate of the deflection line to the right of point D, y_D^{right}, is given by the first area moment of the hatched part of the curvature diagram, with respect to a vertical line passing through point D (second moment-area theorem), yielding

$$y_D^{\text{right}} = \frac{3EI}{2l^2}\left(\frac{l}{2}\times\frac{l}{4}+\frac{1}{2}\frac{l}{2}\times\frac{2}{3}\frac{l}{2}\right)\frac{1}{EI} = \frac{5}{16} = 0.3125\,.$$

This value is smaller than the ordinate of the influence line in point A (0.5). Thus, the maximum positive value of the shear force is obtained with the load at point A, taking the value $V_D^{\text{max}+} = \frac{P}{2}$. The ordinate of the influence line on the left side of point D is the maximum negative value. The maximum negative shear force is therefore obtained with load P on the left side of point D, taking the value

$$y_D^{\text{left}} = 1 - y_D^{\text{right}} = 0.6875 \Rightarrow V_D^{\text{max}-} = 0.6875P\,.$$

XII.11. Verify Maxwell's theorem for the two coordinates represented in the structure depicted in Fig. XII.11-a.

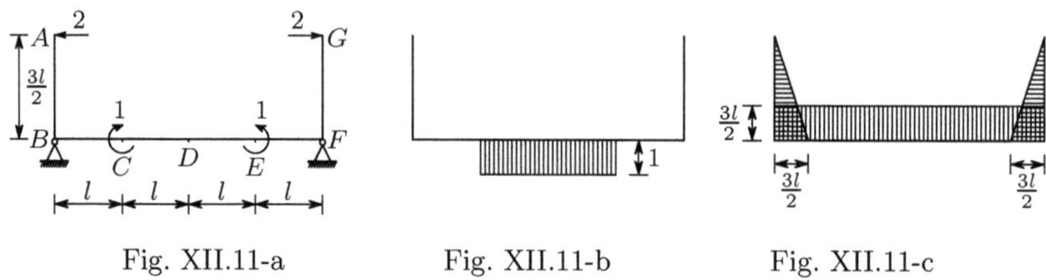

Fig. XII.11-a   Fig. XII.11-b   Fig. XII.11-c

Resolution

The bending moment diagram introduced by a pair of unit moments applied at coordinate 1 takes the form shown in Fig. XII.11-b. The difference between the rotations of cross-sections D and F is equal to the rotation of cross-section F, since, for symmetry reasons, cross-section D does not rotate. Using the first moment-area theorem, we get for the rotation of cross-section F and for the displacement of point G the values

$$\theta_F = \frac{l}{EI} \Rightarrow \delta_G = \frac{3}{2}l \times \theta_F = \frac{3}{2}\frac{l^2}{EI}\,.$$

According to the diagram of bending moment, point G moves to the left. The symmetry of the structure leads to the conclusion that point A suffers an equal displacement to the right. Thus, the distance between points A and G is reduced which corresponds to a negative displacement in coordinate 2 with the value $\delta_2 = -\frac{3l^2}{EI}$.

Applying a pair of unit forces in coordinate 2, we get the bending moment diagram represented in Fig. XII.11-c. The displacement in coordinate 1 is the difference between the rotations of cross-sections C and E. Since in segment CE the upper fibres are tensioned, the concavity of the deflection line points downwards, which corresponds to a negative displacement in coordinate 1. Since the first moment-area theorem gives the absolute value of the relative rotation of two cross-sections directly, we conclude that the displacement in coordinate 1 is given by the area of the curvature diagram between cross-sections C and E, that is

$$\delta_1 = -\frac{3l}{2EI} \times 2l = -\frac{3l^2}{EI} \ .$$

This value coincides with the value obtained for the displacement in coordinate 2, when a unit force acts in coordinate 1, as required by Maxwell's theorem.

XII.12. Using the theorem of virtual displacements, find the values of P which correspond to the collapse mechanisms considered in the resolution of example IX.13 by the kinematic method (Fig. IX.13-c).

Resolution

Denoting by θ the angles between the dashed lines and the horizontal, and applying the theorem of virtual displacements to the first mechanism, we get the following value for force P

$$\begin{cases} dW = 2P\dfrac{\theta l}{2} \\ dU = M_p\theta + M_p2\theta \end{cases} \quad ; \quad dU = dW \ \Rightarrow \ P = 3\frac{M_p}{l} \ .$$

In the same way, we get for the second collapse configuration

$$\begin{cases} dW = P\dfrac{\theta l}{2} \\ dU = M_p\theta + M_p2\theta \end{cases} \quad ; \quad dU = dW \ \Rightarrow \ P = 6\frac{M_p}{l}$$

and for the third

$$\begin{cases} dW = -P\dfrac{\theta l}{2} + 2P\dfrac{\theta l}{2} \\ dU = M_p2\theta + M_p2\theta \end{cases} ; \quad dU = dW \ \Rightarrow \ P = 8\frac{M_p}{l} \ .$$

These values confirm those obtained by means of equilibrium considerations.

XII.13. Using the theorem of virtual displacements, determine the load corresponding to the collapse configuration represented in Fig. IX.14-d.

Resolution

Considering the collapse configuration represented in Fig. IX.14-d and denoting by δ the infinitesimal displacement of cross-section C, we conclude that the work done by the external forces in this virtual deformation takes the value

$$\mathrm{d}W = \int_0^l py\,\mathrm{d}z = p\int_0^l y\,\mathrm{d}z = p\frac{1}{2}\delta l \;,$$

where y represents the displacement of a cross-section located at a distance z of point A. The rotations of segments AC and BC in the deformation defined by δ are, respectively,

$$\theta_A = \frac{\delta}{(1-\alpha)\,l} \;; \qquad \theta_B = \frac{\delta}{\alpha l} \;.$$

Thus, the energy dissipated in the plastic hinges takes the value

$$\mathrm{d}U = M_p\theta_A + M_p\left(\theta_A + \theta_B\right) = M_p\left[\frac{\delta}{(1-\alpha)\,l} \times 2 + \frac{\delta}{\alpha l}\right] = \frac{M_p}{l}\frac{\alpha+1}{\alpha-\alpha^2}\delta \;.$$

Finally, the theorem of virtual displacements yields a relation between p and α

$$\mathrm{d}W = \mathrm{d}U \;\Rightarrow\; p\frac{1}{2}\delta l = \frac{M_p}{l}\frac{\alpha+1}{\alpha-\alpha^2}\delta \;\Rightarrow\; p = \frac{M_p}{l^2}\frac{2\alpha+2}{\alpha-\alpha^2} \;,$$

which is the same as that obtained using equilibrium considerations (example! IX.14).

XII.14. Without using (297), determine the displacement caused by the impact of the weight P on the cross-section B of the beam represented in Fig. XII.14. The beam has bending stiffness EI. The deflection caused by the shear force may be disregarded.

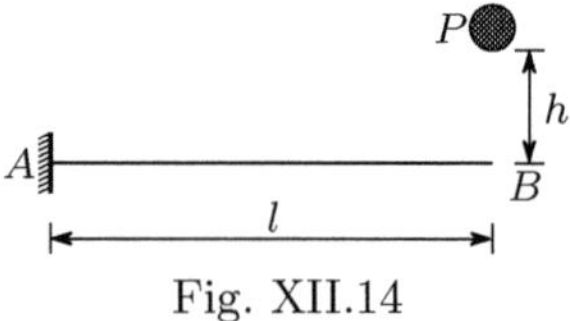

Fig. XII.14

Resolution

The problem may be solved by expressing the elastic potential energy stored by the beam as a function of the required displacement. Since a force F

applied in the free end of a cantilever beam causes the displacement $\delta = \frac{Fl^3}{3EI}$, Clapeyron's theorem yields the expression

$$\delta = \frac{Fl^3}{3EI} \Rightarrow F = \frac{3EI}{l^3}\delta \Rightarrow U = \frac{1}{2}F\delta = \frac{3EI}{2l^3}\delta^2 \;.$$

The work done by the weight P in the fall, until it stops, is given by the expression $W = P(h + \delta)$. The law of energy conservation then allows the computation of the value of δ, yielding, after some manipulation

$$W = U \Rightarrow P(h + \delta) = \frac{3EI}{2l^3}\delta^2 \Rightarrow \delta = \frac{Pl^3}{3EI}\left(1 + \sqrt{1 + 6\frac{EIh}{Pl^3}}\right)\;.$$

Equation (297) would obviously yield the same result, since we have $U_k = Ph$ and $U_s = \frac{1}{2}P\frac{Pl^3}{3EI}$.

XII.15. Determine the bending moment caused in the built-in end of the bar represented in Fig. XII.15 by the impact of the weight P.

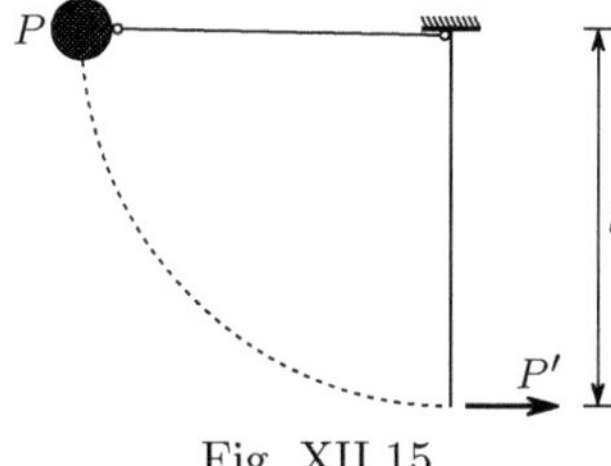

Fig. XII.15

Resolution

The problem may be solved by the direct application of (298). The kinetic energy of the mass with weight P at the moment of the impact is equal to the work done by force P in the fall, that is, $U_k = Pl$. As seen in earlier examples, the stiffness at the free end of a cantilever beam with length l and bending stiffness EI takes the value $K = \frac{3EI}{l^3}$. Thus, the equivalent load P' which would cause the same internal forces as the impact takes the value (298)

$$\begin{cases} U_k = Pl \\ K = \frac{3EI}{l^3} \end{cases} \Rightarrow \quad P' = \sqrt{2KU_k} = \sqrt{2\frac{3EI}{l^3}Pl}$$

$$= \sqrt{6\frac{PEI}{l^2}} \quad \Rightarrow \quad M = P'l = \sqrt{6PEI}\;.$$

XII.16. Figure XII.16 represents a mass m animated with a velocity v that collides with the column AC at point B. The cross-section of the column has a bending stiffness EI. Determine the bending moment caused by the impact in the column, particularizing the results by considering a profile section HEB 300 [9] for the column, $m = 1000\,kg$, $l = 4\,m$ and $v = 5\,m/s$.

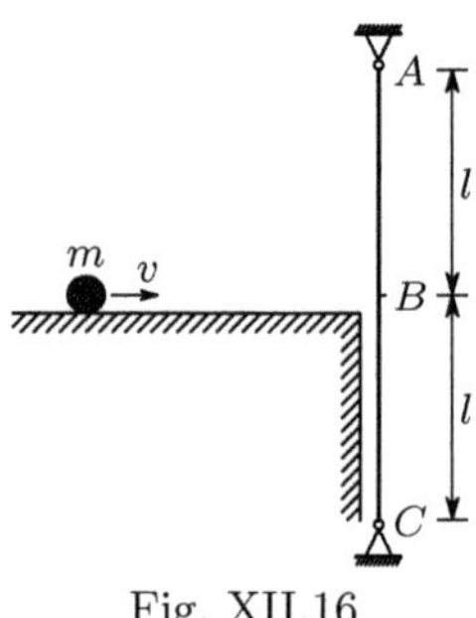

Fig. XII.16

Resolution

The problem may be solved using (298). The stiffness of the column for a horizontal displacement of point B is $K = \frac{6EI}{l^3}$. Thus, we have

$$\begin{cases} K = \frac{6EI}{l^3} \\ U_k = \frac{1}{2}mv^2 \end{cases} \Rightarrow \begin{vmatrix} P' = \sqrt{2KU_k} = \sqrt{2\frac{6EI}{l^3}\frac{1}{2}mv^2} = \frac{v}{l}\sqrt{\frac{6EIm}{l}} \\[2mm] \Rightarrow M_B = \frac{P'l}{2} = \frac{v}{2}\sqrt{\frac{6EIm}{l}}\ . \end{vmatrix}$$

In the specific situation given we get [9]

$$\begin{cases} l = 4\,m \\ E = 206 \times 10^9\,N/m^2 \\ I = 25166 \times 10^{-8}\,m^4 \\ m = 1000\,kg \\ v = 5\,m/s \end{cases} \Rightarrow \begin{vmatrix} M_B = \frac{5}{2}\sqrt{\frac{6\times 206\times 10^9\times 25166\times 10^{-8}\times 1000}{4}} \\[3mm] \approx 697\,150\,Nm\ . \end{vmatrix}$$

XII.17. The structure represented in Fig. XII.17 is made of a material with an elasticity modulus E. Cables AB and CD have a cross-section area Ω. Beam BC has a bending stiffness EI. Determine the displacement of the mid cross-section of the beam caused by the impact of weight P when it falls from the indicated height $(\frac{l}{2})$.

Resolution

The kinetic energy of the weight P at the moment of impact is equal to the work done by force P in the fall, that is $U_k = \frac{Pl}{2}$. The elastic potential

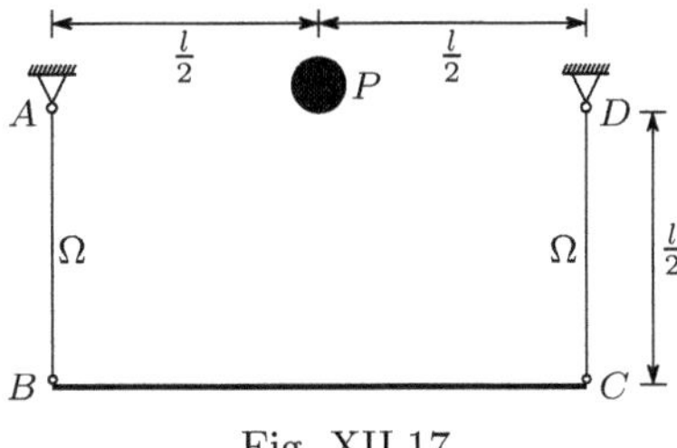

Fig. XII.17

energy that would be stored by the structure if force P were to be applied statically may be computed by means of Clapeyron's theorem. To this end, it is necessary to compute the displacement that would be caused by force P at the mid cross-section. This displacement results from the axial deformation of the cables and the bending deformation of the beam. Since the displacement caused by a concentrated load on the mid cross-section of a simply-supported beam is $\delta = \frac{Pl^3}{48EI}$, we have

$$\delta_s = \frac{\frac{P}{2}\frac{l}{2}}{E\Omega} + \frac{Pl^3}{48EI} = \frac{Pl}{4E}\left(\frac{1}{\Omega} + \frac{l^2}{12I}\right) \Rightarrow U_s = \frac{1}{2}P\delta_s = \frac{P^2 l}{8E}\frac{12I + \Omega l^2}{12I\Omega}.$$

Substituting in (297), we get

$$\delta = \delta_s \psi = \delta_s \left(1 + \sqrt{1 + \frac{U_k}{U_s}}\right) = \frac{Pl}{4E}\frac{12I + \Omega l^2}{12I\Omega}\left(1 + \sqrt{1 + \frac{4E}{P}\frac{12I\Omega}{12I + \Omega l^2}}\right).$$

XII.18. Figure XII.18 represents a structure which is in a horizontal plane. The bars have a rectangular cross-section with base b and height $2b$ and are made of a material with an elasticity modulus E and Poisson's coefficient $\nu = 0.25$. Disregarding the deformation caused by the shear force, determine the vertical displacement of point A when a weight P falls on it from a height $2l$.

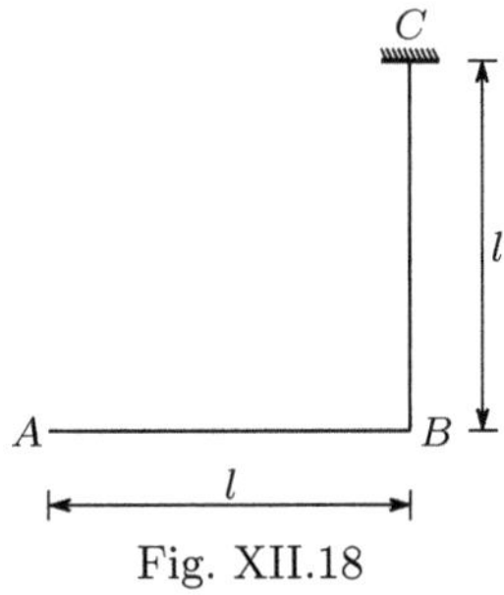

Fig. XII.18

Resolution

The bending and torsional stiffnesses of the cross-section take the values (see Subsect. X.4.d)

$$EI = E\frac{b\,(2b)^3}{12} = E\frac{2}{3}b^4 = 0.6667Eb^4 \; ;$$

$$GJ = Gc_1hb^3 = \frac{E}{2\,(1+\nu)}0.2287 \times 2b \times b^3 = 0.18296Eb^4 \; .$$

If the load P were to be applied statically at point A, bar AB would have a bending moment M^{AB} with a linear variation between zero (cross-section A) and Pl (cross-section B) and a vanishing twisting moment. Bar BC would have a bending moment M^{BC} with a linear variation between zero (cross-section B) and Pl (cross-section C) and a constant positive twisting moment T^{BC} with the value Pl. These internal forces cause the following displacements in cross-section A

$$M^{AB} \quad \longrightarrow \quad \delta_1 = \frac{Pl^3}{3EI} = \frac{Pl^3}{2Eb^4} \; ;$$

$$M^{BC} \quad \longrightarrow \quad \delta_2 = \delta_1 \; ;$$

$$T^{BC} \quad \longrightarrow \quad \delta_3 = \underbrace{\frac{T}{GJ}}_{\varphi} \times l \times l = \frac{Pl}{0.18296Eb^4}l^2 = 5.4657\frac{Pl^3}{Eb^4} \; .$$

Thus, the energies U_s and U_k take the values

$$\delta = 2\delta_1 + \delta_3 = 6.4657\frac{Pl^3}{Eb^4} \Rightarrow U_s = \frac{1}{2}P\delta = 3.2328\frac{P^2l^3}{Eb^4} \; ; \qquad U_k = 2Pl \; .$$

The required displacement is then

$$\frac{U_k}{U_s} = 0.6187\frac{Eb^4}{Pl^2} \Rightarrow \delta_A = \delta\psi = 6.4657\frac{Pl^3}{Eb^4}\left(1 + \sqrt{1 + 0.6187\frac{Eb^4}{Pl^2}}\right) \; .$$

XII.19. The cross-section of beam AB of the structure represented in Fig. XII.19 has a moment of inertia $I = 100\,cm^4$. The cable CD has a cross-section area $\Omega = 0.2\,cm^2$. The structure is made of steel ($E = 206\,GPa$). Determine the maximum value of the bending moment introduced into the beam by a weight $P = 1\,kN$ falling from the indicated height, considering:
(a) cable CD as non-deformable;
(b) cable CD as deformable.

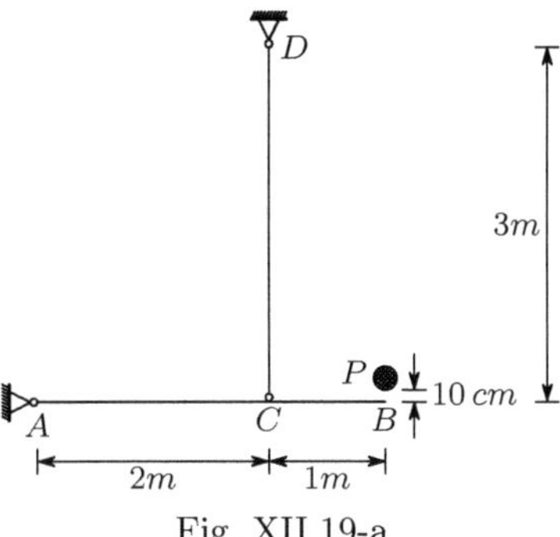

Fig. XII.19-a

Resolution

(a) A statically applied load $P = 1\,kN$ on point B introduces a bending moment $M_C = 1\,kNm$ into cross-section C. The displacement of point B, with an undeformed cable, may be computed using Castigliano's theorem, yielding

$$\delta_B = \frac{1000}{EI} = \frac{1000}{206 \times 10^9 \times 100 \times 10^{-8}} \approx 4.85437 \times 10^{-3} m\ .$$

The elastic potential energy stored by the structure under these conditions, U_s, the kinetic energy of the mass at the moment of impact, U_k, and the dynamic coefficient, ψ, take the values

$$\begin{cases} U_s = \dfrac{1}{2} P \delta_B \approx 2.42718\,J \\ U_k = 0.1\,P = 100\,J \end{cases} \Rightarrow \psi = 1 + \sqrt{1 + \frac{U_k}{U_s}} \approx 7.49615\ .$$

The maximum value of the bending moment is then

$$M_C = 1000\psi = 7\,496.15\,Nm\ .$$

(b) Considering the axial deformation of the cable, the displacement of point B caused by a statically applied load P would be ($N_{CD} = 1500\,N$)

$$\delta_B = \frac{1000}{EI} + \frac{3}{2}\frac{1500 \times 3}{E\Omega} \approx 4.85437 \times 10^{-3} + 1.63835 \times 10^{-3} \approx 6.49272 \times 10^{-3} m\ .$$

The elastic potential energy, the dynamic coefficient and the bending moment in cross-section C take then the values

$$U_s = \frac{1}{2} 1000 \times 6.49272 \times 10^{-3} \approx 3.24636\,J$$

$$\Rightarrow \psi = 1 + \sqrt{1 + \frac{U_k}{3.24636}} \approx 6.63948 \Rightarrow M_C = 6\,639.48\,Nm\ .$$

XII.20. Using Maxwell's theorem, determine the influence line of the displacement of cross-section C of the beam represented in Fig. XII.20.

XII.21. Verify the validity of Maxwell's theorem in the two coordinates represented in the beam depicted in Fig. XII.21.

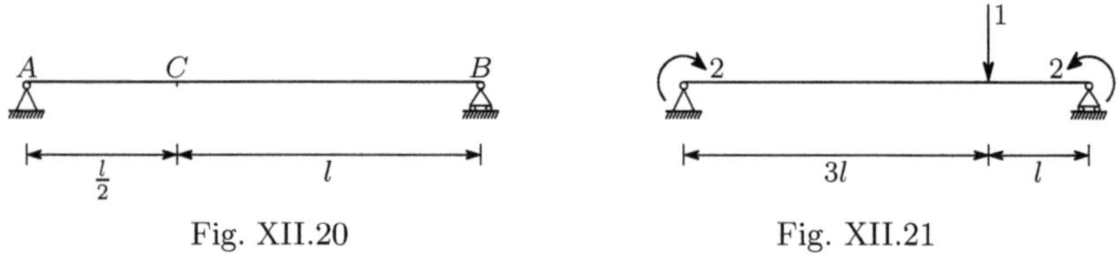

Fig. XII.20   Fig. XII.21

XII.22. In the beam represented in Fig. XII.22, identify the approximate position of the points where load P should be placed in order to get the maximum negative and positive values of the bending moment in cross-section B.

XII.23. The beam represented in Fig. XII.23 supports a uniformly distributed load p. Where should this load be considered so as to get the maximum absolute value of the shear force in cross-section A?

XII.24. In the beam represented in Fig. XII.24, identify the approximate position of load P in order to get the maximum bending moment in cross-section B.

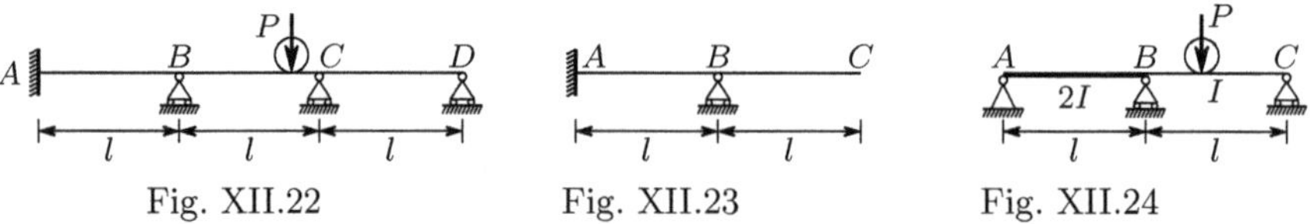

Fig. XII.22   Fig. XII.23   Fig. XII.24

XII.25. In the symmetric beam represented in Fig. XII.25, where should the action of a uniformly distributed load p be considered in order to get the maximum absolute value of the shear force in the cross-section located immediately to the right of support C ?

Answer the same question, with respect to the maximum bending moment in the mid cross-section of beam segment CD.

XII.26. In the beam represented in Fig. XII.26, where should the action of force P be considered in order to get the maximum absolute value of the bending moment in cross-section A ?

Answer the same question, with respect to the shear force in the cross-section located immediately to the left of support B.

XII.27. Using Maxwell's theorem show that, in the structure represented in Fig. XII.27, load P never causes tension in the cable AB, regardless of its position on beam CD.

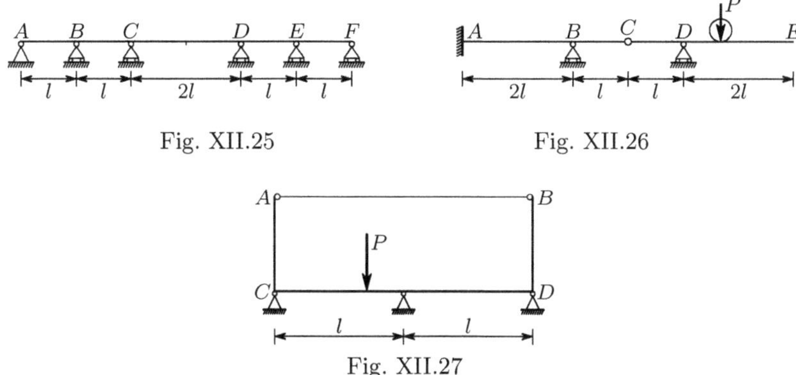

Fig. XII.25   Fig. XII.26

Fig. XII.27

XII.28. The beam represented in Fig. XII.28 is made of a material with linear
elastic behaviour with an elasticity modulus E. Its cross-section has
a moment of inertia I. On this beam a weight P falls from a height
$h = \frac{l}{2}$. The impact may occur on any cross-section of the beam.
  (a) Determine the point of the beam where a static load P must be
       applied, in order to get the maximum bending moment in cross-
       section B.
  (b) Determine the bending moment induced in cross-section B by the
       fall of the weight on that cross-section.
  (c) Does the fact that the deformation caused by the shear force
       is not considered introduce an advantageous or disadvantageous
       error for the safety of the structure ?
XII.29. In which of the two beams represented in Fig. XII.29 does the impact
of the weight cause higher internal forces ? Justify the answer.

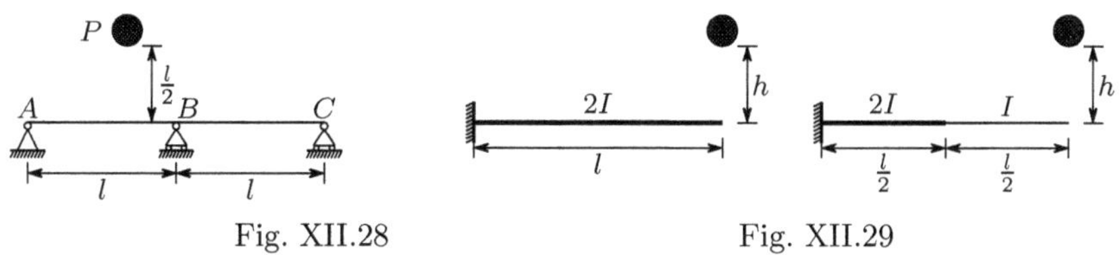

Fig. XII.28   Fig. XII.29

XII.30. On point A of the beam represented in Fig. XII.30 a weight P falls. If
an additional support is placed at point C, does the bending moment
caused by the impact on cross-section B increase or decrease ? Justify
the answer.
XII.31. The structure represented in Fig. XII.31 is made of a material with linear elastic behaviour with an elasticity modulus E. The cross-sections
of bars AB and BC have the moments of inertia $2I$ and I, respectively.

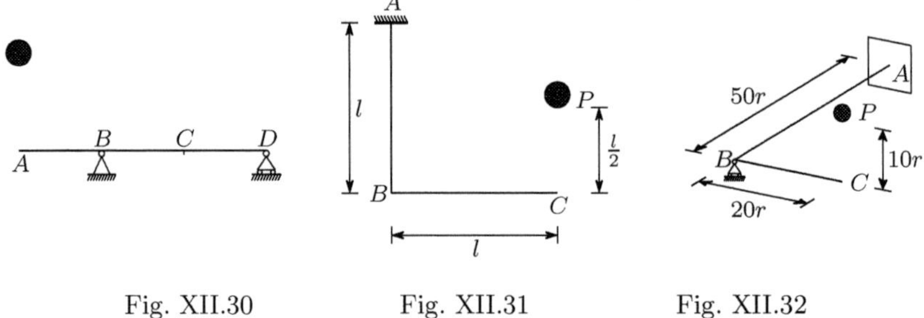

Fig. XII.30   Fig. XII.31   Fig. XII.32

(a) Determine the bending moments introduced into the structure by the impact of the weight P.

(b) If the moment of inertia of the cross-section of bar AB were to be reduced to the half, would those bending moments increase or decrease ? Justify the answer.

XII.32. The structure represented in Fig. XII.32 is contained in a horizontal plane. The two bars are perpendicular to each other and have a circular cross-section with a radius r. The support at point B prevents the vertical displacement.

(a) Determine the internal forces caused in the structure by the impact of the weight P when it falls from the given height (consider $G = 0.4E$).

(b) If the length of bar AB were to increase to double ($100r$), would the internal forces increase or decrease ? Justify the answer.

Answers to Proposed Exercises

Chapter II

II.3. (a) Substituting the given expressions into the differential equations of equilibrium (5) it is immediately seen that they are not satisfied.

(b) $X = \frac{\eta\rho}{\lambda}$, $Y = \eta\rho$.

Chapter III

III.1.
$$\varepsilon_x = 10x + 3y \qquad \varepsilon_{xy} = \tfrac{3}{2}x + 7y + \tfrac{3}{2}z \qquad \omega_{xy} = -\tfrac{3}{2}x - y - \tfrac{3}{2}z$$
$$\varepsilon_y = 6x + 8y \qquad \varepsilon_{xz} = \tfrac{3}{2}y + 2z \qquad \omega_{xz} = \tfrac{3}{2}y - 2z$$
$$\varepsilon_z = 4x + 12y \qquad \varepsilon_{yz} = 2y + 2z + \tfrac{3}{2} \qquad \omega_{yz} = 2y - 2z + \tfrac{3}{2}$$

III.2. $\varepsilon_x = 12A + 72A^2 + 72D^2 + 312.5G^2$

III.3. (a) 4; (b) 4; (c) 2.

Chapter IV

IV.7. *Monotropic material*: revolution ellipsoid with the revolution axis parallel to the monotropy direction;

Orthotropic material: ellipsoid with principal axes parallel to the material orthotropy directions.

IV.8. (a) $\varepsilon_v = \frac{\sigma_x}{E_x}\left(1 - \nu_{xy} - \nu_{xz}\right) + \frac{\sigma_y}{E_y}\left(1 - \nu_{yx} - \nu_{yz}\right) + \frac{\sigma_z}{E_z}\left(1 - \nu_{zx} - \nu_{zy}\right)$

(x, y and z are the orthotropy directions)

(b) $\varepsilon_v = \frac{\sigma_x}{E_l}(1 - 2\nu_l) + (\sigma_y + \sigma_z)\left(\frac{1-\nu_t}{E_t} - \frac{1-\nu_l}{E_l}\right)$

(x is the monotropy direction)

IV.9. Ellipsoid with principal axes $R_x = \left(1 + \frac{1}{30}\frac{\sigma}{E}\right)R$, $R_y = \left(1 + 0.15\frac{\sigma}{E}\right)R$ and $R_z = \left(1 + 0.775\frac{\sigma}{E}\right)R$.

IV.10. (a) $U_d = \frac{\sigma_0^2 \Delta t}{\eta}$ (b) $U_d = \frac{3}{8}E\varepsilon_0^2$.

IV.11. Axisymmetric stress state where the symmetry axis has the direction of the material symmetry direction.

IV.12. Nine, since it is an orthotropic material with linear elastic behaviour: the material has symmetric rheological properties with respect to the plane of the layers and with respect to the two other planes that are perpendicular to it and are parallel to the fibres in the 1st, 5th, 9th, etc., and in the 2nd, 6th, 10th, etc., layers, respectively.

IV.13. (a) $\varepsilon_v = \left(\frac{1-\nu_{xy}-\nu_{xz}}{E_x} + \frac{1-\nu_{yx}-\nu_{yz}}{E_y} + \frac{1-\nu_{zx}-\nu_{zy}}{E_z}\right)\sigma_m$.

(b) $\varepsilon_v = \left(\frac{1-4\nu_l}{E_l} + \frac{2-2\nu_t}{E_t}\right)\sigma_m$ (σ_m is the isotropic stress).

IV.14. $[l] = \begin{bmatrix} l_1 & m_1 & n_1 \\ l_2 & m_2 & n_2 \\ l_3 & m_3 & n_3 \end{bmatrix}$ $[\sigma_p] = \begin{bmatrix} \sigma_1 & 0 & 0 \\ 0 & \sigma_2 & 0 \\ 0 & 0 & \sigma_3 \end{bmatrix}$.

(a) 1 - $[\sigma] = [l]^t [\sigma_p][l] = \begin{bmatrix} \sigma_x & \tau_{yx} & \tau_{zx} \\ \tau_{xy} & \sigma_y & \tau_{zy} \\ \tau_{xz} & \tau_{yz} & \sigma_z \end{bmatrix}$;

2 - $[\sigma] \longrightarrow [\varepsilon] = \begin{bmatrix} \varepsilon_x & \varepsilon_{yx} & \varepsilon_{zx} \\ \varepsilon_{xy} & \varepsilon_y & \varepsilon_{zy} \\ \varepsilon_{xz} & \varepsilon_{yz} & \varepsilon_z \end{bmatrix}$ (85);

3 - $\varepsilon_v = \varepsilon_x + \varepsilon_y + \varepsilon_z$ (invariant).

(b) Operations 1 and 2 equal to answer (a);

3 - $[\varepsilon'] = [l][\varepsilon][l]^t = \begin{bmatrix} \varepsilon_{x'} & \varepsilon_{y'x'} & \varepsilon_{z'x'} \\ \varepsilon_{x'y'} & \varepsilon_{y'} & \varepsilon_{z'y'} \\ \varepsilon_{x'z'} & \varepsilon_{y'z'} & \varepsilon_{z'} \end{bmatrix}$

($x' \equiv 1$, $y' \equiv 2$, $z' \equiv 3$ – principal directions of the stress state).

IV.15. The relaxation curve is similar to that corresponding to the Maxwell model (Fig. 33), except in relation to the asymptote, since, after relaxation ($t \to \infty$) a residual stress remains which is necessary to keep the additional spring deformed.

IV.16. (a) Monotropy direction and all directions of the isotropy plane.

(b) Orthotropy directions.

IV.17. $U = \frac{1}{2}\left(\frac{\sigma_x^2}{E_l} + \frac{\sigma_y^2+\sigma_z^2}{E_t}\right) - \frac{\nu_l}{E_l}(\sigma_x\sigma_y + \sigma_x\sigma_z) - \frac{\nu_t}{E_t}\sigma_y\sigma_z + \frac{\tau_{xy}^2+\tau_{xz}^2}{2G_t} + \frac{1+\nu_t}{E_t}\tau_{yz}^2$.

IV.18. $U_e = \frac{\sigma_0}{2\alpha}\left(1 - 2e^{-\frac{\alpha}{100}} + e^{-\frac{\alpha}{50}}\right)$; $U_d = \frac{\sigma_0}{\alpha}\left(\frac{\alpha}{100} - \frac{3}{2} + 2e^{-\frac{\alpha}{100}} - \frac{1}{2}e^{-\frac{\alpha}{50}}\right)$.

IV.19. $U_e = \frac{1}{2}E(\dot{\varepsilon}\Delta t)^2$; $U_d = \eta\dot{\varepsilon}^2\Delta t$.

Chapter VI

VI.19. $N_{AB} = \frac{66}{29}P \qquad N_{BC} = \frac{8}{29}P \qquad N_{CD} = -\frac{108}{29}P.$

VI.20. $\sigma_a = -\frac{20}{7}E\alpha\Delta t$ and $\sigma_b = \frac{20}{7}E\alpha\Delta t$. (135) may be used to compute the stresses, although the cross-section does not have symmetry axes. In fact, because of the way as the areas occupied by the two materials are distributed in the cross-section, the resulting moment of the stresses obtained by means of (135) vanishes. This means that the equilibrium conditions are verified in the conditions for which (135) has been deduced: constant strain throughout the cross-section, that is, with the bar axis remaining straight during the deformation.

VI.21. $e = \frac{\sqrt{5}-1}{2}c \approx 0.618c$. This value is computed by equating to zero the resulting moment of the stresses caused by a temperature variation, when the bar axis remains straight (constant strain in the cross-section).

XII.8 Chapter VII

VI.3.   (a) $\frac{I}{v} = \frac{bh^2}{6}$;   (b) $\frac{I}{v} = 2\frac{bh^2}{6}$;   (c) $\frac{I}{v} = \frac{1}{2}\frac{bh^2}{6}$.

VII.10. (a) rhombus with height $\frac{h}{3}$ and width $\frac{b}{3}$;
      (b) circle of radius $\frac{r}{4}$;
      (c) rectangle of base $\frac{b}{6}$ and height $\frac{h}{6}$;
      (d) equilateral triangle with side length $\frac{a}{4}$;
      (e) ellipse with semi-axes lengths $\frac{b}{4}$ and $\frac{h}{4}$.

VII.18. $M_{el}^{\max} = 0.16\,\sigma_c b^3$, $M_p^{\max} = 0.264\,\sigma_c b^3$, $\varphi = 1.65$.

VII.19. Width $= a \Rightarrow D \geq 3a\left(\frac{1}{\varepsilon_c}-1\right)$,   Width $= 3a \Rightarrow D \geq a\left(\frac{1}{\varepsilon_c}-1\right)$
      D – diameter of the interior face of the winding (drum's diameter).

VII.20. (a) $\Delta l = 2.4\,\alpha\Delta t\, l$;
      (b) $\frac{1}{\rho} = \frac{3}{34}\frac{M}{c^4 E}$;
      (c) $y = 0 \Rightarrow \sigma = 1.2E\alpha\Delta T$,
            $y = \pm c \Rightarrow \sigma_b = \pm\frac{9}{34}\frac{M}{c^3} + 1.2E\alpha\Delta T$,
            $y = \pm c \Rightarrow \sigma_a = \pm\frac{6}{34}\frac{m}{c^3} - 1.2E\alpha\Delta T$,
            $y = \pm 2c \Rightarrow \sigma_a = \pm\frac{12}{34}\frac{M}{c^3} - 1.2E\alpha\Delta T$.

VII.21. (a) $\frac{1}{\rho_{el}} = \frac{3}{568}\frac{M\varepsilon_0}{c^4\sigma_0}$;
      (b) Material b, because the yielding strain of this material is smaller than that of material a and the distance of the farthest fibres from the neutral axis is greater in material b than in material a;
      $\sigma_b = \frac{9}{284}\frac{M}{c^3}$.

VII.22. The section modulus increases 19.2%. The bending strength of the bar increases by the same amount, for the same reasons as explained in the resolution of example VII.6.

In the case of a ductile material, removing the small rectangles increases the section modulus, but decreases the plastic modulus, that is, the plastic moment. In fact, this modulus increases whenever area is added to the cross-section. Removing the small rectangles causes a decrease $\Delta M_p = \frac{3}{125} c^3 \sigma_c$ in the plastic moment of the cross-section.

VII.23. (a) $M = \frac{29}{3}\,\sigma_c a^3$;

(b) $y = \pm a \Rightarrow \sigma = \mp\frac{15}{44}\sigma_c, \qquad y = \pm 2a \Rightarrow \sigma = \pm\frac{14}{44}\sigma_c$;

(c) $\frac{1}{\rho_{re}} = \frac{15}{44}\frac{\sigma_c}{Ea}$;

VII.24. $\sigma_b^{\max} = \frac{192}{185}\frac{M}{c^3}, \; \sigma_a^{\max} = \frac{576}{185}\frac{M}{c^3}$.

VII.25. $d = \sqrt{\frac{3}{2}}\,a$. With this value of d the two principal moments of inertia are equal. When this happens, any axis is a principal axis, that is bending is plane for any direction of the action axis of the bending moment.

Chapter VIII

VIII.18. $\left(\frac{\mathrm{d}E}{\mathrm{d}z}\right)_{\max} = \frac{12}{37}\frac{pl}{c}$.

VIII.19. (a) $\left(\frac{\mathrm{d}E}{\mathrm{d}z}\right)_{\max} = 9.7455p$.

(b) The plane containing the loading should pass through the intersection point of the centre lines of the two rectangles.

XII.9 Chapter IX

IX.17. (a) $\theta_D = \frac{Ml}{6EI}$;   (b) $\delta = \frac{Pl^3}{4EI}$;   (c) $\delta = \frac{Pl^3}{EI}$;   (d) $\delta_B = -\frac{Ml^2}{3EI}$.

IX.18. (a) $M_A = \frac{M}{3}$.

(b) $M_B = \frac{Pl}{20} - \frac{7pl^2}{15}, \; M_A = -\frac{Pl}{5} - \frac{2pl^2}{15}$.

IX.19. (a) $P_{col} = \frac{M_p}{l}$;   (b) $P_{col} = \frac{2M_p}{3l}$;   (c) $P_{col} = \frac{7}{4}\frac{M_p}{l}$;

(d) $p_{col} = 6.9641\frac{M_p}{l^2}$.

Chapter X

X.15. (a) $f = \frac{T}{2\pi r^2}$ (shear flow in the cross-section).

(b) $\theta = \frac{25T}{12\pi r^4 G} \approx \frac{2.083T}{\pi r^4 G}$ (result obtained by means of the expressions deduced in example X.6-a).

X.16. (a) $\tau_a \approx 4.081 \frac{T}{a^3}$ $\tau_b \approx 6.122 \frac{T}{a^3}$.

(b) $\theta \approx 4.122 \frac{T}{Ga^4}$ (this result and that of Sub-question X.16.a have been obtained by means of the expressions deduced in example X.6-b; note that $A = (0.99a)^2$).

X.17. (a) Open cross section: $T^a_{\text{all}} = \frac{179}{3}a^3\tau_{\text{all}}$, $J_a = \frac{358}{3}a^4$;

Closed cross section: $T^f_{\text{all}} = 800a^3\tau_{\text{all}}$, $J_f = \frac{32000}{3}a^4$;

$\frac{T^f_{\text{all}}}{T^a_{\text{all}}} = 13.4078$, $\frac{J_f}{J_a} = 89.3855$.

(b) $f = \tau_{\text{all}}\, a$.

X.18. Considering only the closed part, we get: $\tau^f_{\max} = \frac{T}{400a^3}$, $\theta^f = \frac{T}{3200Ga^4}$; with the whole cross-section, we get: $\tau^t_{\max} \approx \frac{T}{426.667a^3}$, $\theta^t \approx \frac{T}{3413.333Ga^4}$;

$\frac{\tau^f_{\max}}{\tau^t_{\max}} = \frac{\theta^f}{\theta^t} \approx 1.0667$.

X.19. (a) $\tau_a = \frac{T}{21546a^3}$; $\tau_b = \frac{T}{10773a^3}$.

(b) $\theta = 1.4023 \times 10^{-6}\frac{T}{Ga^4}$.

X.20. $\theta = \frac{\gamma_0}{6a}$.

X.21. (a) $\frac{-0.01\ln 0.12}{2r} = \frac{0.0106}{r}$;   (b) $M = \frac{1.6\tau_0 r^3\pi}{20}$.

X.22. $\delta_M = \frac{80000}{3}\frac{P}{\pi Er}$; $\delta_V = 1500\frac{P}{\pi Er}$; $\delta_T = 75000\frac{P}{\pi Er}$.

If the warping of the built-in cross-section is prevented, the last component (δ_T) will be smaller, as explained in Footnote 75.

Chapter XI

XI.18. $e = \frac{1}{6}\alpha b$.

XI.19. $p_{cr} = 5.12EL$

XI.20. $\Omega > \frac{125\pi^2}{576}\frac{I}{l^2} \approx 2.14184\frac{I}{l^2}$

XI.21. The spring is necessary; $P_{cr} = \frac{2}{3}El$.

XI.22. $p_{cr} = 80.568\frac{EI}{l^3}$.

XI.23. The problem is solved in the same way as example XI.14. Considering the same reference system, the equation of the deflection curve is:

$$y = \frac{Q}{2Pk\cos\left(\frac{kl}{2}\right)}\sin(kz) - \frac{Q}{2P}z, \text{ with } k^2 = \frac{P}{EI}.$$

The maximum value of the bending moment is : $M_{\max} = \frac{Q}{2k}\tan\left(\frac{kl}{2}\right)$.

Chapter XII

XII.20. The influence line of the displacement in cross-section C is the deflection line of the beam when a unit load is placed on point C. In fact, Maxwell's theorem leads to the conclusion that the displacement in cross-section C, caused by a unit load acting on any given cross-section, is equal to the displacement in that cross-section, caused by a unit load acting on point C.

XII.21. A unit load applied in coordinate 1 causes in coordinate 2 a generalized displacement (relative rotation of the end cross-sections) equal to $\frac{3}{2}\frac{l^2}{EI}$; a generalized unit force in in coordinate 2 (a unit moment at each end of the beam) causes a displacement with equal value in coordinate 1.

XII.22. Drawing the approximate shape of the influence line, we conclude that the maximum negative ordinate occurs in beam segment BC, slightly to the left of the mid point of this segment, since cross-section A cannot rotate, while a small rotation of cross-section C takes place; thus, the deformation of segment BC is larger than the deformation of segment AB; the maximum positive ordinate occurs in segment CD at the distance $\frac{1}{\sqrt{3}}l \approx 0.577l$ of support D (see example XII.10-a).

XII.23. The load should be placed on the whole segment BC, since the influence line has positive ordinates in segment AB and negative ordinates in segment BC; the area of the negative part of the influence line is clearly larger than that of the positive part.

XII.24. The load should be placed slightly to the left of the middle of beam segment BC (at the distance $0.577l$ of support C, cf. example XII.10-a), since the ordinates of the influence line in this segment are greater than in segment AB, because the bending stiffness is smaller.

XII.25. $V_{\max}$ in the right section of support C: segments BC, CD and EF; $M_{\max}$ in the middle of segment CD: segments AB, CD and EF.

XII.26. $M_A^{\max}$: load on point E; $V_{B-\text{left}}^{\max}$: load on point E, since the slope of the influence line has the same absolute value in cross-sections B and D.

XII.27. Considering the structure without the cable, with a coordinate 1 composed by a pair of horizontal forces applied to points A and B, positive when they cause an increase of the distance AB, and a vertical downwards coordinate 2 in any point of the beam CD, we conclude easily that a positive force (pair of forces) in coordinate 1 causes a negative displacement in coordinate 2. Maxwell's theorem leads to the conclusion that a positive force in coordinate 2 causes a negative displacement in coordinate 1, that is the distance between points A and B decreases.

XII.28. (a) The load should fall on the point at the distance $(1 - \frac{1}{\sqrt{3}})l \approx 0.4226l$ of support B.

(b) $M_B = 0.0962 Pl \left(1 + \sqrt{1 + \frac{73.47EI}{Pl^2}}\right)$.

(c) The error is advantageous to safety, since a stiffer beam is considered. This increases the dynamic coefficient, and so higher internal forces than the actual ones are computed.

XII.29. In the left beam, since it is stiffer.

XII.30. The support increases the stiffness of the structure in the vertical displacement of point A which reduces the value of U_s, increasing the value of the dynamic coefficient. As segment AB is statically determinate, the bending moment in cross-section B depends only on the weight and the dynamic coefficient, so that that moment increases.

XII.31. (a) $M_B = Pl \left(1 + \sqrt{1 + \frac{6EI}{5Pl^2}}\right)$.

(b) The moments would decrease, since this geometry change does not change the internal forces if the load is statically applied, but it does reduce the dynamic coefficient, as the rotation of cross-section B increases.

XII.32. (a) Twisting moment in AB: $20rP\psi$;

Maximum bending moment in BC: $20rP\psi$;

$\psi = 1 + \sqrt{1 + \frac{15E\pi r^2}{79000P}}$.

(b) The internal forces would be reduced (same justification as in answer XII.31.b).

References

1. A.J. Durelli, E.A. Phillips, C.H. Tsao, *Introduction to the Theoretical and Experimental Analysis of Stress and Strain*, McGRAW-HILL BOOK COMPANY, INC., 1958.
2. Y.C. Fung, *Foundations of Solid Mechanics*, PRENTICE-HALL, INC., Englewood Cliffs, New Jersey, 1965.
3. Ch. Massonnet, S. Cescotto, *Mécanique des Matériaux*, EYROLLES, Paris, 1980.
4. S.P. Timoshenko, J.N. Goodier, *Theory of Elasticity*, third edition, McGraw-Hill Book Company, 1970.
5. Bronstein, Semendjajew, *Taschenbuch der Mathematik*, 24th edition, Verlag Harri Deutsch Thun und Frankfurt/Main, 1989.
6. D.R.J. Owen, E. Hinton, *Finite Elements in Plasticity - Theory and Practice*, Pineridge Press Limited, Swansea, U.K., 1980.
7. Odone Belluzzi, *Ciencia de la Construccion*, Aguilar s a de ediciones, 1973.
8. Charles Massonnet, *Résistance des Matériaux*, II volume, DUNOD, Paris, 1965.
9. J.S. Farinha, A. Correia dos Reis, *Tabelas Técnicas*, edição P.O.B., Setúbal, 1993 (in Portuguese).
10. prEN 1993-3: 20xx, *Eurocode 3: Design of steel structures: Part 1-1: General structural rules*, 2001.
11. S.P. Timoshenko, J.M. Gere, *Theory of Elastic Stability*, second edition, McGraw-Hill Book Company, 1961.
12. Curt F. Kollbrunner, Konrad Basler, *Torsion*, Springer-Verlag, Berlin and Heidelberg, 1966.
13. W.T. Koiter, General theorems for elastic-plastic solids, in *Progress in Soild Mechanics*, Vol. 1, I.N. Snedon and R. Hill (Eds.), Chapter 4, North--Holland, Amesterdam, 1960.
14. V. Dias da Silva, *Introdução à Análise Não-Linear de Estruturas*, Departamento de Engenharia Civil, Faculdade de Ciências e Tecnologia, Universidade de Coimbra, 2002 (in Portuguese).
15. Ted Belytschko, Wing Kam Liu, Brian Moran, *Nonlinear Finite Elements for Continua and Structures*, John Wiley & Sons Ltd., 2000.

Index

If you have any concerns about our products,
you can contact us on
ProductSafety@springernature.com

In case Publisher is established outside the EU,
the EU authorized representative is:
Springer Nature Customer Service Center GmbH
Europaplatz 3, 69115 Heidelberg, Germany

Printed by Libri Plureos GmbH
in Hamburg, Germany